T0213564

# Lecture Notes in Artificial Intelligence    10536

Subseries of Lecture Notes in Computer Science

More information about this series at http://www.springer.com/series/1244

Yasemin Altun · Kamalika Das
Taneli Mielikäinen · Donato Malerba
Jerzy Stefanowski · Jesse Read
Marinka Žitnik · Michelangelo Ceci
Sašo Džeroski (Eds.)

# Machine Learning and Knowledge Discovery in Databases

European Conference, ECML PKDD 2017
Skopje, Macedonia, September 18–22, 2017
Proceedings, Part III

 Springer

*Editors*
Yasemin Altun
Google Research
Google Inc.
Zurich
Switzerland

Kamalika Das
NASA Ames Research Center
Mountain View
USA

Taneli Mielikäinen
Oath
Sunnyvale
USA

Donato Malerba
Department of Computer Science
University of Bari Aldo Moro
Bari
Italy

Jerzy Stefanowski
Institute of Computing Science
Poznan University of Technology
Poznan
Poland

Jesse Read
Laboratoire d' Informatique (LIX)
École Polytechnique
Palaiseau
France

Marinka Žitnik
Department of Computer Science
Stanford University
Stanford
USA

Michelangelo Ceci ⓘ
Università degli Studi di Bari Aldo Moro
Bari
Italy

Sašo Džeroski
Jožef Stefan Institute
Ljubljana
Slovenia

ISSN 0302-9743          ISSN 1611-3349  (electronic)
Lecture Notes in Artificial Intelligence
ISBN 978-3-319-71272-7        ISBN 978-3-319-71273-4  (eBook)
https://doi.org/10.1007/978-3-319-71273-4

Library of Congress Control Number: 2017961799

LNCS Sublibrary: SL7 – Artificial Intelligence

Printed on acid-free paper

This Springer imprint is published by Springer Nature
The registered company is Springer International Publishing AG
The registered company address is: Gewerbestrasse 11, 6330 Cham, Switzerland

# Preface

This year was the 10th edition of ECML PKDD as a single conference. While ECML and PKDD have been organized jointly since 2001, they only officially merged in 2008. Following the growth of the field and the community, the conference has diversified and expanded over the past decade in terms of content, form, and attendance. This year, ECML PKDD attracted over 600 participants.

We were proud to present a rich scientific program, including high-profile keynotes and many technical presentations in different tracks (research, journal, applied data science, nectar, and demo), fora (EU projects, PhD), workshops, tutorials, and discovery challenges. We hope that this provided ample opportunities for exciting exchanges of ideas and pleasurable networking.

Many people put in countless hours of work to make this event happen: To them we express our heartfelt thanks. This includes the organization team, i.e., the program chairs of the different tracks and fora, workshops and tutorials, and discovery challenges, as well as the awards committee, production and public relations chairs, local organizers, sponsorship chairs, and proceedings chairs. In addition, we would like to thank the program committees of the different conference tracks, the organizers of the workshops and their respective committees, the Cankarjev Dom congress agency, and the student volunteers. Furthermore, many thanks to our sponsors for their generous financial support. We would also like to thank Springer for their continuous support, Microsoft for allowing us to use their CMT software for conference management, the European project MAESTRA (ICT-2013-612944), as well as the ECML PKDD Steering Committee (for their suggestions and advice). We would like to thank the organizing institutions: the Jožef Stefan Institute (Slovenia), the Ss. Cyril and Methodius University in Skopje (Macedonia), and the University of Bari Aldo Moro (Italy).

Finally, thanks to all authors who submitted their work for presentation at ECML PKDD 2017. Last, but certainly not least, we would like to thank the conference participants who helped us make it a memorable event.

September 2017

Sašo Džeroski  
Michelangelo Ceci

# Foreword to the ECML PKDD 2017
# Applied Data Science Track

We are pleased to present the proceedings of the Applied Data Science (ADS) Track of ECML PKDD 2017. This track aims to bring together participants from academia, industry, governments, and NGOs (non-governmental organizations) in a venue that highlights practical and real-world studies of machine learning, knowledge discovery, and data mining. Novel and practical ideas, open problems in applied data science, description of application-specific challenges, and unique solutions adopted in bridging the gap between research and practice are some of the relevant topics for which papers have been submitted and accepted in this track. This year's Applied Data Science Track included 27 accepted paper presentations distributed across six sessions. Given a total of 93 submissions, this year's track was highly selective: Only 27 papers could be accepted for publication and for presentation at the conference, corresponding to an acceptance rate of 29%. Each of the 93 submissions was thoroughly reviewed, and accepted papers were chosen both for their originality and for the application they promoted. The accepted papers focus on topics ranging from machine-learning methods and data science processes to dedicated applications. Topics covered include deep learning, time series mining, text mining, for a variety of applications such as e-commerce, fraud detection, social good, ecology, experiment design, and social network analysis. We thank all the authors who submitted the 93 papers for their work and effort to bring machine learning to solve many interesting problems. We also thank all the Program Committee members of the ADS track for their substantial efforts to guarantee the quality of these proceedings. We hope that this program was enjoyable to both academics and practitioners alike, and fostered the beginning of new industry–academia collaborations.

September 2017

Yasemin Altun
Kamalika Das
Taneli Mielikäinen

# Foreword to the ECML PKDD 2017 Nectar Track

We are pleased to present the proceedings of the Nectar Track of the ECML PKDD 2017 conference held in Skopje. This track, which started in 2012, provides a forum for the discussion of recent high-quality research results at the frontier of machine learning and data mining with other disciplines, which have been already published in related conferences and journals. For researchers from the other disciplines, the Nectar Track offers a place to present their work to the ECML PKDD community and to raise the community's awareness of data analysis results and open problems in their field. Particularly welcome were papers illustrating the pervasiveness of data-driven exploration and modelling in science, technology, and society, as well as innovative applications, and also theoretical results. Authors were invited to submit four-page summaries of their previously published work.

We received 25 submissions and each of them was thoroughly reviewed by two Program Committee (PC) members. Finally, ten papers were selected for publication in the proceedings and presentation during the conference. The accepted papers cover a wide range of machine learning and data mining methods, as well as quite diverse domains of applications. The topics cover, among others, automatic music generation, music chord prediction, phenotype inference from biomedical texts and genomic databases, new data-driven approaches for finding a parking space in cities, process-based modelling to construct dynamical systems, advances in kernel-based graph classification, user interactions and influence in social networks, efficient exploitation of tree ensembles in Web search and document ranking, data cleaning with AI planning solvers, and applications of predictive clustering trees to image analysis.

We take this opportunity to thank all authors for submitting their papers to the Nectar Track. We also wish to express our gratitude to all PC members who helped us in the reviewing process, providing insightful feedback that helped the authors of accepted papers to prepare good presentations during the conference. Finally, we would like to thank the ECML PKDD general chairs and the other members of the Organizing Committee for their excellent co-operation and support for all our efforts. We hope that the readers will enjoy these short papers and that the papers, conference presentations, and discussions will inspire further interesting research at the boundaries of machine learning and data mining with many other interesting fields.

September 2017

Donato Malerba
Jerzy Stefanowski

# Foreword to the ECML PKDD 2017 Demo Track

We present, with great pleasure, the Demo Track of ECML PKDD 2017. Since its inception, this Demo Track is among the major forums in the field for presenting state-of-the-art data mining and machine learning systems and research prototypes, and for disseminating new methods and techniques in a variety of application domains. Each selected demo was presented at the conference and allocated a four-page paper in the proceedings.

The evaluation criteria encompassed innovation and technical advances, meeting novel challenges, and the potential impact and interest for researchers and practitioners in the machine learning and data-mining community. Each submission was first reviewed by at least two expert referees, with a majority receiving three reviews. Consensus on each paper was reached through discussion between the demo chairs. In total, 52 reviews were made, and from 17 original submissions 10 were accepted for publication in the conference proceedings and presentation at the demo sessions during the conference in Skopje. The accepted demonstration papers cover a wide range of machine learning and data mining techniques, as well as a very diverse set of real-world application domains. We believe the review system was successful in ensuring that the accepted work is of high quality and suited for publication in the track.

We thank all authors for submitting their work, without which this track would not be possible. We are deeply grateful to our Program Committee for volunteering their time and expertise. Their contribution is at the core of the scientific quality of the Demo Track. The expert Program Committee included a mix of experienced individuals from previous years as well as experts newly recruited to ensure broad technical expertise and to promote inclusivity of various data mining and machine learning research areas. Finally, we wish to thank the general chairs and the program chairs for entrusting us with this track and providing us with their expert advice. We hope that the readers will enjoy this set of short papers and that the demonstrated systems, prototypes, and libraries of this track will inspire interaction and discussion that will be valuable to both the authors and the community at large.

September 2017

Jesse Read
Marinka Zitnik

# Organization

## ECML PKDD 2017 Organization

### Conference Chairs

| | |
|---|---|
| Michelangelo Ceci | University of Bari Aldo Moro, Italy |
| Sašo Džeroski | Jožef Stefan Institute, Slovenia |

### Program Chairs

| | |
|---|---|
| Michelangelo Ceci | University of Bari Aldo Moro, Italy |
| Jaakko Hollmén | Aalto University, Finland |
| Ljupčo Todorovski | University of Ljubljana, Slovenia |
| Celine Vens | KU Leuven Kulak, Belgium |

### Journal Track Chairs

| | |
|---|---|
| Kurt Driessens | Maastricht University, The Netherlands |
| Dragi Kocev | Jožef Stefan Institute, Slovenia |
| Marko Robnik-Šikonja | University of Ljubljana, Slovenia |
| Myra Spiliopoulu | Magdeburg University, Germany |

### Applied Data Science Track Chairs

| | |
|---|---|
| Yasemin Altun | Google Research, Switzerland |
| Kamalika Das | NASA Ames Research Center, USA |
| Taneli Mielikäinen | Yahoo! USA |

### Local Organization Chairs

| | |
|---|---|
| Ivica Dimitrovski | Ss. Cyril and Methodius University, Macedonia |
| Tina Anžič | Jožef Stefan Institute, Slovenia |
| Mili Bauer | Jožef Stefan Institute, Slovenia |
| Gjorgji Madjarov | Ss. Cyril and Methodius University, Macedonia |

### Workshops and Tutorials Chairs

| | |
|---|---|
| Nathalie Japkowicz | American University, USA |
| Pance Panov | Jožef Stefan Institute, Slovenia |

## Awards Committee

Peter Flach                 University of Bristol, UK
Rosa Meo                    University of Turin, Italy
Indrė Žliobaitė             University of Helsinki, Finland

## Nectar Track Chairs

Donato Malerba              University of Bari Aldo Moro, Italy
Jerzy Stefanowski           Poznan University of Technology, Poland

## Demo Track Chairs

Jesse Read                  École Polytechnique, France
Marinka Žitnik              Stanford University, USA

## PhD Forum Chairs

Tomislav Šmuc               Rudjer Bošković Institute, Croatia
Bernard Ženko               Jožef Stefan Institute, Slovenia

## EU Projects Forum Chairs

Petra Kralj Novak           Jožef Stefan Institute, Slovenia
Nada Lavrač                 Jožef Stefan Institute, Slovenia

## Proceedings Chairs

Jurica Levatić              Jožef Stefan Institute, Slovenia
Gianvito Pio                University of Bari Aldo Moro, Italy

## Discovery Challenge Chair

Dino Ienco                  IRSTEA - UMR TETIS, France

## Sponsorship Chairs

Albert Bifet                Télécom ParisTech, France
Pance Panov                 Jožef Stefan Institute, Slovenia

## Production and Public Relations Chairs

Dragi Kocev                 Jožef Stefan Institute, Slovenia
Nikola Simidjievski         Jožef Stefan Institute, Slovenia

# ECML PKDD Steering Committee

| | |
|---|---|
| Michele Sebag | Université Paris Sud, France |
| Francesco Bonchi | ISI Foundation, Italy |
| Albert Bifet | Télécom ParisTech, France |
| Hendrik Blockeel | KU Leuven, Belgium and Leiden University, The Netherlands |
| Katharina Morik | University of Dortmund, Germany |
| Arno Siebes | Utrecht University, The Netherlands |
| Siegfried Nijssen | LIACS, Leiden University, The Netherlands |
| Chedy Raïssi | Inria Nancy Grand-Est, France |
| Rosa Meo | Università di Torino, Italy |
| Toon Calders | Eindhoven University of Technology, The Netherlands |
| João Gama | FCUP, University of Porto/LIAAD, INESC Porto L.A., Portugal |
| Annalisa Appice | University of Bari Aldo Moro, Italy |
| Indré Žliobaité | University of Helsinki, Finland |
| Andrea Passerini | University of Trento, Italy |
| Paolo Frasconi | University of Florence, Italy |
| Céline Robardet | National Institute of Applied Science in Lyon, France |
| Jilles Vreeken | Saarland University, Max Planck Institute for Informatics, Germany |

# Applied Data Science Track Program Committee

Michele Berlingerio
Michael Berthold
Kanishka Bhaduri
Berkant Barla Cambazoglu
Soumyadeep Chatterjee
Abon Chaudhuri
Debasish Das
Mahashweta Das
Dinesh Garg
Guillermo Garrido
Rumi Ghosh
Slawek Goryczka
Francesco Gullo
Georges Hebrail
Hongxia Jin
Anuradha Kodali

Deguang Kong
Mikhail Kozhevnikov
Hardy Kremer
Sricharan Kumar
Mounia Lalmas
Zhenhui Li
Jiebo Luo
Arun Maiya
Silviu Maniu
Luis Matias
Dimitrios Mavroeidis
Thomas Meyer
Daniil Mirylenka
Xia Ning
Nikunj Oza
Daniele Pighin

Fabio Pinelli
Elizeu Santos-Neto
Manali Sharma
Alkis Simitsis
Siqi Sun
Maguelonne Teisseire
Ingo Thon
Antti Ukkonen
Ranga Vatsavai
Pinghui Wang
Xiang Wang
Ding Wei
Cheng Weiwei
Yanchang Zhao

## Nectar Track Program Committee

| | | |
|---|---|---|
| Annalisa Appice | Peter Flach | Pauli Miettinen |
| Hendrik Blockeel | João Gama | Ernestina Menasalvas |
| Toon Calders | Kristian Kersting | Celine Robardet |
| Tijl De Bie | Stan Matwin | Bernhard Pfahringer |

## Demo Track Program Committee

| | | |
|---|---|---|
| Monica Agrawal | Vladimir Gligorijevic | Noel Malod Dognin |
| Albert Bifet | Francois Jacquenet | Olivier Pallanca |
| Aleksandar Dimitriev | Isak Karlsson | Joao Papa |
| Elisa Fromont | Mark Last | Mykola Pechenizkiy |
| Ricard Gavalda | Noel Malod | Bo Wang |

## Sponsors

### Gold Sponsors

| | |
|---|---|
| Deutsche Post DHL Group | http://www.dpdhl.com/ |
| Google | https://research.google.com/ |

### Silver Sponsors

| | |
|---|---|
| AGT | http://www.agtinternational.com/ |
| ASML | https://www.workingatasml.com/ |
| Deloitte | https://www2.deloitte.com/global/en.html |
| NEC Europe Ltd. | http://www.neclab.eu/ |
| Siemens | https://www.siemens.com/ |

### Bronze Sponsors

| | |
|---|---|
| Cambridge University Press | http://www.cambridge.org/wm-ecommerce-web/academic/landingPage/KDD17 |
| IEEE/CAA Journal of Automatica Sinica | http://www.ieee-jas.org/ |

### Awards Sponsors

| | |
|---|---|
| Machine Learning | http://link.springer.com/journal/10994 |
| Data Mining and Knowledge Discovery | http://link.springer.com/journal/10618 |
| Deloitte | http://www2.deloitte.com/ |

**Lanyards Sponsor**

KNIME                          http://www.knime.org/

**Publishing Partner and Sponsor**

Springer                       http://www.springer.com/gp/

**PhD Forum Sponsor**

IBM Research                   http://researchweb.watson.ibm.com/

**Invited Talk Sponsors**

EurAi                          https://www.eurai.org/
GrabIT                         https://www.grabit.mk/

# Contents – Part III

# Contents – Part I

## Ensembles and Meta Learning

## Feature Selection and Extraction

## Kernel Methods

**Networks and Graphs**

## Neural Networks and Deep Learning

# Contents – Part II

## Recommendation

## Regression

# Applied Data Science Track

# A Novel Framework for Online Sales Burst Prediction

Rui Chen and Jiajun Liu[✉]

Big Data Analytics and Intelligence Lab, School of Information,
Renmin University of China, Beijing, China
{r_chen,jiajunliu}@ruc.edu.cn

**Abstract.** With the rapid growth of e-commerce, a large number of online transactions are processed every day. In this paper, we take the initiative to conduct a systematic study of the challenging prediction problems of sales bursts. Here, we propose a novel model to detect bursts, find the bursty features, namely the start time of the burst, the peak value of the burst and the off-burst value, and predict the entire burst shape. Our model analyzes the features of similar sales bursts in the same category, and applies them to generate the prediction. We argue that the framework is capable of capturing the seasonal and categorical features of sales burst. Based on the real data from JD.com, we conduct extensive experiments and discover that the proposed model makes a relative MSE improvement of 71% and 30% over LSTM and ARMA.

**Keywords:** Burst prediction · E-commerce

## 1 Introduction

E-commerce websites have become an ubiquitous mechanism for online shopping. Devendra first defined that electronic commerce, commonly known as e-commerce or eCommerce, consists of the buying and selling of products or services over electronic system such as internet and other computer network [1]. The effects of e-commerce have reached all areas of business, from customer service to new product design [2].

According to statistics in 2015, turnover of E-commerce has reached 18.3 trillion in China, up by 36.5% [4]. This can also be seen from the huge trading volume on some special shopping festivals in China. In the shopping festival of 11th Nov, 2016, the single-day merchandise trade of tmall.com reached 912.17 billion yuan, up by 61%. For the logistics industry, the number of orders will have a corresponding explosive growth. Nowadays, the e-commerce platforms usually launch promotional activities on a particular category of products. Hence the products in the same category always show similar sales changes.

In e-commerce field, time series prediction is wildly used. A large number of methods have been developed and applied to time series forecasting problems, such as sophisticated statistical methods [16] and neural network [15]. For any

© Springer International Publishing AG 2017
Y. Altun et al. (Eds.): ECML PKDD 2017, Part III, LNAI 10536, pp. 3–14, 2017.
https://doi.org/10.1007/978-3-319-71273-4_1

product, the sales series can increase or fall sharply, which we called spikes or bursts. The prediction of bursts is beneficial in several aspects, such as the storage optimization of the suppliers and the stability maintanance of the e-commerce website. In the existing studies about burst prediction, most of them are concerned about predicting the bursty features. Few of them focus on the prediction of the entire sales burst shape.

In this paper, we study the task of analyzing the bursty features of product sales, and propose a model framework to predict sales burst. The major contributions of this paper include:

1. We define a new problem of time series prediction that mainly concerns about the bursts. To formulate this problem, we split it into three parts: detecting bursts, predicting the features and predicting the shape of bursts.
2. We propose a novel framework to capture the seasonal and categorical features for predicting the entire burst series. We also take the initiative to use the reshaped nearest neighbors to simulate the burst shape.
3. We conduct extensive experiments on real datasets from sale records on JD.com and evaluate the advantages and characteristics of the proposed model. The results show that our model has a significant improvement.

The rest of the paper is organized as follows. In Sect. 2 we present the background and relevent literature of the problem studied. In Sect. 3 we give the problem formulation and notations. Datasets with the preprocessing steps are shown in Sect. 4. In Sect. 5 we give the intuition of our model and describe the layers of our model framework. Experimental results on real life data are shown in Sect. 6. Finally, we conclude our work in Sect. 7.

## 2   Related Work

In recent years, many methods have been developed and applied to time series forecasting problems, such as sophisticated statistical methods [16] and neural networks [15]. Schaidnagel et al. [14] presented a parametrized time series algorithm that predicts sales volumes with variable product prices and low data support.

Burst, defined as "a brief period of intensive activity followed by long period of nothingness" [3], is a common phenomenon in time series. Most existing studies about burst prediction, mainly applied in social networks. Lin et al. [11] proposed a framework to capture dynamics of hashtags based on their topicality, interactivity, diversity, and prominence. Ma et al. [12] predicted the popularity of hashtags on daily basis. [8,9] introduced a method to predict the burst of Twitter hashtags before they actually burst. Bursts of topics, sentiments and questions have been demonstrated to have a predictive power of product sales. Gruhl et al. [6] proposed to use online postings to predict spikes in sales rank. Such methods have been proved to be effective and achieved good performance. However, how to design a framework that can predict the entire burst shape, is yet to be answered. Inspired by the success of these methods, we divided the problem into several parts and investigate using a framework with three layers.

# 3   Problem Formulation and Notations

The sales of a product can be formed a time series $<x_1, x_2, ...x_t, ...>$. $x_t$ donates the count of sales at the $t$-th time interval. Features of a sales burst contains the start time, start value, burst time, peak value, period, off-burst time, etc. Given a product $p$ in category $c$, we consider a prediction scenario: a product $p$ has a historical sales series $x$, at time $t$, will the sales of product burst in the near future? If the product will be a bursting one, what's the shape of the entire burst?

In the rest of the paper, the technical details of the algorithms will be described mostly in vector forms. All the assumptions and notations mentioned in this paper are listed in Table 1.

**Table 1.** Table of notations

| Notation | Description |
|---|---|
| $X_c$ | The set of time series in category $c$ |
| $\bar{X}_c$ | The set of reshaped burst series in category $c$ |
| $x \in \mathbb{R}^d$ | The time series of length d |
| $\bar{x} \in \mathbb{R}^m$ | The reshaped burst series of length m |
| $w \in \mathbb{R}^k$ | The window with length k |
| $s \in \mathbb{R}^k \sim w$ | The time sequence within the window with length k |
| $lmaxs$ | The set of local max points in a time series |
| $lmins$ | The set of local min points in a time series |
| $B_c$ | The set of bursty feature vector for category $c$ |
| $b = (s, b, p, T, e)$ | The bursty feature vector: start time, burst time, peak value, period, off-burst value |

# 4   Dataset and Set up

## 4.1   Dataset Description

To support the comprehensive experiments, we collected product information and transaction records from a real-life e-commerce website, JD.com. The original dataset contains two parts. The first part is a 139 million set of transaction-records from 1/1/2008 to 12/1/2013, including user ID, product ID, and the purchasing date. Single-day sale of one product ranges from 0 to 7802. The second part is a 0.25 million set of product information, including product ID, categories, brand and product name. There are 167 categories in total.

## 4.2   Preprocess

In order to prepare the datasets required in the framework, we conducted the following steps.

**Filtering.** Since random factors have a great impact on time series when the whole number of records is small, we select records from 2010 to 2013, which constitute 90% of all sales. Besides, we choose products with good selling frequency, using a threshold of 100 records in total.

**Time Series Generation.** For each product, We calculate the daily sales volume as the value of each node in the time series with the transaction records.

**Smoothing.** After the above process, we will get plenty of raw time series with a lot of sharp rise and fall. To detect bursts, we smooth the time series. In specific, we perform discrete kalman filter [7] with exponentially weighted moving average (EWMA) on each time series. The smoothed series in each category will form a new time series set.

**Sample Bursts Extraction.** For category $c$, we have time series set $X_c$. Suppose $X \in X_c, X = x_1, x_2, ...x_d$, the sample bursts are extracted through the following steps. First, we find all the local maximums $lmaxs$ and local minimums $lmins$ for $X$. For the sequence between each local minimal pair, $(lmin_i, lmin_{i+1})$, there must be a $lmax_i$. So we judge whether the sequence contains a burst by the slope of $(lmin_i, lmax_i)$. The threshold $delta$ is computed by the standard deviation of $X$ and window size $k$.

In the process of rise and fall, sales series may have small fluctuations. Once a new burst $b$ is detected, we will judge whether it should be merged with the previous burst $b'$ with the features of $b$ and $b'$. In specific, there are two situations when we choose to merge two bursts:

- Burst $b'$ appeared in the falling process of burst $b$: merge $b$ and $b'$ as a new burst $b'' = (s, p, b, T + T', e')$
- Burst $b$ appeared in the rising process of burst $b'$: merge $b$ and $b'$ as a new burst $b'' = (s, p', b', T + T', e')$

**Reshaping.** Based on the assumption that products in the same category have similar bursty features, we propose to generate a new type of dataset for each category with the burst series. The burst series with period $T$ in each category are reshaped into a fixed length $m$:

$$\bar{X} = \bar{x}_1, \bar{x}_2, ..., \bar{x}_m, \bar{x}_i = x_k + (x_{k+1} - x_k) \times (i \times \frac{T}{m} - k), k = \lfloor i \times \frac{T}{m} \rfloor$$

All reshaped bursts will form a new dataset $\bar{X}_c$.

Hence, for each category $c$, we have three datasets: the smoothed time series dataset $X_c$, the reshaped bursts dataset $\bar{X}_c$, and the bursty feature dataset $B_c$. The categories with less than $N$ products (we set $N$ as 100) are ignored in this process, and we keep 64 categories in total. Figure 1 shows the distribution of the bursty feature datasets.

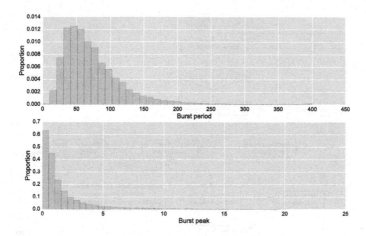

**Fig. 1.** The distribution of bursty features

# 5   The Model

## 5.1   Intuition

In the existing studies about burst prediction, most of them focus on how to predict the bursty features, rather than the entire burst. The performance of models are improved by the usage of various features. For sales time series, the extra information is much less than topics or hashtags on websites. Most of the time, we just have the transaction records and basic information about the products. When we are faced with a new product with no historical data, the prediction problem will become more difficult. In the absence of sufficient information and data, we need to figure out a proper approach to predict the entire burst shape. Next we will present an overview of the model framework.

## 5.2   Model Overview

We propose a model framework to predict the entire burst shape of online products, shown in Fig. 2. The model has three layers, namely the burst detection layer, bursty feature prediction layer, and the burst shape prediction layer. The final layer will generate the output prediction of the whole model. The model implements the following work flow:

## 5.3   Burst Detection

The burst detection layer aims to detect the start of a burst. For time series $X$, we let the window $w$ slide with the timeline from the start point of $X$. At time $t$, we get a sequence $s_t = x_t, x_{t+1}, ..., x_{t+d}$, and we will use this sequence as the input of the classification model. The sliding window is to simulate and solve the real situation in e-commercial websites.

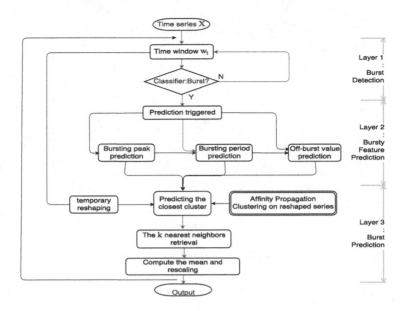

**Fig. 2.** The framework of burst shape prediction model

For the classification part, we apply a SVM-based method to solve this problem. We first define a feature space based on the time series features in [10], and add some new features such as $max_value$, $min_value$, $id_max$ and $id_min$. The feature vector $f$ is the input of the SVM model, and we will get the class label as output, indicating whether it will be a bursting one.

### 5.4 Bursty Feature Prediction

Once a burst has been detected, we want to know its bursty features, such as how high the peak will reach, when it will be off-burst and what the off-burst value is. In this layer, these features are predicted by three different regression model, named $M^{(1)}, M^{(2)}$ and $M^{(3)}$. The predicted results in $M^{(i-1)}$ and $f$ will form a new vector, as the input for $M^{(i)}$. For each category $c$, We will train several different types of $M^{(i)}$, and choose the one with smallest mean square error (MSE).

### 5.5 Burst Shape Prediction

The burst shape prediction layer aims to forecast the whole burst shape using the bursty features $b$ and the time sequence $s_t$. For each category $c$, a clustering model is pre-trained with the reshaped bursts dataset $\bar{X}_c$.

The key idea is to use the corresponding part and the bursty features as the measure of similarities. First, with the predicted period $T$, the time sequence $s_t$ will be reshaped into a new sequence of fixed length $l = k \times (d/m)$,

$s'_t = s_1, s_2, ..., s_l$. For each cluster center of reshaped bursts $\bar{x} = \bar{x}_1, \bar{x}_2, ..., \bar{x}_m$, we only use the same part $\bar{x}_1, \bar{x}_2, ..., \bar{x}_l$ to calculate the similarity of two sequences and find the predicted cluster $s_t$. There are plenty of methods to measure similarity, such as Euclidean distance, cosine similarity and so on. For time series, Dynamic Time Warping (DTW) [13] is good alternative, which is designed to find an optimal alignment between two given (time-dependent) sequences. Then, we can find all reshaped bursts in $s_t$, and calculate new similarities between $s'_t$ and their corresponding parts. Here, the new similarity contains sequence similarity, absolute value of period similarity, peak similarity and off-burst value similarity. The weight of the first part is highest and for other parts we assign same weights. We choose the highest ranking $k$ sequences, calculate its mean series $\bar{x}'$, and rescale it with the period $T$ as the output prediction. The rescaling method uses the reverse process of reshaping: $X = x_1, x_2, ..., x_T, x_i = \bar{x}'_k + (\bar{x}'_{k+1} - \bar{x}'_k) \times (i \times \frac{m}{T} - k), k = \lfloor i \times \frac{m}{T} \rfloor$.

## 6   Experiments and Evaluation

**Setup.** For each individual dataset, we randomly divide the samples into three folds: the training set, the validation set and the test set, with the proportion of 3:1:1. The model is trained using the training set, and is then tested on the validation set. Such cross-validation is performed on the same individual dataset for ten times with random splits, and the reported performance is the averaged value cross the ten iterations. Finally we test the model on the test set and report the performance.

### 6.1   Evaluation of the Burst Detection Layer

In this layer, we set a maximum number of samples as 30000 to reduce training time, and randomly select the same number of positive and negative samples from the training dataset. Before the training, we evaluate the contributions of features with a L1-based linear SVC model, and then reduce the dimensionality of input feature vector of the classifier, a SVC model with rbf kernel. Precision, Recall and F1-score are reasonable metrics for the evaluation of this layer.

Table 2 and Fig. 3 show the performance comparison of the first layer, detecting bursts. It can be observed that our model shows the best performance of F1-score for 90% categories listed in the table, and achieves the highest F1-score of 0.77 on average. Besides, our model achieves relatively high precision and recall scores on average, both scoring over 0.72. The performance of K-Nearest Neighbors is quite different, which achieves the best recall score 0.84 and the worst precision 0.63. SVM models with sigmoid kernel is the most ineffective and unstable one, with the deviations of all scores over 0.15. SVM models with linear kernel and rbf kernel always have similar performance.

In practical applications, there may be different requirements on precision and recall. These requirements can be satisfied by training optimal prediction models using different types of F-scores, and select the one with best score.

Table 2. Performance comparision of detecting bursts (partial data)

| ID | F1-score | | | | | |
|---|---|---|---|---|---|---|
| | SVM(linear) | SVM(rbf) | SVM(sigmoid) | KNeighbors | DecisionTree | Our method |
| 0 | 0.78 | 0.78 | 0.31 | 0.69 | 0.77 | **0.85** |
| 1 | 0.81 | 0.83 | 0.61 | 0.63 | 0.84 | **0.85** |
| 2 | 0.78 | 0.78 | 0.77 | 0.68 | 0.79 | **0.80** |
| 3 | **0.75** | **0.75** | 0.63 | 0.87 | **0.75** | 0.75 |
| 4 | 0.70 | 0.69 | 0.69 | **0.73** | 0.70 | 0.69 |
| 5 | 0.59 | 0.59 | 0.48 | 0.59 | 0.57 | **0.64** |
| 6 | 0.73 | 0.73 | 0.65 | 0.70 | **0.74** | **0.74** |
| 7 | 0.87 | 0.89 | 0.61 | 0.70 | 0.90 | **0.91** |
| 8 | 0.73 | **0.74** | 0.68 | 0.69 | 0.73 | **0.74** |
| 9 | **0.77** | **0.77** | 0.71 | 0.67 | **0.77** | 0.77 |

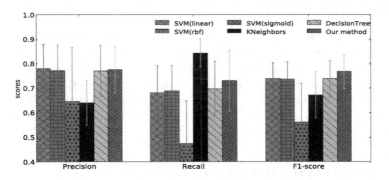

**Fig. 3.** Mean average classification performance with standard deviation (Precision/Recall/F1-score: the higher the better)

## 6.2   Evaluation of the Bursty Features Prediction Layer

For each category $c$, we train multiple regression models, select the best one, that is, the one with smallest MSE, as prediction model for current $c$. Using features we generated, we can find typical features for each $c$. In specific, we use the feature selection algorithm $\chi^2$, to compute the score of each feature, and select the $K = 10$ highest scoring features.

Table 3 studies the performance of the proposed method and the competitive methods on 64 individual datasets of different category, evaluated by average HitRate (HR)@20% and 50%, Mean Squared Error (MSE) and Mean absolute Relative Error (MRE). HitRate is the percentage of times when the relative error is within a specific range. For example, for a set of targets that equal 10, if 50% of time the predicted value is within [6–14], the HitRate@40% is 0.5.

The highlighted numbers in red, blue and black indicate the winners on each model under the corresponding metric. To compare the methods quantitatively,

**Table 3.** Performance comparision on scores of predicting peak value, period and off-burst value of bursts (HR: the higher the better, MSE/MRE: the lower the better)

| Method | Avg MSE | Avg MRE | Avg HitRate@20% | Avg HitRate@50% |
|---|---|---|---|---|
| Linear SVR | 29.91—12096.26—3.88 | 1.91—50.90—0.69 | 0.14—0.27—0.12 | 0.37—0.61—0.26 |
| SVR | 26.76—10573.25—7.29 | 1.88—49.65—0.94 | 0.16—0.27—0.09 | 0.38—0.61—0.22 |
| Linear Regression | 30.95—12861.00—4.98 | 2.23—62.28—0.57 | 0.12—0.22—0.14 | 0.30—0.50—0.30 |
| Bayes | 30.22—12399.39—4.97 | 2.22—61.72—0.57 | 0.12—0.22—0.14 | 0.30—0.50—0.30 |
| CART | 42.95—15538.12—7.53 | 2.49—63.28—0.70 | 0.13—0.25—0.11 | 0.31—0.54—0.26 |
| Our method | **25.66—10032.34—3.78** | **1.87—47.85—0.56** | **0.16—0.28—0.15** | **0.40—0.68—0.38** |

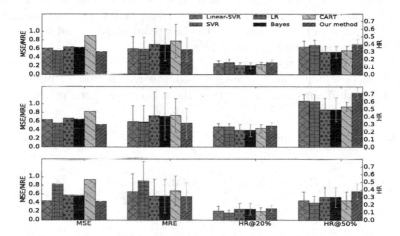

**Fig. 4.** Mean average prediction performance with standard deviation. The three sub-figures represent the task of predicting the peak value, period and off-burst value) (HR: the higher the better, MSE/MRE: the lower the better)

we also provide Fig. 4 (MRE/MSE is normalized with the maximum MRE/MSE among the methods in each entry).

Our model achieves the best performance of MSE and MRE and the highest score of HR@20% and HR@50% with a relatively low deviation. For example, in the task of the peak value prediction, we find our model's performance and the average of other methods' performance under MSE, MRE, HR@20% and HR@50% are 25.66 vs. 32.16 (unnormalized), 1.87 vs. 2.15 (unnormalized), 0.16 vs. 0.13 and 0.40 vs. 0.33 respectively, showing that our model makes a relative improvement of 20%, 13%, 23% and 21% respectively. In the evaluation of forecasting the burst period and off-burst value, our model has a optimal performance in HR@50%, reaching 0.68 and 0.38 on average.

It can also be observed that Linear Regression and Bayes Ridge Regression show similar performance on HR@20% and HR@50%. Linear SVR and SVR with rbf kernel have different performance under MSE and MRE. The CART method have the worst performance on MSE and MRE among all the methods.

## 6.3    Evaluation of the Burst Shape Prediction Layer

For this prediction task, we compare our model with two different models commonly used for time series prediction. The metric mean square error (MSE) is used to evaluate the prediction performance. We perform the Affinity Propagation Clustering [5] on the dataset of reshaped bursts. Euclidean distance is applied to calculate the input similarity matrix of the training set, and we choose DTW distance as the measure of similarity when forecasting. Besides, when the start value of the forecasted result and the last value in the window differ greatly, we apply a decay function to smooth the 30% of the predicted burst series, namely $y_{30\%}$: $y = \alpha f + (1 - \alpha)g$. Here, we set $f$ as the straight line from the last point in the window $w$ to the last point of $y_{30\%}$, and set $g$ as $y_{30\%}$. $\alpha$ is set as the exponential function $e^{(-1/3)t}$.

Table 4 and Fig. 5 shows the results of performance comparison. For the first two methods, we judge the performance on the real period of bursts. If the predicted period of our model is less than the real one, we set the left part with

**Table 4.** Performance comparision on the entire period of bursts (MSE score, partial data)

| Category ID | LSTM | ARMA | Our model |
|---|---|---|---|
| 0 | 9.48 | 8.39 | **5.94** |
| 1 | 10.45 | 8.32 | **6.43** |
| 2 | 1.35 | 0.97 | **0.78** |
| 3 | 40.15 | 32.70 | **31.72** |
| 4 | 15.62 | **5.01** | 14.39 |
| 5 | 32.66 | 5.61 | **2.82** |
| 6 | 0.83 | 0.61 | **0.28** |
| 7 | 13.91 | 8.48 | **6.51** |
| 8 | 21.67 | 2.64 | **0.86** |
| 9 | 14.10 | 7.55 | **4.69** |

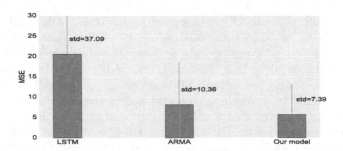

**Fig. 5.** Mean average burst series prediction MSE scores with standard deviation

the average value of our prediction. We also draw some of the predicted results into Fig. 6. Conclusions can be drawn as follows:

- The LSTM model shows the worst performance on average MSE, scoring 20.58. Besides, the performance of LSTM are not that stable, with relatively high deviations of 37.09.
- Take all categories into account, our model wins 57 out of 64 times on the score of MSE. The average MSE of our method is 5.88, which significantly outperforms the other related methods, showing a relative improvement of 71% and 30%. The MSE standard deviations of our model and other methods are 37.10, 10.36 and 7.57 on average, indicating that our model makes a significant improvement by 80% and 29%.

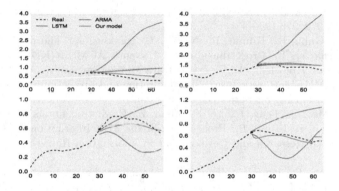

**Fig. 6.** Samples of the predicted burst results

# 7   Conclusion

In this paper, we take the initiative to propose a burst prediction framework of online product sales. The framework includes three layers: a burst detection layer, a bursty feature prediction layer and a burst shape prediction layer. The burst detection layer detect the start of a burst with a sliding window and a optimized classification model. The three bursty features, the burst peak, period and off-burst value, are predicted by different regression model with the best training score. The entire burst shape is generated by the burst series with similar seasonal features in the same category. Extensive experiments are conducted on real datasets from JD.com. We find that in average our framework achieves 4% to 45% advantage of F1-score in the classification and up to 73% improvement of HitRate@50% on the feature prediction against other methods. The result shows that the proposed solutions are effective to the burst prediction task, with an average improvement of 71% and 30% on MSE. We expect our framework to be of great value in e-commerce field.

**Acknowledgments.** This work was supported by the National Natural Science Foundation of China (No. 61602487), the Fundamental Research Funds for the Central Universities, and the Research Funds of Renmin University of China (No. 2015030275).

# References

1. Agrawal, D., Agrawal, R.P., Singh, J.B., Tripathi, S.P.: E-commerce: true indian picture. J. Adv. IT **3**(4), 250–257 (2012)
2. Avery, S.: Online tool removes costs from process. Purchasing **123**(6), 79–81 (1997)
3. Barabási, A.-L.: Bursts: The Hidden Patterns Behind Everything we do, from Your E-mail to Bloody Crusades. Penguin, New York (2010)
4. Cao, L., Zhang, Z.: The 2015 Annual China Electronic Commerce Market Data Monitoring Report. China Electronic Commerce Research Center, Hangzhou (2016)
5. Frey, B.J., Dueck, D.: Clustering by passing messages between data points. Science **315**(5814), 972–976 (2007)
6. Gruhl, D., Guha, R., Kumar, R., Novak, J., Tomkins, A.: The predictive power of online chatter. In: Proceedings of the Eleventh ACM SIGKDD International Conference on Knowledge Discovery in Data Mining, pp. 78–87. ACM (2005)
7. Kalman, R.E.: A new approach to linear filtering and prediction problems. Trans. ASME-J. Basic Eng. **82**(Series D), 35–45 (1960)
8. Kong, S., Mei, Q., Feng, L., Ye, F., Zhao, Z.: Predicting bursts and popularity of hashtags in real-time. In: Proceedings of the 37th International ACM SIGIR Conference on Research and Development in Information Retrieval, pp. 927–930. ACM (2014)
9. Kong, S., Mei, Q., Feng, L., Zhao, Z.: Real-time predicting bursting hashtags on Twitter. In: Li, F., Li, G., Hwang, S., Yao, B., Zhang, Z. (eds.) WAIM 2014. LNCS, vol. 8485, pp. 268–271. Springer, Cham (2014). https://doi.org/10.1007/978-3-319-08010-9_29
10. Kong, S., Mei, Q., Feng, L., Zhao, Z., Ye, F.: On the real-time prediction problems of bursting hashtags in twitter. arXiv preprint arXiv:1401.2018 (2014)
11. Lin, Y.-R., Margolin, D., Keegan, B., Baronchelli, A., Lazer, D.: # bigbirds never die: understanding social dynamics of emergent hashtag. arXiv preprint arXiv:1303.7144 (2013)
12. Ma, Z., Sun, A., Cong, G.: On predicting the popularity of newly emerging hashtags in Twitter. J. Am. Soc. Inf. Sci. Technol. **64**(7), 1399–1410 (2013)
13. Ratanamahatana, C.A., Keogh, E.: Making time-series classification more accurate using learned constraints. In: Proceedings of the 2004 SIAM International Conference on Data Mining, pp. 11–22. SIAM (2004)
14. Schaidnagel, M., Abele, C., Laux, F., Petrov, I.: Sales prediction with parametrized time series analysis. In: Proceedings DBKDA, pp. 166–173 (2013)
15. Thiesing, F.M., Vornberger, O.: Sales forecasting using neural networks. In: 1997 International Conference on Neural Networks, vol. 4, pp. 2125–2128. IEEE (1997)
16. Weigend, A.S.: Time Series Prediction: Forecasting the Future and Understanding the Past. Santa Fe Institute Studies in the Sciences of Complexity (1994)

# Analyzing Granger Causality in Climate Data with Time Series Classification Methods

Christina Papagiannopoulou[1]($\boxtimes$), Stijn Decubber[1], Diego G. Miralles[2,3], Matthias Demuzere[2], Niko E. C. Verhoest[2], and Willem Waegeman[1]

[1] Department of Mathematical Modelling, Statistics and Bioinformatics, Ghent University, Ghent, Belgium
{christina.papagiannopoulou,stijn.decubber,willem.waegeman}@ugent.be
[2] Laboratory of Hydrology and Water Management, Ghent University, Ghent, Belgium
{matthias.demuzere,niko.verhoest,diego.miralles}@ugent.be
[3] Department of Earth Sciences, VU University Amsterdam, Amsterdam, The Netherlands

**Abstract.** Attribution studies in climate science aim for scientifically ascertaining the influence of climatic variations on natural or anthropogenic factors. Many of those studies adopt the concept of Granger causality to infer statistical cause-effect relationships, while utilizing traditional autoregressive models. In this article, we investigate the potential of state-of-the-art time series classification techniques to enhance causal inference in climate science. We conduct a comparative experimental study of different types of algorithms on a large test suite that comprises a unique collection of datasets from the area of climate-vegetation dynamics. The results indicate that specialized time series classification methods are able to improve existing inference procedures. Substantial differences are observed among the methods that were tested.

**Keywords:** Climate science · Attribution studies · Causal inference Granger causality · Time series classification

## 1 Introduction

Research questions in climate change research are mostly related to either *climate projection* or to *climate change attribution*. Climate projection or forecasting aims at predicting the future state of the climatic system, typically over the next decades. The goal of climatic attribution on the other hand is to identify and quantify cause-effect relationships between climate variables and natural or anthropogenic factors. A well-studied example, both for projection and attribution, is the effect of human greenhouse gas emissions on global temperature.

The standard approach in the field of climate science is based on simulation studies with mechanistic climate models, which have been developed, expanded and extensively studied over the last decades. Data-driven models, in contrast to mechanistic models, assume no underlying physical representation of reality

© Springer International Publishing AG 2017
Y. Altun et al. (Eds.): ECML PKDD 2017, Part III, LNAI 10536, pp. 15–26, 2017.
https://doi.org/10.1007/978-3-319-71273-4_2

but directly model the phenomenon of interest by learning a more or less flexible function of some set of input data. Climate science is one of the most data-rich research domains. With global observations on ever finer spatial and temporal resolutions from both satellite and *in-situ* measurements, the amount of (publicly available) climatic data sets has vastly grown over the last decades. It goes without any doubt that there is a big potential for making progress in climate science with advanced machine learning models.

The most common data-driven approach for identifying causal relationships in climate science consists of Granger causality modelling [17]. Analyses of this kind have been applied to investigate the influence of one climatic variable on another, e.g., the Granger causal effect of $CO_2$ on global temperature [1,20], of vegetation and snow coverage on temperature [19], of sea surface temperatures on the North Atlantic Oscillation [26], or of the El Niño Southern Oscillation on the Indian monsoon [25]. In Granger causality studies, one assumes that a time series A Granger-causes a time series B, if the past of A is helpful in predicting the future of B. The underlying predictive model that is commonly considered in such a context is a linear vector autoregressive model [8,32]. Similar to other statistical inference procedures, conclusions are only valid as long as all potential confounders are incorporated in the analysis. The concept of Granger-causality will be reviewed in Sect. 2.

In recent work, we have shown that causal inference in climate science can be substantially improved by replacing traditional statistical models with non-linear autoregressive methods that incorporate hand-crafted higher-level features of raw time series [27]. However, approaches of that kind require a lot of domain knowledge about the working of our planet. Moreover, higher-level features that are included in the models often originate from rather arbitrary decisions. In this article, we postulate that causal inference in climate science can be further improved by using automated feature construction methods for time series. In recent years, methods of that kind have shown to yield substantial performance gains in the area of time series classification. However, we believe that some of those methods also have a lot of potential to improve causal inference in climate science, and the goal of this paper is to provide experimental evidence for that. We experimentally compare a large number of time series classification methods – an overview and more discussion of these methods will be given in Sect. 3.

Most attribution studies in climate science infer causal relationships between time series of continuous measurements, leading to regression settings. However, classification settings arise when targeting extreme events, such as heatwaves, droughts or floods. We will conduct an experimental study in the area of investigating climate-vegetation dynamics, where such a classification setting naturally arises. This is an interesting application domain for testing time series classification methods, due to the availability of large and complex datasets with worldwide coverage. It is also a practically-relevant setting, because extremes in vegetation can reveal the vulnerability of ecosystems w.r.t. climate change [23]. A more precise description of this application domain and the experimental setup will be provided in Sect. 4. In Sect. 5, we will present the main results, which will allow us to formulate conclusions concerning which methods are more appropriate in the area of climate sciences.

## 2   Granger Causality for Attribution in Climate Science

Granger causality [17] can be seen as a predictive notion of causality between time series. In the bivariate case, when two time series are considered, one compares the forecasts of two models; a baseline model that includes only information from the target time series (which resembles the effect) and a so-called full model that includes also the history of the second time series (which resembles the cause). Given two time series $x = [x_1, x_2, ..., x_N]$ and $y = [y_1, y_2, ..., y_N]$, with $N$ being the length of the time series, one says that the time series $x$ Granger-causes the time series $y$ if the forecast of $y$ at a specific time stamp $t$ improves when information of the history of $x$ is included in the model.

In this paper we will limit our analysis to situations where the target time series $y$ consists of $\{0, 1\}$-measurements that denote the presence or absence of an extreme event at time stamp $t$. As such, one ends up with solving two classification problems, one for the baseline and one for the full model. We will work with the Area Under the Curve (AUC) as performance measure, because the class distribution will be heavily imbalanced, as a natural result of modelling extreme events. Let $\hat{y}$ denote the new time series that originates as the one-step ahead forecast of $y$ using either the baseline or the full model, then Granger causality can be formally formulated as follows:

**Definition 1.** *A time series $x$ Granger-causes $y$ if $AUC(y, \hat{y})$ increases when $x_{t-1}, x_{t-2}, ..., x_{t-P}$ are considered for predicting $y_t$, in contrast to considering $y_{t-1}, y_{t-2}, ..., y_{t-P}$ only, where $P$ is the lag-time moving window.*

Granger causality studies might yield incorrect conclusions when additional (confounding) effects exerted by other climatic or environmental variables are not taken into account [13]. The problem can be mitigated by considering time series of additional variables. For example, let us assume one has observed a third variable $w$, which might act as a confounder in deciding whether $x$ Granger-causes $y$. The above definition then naturally extends as follows.

**Definition 2.** *We say that time series $x$ Granger-causes $y$ conditioned on time series $w$ if $AUC(y, \hat{y})$ increases when $x_{t-1}, x_{t-2}, ..., x_{t-P}$ are included in the prediction of $y_t$, in contrast to considering $y_{t-1}, y_{t-2}, ..., y_{t-P}$ and $w_{t-1}, w_{t-2}, ..., w_{t-P}$ only, where $P$ is the lag-time moving window.*

An extension to more than three time series is straightforward. In our experiments, $y$ will represent the vegetation extremes at a given location, whereas $x$ and $w$ can be the time series of any climatic variable at that location (e.g., temperature, precipitation or radiation).

Generally, the null hypothesis ($H_0$) of Granger causality is that the baseline model has equal prediction error as the full model. Alternatively, if the full model predicts the target variable $y$ significantly better than the baseline model, $H_0$ is rejected. In most applications, inference is drawn in vector autoregressive models by testing for significance of individual model parameters. Other studies have used likelihood-ratio tests, in which the full and baseline models are nested models [26]. Those procedures have a number of important shortcomings: (1) existing statistical tests only apply to stationary time series, which is an unrealistic assumption for attribution studies in climate science, (2) most tests are based on linear models, whereas cause-effect relationships can be non-linear, and (3) the models used for such tests are trained and evaluated on in-sample data, which will typically result in overfitting when the dimensionality or the model complexity increases.

In recent work, we have introduced an alternative way of assessing Granger-causality, by focussing on quantitative instead of qualitative differences in performance between baseline and full models [27]. In this way, traditional linear models can be replaced by more accurate machine learning models. If both the baseline and the full model give evidence of better predictions, one can draw stronger conclusions w.r.t. cause-effect relationships. To this end, no statistical tests are computed, but the differences between the two types of models is visualized and interpreted in a quantitative way.

## 3    From Granger Causality to Time Series Classification

In the general framework that we presented in [27] we constructed hand-crafted features based on knowledge that has been described in the climate literature [12]. These features include lagged variables, cumulative variables as well as extreme indices. Therefore, we ended up with in total ~360 features extracted from one time series. Our previous study has shown that incorporating those features in any classical regression or classification algorithm might lead to a substantial increase in performance (for both the baseline and the full model).

In this article, we investigate whether this feature construction process can be automated using time series classification methods. Due to the increased public availability of datasets from various domains, many novel time series classification algorithms have been proposed in recent years. All those methods either try to find higher-level features that represent discriminative patterns or similarity measures that define an appropriate notion of relatedness between two time series [2,11,21]. The following categories can be distinguished:

(a) Algorithms that use the whole series or the raw data for classification. To this family of algorithms belongs the one nearest neighbour (1-NN) classifier with different distance measures such as the dynamic time warping (DTW) [29], which is usually the standard benchmark measure, and variations of it, the complexity invariant distance (CID) [3], the derivative DTW [14], the derivative transform distance (DTD) [15] and the Move-split-merge (MSM) [33] distance.
(b) Algorithms that are based on sub-intervals of the original time series. They usually use summary measures of these intervals as features. Typical algorithms in this category are the time series forest (TSF) [10], the time series bag of features (TSBF) [5] and the learned pattern similarity (LPS) [4].

(c) Algorithms that are attempting to find informative patterns, called shapelets, in the data. An informative shapelet is a pattern that helps in distinguishing the classes by its presence or absence. Representative algorithms of this class are the Fast shapelets (FS) [28], the Shapelet transform (ST) [18] and the Learned shapelets (LS) [16].

(d) Algorithms that are based on the frequency of the patterns in a time series. These algorithms build a vocabulary of patterns and form a histogram for each observation by using this vocabulary. Algorithms such as the Bag of patterns (BOP) [22], the Symbolic aggregate approximation-vector space model (SAXVSM) [31] and the Bag of SFA symbols (BOSS) [30] are based on the idea of a pattern vocabulary.

(e) Finally, there are approaches that combine more than one from the above techniques, forming ensemble models. A recently proposed algorithm named Collection of transformation ensembles (COTE) combines a large number of classifiers constructed in the time, frequency, and shapelet transformation domains.

In our comparative study, we run algorithms from the first four different groups. The main criteria for including a particular algorithm in our analysis are (1) availability of source code, (2) running time for the datasets that we consider, and (3) interpretability of the extracted features. Since we have collected multiple time series for a large part of the world (3,536 locations in total), the algorithms should run in a reasonable amount of time. Several algorithms had problems to finish within 3 days.

## 4    Experimental Setup

In order to evaluate the above-mentioned time series classification methods for causal inference, we adopt an experimental setup that is similar to [27]. The non-linear Granger causality framework is adopted to explore the influence of past-time climate variability on vegetation dynamics. To this end, data sets of observational nature were collected to construct climatic time series that are then used to predict vegetation extremes. Data sets have been selected on the basis of meeting the following requirements: (a) an expected relevance of the variable for vegetation dynamics, (b) the availability of multi-decadal records, and (c) the availability of an adequate spatial and daily temporal resolution. In our previous work, we collected in this way in total 21 datasets. For the present study, we retained three of them, while covering the three basic climatic variables: water availability, temperature, and radiation. The main reason for making this restriction was that in that way the running time of the different time series classification algorithms could be substantially reduced. Specifically, we collected one precipitation dataset, which is coming from a combination of *in-situ*, satellite data, and reanalysis outputs, called Multi-Source Weighted-Ensemble Precipitation (MSWEP) [7]. We include one temperature dataset, which is a reanalysis data set, and one radiation dataset from the European Centre for Medium-Range Weather Forecasts (ECMWF) ERA-Interim [9].

In addition to those three climatic datasets, we also collected a vegetation dataset. We use the satellite-based Normalized Difference Vegetation Index (NDVI) [34], which is a commonly used monthly long-term global indicator of vegetation [6]. Roughly speaking, NDVI is a graphical greenness indicator which measures how green is a specific point on the Earth at a specific time stamp. The study period starting from 1981–2010 is set by the length of the NDVI record. The dataset is converted to a 1° spatial resolution to match with the climatic datasets.

For most locations on Earth, NDVI time series exhibit a clear seasonal cycle and trend – see top panel of Fig. 1 for a representative example. However, in climate science, the interesting part of such a time series is the residual component, usually referred to as seasonal anomalies. In a statistical sense, climatic data can only be useful to predict this residual component, as both the seasonal cycle and the trend can be modelled with pure autoregressive features. Similarly as in [27], we isolate the residual component using time series decomposition methods, and we work further with this residual component – see bottom panel of Fig. 1 for an illustration. In a next step, extremes are obtained from the residuals, while taking the spatial distribution of those extremes into account. The most straightforward way is setting a fixed threshold per location, such as the 10% percentile of the residuals. However, this leads to spatial distributions that are physically not plausible, because one cannot expect that the same number of vegetation extremes is observed in all locations on Earth. At some locations, vegetation extremes are more probable to happen. For this reason, we group the location pixels into areas with the same vegetation type, by using the global vegetation classification scheme of the International Geosphere-Biosphere Program (IGBP) [24], which is generically used throughout a range of communities. We selected the map of the year 2001 (closer to the middle of our period of interest). In order to end up with coherent regions that have similar climatic and vegetation characteristics, we further divided the vegetation groups into areas in which only neighboring pixels can belong to the same group. That way, we create 27 different pixel groups in America, see Fig. 2. We limit the study to America because some of the time series classification methods that we analyse have a long running time. Once we know which of those methods perform well, the study can of course be further extended to other regions, under the assumption that the same methods are favored for those regions.

The vegetation extremes are then defined by applying a $10^{th}$ percentile threshold on the seasonal anomalies of each region. This is a common threshold in defining extremes in vegetation [35]. Applying a lower threshold would result in extreme events that are extremely rare, making it impossible to train predictive models. In this way, we produce the target variable of our time series classification task. The presence of an extreme is denoted with a '1' and the absence with a '0'. Unsurprisingly, the distribution of the vegetation extremes in time indicates that many more extremes occur in recent years, which means that a clear trend appears again in the time series of extreme events, even though the initial time series was detrended. This makes the time series highly

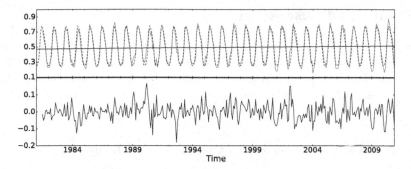

**Fig. 1.** The three components of an NDVI time series visualized for one particular location. The top panel shows the linear trend (black continuous line) and the seasonal cycle (dashed black line) that are obtained from the raw time series (red). The bottom panel visualizes the residuals, which are obtained by subtracting the seasonal cycle and the linear trend from the raw data. Only the residuals are further used to define extreme events. (Color figure online)

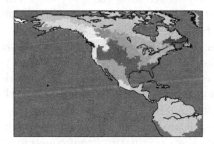

**Fig. 2.** Groups of pixels that are regions with similar climatic and vegetation characteristics. Based on the time series of each region we calculate the vegetation extremes for the pixels of that region.

non-stationary. Moreover, also a seasonal cycle typically re-appears, as one observes more extremes in certain months. Correctly identifying those two components (trend and seasonality) is essential when inferring causal relationships between vegetation extremes and climate.

As discussed in Sect. 2, a baseline model only includes information from the target time series (i.e. previous time stamps). We both consider the residuals as well as their binarized extreme counterparts as features for the baseline model. However, due to the existence of seasonal cycles and trends when considering binary time series of extreme vegetation, we also include 12 dummy variables which indicate the month of the observation and a variable for the year of this observation. These last two components are necessary because the baseline model should tackle as good as possible the seasonality and the trend that exists in the time series of NDVI extremes. In this paper, we perform a general test for causal relationships between climatic time series and vegetation. As such, the full model extends the baseline model with the above-mentioned climatic variables.

# 5    Results and Discussion

We present two types of experimental results. First, we analyze the predictive performance of various time series classification methods as representatives for the full model in a Granger-causality context. Subsequently, we select the best-performing algorithm for a Granger causality test, in which a baseline and a full model are compared.

## 5.1    Comparison of Time Series Classification Methods

For the first step we performed a straightforward comparison of the performance of the following algorithms: CID [3], LPS [4], TSF [10], SAXVSM [31], BOP [22], BOSS [30], FS [28] and hand-crafted features in combination with a classification algorithm [27]. In this setting, our dataset consists of monthly observations (there are in total 360 observations per pixel), and the input time series for each observation includes the 365 past daily values of precipitation time series before the month of interest (excluding the daily values of the current month). Only the precipitation time series is used, as some of the methods are unable to handle multivariate time series as input. We train the models per region by concatenating the observations of the pixels. The evaluation is performed per pixel by using random 3-fold cross-validation and AUC as performance measure.

Figure 3 shows the results. The vocabulary-based algorithms outperform the other representations, which implies that the frequency of the patterns makes the two classes of our dataset more distinguishable. Algorithms which distinguish the observations according to a presence or an absence of a shapelet perform poor, probably because observations originating from consecutive time windows have similar shapelets (the daily values of the next month is added for the next observation). In addition, the shapelet-based FS algorithm is also not very efficient in terms of memory space for large datasets. For this reason, we could not obtain results for the 4 largest regions of our dataset – see Table 1. For the algorithms that compare the whole raw time series by using a distance measure (i.e., CID) one can observe that the performance is also very low, probably also due to the strong similarity between consecutive observations. Similarly, algorithms that attempt to form a characteristic vector for each class fail since the patterns in both classes are very similar (i.e., SAXVSM). On the other hand, from the algorithms that use sub-intervals of time series, LPS has a similar performance as the vocabulary-based algorithms, because it takes local patterns and their relationships into account and forms a histogram out of them, while TSF fails in capturing useful information. We note that the LPS algorithm includes randomness so in each run it extracts different patterns from the data and also it is more time and space inefficient by comparison with the vocabulary-based algorithms. Finally, the hand-crafted features are not able to extract local patterns of the raw daily time series and are mostly based on statistic measurements. Table 1 presents the arithmetical results for the 9 largest regions. As one can observe, the results of BOP and BOSS are very similar. In most regions they give rise to substantially better results than the other methods that were tested.

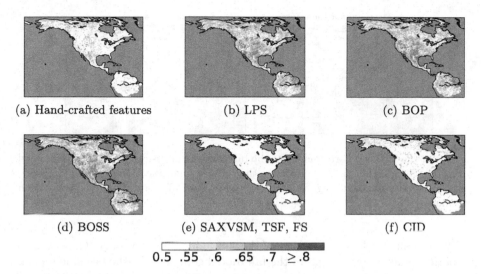

(a) Hand-crafted features      (b) LPS      (c) BOP

(d) BOSS      (e) SAXVSM, TSF, FS      (f) CID

0.5 .55  .6  .65  .7  ≥ .8

**Fig. 3.** Performance comparison in terms of AUC of the time series classification algorithms in the univariate time series classification setting on climate data.

**Table 1.** Mean and standard deviation of the AUC for areas which include more than 100 pixels. The vocabulary-based algorithms as well as the LPS algorithm perform very similar. Results of the algorithms SAXVSM and TSF are omitted due to their low performance.

| Algorithm | Reg 1 | Reg 2 | Reg 3 | Reg 4 | Reg 5 | Reg 6 | Reg 7 | Reg 8 | Reg 9 |
|---|---|---|---|---|---|---|---|---|---|
| Hand-crafted | 0.50±0.01 | 0.50±0.00 | 0.54±0.05 | 0.52±0.03 | 0.51±0.02 | 0.50±0.00 | 0.50±0.00 | 0.50±0.01 | 0.51±0.01 |
| LPS | 0.59±0.06 | **0.56±0.04** | **0.65±0.09** | **0.65±0.07** | 0.61±0.06 | **0.62±0.05** | 0.60±0.05 | 0.65±0.07 | 0.59±0.05 |
| BOP | **0.60±0.07** | **0.56±0.05** | **0.65±0.08** | 0.64±0.07 | 0.60±0.06 | 0.61±0.05 | **0.61±0.06** | 0.66±0.07 | **0.60±0.05** |
| BOSS | **0.60±0.06** | **0.56±0.04** | 0.64±0.08 | **0.65±0.07** | 0.61±0.05 | 0.61±0.05 | **0.61±0.05** | **0.67±0.07** | 0.59±0.05 |
| CID | 0.50±0.03 | 0.50±0.02 | 0.51±0.05 | 0.51±0.04 | 0.50±0.03 | 0.54±0.04 | 0.53±0.03 | 0.55±0.05 | 0.51±0.03 |
| FS | – | 0.50±0.00 | 0.50±0.00 | – | 0.50±0.00 | – | 0.50±0.00 | – | 0.50±0.00 |

## 5.2    Quantification of Granger Causality

In a second step, we combine the best representation coming from the time series classification algorithms and we apply it to the non-linear Granger causality framework in order to test causal effects of climate on vegetation extremes. Our main goal is to replace the hand-crafted features constructed in [27]. As the BOSS algorithm has the best performance compared to the other time series algorithms, we use the vocabulary of patterns that BOSS automatically extracts from the climatic time series as features. To evaluate Granger causality, the baseline model includes information from the NDVI extremes, while the full model includes also the automatically-extracted features from the climatic time series. In contrast to the previous set of experiments, we now include three climatic time series instead of only the precipitation time series.

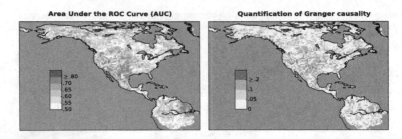

**Fig. 4.** On the left, the performance of the full model that uses the patterns extracted by the BOSS algorithm as predictors. On the right, a quantification of Granger causality; positive values indicate regions with Granger-causal effects of climate on vegetation extremes.

Figure 4 shows the performance of the full model in terms of AUC, as well as the performance improvement of the full model compared to the baseline model. It is clear that by using information from climatic time series the prediction of vegetation extremes improves in most of the regions. Therefore, one can conclude that – while not bearing into consideration all potential control variables in our analysis – climate dynamics indeed Granger-cause vegetation extremes in most of the continental land surface of North and Central America.

As results of that kind could not be obtained with hand-crafted feature representations, we do conclude that more specialized time series classification methods such as BOSS have the potential of enhancing causal inference in climate science. While this paper presents particular results for the case of climate-vegetation dynamics, we believe that the approach might be useful in other causal inference studies, too.

**Acknowledgements.** This work is funded by the Belgian Science Policy Office (BEL-SPO) in the framework of the STEREO III programme, project SAT-EX (SR/00/306). D. G. Miralles acknowledges support from the European Research Council (ERC) under grant agreement n° 715254 (DRY-2-DRY). The data used in this manuscript can be accessed using http://www.SAT-EX.ugent.be as gateway.

# References

1. Attanasio, A.: Testing for linear granger causality from natural/anthropogenic forcings to global temperature anomalies. Theor. Appl. Climatol. **110**(1–2), 281–289 (2012)
2. Bagnall, A., Lines, J., Bostrom, A., Large, J., Keogh, E.: The great time series classification bake off : a review and experimental evaluation of recent algorithmic advances. Data Min. Knowl. Disc. **31**(3), 606–660 (2017). https://doi.org/10.1007/s10618-016-0483-9. ISSN: 1573-756X
3. Batista, G.E.A.P.A., Keogh, E.J., Tataw, O.M., De Souza, V.M.A.: CID: an efficient complexity-invariant distance for time series. Data Min. Knowl. Discov. **28**(3), 634–669 (2014)

4. Baydogan, M.G., Runger, G.: Time series representation and similarity based on local autopatterns. Data Min. Knowl. Discov. **30**(2), 476–509 (2016)
5. Baydogan, M.G., Runger, G., Tuv, E.: A bag-of-features framework to classify time series. IEEE Trans. Patt. Anal. Mach. Intell. **35**(11), 2796–2802 (2013)
6. Beck, H.E., McVicar, T.R., van Dijk, A.I.J.M., Schellekens, J., de Jeu, R.A.M., Bruijnzeel, L.A.: Global evaluation of four AVHRR-NDVI data sets: intercomparison and assessment against landsat imagery. Remote Sens. Environ. **115**(10), 2547–2563 (2011)
7. Beck, H.E., van Dijk, A.I.J.M., Levizzani, V., Schellekens, J., Miralles, D.G., Martens, B., de Roo, A.: MSWEP: 3-hourly 0.25° global gridded precipitation (1979–2015) by merging gauge, satellite, and reanalysis data. Hydrol. Earth Syst. Sci. Discuss. **2016**, 1–38 (2016)
8. Chapman, D., Cane, M.A., Henderson, N., Lee, D.E., Chen, C.: A vector autoregressive ENSO prediction model. J. Clim. **28**(21), 8511–8520 (2015)
9. Dee, D.P., Uppala, S.M., Simmons, A.J., Berrisford, P., Poli, P., Kobayashi, S., Andrae, U., Balmaseda, M.A., Balsamo, G., Bauer, P., et al.: The ERA-Interim reanalysis: configuration and performance of the data assimilation system. Q. J. Royal Meteorol. Soc. **137**(656), 553–597 (2011)
10. Deng, H., Runger, G., Tuv, E., Vladimir, M.: A time series forest for classification and feature extraction. Inf. Sci. **239**, 142–153 (2013)
11. Ding, H., Trajcevski, G., Scheuermann, P., Wang, X., Keogh, E.: Querying and mining of time series data: experimental comparison of representations and distance measures. Proc. VLDB Endowment **1**(2), 1542–1552 (2008)
12. Donat, M.G., Alexander, L.V., Yang, H., Durre, I., Vose, R., Dunn, R.J.H., Willett, K.M., Aguilar, E., Brunet, M., Caesar, J., et al.: Updated analyses of temperature and precipitation extreme indices since the beginning of the twentieth century: the HadEX2 dataset. J. Geophys. Res.: Atmos. **118**(5), 2098–2118 (2013)
13. Geiger, P., Zhang, K., Gong, M., Janzing, D., Schölkopf, B.: Causal inference by identification of vector autoregressive processes with hidden components. In Proceedings of 32th International Conference on Machine Learning (ICML 2015) (2015)
14. Górecki, T., Łuczak, M.: Using derivatives in time series classification. Data Min. Knowl. Disc. **26**(2), 310–331 (2013). https://doi.org/10.1007/s10618-012-0251-4. ISSN: 1573-756X
15. Górecki, T., Łuczak, M.: Non-isometric transforms in time series classification using DTW. Knowl.-Based Syst. **61**, 98–108 (2014)
16. Grabocka, J., Schilling, N., Wistuba, M., Schmidt-Thieme, L.: Learning time-series shapelets. In: Proceedings of the 20th ACM SIGKDD International Conference on Knowledge Discovery and Data Mining, pp. 392–401. ACM (2014)
17. Granger, C.W.J.: Investigating causal relations by econometric models and cross-spectral methods. Econometrica: J. Econ. Soc. **37**, 424–438 (1969)
18. Hills, J., Lines, J., Baranauskas, E., Mapp, J., Bagnall, A.: Classification of time series by shapelet transformation. Data Min. Knowl. Discov. **28**(4), 851–881 (2014)
19. Kaufmann, R.K., Zhou, L., Myneni, R.B., Tucker, C.J., Slayback, D., Shabanov, N.V., Pinzon, J.: The effect of vegetation on surface temperature: a statistical analysis of NDVI and climate data. Geophys. Res. Lett. **30**(22) (2003)
20. Kodra, E., Chatterjee, S., Ganguly, A.R.: Exploring granger causality between global average observed time series of carbon dioxide and temperature. Theor. Appl. Climatol. **104**(3–4), 325–335 (2011)
21. Liao, T.W.: Clustering of time series data a survey. Patt. Recogn. **38**(11), 1857–1874 (2005)

22. Lin, J., Khade, R., Li, Y.: Rotation-invariant similarity in time series using bag-of-patterns representation. J. Intell. Inf. Syst. **39**(2), 287–315 (2012)
23. Liu, G., Liu, H., Yin, Y.: Global patterns of NDVI-indicated vegetation extremes and their sensitivity to climate extremes. Environ. Res. Lett. **8**(2), 025009 (2013)
24. Loveland, T.R., Belward, A.S.: The IGBP-DIS global 1km land cover data set, discover: first results. Int. J. Remote Sens. **18**(15), 3289–3295 (1997)
25. Mokhov, I.I., Smirnov, D.A., Nakonechny, P.I., Kozlenko, S.S., Seleznev, E.P., Kurths, J.: Alternating mutual influence of El-Niño/southern oscillation and Indian monsoon. Geophys. Res. Lett. **38**(8) (2011)
26. Mosedale, T.J., Stephenson, D.B., Collins, M., Mills, T.C.: Granger causality of coupled climate processes: ocean feedback on the North Atlantic Oscillation. J. Clim. **19**(7), 1182–1194 (2006)
27. Papagiannopoulou, C., Miralles, D.G., Verhoest, N.E.C., Dorigo, W.A., Waegeman, W.: A non-linear Granger causality framework to investigate climate-vegetation dynamics. Geosci. Model Dev. **10**, 1–24 (2017)
28. Rakthanmanon, T., Keogh, E.: Fast shapelets: a scalable algorithm for discovering time series shapelets. In: Proceedings of the 2013 SIAM International Conference on Data Mining, pp. 668–676. SIAM (2013)
29. Sakoe, H., Chiba, S.: Dynamic programming algorithm optimization for spoken word recognition. IEEE Trans. Acoust. Speech Sig. Process. **26**(1), 43–49 (1978)
30. Schäfer, P.: The BOSS is concerned with time series classification in the presence of noise. Data Min. Knowl. Discov. **29**(6), 1505–1530 (2015)
31. Senin, P., Malinchik, S.: SAX-VSM: interpretable time series classification using sax and vector space model. In: 2013 IEEE 13th International Conference on Data Mining (ICDM), pp. 1175–1180. IEEE (2013)
32. Shahin, M.A., Ali, M.A., Ali, A.B.M.S.: Vector Autoregression (VAR) modeling and forecasting of temperature, humidity, and cloud coverage. In: Islam, T., Srivastava, P.K., Gupta, M., Zhu, X., Mukherjee, S. (eds.) Computational Intelligence Techniques in Earth and Environmental Sciences, pp. 29–51. Springer, Dordrecht (2014). https://doi.org/10.1007/978-94-017-8642-3_2
33. Stefan, A., Athitsos, V., Das, G.: The move-split-merge metric for time series. IEEE Trans. Knowl. Data Eng. **25**(6), 1425–1438 (2013)
34. Tucker, C.J., Pinzon, J.E., Brown, M.E., Slayback, D.A., Pak, E.W., Mahoney, R., Vermote, E.F., El Saleous, N.: An extended AVHRR 8-km NDVI dataset compatible with MODIS and SPOT vegetation NDVI data. Int. J. Remote Sens. **26**(20), 4485–4498 (2005)
35. Zscheischler, J., Mahecha, M.D., Harmeling, S., Reichstein, M.: Detection and attribution of large spatiotemporal extreme events in Earth observation data. Ecol. Inform. **15**, 66–73 (2013)

# Automatic Detection and Recognition of Individuals in Patterned Species

Gullal Singh Cheema$^{(\boxtimes)}$ and Saket Anand

IIIT-Delhi, New Delhi, India
{gullal1408,anands}@iiitd.ac.in

**Abstract.** Visual animal biometrics is rapidly gaining popularity as it enables a non-invasive and cost-effective approach for wildlife monitoring applications. Widespread usage of camera traps has led to large volumes of collected images, making manual processing of visual content hard to manage. In this work, we develop a framework for automatic detection and recognition of individuals in different patterned species like tigers, zebras and jaguars. Most existing systems primarily rely on manual input for localizing the animal, which does not scale well to large datasets. In order to automate the detection process while retaining robustness to blur, partial occlusion, illumination and pose variations, we use the recently proposed Faster-RCNN object detection framework to efficiently detect animals in images. We further extract features from AlexNet of the animal's flank and train a logistic regression (or Linear SVM) classifier to recognize the individuals. We primarily test and evaluate our framework on a camera trap tiger image dataset that contains images that vary in overall image quality, animal pose, scale and lighting. We also evaluate our recognition system on zebra and jaguar images to show generalization to other patterned species. Our framework gives perfect detection results in camera trapped tiger images and a similar or better individual recognition performance when compared with state-of-the-art recognition techniques.

**Keywords:** Animal biometrics · Wildlife monitoring
Detection · Recognition · Convolutional neural network
Computer vision

## 1 Introduction

Over the past two decades, advances in visual pattern recognition have led to many efficient visual biometric systems for identifying human individuals through various modalities like iris images [7,27], facial images [1,28] and fingerprints [13,14]. Since the identification process relies on visual pattern matching, it is convenient and minimally invasive, which in turn makes it amenable to use with non-cooperative subjects as well. Consequently, visual biometrics has been applied to wild animals, where non-invasive techniques provide a huge advantage

© Springer International Publishing AG 2017
Y. Altun et al. (Eds.): ECML PKDD 2017, Part III, LNAI 10536, pp. 27–38, 2017.
https://doi.org/10.1007/978-3-319-71273-4_3

in terms of cost, safety and convenience. Apart from identifying or recognizing an individual, visual pattern matching is also used to classify species, detect occurrence or variation in behavior and also morphological traits.

Historically, since mid-1900s, ecologists and evolutionary researchers have used sketch collections [24] and photographic records [15,19] to study, document and index animal appearance [22]. This is due to the fact that a large variety of animal species carry unique coat patterns like stripes on zebras and spots on jaguar. Even though the earlier studies provided ways to formalize unique animal appearance, manually identifying individuals is tedious and requires a human expert with specific skills, making the identification process prone to subjective bias. Moreover, as the volume of images increase, manual processing becomes prohibitively expensive.

With the advancement of computer vision techniques like object detection and localization [8], pose estimation [31] and facial expression recognition [5], ecologists and researchers have an opportunity to systematically apply visual pattern matching techniques to automate wildlife monitoring. As opposed to traditional approaches like in field manual monitoring, radio collaring and GPS tracking, these approaches minimize the subjective bias, are repeatable, cost-effective, safer and less stressful for the human as well as the animal. However, unlike in the human case, there is little control over environmental factors during data acquisition of wild animals. Specifically, in the case of land animals, most of the image data is collected using camera traps fixed at probable locations where the animal of interest can be located. Due to these reasons, the recognition systems have to be robust enough to work on images with drastic illumination changes, blurring and occlusion due to vegetation. For example, some of the challenging images in our tiger dataset can be seen in the Fig. 1.

In recent years, organizations like WWF-India and projects like *Snapshot Serengeti* project [26] have gathered and cataloged millions of images through hundreds of camera trap sites spanning large geographical areas. With this unprecedented increase in quantity of camera trap images, there is a requirement

**Fig. 1.** Sample challenging camera trapped images of tiger

of visual monitoring systems that can automatically sort and organize images based on a desired category (species/individual level) in short amount of time. Also, many of the animal species are endangered and require continuous monitoring, especially in areas where they are vulnerable to poaching, predators and already less in number. Such monitoring efforts can help to protect animals, maintain population across different geographical areas and also protect the local ecosystem.

In this work, we develop a framework for detecting and recognizing individual patterned species that have unique coat patterns such as stripes on zebras, tigers and spots on Jaguars. State-of-the-art systems such as Extract-Compare [12] by Hiby *et al.* and HotSpotter [6] by Crall *et al.* work well, but require user input for every image and hence fail to scale to large datasets. Automatic detection methods proposed in [3,4] detect smaller patches on animals but not the complete animal, and are sensitive to lighting conditions and multiple instances of animals in the same image. In this work, we use the recently proposed convolutional neural network (CNN) based detector, Faster-RCNN [23] by Ren *et al.* that is able to detect different objects at multiple scales. The advantage of using a deep CNN based architecture is robustness to illumination and pose variations as well as location invariance, which proves to be very effective for localizing animals in images in uncontrolled environments. We use Faster-RCNN to detect the body and flank region of the animal and pass it through a pre-trained AlexNet [16] to extract discriminatory features, which is used by a logistic regression classifier for individual recognition.

The remainder of the paper is structured as follows. In Sect. 2, we will briefly talk about recent related work in animal detection and individual animal recognition. Section 3 lays the groundwork for the description of our proposed framework in Sect. 4. We then present empirical results on data sets of various patterned species, and report performance comparisons in Sect. 5 before concluding in Sect. 6.

## 2   Related Work

In this section we briefly discuss recent advances in animal species and individual identification, focusing on land animals exhibiting unique coat patterns.

### 2.1   Animal Detection

One of the earliest works on automatic animal detection [3,4] uses Haar-like features and a low-level feature tracker to detect a lion's face and extract information to predict its activity like still, walking or trotting. The system works in real time and is able to detect faces at multiple scales although only with *slight* pose variations. Zhang *et al.* [30] detect heads of animals like tiger, cat, dog, cheetah, etc. by using shape and texture features to improve image retrieval. The approach relies on prominent 'pointed' ear shapes in frontal poses which makes it sensitive to head-pose variations. These approaches rely on identifying different parts of the animal to detect and track an individual, but are likely to fail in case of occlusion or significant pose change.

CNNs are known to be robust to occlusion and pose variations and have also been used to automatically learn discriminatory features from the data to localize Chimpanzee faces [9]. Also, recently Norouzzadeh et al. [20] used various CNN architectures like Alexnet [16], VGGnet [25] and ResNet [11] to classify 48 animal species using the Snapshot Serengeti [26] dataset with 3.2 million camera trap images and achieved ~96% classification accuracy.

## 2.2  Individual Animal Recognition

Hiby et al. [12] developed 'Extract-Compare', one of the first interactive software tools for recognizing individuals by matching coat patterns of species like tiger, cheetah, giraffe, frogs, etc. The tool works in a retrieval framework, where a user inputs a query image and individuals with similar coat patterns are retrieved from a database for final verification by the user. Prior to the pattern matching, a coarsely parametric 3D surface model is fit on the animal's body, e.g., around the flank of a tiger, or the head of an armadillo. This surface model fitting makes the pattern matching robust to animal and camera pose. However, in order to fit the 3D surface model, the user has to carefully mark several key points like the head, tail, elbows, knees, etc. While this approach works well in terms of accuracy, it is not scalable to a large number of images as the manual processing time for an image could be as high as 30 s.

Lahiri et al. introduced StripeSpotter [17] that extracts features from flanks of a Zebra as 2D arrays of binary values. This 2D array depicts the white and black stripe pattern which can be used to uniquely identify a zebra. The algorithm uses a dynamic programming approach to calculate a value similar to edit distance between two strings. Again, the flank region is extracted manually and each query image is matched against every other image in the database.

HotSpotter [6] and Wild-ID [2] use SIFT [18] features to match query image of the animal with a database of existing animals. Both the tools require a manual input for selecting the region of interest so that SIFT features are unaffected by background clutter in the image. In addition to matching each query image descriptor against each database image separately, Hotspotter also uses *one vs. many* approach by matching each query image descriptor with all the database descriptors. It uses efficient data structure such as a forest of kd-trees and different scoring criterion to efficiently find the approximate nearest neighbor. Hotspotter also performs spatial re-ranking to filter out any spatially inconsistent descriptor matches by using RANSAC solution used in [21]. However, spatial re-ranking does not perform better than simple *one vs. many* matching.

## 3  Background

In this section we briefly describe the deep neural network architectures that we employ in our animal detection and individual recognition framework.

## 3.1   Faster-RCNN

Faster-RCNN [23] by Ren *et al.* is a recently proposed object detection technique that is composed of two modules in a single unified network. The first module is a deep CNN that works as a Region Proposal Network (RPN) and proposes regions of interest (ROI), while the second module is a Fast R-CNN [10] detector that categorizes each of the proposed ROIs. This unification of RPN with the detector lowers test-time computation without noticeable loss in detection performance.

The RPN takes input an image of any size and outputs a set of rectangular object proposals, each with an objectness (object vs. background) score. In addition to shared convolutional layers, RPN has an additional small network with one $n \times n$ convolutional layer and two sibling fully connected layers (one for box-regression and one for box-classification). At each sliding-window location for the $n \times n$ convolution layer, multiple region proposals (called anchors) are predicted with varying scales and aspect ratios. Each output is then mapped to lower-dimensional feature which is then fed into the two sibling layers.

The Fast R-CNN detection network on the other hand can be a ZF [29] or VGG [25] net which, in addition to shared convolution layers, has two fully connected (fc6 and fc7) layers and two sibling class score and bounding box prediction fully connected layers. For further details on cost functions and training of Faster-RCNN, see [23]. We discuss training, hyperparameter setting and implementation details specific to tiger detection in Sects. 4 and 5.

## 3.2   AlexNet

AlexNet was proposed in [16] with five convolutional and three fully-connected layers. With 60 million parameters, the network was trained using a subset of about 1.2 million images from the ImageNet dataset for classifying about 150,000 images into 1000 different categories. The success of AlexNet on a large-scale image classification problem led to several works that used *pre-trained* networks for feature representations which are fed to an application specific classifier. We follow a similar approach for recognition of individuals in patterned species, with a modification of the input size and consequently the feature map dimensions.

# 4   Methodology

In this work we address two problems in animal monitoring: First is to detect and localize the patterned species in a camera trap image, and the second is to uniquely identify the detected animal against an existing database of the same species. The proposed framework can be seen in the Fig. 2.

## 4.1   Data Augmentation

To increase the number of images for training phase and avoid over-fitting, we augment the given training data for both detection and individual recognition.

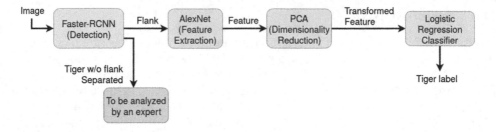

**Fig. 2.** Proposed framework for animal detection and individual recognition

For detection, we double the number of training images by horizontally flipping (mirroring) each image while training the Faster-RCNN.

In case of recognizing individuals, the number of training samples is very small because of relatively few side pose captures per tiger. Therefore, in order to learn to classify individual animals, we need stronger data augmentation techniques. We use contrast enhancement and random filtering (Gaussian or median) for each training image increasing our training set to thrice the number of training images originally.

### 4.2   Detection Using Faster-RCNN

We detect both the tiger and the flank region using Faster-RCNN. During training, both the image and the bounding boxes (tiger and flank) are input to the network. The bounding boxes for the flanks are given for only those images in which the flank is not occluded and distorted due to the pose of the tiger. The network is trained to detect 3 classes: tiger, flank and the background. All the parameters used for training are as used in the original implementation.

For training, the whole network is trained with 4-step alternating training mentioned in [23]. We use ZF [29] net in our framework which has five shareable convolutional layers. In the first step, RPN is trained end-to-end for the region proposals task by initializing the network with an ImageNet-pre-trained model. Fast R-CNN is then trained in the second step with weights initialized by ImageNet-pre-trained model and by using the proposals generated by step-1 RPN. Weight sharing is performed in the third and fourth step, where RPN training is initialized with detector network and by fixing the shared convolutional layers, only layers unique to RPN are fine-tuned. Similarly, Fast R-CNN is trained in the fourth step by fixing the shared layers and fine-tuning only the unique layers of the detector. Additionally, we also fix first two convolutional layers in first two steps of the training for tiger detection as the initial layers are already fine-tuned to detect low-level features like edges.

During testing, only an image is input to the network and it outputs the bounding boxes and the corresponding objectness scores. As Faster-RCNN outputs multiple bounding boxes per category, some of which are highly overlapping, non-maximum suppression (NMS) is applied to reduce the redundant boxes.

Because the convolution layers are shared, we can test the image in one go in very less time (0.3–0.6 s/image on a GPU).

### 4.3  Identification

For identification, we only use flank regions because they contain the discriminatory information to uniquely identify the patterned animals. The images in which the tiger is detected, but the flank is not, are separated to be analyzed by the expert. A tool such as Extract-Compare [12] can be used for difficult cases with extreme pose or occlusion.

We use ImageNet-pre-trained AlexNet [16] to extract features from the flank region and train a logistic regression classifier to recognize the individuals. While this deviates from the end-to-end framework, typical of deep networks, our choice of this approach was to resolve the problem of very low training data for identifying individuals. We tried fine-tuning AlexNet with our data, however, the model overfitted the training set. For feature representation, we used different convolutional layers and fully connected layers to train our classifier and obtained the best results with the third convolutional layer (conv3). Since ImageNet is a large-scale dataset, the pre-trained weights of AlexNet in the higher layers are not optimized for a fine-grained task such as individual animal recognition. On the other hand, the middle layers (like conv3) capture interactions between edges and are discriminative enough to give good results for our problem.

To minimize distortion introduced by resizing the detected flank region to a unit aspect ratio of AlexNet ($227 \times 227$), we modify the size of input to AlexNet and hence the subsequent feature maps. Since the conv3 feature maps are high dimensional, we apply a PCA (Principal Component Analysis) based dimensionality reduction and use principal components that explain 99% of the energy.

## 5  Experiments

All experiments are carried out with Python (and PyCaffe) running on i7-4720HQ 3.6 GHz processor and Nvidia GTX-950M GPU. For Faster-RCNN [23] training, we use a server with Nvidia GTX-980 GPU. We used the python implementation of Faster-RCNN[1] and labelImg[2] annotation tool for annotating the tiger and jaguar images. We also use python's sklearn library for logistic regression classifier. Over three different datasets, we compare our results with HotSpotter [6], which showed superior performance as compared to Wild-ID [2] and StripeSpotter [17].

---

[1] https://github.com/rbgirshick/py-faster-rcnn.
[2] https://github.com/tzutalin/labelImg.

## 5.1   Datasets

**Tiger Dataset:** The dataset is provided by Wildlife Institute of India (WII) and contains about 770 images captured from camera traps. The images as shown in Fig. 1 are very challenging due to severe viewpoint and illumination changes, motion blur and occlusions. We use this for both detection and individual recognition.

**Plains Zebra Dataset**[3] was used in StripeSpotter [17]. The stripe patterns are less discriminative than tigers, however, the images in this dataset have little viewpoint and appearance variations as most images were taken within seconds of each other. We use the cropped flank regions provided in the dataset for comparison with hotspotter.

**Jaguar Dataset**[4] is a smaller dataset also obtained from camera traps, but have poorer image quality (mostly night images), and moderate viewpoint variations (Fig. 3).

**Fig. 3.** Sample images from the other two datasets. Row 1: Jaguars. Row 2: Plains Zebra.

We summarize the three datasets and our model parameters for the individual recognition task in Table 1.

**Table 1.** Dataset statistics and model parameters. **C** is the inverse of regularization strength used in logistic regression classifier.

| Species | #Images | #Labels | Feature size conv3 | Feature size after PCA | C |
|---|---|---|---|---|---|
| Tiger | 260 | 44 | 63360 | ~180 | 1e6 |
| Plains Zebra | 821 | 83 | 40320 | ~460 | 1e6 |
| Jaguar | 112 | 37 | 63360 | ~70 | 1e5 |

---

[3] http://compbio.cs.uic.edu/~stripespotter/.
[4] Provided by Marcella J Kelly upon request: http://www.mjkelly.info/.

## 5.2    Detection

We use 687 tiger images for training and testing the detection system after removing the ones in which the tiger is hardly visible (only tail) and a few very poor quality images (very high contrast due to flash/sun rays). We divide the data for training and testing with a split of 75%/25% respectively into a disjoint set of tigers. With data augmentation, we have a total of 1032 ($516 \times 2$) images in the training set and 171 in the testing set.

For training Faster-RCNN, we randomly initialize all new layers by drawing weights from a zero-mean Gaussian distribution with standard deviation 0.01. We fine-tune RPN in both step 1 and 3 for 12000 iterations and Fast-RCNN in both step 2 and 4 for 10000 iterations. We use a learning rate of 0.001 for 10k and 8k mini-batches respectively, and 0.0001 for the next 2k mini-batches. We use a mini-batch size 1 (RPN) and 2 (Fast-RCNN) images, momentum of 0.9 and a weight decay of 0.0005 as used in [23]. For applying non-maximum suppression (NMS), we fix the NMS threshold at 0.3 (best) on predicted boxes with objectness score more than 0.8, such that all the boxes with IoU greater than the threshold are suppressed.

We report Average Precision (AP) and mean AP for tiger and flank detection, which is a popular metric used for object detection. The results for tiger and flank detection with varying NMS threshold are reported in Table 2. With increasing

**Table 2.** Results for tiger and flank detection

| Object/NMS threshold | 0.2 | 0.3 | 0.4 | 0.5 | 0.6 | 0.7 | 0.8 | 0.9 |
|---|---|---|---|---|---|---|---|---|
| Tiger | 90.7 | **90.9** | 90.6 | 90.3 | 88.9 | 85.3 | 73.5 | 45.2 |
| Flank | 90.6 | **90.6** | 90.4 | 89.4 | 87.2 | 76.9 | 57.0 | 41.9 |
| Mean AP | 90.6 | **90.7** | 90.5 | 89.9 | 88.0 | 81.1 | 65.2 | 43.6 |

**Fig. 4.** Qualitative detection results on images taken from the Internet. The detected boxes are labeled as (Label: Objectness score).

NMS threshold, number of output bounding boxes also increase which leads to poor detection results. We also show some qualitative results on tiger images taken from the Internet, which are quite different in quality and background when compared to the camera trap images as shown in Fig. 4.

### 5.3  Individual Recognition

We use conv3 features of AlexNet for training a logistic regression classifier to classify individuals. For each dataset, we generate five random splits with 75% for training and 25% for testing. For our framework, flanks of tiger and jaguar are resized to $256 \times 192$ and for the zebra to $256 \times 128$ which is equivalent to average size of flank images for the respective dataset. We learn a logistic regression model with $\ell_1$ regularization and perform grid search to find the parameter C. Specific data statistics and model parameters are reported in Table 1. We compare our results with HotSpotter and report the average rank 1 accuracy for all the datasets in Table 3. In Fig. 5, we show the Cumulative Match Characteristic (CMC) curves from rank 1 to rank 5 for our method compared with Hotspotter over all the datasets. The CMC curves indicate that the CNN based architecture clearly works better than HotSpotter in case of stripe patterns, even as we compare lower-rank accuracies. In the jaguar dataset, Hotspotter has a much higher rank-1 accuracy, but we observe a rising trend of our deep learning based approach as we compare lower-rank accuracies. We conjecture that the pre-trained AlexNet feature representation is not as discriminative for spots in jaguars as in case of stripes in tigers or zebras.

**Table 3.** Average rank 1 accuracy comparison

| Dataset | Ours ($227 \times 227$) | Ours (resized) | HotSpotter |
|---------|------------------------|----------------|------------|
| Tiger   | $76.5 \pm 2.2$         | $\mathbf{80.5 \pm 2.1}$ | $75.3 \pm 1.2$ |
| Jaguar  | $73.5 \pm 1.8$         | $78.6 \pm 2.3$ | $\mathbf{92.4 \pm 1.1}$ |
| Zebra   | $91.1 \pm 1.2$         | $\mathbf{93.2 \pm 1.4}$ | $90.9 \pm 0.8$ |

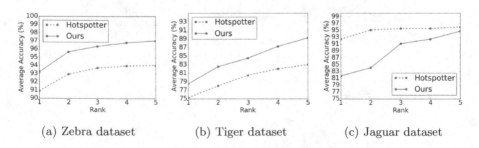

(a) Zebra dataset          (b) Tiger dataset          (c) Jaguar dataset

**Fig. 5.** CMC curve comparison

# 6    Conclusion

In this paper, we proposed a framework for automatic detection and individual recognition in patterned animal species. We used the state-of-the-art CNN based object detector Faster-RCNN [23] and fine-tuned it for the purpose of detecting the whole body and the flank of the tiger. We then used the detected flanks and extracted features from a pre-trained AlexNet [16] to train a logistic regression classifier for classifying individual tigers. We also performed individual recognition task on zebras and jaguars. We get perfect results for tiger detection and perform better than Hotspotter [6] while comparing rank-1 accuracy for individual recognition for tiger and zebra images. Even though AlexNet [16] features used for individual recognition are trained on Imagenet data, they seem to be as robust as SIFT [18] features as shown by our quantitative results. We plan do a thorough comparison in future with larger datasets to obtain deeper insights. For jaguar images, Hotspotter works better at rank-1 accuracy, but the proposed method shows improving trends as we compare lower-rank accuracies.

**Acknowledgments.** The authors would like to thank WII for providing the tiger data, Infosys Center for AI at IIIT-Delhi for computing resources and the anonymous reviewers for their invaluable comments.

## References

1. Ahonen, T., Hadid, A., Pietikainen, M.: Face description with local binary patterns: application to face recognition. IEEE TPAMI **28**(12), 2037–2041 (2006)
2. Bolger, D.T., Morrison, T.A., Vance, B., Lee, D., Farid, H.: A computer-assisted system for photographic mark-recapture analysis. Methods Ecol. Evol. **3**(5), 813–822 (2012)
3. Burghardt, T., Calic, J.: Real-time face detection and tracking of animals. In: Neural Network Applications in Electrical Engineering, pp. 27–32. IEEE (2006)
4. Burghardt, T., Calic, J., Thomas, B.T.: Tracking animals in wildlife videos using face detection. In: EWIMT (2004)
5. Cohen, I., et al.: Facial expression recognition from video sequences: temporal and static modeling. CVIU **91**(1), 160–187 (2003)
6. Crall, J.P., et al.: Hotspotter - patterned species instance recognition. In: WACV, pp. 230–237. IEEE (2013)
7. Daugman, J.: How iris recognition works. IEEE Trans. Circuits Syst. Video Technol. **14**(1), 21–30 (2004)
8. Felzenszwalb, P.F., Girshick, R.B., McAllester, D., Ramanan, D.: Object detection with discriminatively trained part-based models. IEEE TPAMI **32**(9), 1627–1645 (2010)
9. Freytag, A., Rodner, E., Simon, M., Loos, A., Kühl, H.S., Denzler, J.: Chimpanzee faces in the wild: log-Euclidean CNNs for predicting identities and attributes of primates. In: Rosenhahn, B., Andres, B. (eds.) GCPR 2016. LNCS, vol. 9796, pp. 51–63. Springer, Cham (2016). https://doi.org/10.1007/978-3-319-45886-1_5
10. Girshick, R.: Fast R-CNN. In: ICCV, pp. 1440–1448 (2015)
11. He, K., Zhang, X., Ren, S., Sun, J.: Deep residual learning for image recognition. CVPR, 770–778 (2016)

12. Hiby, L., Lovell, P., Patil, N., Kumar, N.S., Gopalaswamy, A.M., Karanth, K.U.: A tiger cannot change its stripes: using a three-dimensional model to match images of living tigers and tiger skins. Biol. Lett. **5**(3), 383–386 (2009)
13. Jain, A.K., Prabhakar, S., Hong, L., Pankanti, S.: Filterbank-based fingerprint matching. IEEE Trans. Image Process. **9**(5), 846–859 (2000)
14. Jiang, X., Yau, W.Y.: Fingerprint minutiae matching based on the local and global structures. In: ICPR, vol. 2, pp. 1038–1041. IEEE (2000)
15. Klingel, A.: Social organization and behavior of grevy's zebra (Equus grevyi). Z. Tierpsychol. **36**, 37–70 (1974)
16. Krizhevsky, A., Sutskever, I., Hinton, G.E.: ImageNet classification with deep convolutional neural networks. In: NIPS, pp. 1097–1105 (2012)
17. Lahiri, M., Tantipathananandh, C., Warungu, R., Rubenstein, D.I., Berger-Wolf, T.Y.: Biometric animal databases from field photographs: identification of individual zebra in the wild. In: International Conference on Multimedia Retrieval, p. 6. ACM (2011)
18. Lowe, D.G.: Distinctive image features from scale-invariant keypoints. Int. J. Comput. Vision **60**(2), 91–110 (2004)
19. Mizroch, S.A., Harkness, S.A.: A test of computer-assisted matching using the North Pacific humpback whale, Megaptera novaeangliae, tail flukes photograph collection. Mar. Fisheries Rev. **65**(3), 25–37 (2003)
20. Norouzzadeh, M.S., Nguyen, A., Kosmala, M., Swanson, A., Packer, C., Clune, J.: Automatically identifying wild animals in camera trap images with deep learning. arXiv preprint arXiv:1703.05830 (2017)
21. Philbin, J., Chum, O., Isard, M., Sivic, J., Zisserman, A.: Object retrieval with large vocabularies and fast spatial matching. In: CVPR, pp. 1–8. IEEE (2007)
22. Prodger, P.: Darwin's Camera: Art and Photography in the Theory of Evolution. Oxford University Press, New York (2009)
23. Ren, S., He, K., Girshick, R., Sun, J.: Faster R-CNN: towards real-time object detection with region proposal networks. In: NIPS, pp. 91–99 (2015)
24. Scott, D.K.: 17 identification of individual Bewick's swans by bill patterns. Recognition Marking of Animals in Research, p. 160 (1978)
25. Simonyan, K., Zisserman, A.: Very deep convolutional networks for large-scale image recognition. arXiv preprint arXiv:1409.1556 (2014)
26. Swanson, A., Kosmala, M., Lintott, C., Simpson, R., Smith, A., Packer, C.: Snapshot serengeti, high-frequency annotated camera trap images of 40 mammalian species in an African Savanna. Sci. Data **2**, 150026 (2015)
27. Tisse, C.l., Martin, L., Torres, L., Robert, M., et al.: Person identification technique using human iris recognition. In: Proceedings of Vision Interface, pp. 294–299 (2002)
28. Turk, M.A., Pentland, A.P.: Face recognition using eigenfaces. In: CVPR, pp. 586–591. IEEE (1991)
29. Zeiler, M.D., Fergus, R.: Visualizing and understanding convolutional networks. In: Fleet, D., Pajdla, T., Schiele, B., Tuytelaars, T. (eds.) ECCV 2014. LNCS, vol. 8689, pp. 818–833. Springer, Cham (2014). https://doi.org/10.1007/978-3-319-10590-1_53
30. Zhang, W., Sun, J., Tang, X.: From tiger to panda: animal head detection. IEEE Trans. Image Process. **20**(6), 1696–1708 (2011)
31. Zhu, X., Ramanan, D.: Face detection, pose estimation, and landmark localization in the wild. In: CVPR, pp. 2879–2886. IEEE (2012)

# Boosting Based Multiple Kernel Learning and Transfer Regression for Electricity Load Forecasting

Di Wu[1][✉], Boyu Wang[2], Doina Precup[1], and Benoit Boulet[1]

[1] McGill University, Montreal, QC H3A 0G4, Canada
di.wu5@mail.mcgill.ca, dprecup@cs.mcgill.ca, benoit.boulet@mcgill.ca
[2] Princeton University, Princeton, NJ 08544, USA
boyuw@princeton.edu

**Abstract.** Accurate electricity load forecasting is of crucial importance for power system operation and smart grid energy management. Different factors, such as weather conditions, lagged values, and day types may affect electricity load consumption. We propose to use multiple kernel learning (MKL) for electricity load forecasting, as it provides more flexibilities than traditional kernel methods. Computation time is an important issue for short-term load forecasting, especially for energy scheduling demand. However, conventional MKL methods usually lead to complicated optimization problems. Another practical aspect of this application is that there may be very few data available to train a reliable forecasting model for a new building, while at the same time we may have prior knowledge learned from other buildings. In this paper, we propose a boosting based framework for MKL regression to deal with the aforementioned issues for short-term load forecasting. In particular, we first adopt boosting to learn an ensemble of multiple kernel regressors, and then extend this framework to the context of transfer learning. Experimental results on residential data sets show the effectiveness of the proposed algorithms.

**Keywords:** Electricity load forecasting · Boosting
Multiple kernel learning · Transfer learning

## 1 Introduction

Electricity load forecasting is very important for the economic operation and security of a power system. The accuracy of electricity load forecasting directly influences the control and planning of power system operation. It is estimated that a 1% increase of forecasting error would bring in a 10 million pounds increase in operating cost per year (in 1984) for the UK power system [4]. Experts believe that this effect could become even stronger, due to the emergence of highly uncertain energy sources, such as solar and wind energy generation. Depending on the

© Springer International Publishing AG 2017
Y. Altun et al. (Eds.): ECML PKDD 2017, Part III, LNAI 10536, pp. 39–51, 2017.
https://doi.org/10.1007/978-3-319-71273-4_4

lead time horizon, electricity load forecasting ranges from short-term forecasting (minutes or hours ahead) to long-term forecasting (years ahead) [13]. With increasingly competitive markets and demand response energy management [15], short-term load forecasting is becoming more and more important [25]. In this paper, therefore, we will focus on tackling this problem.

Electricity load forecasting is a very difficult task since the load is influenced by many uncertain factors. Various methods have been proposed for electricity load forecasting including statistical methods, time series analysis, and machine learning algorithms [21]. Some recent work uses multiple kernels to build prediction models for electricity load forecasting. For example, in [1], Gaussian kernels with different parameters are applied to learn peak power consumption. In [8], different types of kernels are used for different features and a multi-task learning algorithm is proposed and applied on low level load consumption data to improve the aggregated load forecasting accuracy. However, all of the existing methods rely on a fixed set of coefficients for the kernels (i.e., simply set to 1), implicitly assuming that all the kernels are equally important for forecasting, which is suboptimal in real world applications.

Multiple kernel learning (MKL) [2], which learns both the kernels and their combination weights for different kernels, could be tailored to this problem. Through MKL, different kernels could have different weights according to their influence on the outputs. However, learning with multiple kernels usually involves a complicated convex optimization problem, which limits their application on large scale problems. Although some progresses have been made in improving the efficiency of the learning algorithms, most of them only focus on classification tasks [23,26]. On the other hand, electricity load forecasting is a regression problem and the computation time is an important issue.

Another practical issue for load forecasting is the lack of data to build a reliable forecasting model. For example, consider the case of a set of newly built houses (target domain) for which we want to predict the load consumption. We may not have enough data to build a prediction model for these new houses, while we have a large amount of data or knowledge from other houses (source domain). The challenge here is to perform transfer learning [18], which relies on the assumption is that there are some common structures or factors that can be shared across the domains. The objective of transfer learning for load forecasting is to improve the forecasting performance by discovering shared knowledge and leveraging it for electricity load prediction for target buildings.

In this paper, we address both challenges within a novel boosting-based MKL framework. In particular, we first propose the *boosting based multiple kernel regression* (BMKR) algorithm to improve the computational efficiency of MKL. Furthermore, we extend BMKR to the context of transfer learning, and propose two variants of BMKR: *kernel-level boosting based transfer multiple kernel regression* (K-BTMKR) and *model-level gradient boosting based transfer multiple kernel regression* (M-BTMKR). Our contribution, from an algorithmic perspective, is two-fold: We propose a boosting based learning framework (1) to learn regression models with multiple kernels efficiently, and (2) to leverage the MKL models learned from other domains. On the application side, this work introduces the use of transfer learning for the load forecasting problem, which opens up potential future work avenues.

## 2    Background

### 2.1    Multiple Kernel Regression

Let $S = \{(x_n, y_n), n = 1, \dots, N\} \in \mathbb{R}^d \times \mathbb{R}$ be the data set with $N$ samples, $\mathcal{K} = \{k_m : \mathbb{R}^d \times \mathbb{R}^d \to \mathbb{R}, m = 1, \dots, M\}$ be $M$ kernel functions. The objective of MKL is to learn a prediction model, which is a linear combination of $M$ kernels, by solving the following optimization problem [11]:

$$\min_{\eta \in \Delta} \min_{F \in \mathcal{H}_K} \frac{1}{2}||F||_K^2 + C \sum_{n=1}^{N} \ell(F(x_n), y_n), \tag{1}$$

where $\Delta = \{\eta \in \mathbb{R}_+ | \sum_{m=1}^{M} \eta_m = 1\}$ is a set of weights, $\mathcal{H}_K$ is the reproducing kernel Hilbert space (RKHS) induced by the kernel $K(x, x_n) = \sum_{m=1}^{M} \eta_m k_m(x, x_n)$ and $\ell(F(x), y)$ is a loss function. In this paper we use the squared loss $\ell(F(x), y) = \frac{1}{2}(F(x) - y)^2$ for the regression problem. The solution of Eq. 1 is of the form[1]

$$F(x) = \sum_{n=1}^{N} \alpha_n K(x, x_n), \tag{2}$$

where the coefficients $\{\alpha_n\}$ and $\{\eta_m\}$ are learned from samples.

Compared with single kernel approaches, MKL algorithms can provide better learning capability and alleviate the burden of designing specific kernels to handle diverse multivariate data.

### 2.2    Gradient Boosting and $\epsilon$-Boosting

Gradient boosting [10,16] is an ensemble learning framework which combines multiple hypotheses by performing gradient descent in function space. More specifically, the model learned by gradient boosting can be expressed as:

$$F(x) = \sum_{t=1}^{T} \rho^t f^t(x), \tag{3}$$

where $T$ is the number of total boosting iterations, and the $t$-th base learner $f^t$ is selected such that the distance between $f^t$ and the negative gradient of the loss function at $F = F^{t-1}$ is minimized:

$$f^t = \arg \min_f \sum_{n=1}^{N} \left( f(x_n) - r_n^t \right)^2, \tag{4}$$

---

[1] We ignore the bias term for simplicity of analysis, but in practice, the regression function can accomodate both the kernel functions and the bias term.

where $r_n^t = -\left[\frac{\partial \ell(F(x_n), y_n)}{\partial F}\right]_{F=F^{t-1}}$, and $\rho^t$ is the step size which can either be fixed or chosen by line search. Plugging in the squared loss we have $r_n^t = y_n - F^{t-1}(x_n)$. In other words, gradient boosting with squared loss essentially fits the residual at each iteration.

Let $\mathcal{F} = \{f_1, \ldots, f_J\}$ be a set of candidate functions, where $J = |\mathcal{F}|$ is the size of the function space, and $f : \mathbb{R}^d \to \mathbb{R}^J$, $f(x) = [f_1(x), \ldots, f_J(x)]^\top$ be the mapping defined by $\mathcal{F}$. Gradient boosting with squared loss usually proceeds in a greedy way: the step size is simply set $\rho^t = 1$ for all iterations. On the other hand, if the step size $\rho^t$ is set to some small constant $\epsilon > 0$, it can be shown that under the monotonicity condition, this example of gradient boosting algorithm, referred to as $\epsilon$-boosting in [20], essentially solves an $\ell_1$-regularized learning problem [12]:

$$\min_{||\beta||_1 \leq \mu} \sum_{n=1}^{N} \frac{1}{N} \ell\left(\beta^\top f(x_n), y_n\right), \tag{5}$$

where $\beta \in \mathbb{R}^J$ is the coefficient vector, and $\mu$ is the regularization parameter, such that $\epsilon T \leq \mu$. In other words, $\epsilon$-boosting implicitly controls the regularization via the number of iterations $T$ rather than $\mu$.

## 2.3 Transfer Learning from Multiple Sources

Let $\mathcal{S}_T = \{(x_n, y_n), n = 1, \ldots, N\}$ be the data set from the target domain, and $\{\mathcal{S}_1, \ldots, \mathcal{S}_S\}$ be the data sets from $S$ source domains, where $\mathcal{S}_s = \{(x_n^s, y_n^s), n = 1, \ldots, N_s\}$ are the samples of the $s$-th source. Let $\{F_1, \ldots, F_S\}$ be the prediction models learned from $S$ source domains. In this work, the $s$-th model $F_s$ is trained by some MKL algorithm (e.g., BMKR), and is of the form:

$$F_s = \sum_{m=1}^{M} \eta_m^s h_m^s(x) = \sum_{m=1}^{M} \eta_m^s \sum_{n=1}^{N_s} \alpha_n^s k_m(x, x_n^s). \tag{6}$$

The objective of transfer learning is to build a model $F$ that has a good generalization ability in the target domain using the data set $\mathcal{S}_T$ (which is typically small) and knowledge learned from sources $\{\mathcal{S}_1, \ldots, \mathcal{S}_S\}$. In this work, we assume that such knowledge has been embedded into $\{F_1, \ldots, F_S\}$, and therefore the problem becomes to explore the model structures that can be transferred to the target domain from various source domains. This type of learning approach is also referred to as *parameter transfer* [18].

## 3 Methods

### 3.1 Boosting Based Multiple Kernel Learning Regression

The idea of BMKR is to learn an ensemble model with multiple kernel regressors using the gradient boosting framework. The starting point of our method is similar to multiple kernel boosting (MKBoost) [23], which adapts AdaBoost [9] for

---

**Algorithm 1.** BMKR: Boosting based Multiple Kernel Regression

---

**Input:** Data set $\mathcal{S}$, kernel functions $\mathcal{K}$, number of iterations $T$

1: Initialize residual: $r_n^1 = y_i, \forall n \in \{1, \ldots, N\}$, and $F = 0$
2: **for** $t = 1, \ldots, T$ **do**
3:    **for** $m = 1, \ldots, M$ **do**
4:       Sample $N'$ data points from $\mathcal{S}$
5:       Train a kernel regression model $f_m^t$ with $k_m$ by fitting the residuals of the
        selected $N'$ samples
6:       Compute the loss: $e_m^t = \frac{1}{2} \sum_{n=1}^N \left( f_m^t(x_n) - r_n^t \right)^2$
7:    **end for**
8:    Select the regression model with the smallest fitting error: $f^t = \arg\min_{f_m^t} e_m^t$
9:    Add $f^t$ to the ensemble: $F \leftarrow F + \epsilon f^t$
10:   Update residuals: $r_n^{t+1} = y_n - F(x_n), \forall n \in \{1, 2, \ldots N\}$
11: **end for**

**Output:** the final multiple kernel function $F(x)$

---

multiple kernel classification. We extend this idea to a more general framework of gradient boosting [10,16], which allows different loss functions for different types of learning problems In this paper, we focus on the regression problem and use the squared loss.

At the $t$-th boosting iteration, for each kernel $k_m, m = 1, \ldots, M$, we first train a kernel regression model such as support vector regression (SVR) by fitting the current residuals, and obtain a solution of the form:

$$f_m^t(x) = \sum_{n=1}^N \alpha_{t,n} k_m(x, x_n). \tag{7}$$

Then we choose from $M$ candidates, the regression model with the smallest fitting error

$$f^t = \arg\min_{f_m^t, m \in \{1, \ldots, M\}} e_m^t, \tag{8}$$

where $e_m^t = \frac{1}{2} \sum_{n=1}^N \left( f_m^t(x_n) - r_n^t \right)^2$, and add it to the ensemble $F$. The final hypothesis of BMKR is expressed as in Eq. 3.

The pseudo-code of BMKR is shown in Algorithm 1. For gradient boosting with squared loss, the step size $\rho^t$ is not strictly necessary [3], and we can either simply set it to 1, or a fixed small value $\epsilon$ as suggested by $\epsilon$-boosting. Note that at each boosting iteration, instead of fitting all $N$ samples, we can select only $N'$ samples for training a SVR model, as suggested in [23], which can substantially reduce the computational complexity of each iteration as $N' \ll N$.

## 3.2   Boosting Based Transfer Regression

As explained in Sect. 1, as we typically have very few data in the target domain, and therefore the model can easily overfit, especially if we train a complicated

MKL model, even with the boosting approach. To deal with this issue, we can implicitly regularize the candidate functions at each boosting iteration by constraining the learning process within the function space spanned by the kernel functions trained on the source domains, rather than training the model in the function space spanned by arbitrary kernels. On the other hand, however, the underlying assumption of this approach is that at least one source domain is closely related to the target domain and therefore the kernel functions learned from the source domains can be reused. If this assumption does not hold, *negative transfer* could hurt the prediction performance. To avoid this situation, we also keep a MKL model which is trained only on the target domain. Consequently, the challenge becomes how to balance the knowledge embedded in the model learned from the source domains and the data fitting in the target domain.

To address this issue in a principled manner, we follow the idea of $\epsilon$-boosting [6,20] and propose the BTMKR algorithm, which is aimed towards transfer learning. There are two levels of transferring the knowledge of models: kernel-level transfer and model-level transfer, denoted by K-BTMKR and M-BTMKR respectively. At each iteration, K-BTMKR selects a single kernel function from $S \times M$ candidate kernels, while M-BTMKR selects a multiple kernel model from $S$ domains. Therefore, K-BTMKR has higher "resolution" and more flexibility, at the price of higher risk of overfitting, as the dimension of its search space is $M$ higher than that of M-BTMKR.

**Kernel-Level Transfer (K-BTMKR).** Let $\mathcal{H} = \{h_1^1, \ldots, h_M^1, \ldots, h_1^S, \ldots, h_M^S\}$ be the set of $MS$ candidate kernel functions learned from $S$ source domains, and $\mathcal{F} = \{f_1, \ldots, f_J\}$ be the set of $J$ candidate kernel functions from the target domain. Note that as the kernel functions from the source domains are fixed, the size of $\mathcal{H}$ is finite, while the size of the function space of the target domain is infinite, since the weights learned by SVR can be arbitrary (i.e., Eq. 7). For simplicity of analysis, we assume $J$ is also finite. Given the mapping $h : \mathbb{R}^d \to \mathbb{R}^{MS}, h(x) = [h_1^1(x), \ldots, h_M^S]^\top$ defined by $\mathcal{H}$ and the mapping $f$ defined by $\mathcal{F}$, we formulate the transfer learning problem as:

$$\min_{\beta_{\mathcal{S}}, \beta_{\mathcal{T}}} \mathcal{L}(\beta_{\mathcal{S}}, \beta_{\mathcal{T}}) \quad \text{s.t. } ||\beta_{\mathcal{S}}||_1 + \lambda||\beta_{\mathcal{T}}||_1 \leq \mu, \tag{9}$$

where $\mathcal{L}(\beta_{\mathcal{S}}, \beta_{\mathcal{T}}) \triangleq \sum_{n=1}^{N} \ell(\beta_{\mathcal{S}}^\top h(x_n) + \beta_{\mathcal{T}}^\top f(x_n), y_n)$, $\beta_{\mathcal{S}} \triangleq [\beta_1^1, \ldots, \beta_M^S]^\top \in \mathbb{R}^{MS}$, $\beta_{\mathcal{T}} \triangleq [\beta_1, \ldots, \beta_J]^\top \in \mathbb{R}^J$ are the coefficient vectors for the source domains and the target domain respectively, and $\lambda$ is a parameter that controls how much we penalize $\beta_{\mathcal{T}}$ against $\beta_{\mathcal{S}}$. Intuitively, if the data from target domain is limited, we should set $\lambda \geq 1$ to favor the model learned from the source domains, in order to avoid overfitting.

Following the idea of $\epsilon$-boosting [12,20], Eq. 9 can be solved by slowly increasing the value of $\mu$ by $\epsilon$, from 0 to a desired value. More specifically, let $g(x) = [h(x)^\top, f(x)^\top]^\top$, and $\beta = [\Delta\beta_{\mathcal{S}}^\top, \Delta\beta_{\mathcal{T}}^\top]^\top$. At the $t$-th boosting iteration, the coefficient vector $\beta$ is updated to $\beta + \Delta\beta$ by solving the following optimization problem:

$$\min_{\Delta\beta} \mathcal{L} \left(\beta + \Delta\beta\right) \quad \text{s.t.} \quad ||\Delta\beta_S||_1 + \lambda||\Delta\beta_T||_1 \le \epsilon \tag{10}$$

As $\epsilon$ is very small, the objective function of Eq. 10 can be expanded by first-order Taylor expansion, which gives

$$\mathcal{L} \left(\beta + \Delta\beta\right) \approx \mathcal{L}\left(\beta\right) + \nabla\mathcal{L}\left(\beta\right)^\top \Delta\beta, \tag{11}$$

where

$$\frac{\partial \mathcal{L}}{\partial \beta_j} = \sum_{n=1}^{N} -r_n^t g_j(x_n), \quad \forall j \in \{1, \dots, MS + J\}. \tag{12}$$

By changing the coefficients $\tilde{\beta}_T \leftarrow \lambda\beta_T$, it can be shown that minimizing Eq. 10 can be (approximately) solved by

$$\Delta\beta_j = \begin{cases} \epsilon, & \text{if } j = \arg\max_j \frac{\sum_{n=1}^{N} r_n^t g_j(x_n)}{\lambda_j} \\ 0, & \text{otherwise} \end{cases}, \tag{13}$$

where $\lambda_j = 1, \forall j \in \{1, \dots, MS\}$, and $\lambda_j = \lambda$, otherwise. In practice, as the size of function space of target domain is infinite, the candidate functions are actually computed by fitting the current residuals, as shown in Algorithm 2.

**Model-Level Transfer (M-BTMKR).** The derivation of M-BTMKR is similar to that of K-BTMKR, and therefore is omitted here.

### 3.3 Computational Complexity

The computational complexity of BMKR, as analyzed in [23], is $\mathcal{O}(TM\xi(N))$, where $\xi(N)$ is the computational complexity of training a single SVR with $N$ samples. Standard learning approaches formulate SVR as a quadratic programming (QP) problem and therefore $\xi(N)$ is $\mathcal{O}(N^3)$. Lower complexity (e.g., about $\mathcal{O}(N^2)$) can be achieved by using other solvers (e.g., LIBSVM [5]). More important, BMKR can adopt stochastic learning approach, as suggested in [23], which only selects $N'$ samples for training a SVR at each boosting iteration. This approach yields a complexity of $\mathcal{O}(TM(N + \xi(N')))$, which makes the algorithm tractable for large-scale problems by choosing $N' \ll N$. The computational complexity of the BTMKR algorithms is $\mathcal{O}(TM(SN + \xi(N)))$. Note that in the context of transfer learning, we use all the samples from the target domain, as the size of data set is usually small.

# 4 Experiments and Simulation Results

In this section, we evaluate the proposed algorithms on the problem of short-term electricity load forecasting for residential houses. Several factors including day types, weather conditions, and the lagged load consumption itself may affect the load profile of a given house. In this paper, we use three kinds of features for load forecasting: lagged load consumption, i.e., electricity consumed in the last three hours, temperature in the last three hours, and weekday/weekend information.

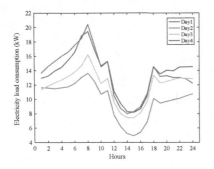

**Fig. 1.** Load data for four winter days       **Fig. 2.** Load data for three houses

## 4.1 Data Description

The historical temperature data are obtained from [14], and the residential house load consumption data are provided by the US Energy department [17]. The data set includes hourly residential house load consumption data for 24 locations in New York state in 2012. For each location, it provides data for three types of houses, based on the house size: low, base, and high. Figure 1 shows load consumption for a base type house for four consecutive winter days. We can see that the load consumption starts to decrease from 8 am and increases very quickly from 4 pm. Figure 2 shows the load consumption for three high load consumption houses in nearby cities for the same winter day. It can be observed that the load consumption for house 1 is similar to house 2 and both are different from house 3.

## 4.2 BMKR for Electricity Load Forecasting

To test the performance of BMKR, we use the data of a high energy consumption house in New York City in 2012. We test the performance of BMKR separately for different seasons, and compare it with single kernel SVR and linear regression. We set the number of boosting iterations for the proposed algorithms to 100, the step-size of $\epsilon$ to 0.05, and the sampling ratio to 0.9. In order to accelerate the learning process, we initialize the model with linear regression. The candidate kernels for BMKR are: Gaussian kernels with 10 different widths $(2^{-4}, 2^{-3}, ..., 2^5)$ and a linear kernel. We repeat the simulation for 10 times, and each time we randomly choose 50% of the data in the season as training data and 50% of the data as testing data.

Table 1 shows the mean and standard deviation (std dev) of the Mean Average Percentage Error (MAPE) measurement for BMKR and the other two baselines. We can see that BMKR achieves the best forecasting performance for all seasons, obtaining 3.3% and 3.8% average MAPE improvements over linear regression and single kernel SVR respectively.

**Table 1.** MAPE (%) performance (mean $\pm$ std dev) for high load consumption houses

| Method | Spring | Summer | Fall | Winter | Average |
|--------|--------|--------|------|--------|---------|
| Linear | $10.42 \pm 0.10$ | $7.78 \pm 0.13$ | $9.21 \pm 0.22$ | $5.81 \pm 0.13$ | $8.30 \pm 0.15$ |
| SVR | $10.95 \pm 0.21$ | $7.73 \pm 0.11$ | $8.82 \pm 0.21$ | $5.88 \pm 0.12$ | $8.34 \pm 0.16$ |
| BMKR | $\mathbf{10.31} \pm 0.17$ | $\mathbf{7.64} \pm 0.02$ | $\mathbf{8.42} \pm 0.11$ | $\mathbf{5.73} \pm 0.07$ | $\mathbf{8.02} \pm 0.10$ |

### 4.3   Transfer Regression for Electricity Load Forecasting

We evaluate the proposed transfer regression algorithms: M-BTMKR and K-BTMKR on high load consumption houses. We randomly pick 6 high load consumption houses as target house and use the remaining 18 high consumption houses as source houses. We repeat the simulation 10 times for each house, and each time we randomly choose 36 samples as the training data, and 100 samples as the testing data for the target house. For source houses, we randomly chose 600 data samples as the training data in each simulation. For K-BTMKR and M-BTMKR, $\lambda$ is chosen by cross validation to balance the model leaned from source house data and the model learned from target house data.

Performance of M-BTMKR and K-BTMKR are compared with linear regression, single kernel SVR and BMKR. The candidate kernels and boosting setting are the same as in Sect. 4.2. For the baselines, the forecasting models are trained only with data from target houses, and the results are shown in Table 2[2], from which it can be observed that the proposed transfer algorithms significantly improve the forecasting performance. For each individual location, the best results are achieved by either K-BTMKR or M-BTMKR, and M-BTMKR shows the best performance on average. The forecasting accuracies of M-BTMKR and K-BTMKR are very close to each other and both are much better than the baseline algorithms without transfer. In other words, with the proposed transfer algorithms, the knowledge learned from the source houses is properly transferred to the target house.

---

[2] Due to the space limitation, we only report the results for high load consumption houses. The results for low and base load consumption houses are similar to the high load consumption houses.

**Table 2.** Transfer learning MAPE (%) performance for high load consumption houses

| Method | Location 1 | Location 2 | Location 3 | Location 4 | Location 5 | Location 6 | Average |
|---|---|---|---|---|---|---|---|
| Linear | $8.02 \pm 0.05$ | $9.11 \pm 0.70$ | $17.39 \pm 1.62$ | $6.05 \pm 0.02$ | $11.43 \pm 0.15$ | $9.42 \pm 0.65$ | $10.24 \pm 0.53$ |
| SVR | $11.53 \pm 0.34$ | $6.82 \pm 0.39$ | $25.90 \pm 0.72$ | $8.24 \pm 0.08$ | $26.31 \pm 1.97$ | $14.00 \pm 0.65$ | $15.47 \pm 0.69$ |
| BMKR | $8.06 \pm 0.03$ | $6.64 \pm 0.54$ | $17.85 \pm 1.31$ | $5.29 \pm 0.01$ | $12.82 \pm 0.21$ | $9.05 \pm 0.57$ | $9.95 \pm 0.45$ |
| M-BTMKR | $\mathbf{5.35} \pm 0.01$ | $5.99 \pm 0.02$ | $\mathbf{5.63} \pm 0.19$ | $\mathbf{5.01} \pm 0.01$ | $9.13 \pm 0.01$ | $\mathbf{5.69} \pm 0.01$ | $\mathbf{6.13} \pm 0.04$ |
| K-BTMKR | $5.38 \pm 0.02$ | $\mathbf{5.46} \pm 0.30$ | $6.97 \pm 0.26$ | $5.55 \pm 0.09$ | $\mathbf{8.96} \pm 0.14$ | $7.31 \pm 0.21$ | $6.60 \pm 0.17$ |

## 4.4 Negative Transfer Analysis

Sometimes the consumption pattern for source houses and target houses can be quite different. We would prefer that the transfer algorithms prevent potential negative transfer for such scenarios. Here we present a case study to show the importance of balancing the knowledge learned from source domains and data fitting in the target domain. We use the same high load target houses as described in Sect. 4.3, but for the source houses, we randomly chose eighteen houses from the low type houses. We repeat the simulation for 10 times and the results are shown in Table 3.

The proposed algorithms are compared with linear regression, single kernel SVR, BMKR, M-BTMKR$_{woT}$, and K-BTMKR$_{woT}$, where M-BTMKR$_{woT}$ and K-BTMKR$_{woT}$ denote the BTMKR algorithms that we do not keep a MKL model trained on the target domain when we learn BTMKR models (i.e., we do not train $f^*$ in Algorithm 2). Simulation results show that, if we do not keep a MKL model trained on the target domain, we would encounter severe negative transfer problem, and the forecasting accuracy would be even much worse than the models learned without transfer. Meanwhile, we can see that the proposed M-BTMKR and K-BTMKR could successfully avoid such negative transfer. In this case, M-BTMKR and K-BTMKR still show better performance than other algorithms, though the forecasting accuracy of K-BTMKR is very close to BMKR. M-BTMKR achieves the best average forecasting performance and provides 14.37% average forecasting accuracy improvements over BMKR. In summary, the BTMKR algorithms can avoid the negative transfer when the data distributions of source domain and target domain are quite different.

**Table 3.** Transfer learning MAPE (%) performance for high load consumption target houses with low load consumption source houses

| Method | Location 1 | Location 2 | Location 3 | Location 4 | Location 5 | Location 6 | Average |
|---|---|---|---|---|---|---|---|
| Linear | $8.02 \pm 0.05$ | $9.11 \pm 0.70$ | $17.39 \pm 1.62$ | $6.05 \pm 0.02$ | $11.43 \pm 0.15$ | $9.42 \pm 0.65$ | $10.24 \pm 0.53$ |
| SVR | $11.53 \pm 0.34$ | $6.82 \pm 0.39$ | $25.90 \pm 0.72$ | $8.24 \pm 0.08$ | $26.31 \pm 1.97$ | $14.00 \pm 0.65$ | $15.47 \pm 0.69$ |
| BMKR | $8.06 \pm 0.03$ | $\mathbf{6.64} \pm 0.54$ | $17.85 \pm 1.31$ | $5.29 \pm 0.01$ | $12.82 \pm 0.21$ | $9.05 \pm 0.57$ | $9.95 \pm 0.45$ |
| M-BTMKR | $\mathbf{7.71} \pm 0.01$ | $8.74 \pm 0.27$ | $\mathbf{8.65} \pm 1.39$ | $6.51 \pm 0.52$ | $\mathbf{11.08} \pm 0.21$ | $8.42 \pm 0.86$ | $\mathbf{8.52} \pm 0.54$ |
| M-BTMKR$_{woT}$ | $57.64 \pm 0.05$ | $59.02 \pm 0.16$ | $59.71 \pm 0.53$ | $46.25 \pm 0.81$ | $38.52 \pm 0.02$ | $56.71 \pm 0.30$ | $52.98 \pm 0.31$ |
| K-BTMKR | $7.80 \pm 0.06$ | $8.60 \pm 0.74$ | $16.27 \pm 2.48$ | $\mathbf{5.77} \pm 0.16$ | $11.42 \pm 0.15$ | $9.33 \pm 0.67$ | $9.87 \pm 0.71$ |
| K-BTMKR$_{woT}$ | $54.81 \pm 0.05$ | $58.31 \pm 0.17$ | $59.00 \pm 0.11$ | $43.95 \pm 0.25$ | $37.49 \pm 0.12$ | $56.81 \pm 0.03$ | $51.73 \pm 0.12$ |

---

**Algorithm 2.** BTMKR: Boosting based Transfer Multiple Kernel Regression

---

**Input:** Data set $\mathcal{S}_T$ from the target domain, number of iterations $T$, regularization parameter $\lambda$, multiple kernel functions $\{F_1, \ldots, F_S\}$ learned from $S$ source domains, where each $F_s$ is given by Eq. 6.

1: Initialize residual: $r_n^1 = y_n, \forall n \in \{1, \ldots, N\}$, and $F = 0$
2: **for** $t = 1, \ldots, T$ **do**
3:   Compute the regression model $f^*$ and $h^*$ (line 8 – 21)
4:   Select the base learner: $f^t = \begin{cases} f^*, & \text{if } \frac{\sum_{n=1}^{N} r_n^t f^*(x_n)}{\lambda} > \sum_{n=1}^{N} r_n^t h^*(x_n) \\ h^*, & \text{otherwise.} \end{cases}$
5:   Add $f^t$ to the ensemble: $F \leftarrow F + \epsilon f^t$
6:   Update residuals: $r_n^{t+1} = y_n - F'(x_n), \ \forall n \in \{1, 2, \ldots N\}$
7: **end for**
**Output:** the final multiple kernel function $F(x)$

---

**K-BTMKR**

8: **for** $s = 1, \ldots, S$ **do**
9:   **for** $m = 1, \ldots, M$ **do**
10:     Fit the current residuals: $\gamma_{s,m}^t = \frac{\sum_{n=1}^{N} r_n^t h_m^s(x_n)}{\sum_{n=1}^{N} h_m^s(x_n)^2}$
11:     Compute the loss of $h_m^s$: $e_{s,m}^t = \frac{1}{2} \sum_{n=1}^{N} \left(\gamma_{s,m}^t h_m^s(x_n) - r_n^t\right)^2$
12:   **end for**
13: **end for**
14: Fit the residuals by training a kernel regressor:
    $f^* = \arg\min_{f \in \mathcal{F}} \frac{1}{2} \sum_{n=1}^{N} \left(f(x_n) - r_n^t\right)$
15: Return the regression models: $f^*$ and $h^* = \arg\min_{\{h_m^s\}} e_{s,m}^t$

---

**M-BTMKR**

16: **for** $s = 1, \ldots, S$ **do**
17:   Fit the current residuals: $\gamma_s^t = \frac{\sum_{n=1}^{N} r_n^t F_s(x_n)}{\sum_{n=1}^{N} F_s(x_n)^2}$
18:   Compute the loss of $F_s$: $e_s^t = \frac{1}{2} \sum_{n=1}^{N} \left(\gamma_s^t F_s(x_n) - r_n^t\right)^2$
19: **end for**
20: Fit the residuals by training a kernel regressor:
    $f^* = \arg\min_{f \in \mathcal{F}} \frac{1}{2} \sum_{n=1}^{N} \left(f(x_n) - r_n^t\right)$
21: Return the regression models: $f^*$ and $h^* = \arg\min_{\{F_s\}} e_s^t$

---

## 5  Related Work

Various techniques have been proposed to efficiently learn MKL models [11], and our BMKR algorithm is originally inspired by [23], which applies the idea of AdaBoost to train a multiple kernel based classifier. BMKR is a more general framework which can adopt different loss functions for different learning tasks. Furthermore, the boosting approach provides a natural approach to solve small sample size problems by leveraging transfer learning techniques. The original work on boosting based transfer learning proposed in [7] introduces a sample-reweighting mechanism based on AdaBoost for classification problem. Later, this approach is generalized to the cases of regression [19], and transferring knowledge from multiple sources [24]. In [6], a gradient boosting based algorithm is proposed for multitask learning, where the assumption is that the model parameters of all the tasks share a common factor. In [22], the transfer boosting and multitask boosting algorithms are generalized to the context of online learning. While both multiple kernel learning and transfer learning have been studied extensively, the effort in simultaneously dealing with these two issues is very limited. Our BTMKR algorithm distinguishes itself from these methods because it deals with these two learning problems in a unified and principled approach. To our best knowledge, this is the first attempt to transfer MKL for regression problem.

# 6    Conclusion

In this paper, we first propose BMKR, a gradient boosting based multiple kernel learning framework for regression, which is suitable for short-term electricity load forecasting problems. Different from the traditional methods for MKL, the proposed BMKR algorithm learns the combination weights for each kernel using a boosting-style algorithm. Simulation results on residential data show that the short-term electricity load forecasting could be improved with BMKR. We further extend the proposed boosting framework to the context of transfer learning and propose two boosting based transfer multiple kernel regression algorithms: K-BTMKR and M-BTMKR. Empirical results suggest that both algorithms can efficiently transfer the knowledge learned from source houses to the target houses and significantly improve the forecasting performance when the target houses and source houses have similar electricity load consumption pattern. We also investigate the effects of negative transfer and show that the proposed algorithms could prevent potential negative transfer when the source houses are quite different from the target houses.

# References

1. Atsawathawichok, P., Teekaput, P., Ploysuwan, T.: Long term peak load forecasting in Thailand using multiple kernel Gaussian process. In: ECTI-CON, pp. 1–4 (2014)
2. Bach, F.R., Lanckriet, G.R., Jordan, M.I.: Multiple kernel learning, conic duality, and the SMO algorithm. In: ICML, pp. 6–13 (2004)
3. Bühlmann, P., Hothorn, T.: Boosting algorithms: regularization, prediction and model fitting. Stat. Sci. **22**, 477–505 (2007)
4. Bunn, D., Farmer, E.D.: Comparative Models for Electrical Load Forecasting. John Wiley and Sons Inc., New York (1985)
5. Chang, C.C., Lin, C.J.: LIBSVM: a library for support vector machines. ACM Trans. Intell. Syst. Technol. **2**(3), 27 (2011)
6. Chapelle, O., Shivaswamy, P., Vadrevu, S., Weinberger, K., Zhang, Y., Tseng, B.: Boosted multi-task learning. Mach. Learn. **85**(1–2), 149–173 (2011)
7. Dai, W., Yang, Q., Xue, G.R., Yu, Y.: Boosting for transfer learning. In: ICML, pp. 193–200 (2007)
8. Fiot, J.B., Dinuzzo, F.: Electricity demand forecasting by multi-task learning. IEEE Trans. Smart Grid **PP**(99), 1 (2016)

9. Freund, Y., Schapire, R.E.: Experiments with a new boosting algorithm. In: ICML, pp. 148–156 (1996)
10. Friedman, J.H.: Greedy function approximation: a gradient boosting machine. Ann. Stat. **29**, 1189–1232 (2001)
11. Gönen, M., Alpaydın, E.: Multiple kernel learning algorithms. J. Mach. Learn. Res. **12**, 2211–2268 (2011)
12. Hastie, T., Tibshirani, R., Friedman, J.: The Elements of Statistical Learning: Data Mining, Inference, and Prediction, 2nd edn. Springer, New York (2009). https://doi.org/10.1007/978-0-387-84858-7
13. Hippert, H.S., Pedreira, C.E., Souza, R.C.: Neural networks for short-term load forecasting: a review and evaluation. IEEE Trans. Power Syst. **16**(1), 44–55 (2001)
14. IEM. https://mesonet.agron.iastate.edu/request/download.phtml
15. Kamyab, F., Amini, M., Sheykhha, S., Hasanpour, M., Jalali, M.M.: Demand response program in smart grid using supply function bidding mechanism. IEEE Trans. Smart Grid **7**(3), 1277–1284 (2016)
16. Mason, L., Baxter, J., Bartlett, P., Frean, M.: Boosting algorithms as gradient descent in function space. In: NIPS, pp. 512–518 (2000)
17. OPENEI. http://en.openei.org/doe-opendata/dataset
18. Pan, S.J., Yang, Q.: A survey on transfer learning. IEEE Trans. Knowl. Data Eng. **22**(10), 1345–1359 (2010)
19. Pardoe, D., Stone, P.: Boosting for regression transfer. In: ICML, pp. 863–870 (2010)
20. Rosset, S., Zhu, J., Hastie, T.: Boosting as a regularized path to a maximum margin classifier. J. Mach. Learn. Res. **5**, 941–973 (2004)
21. Soliman, S.A.H., Al-Kandari, A.M.: Electrical Load Forecasting: Modeling and Model Construction. Elsevier, New York (2010)
22. Wang, B., Pineau, J.: Online boosting algorithms for anytime transfer and multi-task learning. In: AAAI, pp. 3038–3044 (2015)
23. Xia, H., Hoi, S.C.: MKBoost: a framework of multiple kernel boosting. IEEE Trans. Knowl. Data Eng. **25**(7), 1574–1586 (2013)
24. Yao, Y., Doretto, G.: Boosting for transfer learning with multiple sources. In: CVPR, pp. 1855–1862 (2010)
25. Zhang, R., Dong, Z.Y., Xu, Y., Meng, K., Wong, K.P.: Short-term load forecasting of Australian National Electricity Market by an ensemble model of extreme learning machine. IET Gener. Transm. Distrib. **7**(4), 391–397 (2013)
26. Zhuang, J., Tsang, I.W., Hoi, S.C.: Two-layer multiple kernel learning. In: AISTATS, pp. 909–917 (2011)

# CREST - Risk Prediction for Clostridium Difficile Infection Using Multimodal Data Mining

Cansu Sen[1(✉)], Thomas Hartvigsen[1], Elke Rundensteiner[1],
and Kajal Claypool[2]

[1] Worcester Polytechnic Institute, Worcester, MA, USA
{csen,twhartvigsen,rundenst}@wpi.edu
[2] Harvard Medical School, Boston, MA, USA
kajal_claypool@hms.harvard.edu

**Abstract.** Clostridium difficile infection (CDI) is a common hospital acquired infection with a \$1B annual price tag that resulted in ~30,000 deaths in 2011. Studies have shown that early detection of CDI significantly improves the prognosis for the individual patient and reduces the overall mortality rates and associated medical costs. In this paper, we present CREST: **C**DI **R**isk **Est**imation, a data-driven framework for *early* and *continuous* detection of CDI in hospitalized patients. CREST uses a three-pronged approach for high accuracy risk prediction. First, CREST builds a rich set of highly predictive features from Electronic Health Records. These features include clinical and non-clinical phenotypes, key biomarkers from the patient's laboratory tests, synopsis features processed from time series vital signs, and medical history mined from clinical notes. Given the inherent multimodality of clinical data, CREST bins these features into three sets: time-invariant, time-variant, and temporal synopsis features. CREST then learns classifiers for each set of features, evaluating their relative effectiveness. Lastly, CREST employs a second-order meta learning process to ensemble these classifiers for optimized estimation of the risk scores. We evaluate the CREST framework using publicly available critical care data collected for over 12 years from Beth Israel Deaconess Medical Center, Boston. Our results demonstrate that CREST predicts the probability of a patient acquiring CDI with an AUC of 0.76 five days prior to diagnosis. This value increases to 0.80 and even 0.82 for prediction two days and one day prior to diagnosis, respectively.

**Keywords:** Clostridium difficile · Risk stratification
Multimodal data mining · Multivariate time series classification
Electronic Health Records

Y. Altun et al. (Eds.): ECML PKDD 2017, Part III, LNAI 10536, pp. 52–63, 2017.
https://doi.org/10.1007/978-3-319-71273-4_5

# 1   Introduction

**Motivation.** Clostridium difficile infection (CDI) is a common hospital acquired infection resulting in gastrointestinal illness with substantial impact on morbidity and mortality. In 2011, nearly half a million CDI infections were identified in the US resulting in 29,000 patient deaths [1,11]. Despite well-known risk factors and the availability of mature clinical practice guidelines [4], the infection and mortality rates of CDI continue to rise with an estimated $1 billion annual price tag [7]. Early detection of CDI has been shown to be significantly correlated with a successful resolution of the infection within a few days, and is projected to save $3.8 billion in medical costs over a period of 5 years [2]. In current practice, a diagnostic test is usually ordered as a confirmation of a highly-suspect case, only after appearance of symptoms[1]. This points to a tremendous opportunity for employing machine learning techniques to develop intelligent systems for early detection of CDI to eradicate this medical crisis.

**State-of-the-Art.** Our literature review shows that there have been some initial efforts to apply machine learning techniques to develop risk score estimation models for CDI. These efforts largely exploit two approaches. The first, a *moment-in-time approach*, uses only the data from one single moment in patient's stay. This moment can be the admission time [14] or the most recent snapshot data at the time of risk estimation [6]. The second, an *independent-days approach*, uses the complete hospital stay, but treats the days of a patient's stay as independent from each other [16,17]. The *complete physiological state* of the patient, *changes in the physiological state*, and *clinical notes* containing past medical information have been left out of the risk prediction process.

**Challenges.** To fill this gap, the following challenges must be addressed:

**Varying Lengths of Patient Stays.** Stay-lengths vary between patients, complicating the application of learning algorithms. Thus, we must design a fixed-length representation of time series patient-stay data. This requires temporal summarization of data such that the most relevant information for the classification task is preserved.

**Incorporating Clinical Notes.** Clinical notes from a patient's EHR contain vital information (e.g., co-morbidities and prior medications). These are often taken in short-hand and largely abbreviated. Mining and analysis of clinical notes is an open research problem, but some application of current techniques is necessary to transform them into a format usable for machine learning algorithms.

**Combining Multimodal Data.** EHR data is typically multimodal, including text, static data and time series data, that require transformation and normalization prior to use in machine learning. The choices made when transforming

---

[1] The authors would like to thank Elizabeth Claypool, RN, Coordinator of Patient Safety at U. Colorado Health for the valuable information she provided.

the data may have significant impact on classification accuracy if key transformations are not appropriate for the domain.

**Our Proposed CREST System.** CREST: CDI Risk Estimation is a novel framework that addresses these challenges and estimates the risk of a patient contracting CDI. Figure 1 gives an overview of CREST. CREST extracts highly predictive features capturing both time-invariant and time-variant aspects of patient histories from multimodal input data (i.e., consisting of clinical and non-clinical phenotypes, biomarkers from lab tests, time series vital signs, and clinical notes) while maintaining temporal characteristics. Feature selection methods are applied to select the features with the highest predictive power. Feeding these selected features into the classification pipeline, multiple models are fit ranging from primary classifiers to meta-learners. Once trained, CREST continuously generates daily risk scores to aid medical professionals by flagging at-risk patients for improved prognoses.

**Contributions.** In summary, our contributions include:

1. **Time-alignment of time series data.** We design two time-alignment methods that solve the varying length of patient's stay problem. This enables us to bring a *multiple-moments-in-time* approach to the task of predicting patient infections.

2. **Multimodal feature combination.** To our knowledge, CREST is the first work to combine clinical notes and multivariate time series data to perform classification for CDI risk prediction. We

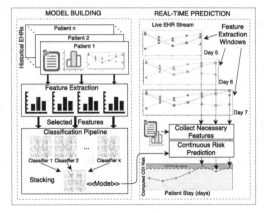

**Fig. 1.** Overview of CREST framework

show that synopsis temporal features from patient time-series data significantly improve classification performance, while achieving interpretable results.

3. **Early detection of the infection.** We evaluate our system with publicly-available critical-care data collected at the Beth Israel Deaconess Intensive Care Unit in Boston, MA [8]. Our evaluation shows that CREST improves the accuracy of predicting high-risk CDI patients by 0.22 one day before and 0.16 five days before the actual diagnosis compared to risk estimated using only admission time data.

## 2    Predictive Features of CREST

We categorize patient EHR information into three feature sets: time-invariant, time-variant, and temporal synopsis. An overview of our feature extraction process is depicted in Fig. 2.

## 2.1    Time-Invariant and Time-Variant Properties of EHR Data

**Time-Invariant Properties.** These represent all data for a patient known at the time of admission which does not change throughout the patient's stay. A number of known CDI risk factors are represented in this data (e.g. age, prior antibiotic usage). To capture these, we extract a set of time-invariant features. *Demographic features* are immutable patient features such as age, gender, and ethnicity. *Stay-specific features* describe a patient's admission such as admission location and insurance type, allowing inference on the patient's condition. These data could be different for the same patient upon readmission. *Medical history features* model historical patient co-morbidities (e.g., diabetes, kidney disease) and medications (e.g., antibiotics, proton-pump inhibitors) associated with increased CDI risk. These are extracted from clinical notes (free-form text files) using text mining. Using the Systematized Nomenclature of Medicine Clinical Terms dictionary (SNOMED CT), synonyms for these diseases and medications are identified to facilitate extraction of said factors from a patient's history.

**Time-Variant Properties.** Throughout the hospital stay of a patient, many observations are recorded continuously such as laboratory results and vital signs, resulting in a collection of time series. A data-driven approach is leveraged to model this data as time-variant features. Additionally, for each day of a patient's stay, we generate multiple binary features flagging the use of antibiotics, H2 antagonists, and proton pump inhibitors, all of which are known to risk factors for CDI. Particularly high risk antibiotics, namely Cephalosporins, Fluoroquinolones, Macrolides, Penicillins, Sulfonamides, and Tetracyclines [9],

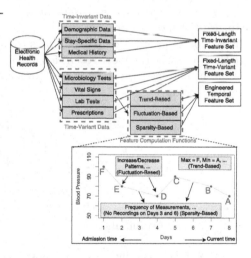

**Fig. 2.** Feature extraction process

are captured by another binary feature flagging the presence of high-risk antibiotics in a patient's body. Using a binary feature avoids one-hot encoding, a method known to dramatically increases dimensionality and sparseness.

## 2.2    Two Strategies for Modeling Variable-Length Time-Series Data

**Time-Alignment for Time-Series Clinical Data.** A patient's stay is recorded as a series of clinical observations that is often characterized as *irregularly spaced time series*. These measurements vary in the frequency at which they are taken (once a day, multiple times a day, etc.). This variation is a function of (a) the observation (a lab test can be taken once a day while a vital sign is measured multiple times), (b) the severity of the patient's condition (patients in more severe conditions must be monitored more closely), and (c) the time of the day (nurses are less likely to wake up patients in the middle of the night). To unify this, we roll up all observations taken more than once a day into evenly sampled averages at the granularity of one day. If there are no measurements for a day, these are considered as missing values and are filled with the median value.

The total number of observations recorded per patient is a function of not only the frequency of observation, but also the length of a patient's stay. After day-based aggregation, we produce a fixed-length feature representation by time-aligning the variable-length feature vectors. This time-alignment can be done by either using the same number of *initial days* since admission or the same number of *most recent days* of each patient's hospital stay. We empirically determine the optimal time-alignment window by evaluating the AUC of the initial days and the most recent days using Random Forests on only time-aligned data. Our results show that AUC using the most recent days was much higher than using the initial days of a patient's stay. We validate our results using SVMs, as shown in Fig. 3. Based on these results, we conclude that when predicting CDI risk on day $p$, the most recent 5 days of the patient stay (i.e. days $p - 5$ to $p - 1$) capture the most critical information. This is consistent with and validated by the incubation period of CDI ($<7$ days with a median of 3 days [4,5]). In CREST, we thus use only the most recent 5 days of each patient's stay as our approach to represent patient vital signs and lab/microbiology tests as continuous numerical feature vectors.

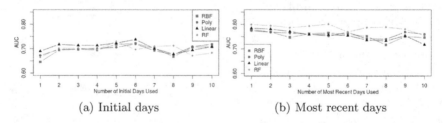

(a) Initial days                    (b) Most recent days

**Fig. 3.** AUC results using initial and most-recent days of patient stays shows that using the most recent 5 days contains the most information about the CDI risk.

**Computing Temporal Synopsis Features.** Time-variant features (e.g., temperature), while capturing the state of the patient for each day of their stay, falsely treat days to be independent from each other. Thus, they do not capture the sequential trends over time inherent in these time series data. For example, the presence or absence of recordings of a time-variant feature may be more informative than the actual values (e.g., *heart rate high alarm* is only measured when a patient has an alarmingly high heart rate). In some cases, the change in an observation (e.g., increase in temperature) may be more important than the actual observed values. To model these trends, in CREST, we introduce feature computation functions, capturing the following temporal synopsis features:

- **Trend-based features** include statistics such as minimum, maximum, and average values. In addition to an equal weighted average, linear and quadratic weighted averages are computed, giving more weight to later days. The relative times of the first and last recordings and of minimum and maximum recordings are also extracted to signal when in a patient's hospital stay these notable events occur.
- **Fluctuation-based features** capture the change characteristic of each time-variant feature. Mean absolute differences, number of increasing and decreasing recordings and the ratio of change in direction are examples of trends we extract to capture these characteristics.
- **Sparsity-based features** model frequency of measurements and proportion of missing values. For example, "heart rate high alarm" is recorded only if a patient's heart rate exceeds the normal threshold.

Figure 2 illustrates the time-variant feature blood pressure for a patient and examples of trends we extract from this time series data.

# 3 Modeling Infection Risk in CREST

## 3.1 Robust Supervised Feature Selection

In CREST, each extracted feature set is fed into a rigorous feature selection module to determine the features that are most relevant to CDI risk. We denote $S_{n \times s}$, $D_{n \times d}$, and $T_{n \times t}$ to be the time-invariant, time-variant, and temporal feature matrices with n instances and $s$, $d$, and $t$ features respectively. For a compact representation, we use $X$ to represent $S$, $D$, and $T$. The goal is to reduce $X_{n \times p}$ into a new feature matrix $X'_{n \times k}$ where $X'_{n \times k} \subset X_{n \times p}$. To achieve this, we combine chi-squared feature selection, a supervised method that tests how features depend on the label vector $Y$, with SVMs. Two issues must be addressed when using this method, namely, determining the optimal cardinality of features, and which features to use.

**Percentile Selection.** We first determine the cardinality of features for each feature set. Using 10-fold cross validation over training data, we select the top K percent of features for $K = (5, 10, 15, \ldots, 100)$ and record the average AUC value by percentile for each of the three feature sets. We then select the percentiles that perform the best.

**Robustness Criterion.** Next, we select as few features as possible while ensuring adequate predictive power. We empirically select which features to use by choosing a robustness criterion, $\gamma$, which we define as "the minimum number of folds in which features must appear to be considered *predictive*". Since we have 10 cross-validation folds, $\gamma \in [1 : 10]$, where $\gamma = 1$ implies all features selected for *any* folds are included in the final feature set (union) and $\gamma = 10$ implies all features selected for *every* fold is included in the final feature set (intersection).

We apply these steps to feature matrices $S$, $D$, and $T$, resulting in reduced feature matrices $S'$, $D'$, and $T'$.

## 3.2   CREST Learning Methodology

We represent a patient's CDI risk as the probability that the patient gets infected with CDI. To compute this probability, we estimate a function $f(X') \rightarrow Y$ using the reduced feature matrix $X'$ (representing $S'$, $D'$, or $T'$) and the label vector $Y$, consisting of binary diagnosis outcomes. The function outputs a vector of predicted probabilities, $\hat{Y}$. In a hospital setting, CREST extracts a feature matrix $X'$ every day of a patient's hospital stay. CREST then employs the classification function on $X'$ (see Fig. 1 for this continuous process). This section describes the process of estimating the function $f$, shown in Fig. 4.

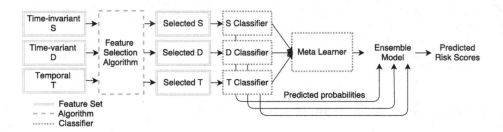

**Fig. 4.** Learning phase of the CRESTframework.

**Type-Specific Classification.** We first train a set of type-specific classifiers built on each of the feature matrices. The task is to estimate $f(X') \rightarrow Y$ which minimizes $|Y - \hat{Y}|$. We use SVMs, Random Forests, and Logistic Regression to estimate $f$. Since imbalanced data is typical in this application domain, CREST uses a modified SVM objective function that includes two cost parameters for positive and negative classes. Thus, a higher misclassification cost is assigned to the minority class. Equation 1 shows the modified SVM objective function we used in CREST and Eq. 2 shows how we choose the cost for positive and negative classes.

$$\text{minimize } (\frac{1}{2}w \cdot w + C^+ \sum_{i \in \mathcal{P}}^{l} \xi_i + C^- \sum_{i \in \mathcal{N}}^{l} \xi_i) \tag{1}$$

$$\text{s.t. } y_i(w \cdot \Phi(x_i) + b) \geq 1 - \xi_i \ \xi_i \geq 0, \quad i = 1 \ldots l.$$

$$C^+ = C \frac{l}{|\mathcal{P}|}, \quad C^- = C \frac{l}{|\mathcal{N}|} \tag{2}$$

where $w$ is a vector of weights, $\mathcal{P}$ is the positive class, $\mathcal{N}$ is the negative class, $l$ is the number of instances, $C$ is the cost, $\xi$ is a set of slack variables, $x_i$ is $i^{th}$ data instance, $\Phi$ is a kernel function, and $b$ is the intercept.

A *static classifier*, trained on feature set $S'$ extracted from admission time data, implies that only the information obtained on admission is necessary to accurately predict risk. This constitutes our baseline as it represents the current practice of measuring risk in hospitals and denotes risk on day 0. A *dynamic classifier*, trained on feature set $D'$, constitutes a multiple-moments-in-time app-roach where the data from many moments in a patient's stay are used as features. This approach allows us to quantify the relationship between the *physiological state* of the patient and their CDI risk. Finally, a *temporal classifier*, trained using feature set $T'$, quantifies the relationship between a patient's *state-change* and their risk, complementing the time-variant features.

**Second-Order Classification.** Since the three type-specific classifiers capture different aspects of a patient's health and hospital stay, we combine them to produce a single continuous prediction based on comprehensive information. We hypothesize that this combination method, termed *second-order classification*, will provide more predictive power. To evaluate this hypothesis, we merge the predicted probability vectors from the type-specific classifiers into a new higher-order feature set $X_{meta} = (\hat{Y}_S, \hat{Y}_D, \hat{Y}_T)$. With this new feature matrix, our task becomes estimating a function $f(X_{meta}) \rightarrow Y$. Beyond naive methods such as model averaging to assign weights to the results produced by the type-specific classifiers, we also develop a stacking-based solution. We train meta learners fusing SVMs with RBF and linear kernels, Random Forests, and Logistic Regression on $X_{meta}$ to learn an integrated ensemble classifier. Henceforth, final predictions are made by these new second-order classifier models.

# 4    Evaluation of CREST Framework

## 4.1    MIMIC-III ICU Dataset and Evaluation Settings

The MIMIC III Database [8], used to evaluate our CRESTFramework, is a pub-licly available critical care database collected from the Beth Israel Deaconess Medical Center Intensive Care Unit (ICU) between 2001 and 2012. The data-base consists of information collected from ~45,000 unique patients and their ~58,000 admissions. Each patient's record consists of laboratory tests, medical procedures, medications given, diagnoses, caregiver notes, etc.

Of the 58,000 admissions in MIMIC, there are 1079 cases of CDI. Approx-imately half of these patients were diagnosed either before or within the first 4 days of their admission. To ensure that CDI cases in our evaluation dataset are contracted during the hospital stay, we exclude patients who test positive for CDI within their first 5 days of hospitalization based on the incubation period of CDI [4,5]. For consistency between CDI and non-CDI patients, we also exclude non-CDI patients whose hospital stay is less than 5 days. As the vast majority of MIMIC consists of patients who do not contract CDI, we end up with an unbalanced dataset (116:1). To overcome this, we randomly subsample from the non-CDI patients to get a 2-to-1 proportion of non-CDI to CDI patients, leaving us with 1328 patient records.

Next, we define the feature extraction window for patients. For CDI patients, it starts on the day of admission and ends $n$ days before the CDI diagnosis, $n \in \{1, \ldots, 5\}$. For non-CDI patients, there are a few alternatives for defining this window. Prior research has used the discharge day as the end of the risk period [6]. However, as the state of the patients can be expected to improve nearing their discharge, this may lead to deceptive results [16]. Instead, we use the halfway point of the non-CDI patient's stay as the end of the risk period or 5 days (minimum length of stay), whichever is greater.

We then split these patients into training and testing subsets with a 70%–30% ratio and maintain these subsets across all experiments. The training set is further split and 5-fold cross-validation is applied to perform hyper-parameter search. We use SVM with linear and RBF kernels, Random Forest and Logistic Regression. All algorithms were implemented using Scikit-Learn in Python.

## 4.2  Classification Results

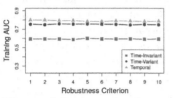

**Fig. 5.** Selection of robustness criterion

Using our feature selection module, we find the best cardinalities to be $K = 20$ for time-invariant, $K = 30$ for time-variant, $K = 90$ for temporal feature sets with robustness-criterion $\gamma = 10$ for all three feature sets. This choice of $\gamma$ is motivated by an almost unchanging validation AUC over all potential $\gamma$ values, as shown in Fig. 5. This shows that mostly the same features are selected for each fold. By choosing $\gamma = 10$, we can be certain that only the features that are strongly related to the response variable are selected.

We first run a set of experiments with type-specific classifiers to determine the predictive power of each type of feature class. We then experiment with ensembles of the type-specific classifiers in two ways: (1) **Equal-weighted model averaging:** We calculate equal weighted averages of the probabilities produced by each type-specific classifier, (2) **Meta-learning:** We train second order meta learners using the outputs of the type-specific classifiers as the input of the meta learners. Table 1 shows the AUC, precision, recall and F-1 scores for each classification method.

Static classifiers constitute our baseline approach. The mean AUC of all static classifiers is 0.60, implying that a risk score can

**Table 1.** Classification results acquired on the test set.

|  |  | AUC | Precision | Recall | F-1 |
|---|---|---|---|---|---|
| Static C. | SVM RBF | 0.544 | 0.57 | 0.62 | 0.58 |
|  | SVM Linear | 0.627 | 0.76 | 0.46 | 0.38 |
|  | Random F | 0.608 | 0.57 | 0.62 | 0.58 |
|  | Logistic R | 0.627 | 0.6 | 0.64 | 0.59 |
|  | **Average** | **0.602** | **0.63** | **0.59** | **0.53** |
| Dynamic C. | SVM RBF | 0.779 | 0.73 | 0.73 | 0.71 |
|  | SVM Linear | 0.756 | 0.71 | 0.72 | 0.69 |
|  | Random F | 0.818 | 0.75 | 0.76 | 0.75 |
|  | Logistic R | 0.758 | 0.72 | 0.73 | 0.71 |
|  | **Average** | **0.778** | **0.73** | **0.74** | **0.72** |
| Temporal C. | SVM RBF | 0.815 | 0.76 | 0.77 | 0.76 |
|  | SVM Linear | 0.817 | 0.76 | 0.72 | 0.72 |
|  | Random F | 0.832 | 0.77 | 0.77 | 0.77 |
|  | Logistic R | 0.809 | 0.75 | 0.76 | 0.75 |
|  | **Average** | **0.818** | **0.76** | **0.76** | **0.75** |
| Model Avg. |  | **0.817** | **0.76** | **0.71** | **0.65** |
| Meta Learn. | SVM RBF | 0.838 | 0.76 | 0.76 | 0.75 |
|  | SVM Linear | 0.833 | 0.76 | 0.73 | 0.74 |
|  | Random F | 0.815 | 0.74 | 0.75 | 0.74 |
|  | Logistic R | 0.831 | 0.76 | 0.77 | 0.76 |
|  | **Average** | **0.829** | **0.76** | **0.75** | **0.75** |

be assigned to a patient at the time of admission. Dynamic classifiers, which use time-variant features, achieve a much higher AUC compared to the static classifiers. This shows that the physiological state of a patient is correlated with the CDI outcome. Among the type-specific classifiers, the temporal classifiers consistently attain the highest AUC. This highlights that patient-state changes are strongly predictive of CDI risk. To the best of our knowledge, ours is the first effort that uses this information to predict CDI risk for patients. Between our two ensemble methods, meta-learners further improve the prediction success over any of the type-specific classifiers, showing that considering all features together is beneficial. The highest AUC is achieved by meta-learners when an SVM with an RBF kernel is used. Figure 6 presents the ROC curves for type-specific classifiers and the meta learners, which show an increasing trend in diagnosis accuracy.

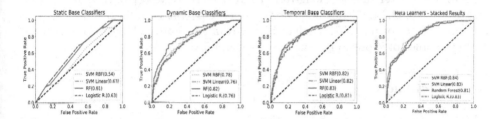

**Fig. 6.** ROC curves for static, dynamic, temporal, and meta classifiers

### 4.3   Early Prediction of CDI

The earlier an accurate prediction can be made, the higher the likelihood that actions can be taken to prevent contraction of CDI. We evaluate the power of our model for early prediction using the best CREST meta learner. Unlike previous experiments, we now train models using the data 1 to 5 days prior to diagnosis. Results indicate that early warnings can maintain high AUC values (Fig. 7). In comparison with the baseline

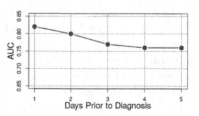

**Fig. 7.** AUC results of early prediction experiments

methods where the mean AUC is 0.60, CREST improves the accuracy of predicting high-risk CDI patients to 0.82 one day prior to diagnosis and to 0.76 five days prior to diagnosis, an improvement of 0.22 and 0.16 over the baseline respectively.

## 5   Related Work

**Feature Extraction from Time Series.** One strategy to deal with clinical time series in machine learning is to extract aggregated features. In healthcare, much work has gone into extracting features from signals such as ECG [13] or EEG [3,10] using methods such as wavelet [13,18] or Fourier [18] transformations. However, EHR time series have largely being ignored. Specially designing feature extraction techniques for EHRs in our model, we demonstrate that prediction accuracy increases using these features over models that do not account for the temporal aspects of the data.

**In-hospital CDI Prediction.** Recent work has begun to investigate prediction models for CDI. [16,17] ignore temporal dependencies in the data and reduce this complex task to univariate time-series classification. [15] while combining time-variant and time-invariant data, neglect the trends in patient records. [12] uses ordered pairs of clinical events to make predictions, missing longer patterns in data. In our work, we apply multivariate time series classification while capturing temporal characteristics and long-term EHR patterns. SVMs [16,17] and Logistic Regression [6,12,14,15] are popular tools for CDI risk prediction models. We apply a variety of models including SVM, Random Forest, Logistic Regression and ensembles of those to produce more comprehensive results.

# 6   Conclusion

CREST is the first system that stratifies a patient's infection risk on a continuous basis throughout their stay and is based on a novel feature extraction and combination method. CREST has been validated for CDI risk using the MIMIC Database. Our experimental results demonstrate that CREST can detect CDI cases with an AUC score of up to 0.84 one day before and 0.76 five days before the actual diagnosis. CDI is a highly contagious disease and early detection of CDI not only greatly improves the prognosis for individual patients by enabling timely precautions but also prevents the spread of the infection within the patient cohort. To our knowledge, this is the first work on multivariate time series classification to predict the risk of CDI. We also demonstrate that our extracted temporal synopsis features improve the AUC by 0.22 over the static classifiers and 0.04 over the dynamic classifiers.

We are in discussion with UCHealth Northern Colorado as well as Brigham and Women's Hospital, part of the Partners Healthcare System in Massachusetts, for the potential deployment of a CREST dashboard integrated with their Electronic Health Records (EPIC). This deployment will be a 4 step process, with the work presented in this paper being the first step. The CREST framework will be independently validated against data from ICUs at these hospitals. Successful validation of CREST will lead to Step 3 - clinical usability of the EPIC-CREST dashboard with a particular ward where daily risk scores produced by CREST will be utilized by the nurses to support diagnosis and early detection. Full scale deployment will be largely dependent on the results of this clinical validation and usability study.

**Acknowledgments.** The authors thank Dr. Richard T. Ellison, III, the head of Infection Control at UMass Memorial Medical Center, Worcester, MA, for his valuable comments that helped us understand the urgency of the CDI crisis. The authors also thank Dr. Alfred DeMaria, Medical Director for the Bureau of Infectious Diseases at Massachusetts Public Health Department for highlighting the effects of this crisis on healthcare systems in Massachusetts and beyond.

# References

1. Centers for Disease Control and Prevention (2017). https://www.cdc.gov/media/releases/2015/p0225-clostridium-difficile.html
2. Centers for Disease Control and Prevention: Antibiotic resistance threats in the United States (2017). https://www.cdc.gov/drugresistance/biggestthreats.html
3. Chaovalitwongse, W.A., Prokopyev, O.A., Pardalos, P.M.: Electroencephalogram (EEG) time series classification: applications in epilepsy. Ann. Oper. Res. **148**(1), 227–250 (2006)
4. Cohen, S.H., et al.: Clinical practice guidelines for Clostridium difficile infection in adults: 2010 update by the Society for Healthcare Epidemiology of America (SHEA) and the Infectious Diseases Society of America (IDSA). Infect. Control Hosp. Epidemiol. **31**(05), 431–455 (2010)
5. Dubberke, E.R., et al.: Hospital-associated Clostridium difficile infection: is it necessary to track community-onset disease? Infect. Control Hosp. Epidemiol. **30**(04), 332–337 (2009)
6. Dubberke, E.R., et al.: Development and validation of a Clostridium difficile infection risk prediction model. Infect. Control Hosp. Epidemiol. **32**(4), 360–366 (2011)
7. Evans, C.T., Safdar, N.: Current trends in the epidemiology and outcomes of Clostridium difficile infection. Clin. Infect. Dis. **60**(suppl 2), S66–S71 (2015)
8. Johnson, A.E., et al.: MIMIC-III, a freely accessible critical care database. Sci. Data **3**, 160035 (2016)
9. Kuntz, J.L., et al.: Incidence of and risk factors for community-associated Clostridium difficile infection: a nested case-control study. BMC Infect. Dis. **11**(1), 194 (2011)
10. Lemm, S., et al.: Spatio-spectral filters for improving the classification of single trial EEG. IEEE Trans. Biomed. Eng. **52**(9), 1541–1548 (2005)
11. Lessa, F.C., et al.: Burden of Clostridium difficile infection in the United States. N. Engl. J. Med. **372**(9), 825–834 (2015)
12. Monsalve, M., et al.: Improving risk prediction of Clostridium difficile infection using temporal event-pairs. In: International Conference on Healthcare Informatics, pp. 140–149. IEEE (2015)
13. Sternickel, K.: Automatic pattern recognition in ECG time series. Comput. Methods Programs Biomed. **68**(2), 109–115 (2002)
14. Tanner, J., et al.: Waterlow score to predict patients at risk of developing Clostridium difficile-associated disease. J. Hosp. Infect. **71**(3), 239–244 (2009)
15. Wiens, J., et al.: Learning data-driven patient risk stratification models for Clostridium difficile. Open Forum Infectious Diseases **1**(2), ofu045 (2014)
16. Wiens, J., et al.: Learning evolving patient risk processes for C. diff colonization. In: ICML Workshop on Machine Learning from Clinical Data (2012)
17. Wiens, J., Horvitz, E., Guttag, J.V.: Patient risk stratification for hospital-associated C. diff as a time-series classification task. In: Advances in Neural Information Processing Systems, pp. 467–475 (2012)
18. Zhang, H., et. al.: Feature extraction for time series classification using disc. wavelet coefficients. In: Advances in Neural Networks. ISNN 2006, pp. 1394–1399 (2006)

# DC-Prophet: Predicting Catastrophic Machine Failures in DataCenters

You-Luen Lee[1], Da-Cheng Juan[2], Xuan-An Tseng[1], Yu-Ting Chen[2],

and Shih-Chieh Chang[1(✉)]

[1] Department of Computer Science, National Tsing Hua University, Hsinchu, Taiwan
peggy199382@gmail.com, killerjack003@gmail.com,
scchang@cs.nthu.edu.tw
[2] Google Inc., Mountain View, CA, USA
dacheng@google.com, yutingchen@google.com

**Abstract.** When will a server fail catastrophically in an industrial datacenter? Is it possible to forecast these failures so preventive actions can be taken to increase the reliability of a datacenter? To answer these questions, we have studied what are probably the largest, publicly available datacenter traces, containing more than *104 million* events from *12,500* machines. Among these samples, we observe and categorize three types of machine failures, all of which are catastrophic and may lead to information loss, or even worse, reliability degradation of a data-center. We further propose a two-stage framework—**DC-Prophet** (DC-Prophet stands for **DataCenter-Prophet**.)—based on One-Class Support Vector Machine and Random Forest. DC-Prophet extracts surprising patterns and accurately pre-dicts the next failure of a machine. Experimental results show that DC-Prophet achieves an AUC of 0.93 in predicting the next machine failure, and a $F_3$-score (The ideal value of $F_3$-score is 1, indicating perfect predictions. Also, the intu-ition behind $F_3$-score is to value "Recall" about three times more than "Precision" [12].) of 0.88 (out of 1). On average, DC-Prophet outperforms other classical machine learning methods by 39.45% in $F_3$-score.

## 1 Introduction

*"When will a server fail catastrophically in an industrial datacenter?" "Is it possible to forecast these failures so preventive actions can be taken to increase the reliability of a datacenter?"* These two questions serve as the motivation for this work.

To meet the increasing demands for cloud computing, Internet companies such as Google, Facebook, and Amazon generally deploy a large fleet of servers in their data-centers. These servers bear heavy workloads and process various, diversified requests [13]. For such a high-availability computing environment, when an unexpected machine failure happens upon a clustered partition, its workload is typically transferred to another machine in the same cluster, which increases the possibility of other failures as a chain effect [11]. Also, this unexpected failure may cause (a) processed data loss, and (b) resource congestion due to machines being suddenly unavailable. In the worst case, these failures may paralyze a datacenter, causing an unplanned outage that requires a very high cost to recover [1]: on average \$9,000/minute, and up to \$17,000/minute.

© Springer International Publishing AG 2017
Y. Altun et al. (Eds.): ECML PKDD 2017, Part III, LNAI 10536, pp. 64–76, 2017.
https://doi.org/10.1007/978-3-319-71273-4_6

To study machine failures in a modern datacenter, we analyze the traces from Google's datacenter [9, 14]; the traces contain more than 104 million events generated by 12,500 machines during 29 days. We observe that approximately 40% of the machines have been removed (due to potential failures or maintenance) at least once during this period. This phenomenon suggests that potential machine failures happen quite frequently, and cannot be simply ignored. Therefore, we want to know: given the trace of a machine, can we accurately predict its next failure, ideally with low computing latency? If the answer is yes, the cloud scheduler (*e.g.*, Borg [17] by Google) can take preventive actions to deal with incoming machine failures, such as by migrating tasks from the machine-to-fail to other machines. In this way, the cost of a machine failure is reduced to the very minimum: only the cost of task migration.

While predicting the next failure of a machine seems to be a feasible and promising solution for improving the reliability of a datacenter, it comes with two major challenges. The first challenge lies in high accuracy being required when making predictions, specifically for reducing false negatives. The false negatives (the machine actually failed but being predicted as normal) may incur a significant recovery cost [1] and should be avoided in Table 1. However, if the objective is set to minimize false negatives, the model will always predict a machine going to fail (so zero false negative), which introduces costs from false positives (the machine actually works but being predicted as failed). Therefore, one major challenge of designing a model is to better trade off between these two costs. The second challenge is the counts between normal events and failure events are highly imbalanced. Among 104 million events, only 8,957 events (less than 1%) are associated with machine failures. In this case, most predictive models will trivially predict every event as normal to achieve a high accuracy (higher than 99%). Consequently, this event-imbalance issue is the second roadblock that needs to be removed.

The contributions of this paper are as follows:

- We analyze probably the largest, publicly-available traces from an industrial datacenter, and categorize three types of machine failures: Immediate-Reboot (IR), Slow-Reboot (SR), and Forcible-Decommission (FD). The frequency and duration of each type of failures categorized by our method further match experts' domain knowledge.
- We propose a two-stage framework: **DC-Prophet** that accurately predicts the occurrence of next failure for a machine. DC-Prophet first applies One-Class SVM to filter out most normal cases to resolve the event-imbalance issue, and then deploys Random Forest to predict the type of failures that might occur for a machine.

**Table 1.** Misprediction issues and the associated costs

|  | Actual: failed | Actual: normal |
| --- | --- | --- |
| Predicted: failed | True positive (correct inference) | False positive: low cost (e.g., extra rescheduling) |
| Predicted: normal | **False negative: high cost (upto $17,000/min)** | True negative (correct inference) |

The experimental results show that DC-Prophet accurately predicts machine failures and achieves an AUC of 0.93 and $F_3$-score of 0.88, both on the test set.

- To understand the effectiveness of DC-Prophet, we also perform a comprehensive study on other widely-used machine learning methods, such as multi-class SVM, Logistic Regression, and Recurrent Neural Network. Experimental results show that, on average, DC-Prophet outperforms other methods by 39.45% in $F_3$-score.
- Finally, we provide a practitioners' guide for using DC-Prophet to predict the next failure of a machine. The latency of invoking DC-Prophet to make one prediction is less 9 ms. Therefore, DC-Prophet can be seamlessly integrated into a scheduling strategy of industrial datacenters to improve the reliability.

The remainder of this paper is organized as follows. Section 2 provides the problem definition, and Sect. 3 details the proposed DC-Prophet framework. Section 4 presents the implementation flow and experimental results, and Sect. 5 provides practitioners' guide. Finally, Sect. 6 concludes this paper.

## 2 Problem Definition

### 2.1 Google Traces Overview

The Google traces [14] consist of the activity logs from 668,000 jobs during 29 days, and each job will spawn one or more tasks to be executed in a 12,500-machine cluster. For each machine, the traces record (a) computing resources consumed by all the tasks running on that machine, and (b) its machine state. Both resource consumption and machine states are recorded with associated time interval of one-microsecond (1 μs) resolution.

We focus on the usage measurements of six types of resources: (a) CPU usage, (b) disk I/O time, (c) disk space usage, (d) memory usage, (e) page cache, and (f) memory access per instruction. All these measurements are normalized by their respective maximum values and thus range from 0 to 1. In this work, the average and peak values during the time interval of 5 min are also calculated for each usage–the interval of 5 min is typically used to report the measured resource footprint of a task in Google's datacenter [14]. Furthermore, resource usages at minute-level provide a more macro view of a machine status [8]. We use $x_{r,t}$ to denote the average usage of resource type $r$ at time interval $t$; similarly, $m_{r,t}$ represents the peak usage. Both $x_{r,t}$ and $m_{r,t}$ are used to construct the training dataset, with further details provided in Sect. 2.4.

In addition, Google traces also contain three types of events to determine machine states: ADD, REMOVE, and UPDATE [14]. In this work, we treat each REMOVE event as an anomaly that could potentially be a machine failure. Detailed analyses are further provided in Sect. 2.3.

### 2.2 Problem Formulation

The problem of predicting the next machine failure is formulated as follows:

*problem 1 (Categorize catastrophic failures).* Given the traces of machine events, categorize the type of each machine failure at time interval $t$ (denoted as $y_t$).

*problem 2 (Forecast catastrophic failures).* Given the traces of resource usages— denoted as $x_{r,t}$ and $m_{r,t}$—up to time interval $\tau - 1$, forecast the next failure and its type at time interval $\tau$ (denoted as $y_\tau$) for each machine. Mathematically, this problem can be expressed as:

$$y_\tau = f(x_{r,t}, m_{r,t}), t = 1 \text{ to } \tau - 1, r \in \text{resources} \tag{1}$$

where $x_{r,t}$ and $m_{r,t}$ represent the respective average and peak usage of resource $r$ at time interval $t$.

We use Fig. 1 to better illustrate the concept in Eq. (1), specifically the temporal relationship among $y_\tau$, $x_{r,t}$ and $m_{r,t}$ for $t = 1$ to $\tau - 1$. One goal here is to find a function $f$ that takes $x_{r,t}$ and $m_{r,t}$ as inputs to predict $y_\tau$.

**Fig. 1.** Relationship among $y_\tau$, $x_{r,t}$ and $m_{r,t}$ for $t = 1$ to $\tau - 1$.

## 2.3 Machine-Failure Analyses

Throughout the 29-day traces, we find a total of 8,957 potential machine failures from the REMOVE events, and Fig. 2(a) illustrates the rank-frequency of these failures. The distribution is power-law-like and heavily skewed: the top-ranked machines failed more than 100 times, whereas the majority of machines (3,397 machines) failed only once. Overall, about 40% (out of 12,500) machines have been removed at least once. We further notice that the resource usages of these most frequently-failing machines are all zeros, indicating a clear abnormal behavior. These machines seem being marked as unavailable internally [2], and hence are apparent anomalies. They are excluded from the analysis later on.

**Observation 1.** *Most frequently-failing machines have failed more than 100 times over 29 days, with usages of all resource types being zero.*

To categorize the type of a failure, we further analyze its duration which is calculated by the time difference between the REMOVE and the following ADD event. Figure 2(b) illustrates the distribution of durations for all machine failures. The failure duration can vary a lot, ranging from few minutes, to few hours, to never back—a machine is never added back to the cluster after its REMOVE event. Furthermore, three "peaks" can be observed in failure durations: $\approx 16$ min, $\approx 2$ h, and never back.

**Observation 2.** *Three "peaks" in the histogram of failure durations correspond to $\approx 16$ min, $\approx 2$ h, and never back.*

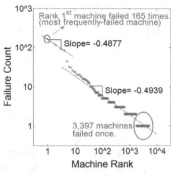 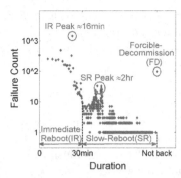

(a) Rank-Frequency plot of machine failures (b) Three "peaks" in the distribution of failure
(log-log scale).                                         durations (log-log scale).

**Fig. 2.** (a) The x-axis represents the rank of each machine sorted based on the number of failures
(high rank means more failures), whereas the y-axis is the number of failures. Both axises are
in logarithmic scale. The distribution is power-law-like: three machines failed more than 100
times, whereas 3,397 machines failed only once. (b) Each dot represents the count of failures at
a specific duration. The x-axis is duration and the y-axis represents the count. Both axises are
in logarithmic scale. Notice the three peaks highlighted by the red circles: $\approx$16 min, $\approx$2 h, and
never back. (Color figure online)

This observation raises an intriguing question: why there are three peaks in failure
durations? We correspond these three peaks ($\approx$16 min, $\approx$2 h, and never back) to three
types of machine failures:

- **Immediate-Reboot (IR).** This type of failures may occur with occasional machine
  errors and these machines can recover themselves in a short duration by rebooting.
  Here, failures of less than 30-min downtime are categorized as IR failures [3].
- **Slow-Reboot (SR).** This type of failures requires more than 30 min to recover.
  According to [3], the causes of slow reboots include file system integrity checks,
  machine hangs that require semiautomatic restart processes, and machine software
  reinstallation and testing. Also, a machine could be removed from a cluster due to
  system upgrades (*e.g.*, automated kernel patching) or network down [7, 10]. We categorize SR failures as the ones with longer than 30-min downtime and will eventually
  be added back to the cluster.
- **Forcible-Decommission (FD).** This type of failures may occur when either a
  machine (*e.g.*, part of hardware) is broken and not repaired before the end of the
  traces, or a machine is taken out from the cluster for some reasons, such as a regular machine retirement (or called "decommission") [2, 3]. We categorize this type of
  failures that a machine is removed permanently from the cluster, as FD failures.

Among 8,771 failure events (186 obvious anomalies are removed beforehand as
Observation 1 described), we summarize 5,894 to be IR failures, 2,783 SR failures, and
94 FD failures. On the other hand, there are 104,644,577 normal operations.

One important goal of this work is to predict the next failure for a machine. If a
failure is mispredicted as a normal operation (a false negative), a high cost can incur.

For example, the user jobs can be killed unexpectedly, leading to processed data loss. If these failures can be predicted accurately in advance, the cloud/cluster scheduler can perform preventive actions such as rescheduling jobs to another available machine to mitigate the negative impacts. Compared to the cost incurred from false negatives, *i.e.*, mispredicting a failure as a normal operation, the cost of "misclassifying" one failure type as another is relatively low. Still, if the right types of failures can be correctly predicted, the cloud/cluster scheduler can plan and arrange the computing resources accordingly.

### 2.4  Construct Training Dataset

We model the prediction of the next machine failure from Eq. (1) as a multi-class classification and construct the training dataset accordingly. Each instance in the dataset consists of a label $y_\tau$ that represents the failure type at time interval $\tau$, and a set of predictive features $x$ (or called a feature vector) extracted from the resource usages up to time interval $\tau - 1$.

The type of a label $y_\tau$ is determined based on the failure duration described in Sect. 2.3. If there is no machine failure at time interval $\tau$, label $y_\tau$ is marked as "normal operation." Therefore, we defined $y_\tau \in \{0, 1, 2, 3\}$, which represents normal operation, IR, SR, and FD, respectively.

For the predictive features $x$, we leverage both the average $x_{r,t}$ and peak values $m_{r,t}$ of six resource types as mentioned in Sect. 2.1. Now the question is: how to select the number of time intervals needed to be included in the dataset for an accurate prediction? We propose to calculate the partial autocorrelation to determine the number of intervals, or called "lags" in time series, to be included in the predictive features $x$. Assume target interval is $\tau$, the interval with "one lag" will be $\tau - 1$ (and the interval with two lags will be $\tau - 2$, etc.). Partial autocorrelation is a type of conditional correlation between $x_{r,\tau}$ and $x_{r,t}$, with the linear dependency of $x_{r,t+1}$ to $x_{r,\tau-1}$ removed [5]. Since the partial autocorrelation can be treated as "the correlation between $x_{r,\tau}$ and $x_{r,t}$, with other linear dependency removed," it suggests how many time intervals (or lags) should be included in the predictive features.

Figure 3(a) illustrates the partial autocorrelation of the CPU usage on one machine, and Fig. 3(b) represents the histogram of partial autocorrelations with certain lags. Both the figures show statistical significance. Notice in general, after 6 lags (30 min), the resource usages are less relevant.

**Observation 3.** *Resource usages from 30 min ago are less relevant to the current usage in terms of partial autocorrelation.*

Based on this observation, we include resource usages within 30 min as features to predict failure type $y_\tau$. In other words, 6 time intervals (lags) are selected for both $x_{r,t}$ and $m_{r,t}$ to construct the predictive features $x_t$. Specifically, $x_t = \{x_{r,t}, m_{r,t}\}$, $r \in$ resources and $t = \tau - j$ where $j = 1$ to 6. Therefore, $x$ has 2 (average and peak usages) × 6 (number of resources) × 6 (intervals) = 72 predictive features.

Now we have constructed the training dataset, and are ready to proceed to the proposed framework. For conciseness, in the rest of this paper each instance will be presented as $(y, x)$ instead of $(y_\tau, x_t)$ with $t = \tau - 1, ..., \tau - 6$.

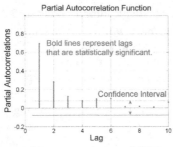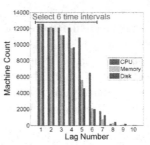

(a) Partial autocorrelation of CPU usage.

(b) Histogram of statistically-significant partial autocorrelations on all machines.

**Fig. 3.** (a) Lags of 1, 2, 3, 5 and 6 correlate with lag 0, *i.e.*, $x_{cpu,\tau}$, and these correlations are statistically significant. (b) For each machine, partial autocorrelations with up to 10 lags are calculated; only statistically-significant lags are reported for plotting this histogram. Notice in general, after 6 lags (or time intervals) the resource usages are less relevant—only few machines report partial autocorrelations with 6+ lags that are statistically significant.

## 3   Methodology

### 3.1   Overview: Two-Stage Framework

Begin immediately, we illustrate the proposed two-stage framework with Fig. 4. In the first stage, One-Class Support Vector Machine (OCSVM) is deployed for anomaly detection. All the detected anomalies are then sent to Random Forest for multi-class classification. Mathematically, DC-Prophet can be expressed as a two-stage framework:

$$f(x) = g(x) \cdot h(x) = \begin{cases} 0, & \text{if } g(x) = 0 \\ h(x), & \text{if } g(x) = 1 \end{cases} \tag{2}$$

where $g(\cdot) \in \{0, 1\}$ is OCSVM and $h(\cdot) \in \{0, 1, 2, 3\}$ is Random Forest. For an incoming instance $x$, it will first be sent to $g(\cdot)$ for anomaly detection. If $x$ is detected

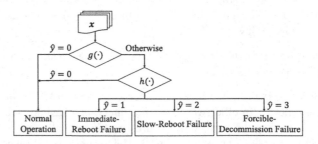

**Fig. 4.** *Flow chart of DC-Prophet: two-stage framework.* At the first stage, a sample $x$ is sent to One-Class SVM $g(\cdot)$ for anomaly detection (*i.e.*, potential machine failure or normal operation). If $x$ is classified as a potential machine failure, then this sample will be further sent to Random Forest $h(\cdot)$ for multi-class (IR, SR, FD, or normal) classification.

as an anomaly, *i.e.*, a potential machine failure, it will be further sent to $h(\cdot)$ for multi-class classification.

In Google traces, the distribution of four label types is extremely unbalanced: 104 millions of normal cases versus 8,771 failures that are treated as anomalies (including all three types of failures). Therefore, OCSVM is applied to filter out most of normal operations and detect anomalies, *i.e.*, potential machine failures. Without doing so, classifiers will be swamped by normal operations, learn only the "normal behaviors," and choose to ignore all the failures. This will cause significant false negatives as mentioned in Table 1 since most machine failures are mispredicted as normal operations.

## 3.2  One-Class SVM

One-class SVM (OCSVM) is often applied for novelty (or outlier) detection [4] and deployed as $g(\cdot)$ in DC-Prophet. OCSVM is trained on instances that have only one class, which is the "normal" class; given a set of normal instances, OCSVM detects the soft boundary of the set, for classifying whether a new incoming instance belongs to that set (*i.e.*, "normal") or not. Specifically, OCSVM computes a non-linear decision boundary, using appropriate kernel functions; in this work, radial basis function (RBF) kernel is used [15]. Equation (3) below show how OCSVM makes an inference:

$$g(\boldsymbol{x}) = \begin{cases} 1, \hat{g}(\boldsymbol{x}) \geq 0 \\ 0, \hat{g}(\boldsymbol{x}) < 1 \end{cases} \text{ where } \hat{g}(\boldsymbol{x}) = \langle \boldsymbol{w}, \phi(\boldsymbol{x}) \rangle + \rho \tag{3}$$

where $\boldsymbol{w}$ and $\rho$ are learnable weights that determine the decision boundary, and the function $\phi(\cdot)$ maps the original feature(s) into a higher dimensional space, to determine the optimal decision boundary. By further modifying the hard-margin SVM to tolerate some misclassifications, we have:

$$\min_{\boldsymbol{w}, \rho} \frac{1}{2} ||\boldsymbol{w}||_2^2 + C \sum_{i}^{n} \xi_i - \rho$$
$$\text{s.t. } \hat{g}(\boldsymbol{x}_i) = \langle \boldsymbol{w}, \phi(\boldsymbol{x}_i) \rangle - \rho \leq \xi_i$$
$$\xi_i \geq 0 \tag{4}$$

where $\xi_i$ represents the classification error of $i^{\text{th}}$ sample, and $C$ represents the weight that trades off between the maximum margin and the error-tolerance.

## 3.3  Random Forest

In the second stage of DC-Prophet, Random Forest [6] is used for multi-class classification. Random Forest is a type of ensemble model that leverages the classification outcomes from several (say $B$) decision trees for making the final classification. In other words, Random Forest is an ensemble of $B$ trees $\{T_1(\boldsymbol{x}), ..., T_B(\boldsymbol{x})\}$, where $\boldsymbol{x}$ is the vector of predictive features described in Sect. 2.4. This ensemble of $B$ trees predicts $B$ outcomes $\{\hat{y}_1 = T_1(\boldsymbol{x}), ..., \hat{y}_B = T_B(\boldsymbol{x})\}$. Then the outcomes of all trees are aggregated for majority voting, and the final prediction $\hat{y}$ is made based on the highest (*i.e.*, most popular) vote. Empirically, Random Forest is robust to overfitting and achieves a very high accuracy.

Given a dataset of $n$ instances $\{(x_1, y_1), ..., (x_n, y_n)\}$, the training procedure of Random Forest is as follows:

1. Randomly sample the training data $\{(x_1, y_1), ..., (x_n, y_n)\}$, and then draw $n$ samples to form a bootstrap batch.
2. Grow a decision tree from the bootstrap batch using the Decision Tree Construction Algorithm [4].
3. Repeat the above two steps until the whole ensemble of $B$ trees $\{T_1(x), ..., T_B(x)\}$ are grown.

After Random Forest is grown, along with the OCSVM in the first stage, DC-Prophet is ready for predicting the type of a machine failure.

## 4    Experimental Results

### 4.1    Experimental Setup

To best compare the proposed DC-Prophet with other machine learning models, we manage to search for the best hyperparameters by using 5-fold cross-validation for all the methods. Then the accuracy of each method is evaluated on the test set. All the experiments are conducted via MATLAB, running on Intel I5 processor (3.20 GHz) with 16 GB of RAM.

For the evaluation metrics, we report $Precision$, $Recall$, $F$-$score$, and $AUC$ (area under ROC curve) to provide a comprehensive study on the performance evaluation for different models. $F$-$score$ is defined as:

$$F_\beta = (1 + \beta^2) \frac{Precision * Recall}{(\beta^2 * Precision) + Recall} \tag{5}$$

where $\beta$ is the parameter representing the relative importance between $Recall$ and $Precision$ [16]. In this work, $\beta$ is selected to be 3, which means $Recall$ is approximately three times more important than $Precision$. Since the false negative (machine failure mispredicted as normal event) is much more costly as mentioned in Table 1, $F_3$-score is used as the main criterion to select the best framework for predicting failure types.

### 4.2    Results Summary

Table 2 shows the experimental results from different methods. We calculate and report $Precision$, $Recall$, $F_3$-score and AUC for comprehensive comparisons. The results demonstrate that the two-stage algorithms have better performance on both $F_3$-score and AUC. It also shows that using One-Class SVM for anomaly detection as the first stage is necessary. Among 8,771 failures, One-Class SVM only mispredicts 11 failures as normal events, which serves as an excellent filter. Furthermore, our proposed framework, DC-Prophet, which combines One-Class SVM and Random Forest, has the best $F_3$-score and AUC among all the two-stage methods.

However, it seems that all the algorithms have very limited capability to recognize FD failures. One reason could be that several FD failures are found to share similar patterns with the other two failure types—IR and SR; also out of 18 FD failures in the test set, 4 failures are predicted and categorized as SR failures. We suspect that for these FD cases, the machines are eventually added back; therefore they should be categorized as SR instead of FD failures. However, the ADD events occur after the end of traces.

We also notice that by simply applying Random Forest algorithm, we can already achieve great results in *Precision*. However, our proposed DC-Prophet still outperforms Random Forest in failure prediction, especially for the IR failures.

To evaluate the capability of DC-Prophet in industrial datacenters during serving, we measure the amortized runtime of one single prediction. Table 2 shows that DC-Prophet only requires 8.7 ms to make one prediction, which is almost negligible for most of the services in datacenters. This short latency allows the cloud scheduler to make preventive actions to deal with possible incoming machine failures. Furthermore, DC-Prophet is memory efficient—only 72 features are stored for making a prediction.

**Table 2.** Experimental result

| Algorithm | $F_3$-score | AUC | Precision | | | | Recall | | | | Runtime (ms) |
|---|---|---|---|---|---|---|---|---|---|---|---|
| | | | Normal | IR | SR | FD | Normal | IR | SR | FD | |
| One-stage method | | | | | | | | | | | |
| DT | 0.846 | 0.920 | 0.995 | 0.663 | 0.438 | 0.222 | 0.995 | 0.684 | 0.423 | 0.111 | 0.002 |
| LR | 0.344 | 0.660 | 0.978 | 0.756 | 0.642 | 0 | 0.999 | 0.336 | 0.077 | 0 | 0.001 |
| SVM | 0.184 | 0.584 | 0.973 | 0.624 | 0.521 | 0 | 0.998 | 0.154 | 0.068 | 0 | 18.62 |
| RNN | 0.505 | 0.740 | 0.983 | 0.742 | 0.689 | 0 | 0.999 | 0.464 | 0.184 | 0 | 0.471 |
| RF | 0.848 | 0.918 | 0.995 | 0.785 | 0.710 | 0 | 0.999 | 0.786 | 0.410 | 0 | 0.117 |
| Two-stage method | | | | | | | | | | | |
| OCSVM + DT | 0.856 | 0.919 | 0.986 | 0.591 | 0.378 | 0.046 | 0.969 | 0.666 | 0.449 | 0.111 | 8.711 |
| OCSVM + LR | 0.442 | 0.707 | 0.940 | 0.735 | 0.640 | 0 | 0.998 | 0.406 | 0.131 | 0 | 8.816 |
| OCSVM + SVM | 0.202 | 0.591 | 0.919 | 0.654 | 0.519 | 0 | 0.996 | 0.173 | 0.074 | 0 | 17.46 |
| OCSVM + RNN | 0.542 | 0.757 | 0.950 | 0.766 | 0.639 | 0 | 0.998 | 0.469 | 0.256 | 0 | 9.247 |
| **OCSVM + RF** | **0.878** | **0.933** | **0.986** | **0.729** | **0.591** | **0.667** | **0.991** | **0.795** | **0.408** | **0.111** | **8.714** |

## 4.3 Feature Analysis

Among all the predictive features, we observe several features to be more discriminative than others. Figure 5 shows how many times a feature in $x$ is selected to be split on in Random Forest. Figure 5(a) shows the number of average-value features $x_{r,t}$ being selected in Random Forest while Fig. 5(b) illustrates the number of peak-value features $m_{r,t}$ being selected. For average-value features, we observe a trend that recent features are more discriminative. In addition, the features related to memory usages are more discriminative than the others.

We also discover that the number of peak-value features is more discriminative than the average-value ones in general. Furthermore, the peak-value features have similar predictive capabilities over six time intervals, as shown in Fig. 5(b). In addition, we

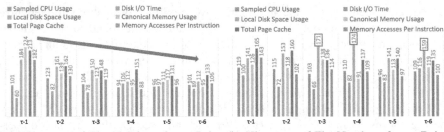

(a) Histogram of The Number of $x_{r,t}$ Being Selected (Average-Value)     (b) Histogram of The Number of $m_{r,t}$ Being Selected (Peak-Value)

**Fig. 5.** *Counts of features selected by Random Forest* : (a) shows the number of average-value features $x_{r,t}$ being selected. We observe a trend that more recent features are more discriminative. (b) shows the number of peak-value features $m_{r,t}$ being selected. (Color figure online)

observe that the peak usage of local disk is an important feature for predicting machine failures (see red circles in Fig. 5(b)).

## 5 Practitioners' Guide

Here we provide the practitioners' guide to applying DC-Prophet for forecasting machine failures in a datacenter:

- **Construct Training Dataset:** Given the traces of machines in a datacenter, extract abnormal events representing potential machine failures, and determine their types based on the observations in Sect. 2.3 for obtaining label $y$. Then calculate the partial autocorrelation for each resource measurement (*e.g.*, CPU usage, disk I/O time, etc.) to determine the number of time intervals (or lags) to be included as the predictive features $x$.
- **One-Class SVM:** After constructing the dataset of $(y, x)$, train OCSVM with the instances labeled as "normal" only, and find the best hyperparameters via grid-search and cross-validation.
- **Random Forest:** After OCSVM is trained, remove the instances detected as normal from the training dataset. Use the rest of dataset (treated as anomalies) to train Random Forest. Choose the number of trees in the ensemble and optimize it by cross-validation.

After both components of DC-Prophet are trained, each new incoming instance will follow the flow in Fig. 4 for failure prediction. Thanks to DC-Prophet's low latency (8.71 ms per invocation), it can be used for both (a) offline analysis in other similar datacenters, and (b) serving as a failure predictor integrated into a cloud/cluster scheduler, with training via historical data offline.

# 6    Conclusion

In this paper, we propose DC-Prophet: a two-stage framework for forecasting machine failures. Thanks to DC-Prophet, we now can answer the two motivational questions: "When will a server fail catastrophically in an industrial datacenter?" "Is it possible to forecast these failures so preventive actions can be taken to increase the reliability of a datacenter?" Experimental results show that DC-Prophet accurately predicts machine failures and achieves an AUC of 0.93 and $F_3$-score of 0.88. Finally, a practitioners' guide is provided for deploying DC-Prophet to predict the next failure of a machine. The latency of invoking DC-Prophet to make one prediction is less 9 ms, and there can be seamlessly integrated into the scheduling strategy of industrial datacenters to improve the reliability.

# References

1. 2016 cost of data center outages report. https://goo.gl/OeNM4U
2. Google cluster data - discussions (2011). https://groups.google.com/forum/#!forum/googleclusterdata-discuss
3. Barroso, L.A., Clidaras, J., Hölzle, U.: The datacenter as a computer: an introduction to the design of warehouse-scale machines. Synth. Lect. Comput. Archit. **8**(3), 1–154 (2013)
4. Bishop, C.: Pattern Recognition and Machine Learning. Information Science and Statistics. Springer, New York (2006)
5. Box, G.E., Jenkins, G.M., Reinsel, G.C., Ljung, G.M.: Time Series Analysis: Forecasting and Control. Wiley, New York (2015)
6. Breiman, L.: Random forests. Mach. Learn. **45**(1), 5–32 (2001)
7. Chen, X., Lu, C.-D., Pattabiraman, K.: Failure analysis of jobs in compute clouds: a Google cluster case study. In: 2014 IEEE 25th International Symposium on Software Reliability Engineering, pp. 167–177. IEEE (2014)
8. Guan, Q., Fu, S.: Adaptive anomaly identification by exploring metric subspace in cloud computing infrastructures. In: 2013 IEEE 32nd International Symposium on Reliable Distributed Systems (SRDS), pp. 205–214. IEEE (2013)
9. Juan, D.-C., Li, L., Peng, H.-K., Marculescu, D., Faloutsos, C.: Beyond poisson: modeling inter-arrival time of requests in a datacenter. In: Tseng, V.S., Ho, T.B., Zhou, Z.-H., Chen, A.L.P., Kao, H.-Y. (eds.) PAKDD 2014. LNCS (LNAI), vol. 8444, pp. 198–209. Springer, Cham (2014). https://doi.org/10.1007/978-3-319-06605-9_17
10. Liu, Z., Cho, S.: Characterizing machines and workloads on a Google cluster. In: 2012 41st International Conference on Parallel Processing Workshops, pp. 397–403. IEEE (2012)
11. Miller, T.D., Crawford Jr., I.L.: Terminating a non-clustered workload in response to a failure of a system with a clustered workload. US Patent 7,653,833, 26 January 2010
12. Powers, D.M.: Evaluation: from precision, recall and f-measure to ROC, informedness, markedness and correlation. J. Mach. Learn. Technol. **2**(1), 37–63 (2011)
13. Reiss, C., Tumanov, A., Ganger, G.R., Katz, R.H., Kozuch, M.A.: Heterogeneity and dynamicity of clouds at scale: Google trace analysis. In: SOCC, p. 7. ACM (2012)
14. Reiss, C., Wilkes, J., Hellerstein, J.L.: Google cluster-usage traces: format + schema. Technical report, Google Inc., Mountain View, CA, USA, version 2.1, November 2011. https://github.com/google/cluster-data. Accessed 17 Nov 2014
15. Scholkopf, B., Sung, K.-K., Burges, C.J., Girosi, F., Niyogi, P., Poggio, T., Vapnik, V.: Comparing support vector machines with Gaussian kernels to radial basis function classifiers. IEEE Trans. Signal Process. **45**(11), 2758–2765 (1997)

16. van Rijsbergen, C.: Information Retrieval, 2nd edn. Butterworths, London (1979)
17. Verma, A., Pedrosa, L., Korupolu, M., Oppenheimer, D., Tune, E., Wilkes, J.: Large-scale cluster management at Google with Borg. In: Proceedings of the Tenth European Conference on Computer Systems, p. 18. ACM (2015)

# Disjoint-Support Factors and Seasonality Estimation in E-Commerce

Abhay Jha[✉]

Facebook, Inc., Menlo Park, CA, USA
abhaykj@fb.com

**Abstract.** Successful inventory management in retail entails accurate demand forecasts for many weeks/months ahead. Forecasting models use *seasonality*: recurring pattern of sales every year, to make this forecast. In e-commerce setting, where the catalog of items is much larger than brick and mortar stores and hence includes a lot of items with short history, it is infeasible to compute seasonality for items individually. It is customary in these cases to use ideas from factor analysis and express seasonality by a few factors/basis vectors computed together for an entire assortment of related items. In this paper, we demonstrate the effectiveness of choosing vectors with disjoint support as basis for seasonality when dealing with a large number of short time-series. We give theoretical results on computation of disjoint support factors that extend the state of the art, and also discuss temporal regularization necessary to make it work on walmart e-commerce dataset. Our experiments demonstrate a marked improvement in forecast accuracy for items with short history.

## 1 Introduction

Seasonality refers to patterns in a time-series that repeat themselves every season. For example, retail sales always increase in November, unemployment drops in December, temperature increases in summer. In general, one is interested in finding the smooth periodic pattern underlying a long univariate time-series which has data for many past seasons. This reduces to some form of regression of observation on the season, for e.g., day/week, as exemplified in a lot of time-series literature [1–3].

In this paper, we will focus on finding the weekly seasonality of sales on an annual basis. We focus on e-commerce, which is a decidedly different and arguably more challenging task, because the assortment of items is larger and more dynamic– this implies there is a large number of time-series and most of them do not have enough data for even one year. This make the traditional approach of regression infeasible, since we cannot estimate a 52 week seasonality from, say only 6 weeks of sales. The problem in this domain is more suited to factor analysis and matrix factorization techniques, which have been successfully used for imputation in other scenarios with a lot of missing data [4]. In this

---

Work done while the author was at @WalmartLabs.

© Springer International Publishing AG 2017
Y. Altun et al. (Eds.): ECML PKDD 2017, Part III, LNAI 10536, pp. 77–88, 2017.
https://doi.org/10.1007/978-3-319-71273-4_7

approach, one computes instead a few orthogonal basis vectors, called seasonal basis for an entire category of related items. Figure 1 illustrates the seasonal basis of a certain group of items when computed on the online sales data over 52 weeks of the year. Seasonality for an item can be evaluated with a regression, generally by a time-series forecasting model with time varying coefficients, on the seasonal basis. We illustrate a simple forecasting model that incorporates seasonal basis in (1). However, this regression can lead to unreliable results for two reasons. First, in the span of a short time-series, individual seasonal basis might not be orthogonal. For e.g., in Fig. 1, basis 2 and 3 from PCA have very similar curve from week 20 to 35 and same with SPCA for weeks 35 to 52, which makes it impossible to disambiguate between them if a time-series only had data for those weeks. One solution is to work with fewer basis; but unless one always works with one basis, there is no guarantee that they would be orthogonal for every segment. This is a big issue when a vast majority of items being forecasted don't even have a year of data.

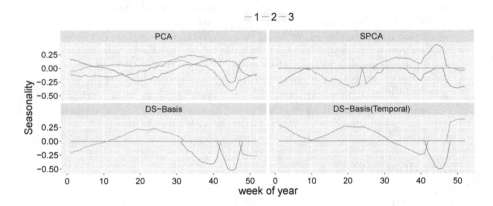

**Fig. 1.** Seasonal factors computed with different methods on a group of items from walmart-ecommerce. Notice how basis 2 and 3 from PCA have very similar curve from week 20 to 35 and same with SPCA for weeks 35 to 52. This makes it impossible to disambiguate between the two basis if a time-series only had data for those weeks. This leads to unreliable estimate of seasonality and hence unreliable forecasts. Unless one works with only one basis, this problem is inevitable; hence the notion of disjoint support basis that lead to orthogonality for every segment. DS-basis only have one non-zero component at a time– the curves sometimes seem to overlap as one basis goes from zero to non-zero and another from non-zero to zero.

Another, a more intuitive problem, is that not all parts of the year are related to each other. For an item, sales in February may have no relation to sales in September, but may be related to sales in December. Hence, we should not be modifying forecasts for the entire year based on the sales during a part, which is what happens in general. Fortunately, both problems lead to one solution. To solve the first problem, we enforce a stricter notion of orthogonality where every segment of two vectors is orthogonal; it can be shown to be equivalent to them having Disjoint Support (*DS*). Figure 1 illustrates *DS*-basis. They solve the second problem as well by segmenting the year into different weeks which exhibit a distinct behavior. However, with disjoint supports there is only one curve to model the variation during any part of the year, which is not always true when one considers a large group of possibly unrelated items. So, one way of viewing *DS*-basis is as a strong regularizer imposed on a group of items which forces their sales to follow one curve during the seasonal events. This is not recommended for the entire catalog together, but for groups of related items in the catalog hierarchy.

In this paper, we will study how to compute *DS*-basis, both with theoretical results, and practical lessons learned in applying it at walmart e-commerce. We show that *DS*-basis for a low rank matrix can be computed in polynomial time. Our proof relies on bounding the number of regions of a low rank hyperplane arrangement. For general matrices, we show that the problem is NP-hard, with a reduction from graph coloring. We also give a constant factor approximation algorithm, and prove hardness of approximation results.

When applying this technique at Walmart E-Commerce, we observed multiple anomalies in the basis computed, compared to our domain knowledge. This is because real world datasets are noisy and there are many factors that lead to variation in sales that cannot be accounted for. We propose certain temporal regularizations that can overcome this noise, by exploiting the fact that our data is a time-series. Computing the basis with these regularization entails learning in Switching State Space Models which is often done with moment-matching Kalman Filters [5]. We propose an alternative faster approach that leverages forward-backward algorithm for estimation in HMM, to achieve the same accuracy, in less execution time.

Our experiments demonstrate that forecast accuracy is markedly improved for items with short history. Further empirical evaluations are done on a synthetic dataset for a more detailed comparison. This paper is organized as follows: Sect. 2 gives some background and discusses the related work, and we describe our approach along with the computational complexity of the problem in Sect. 3, and Sect. 4, which focuses on a more general problem with temporal regularizations. Section 5 has the empirical evaluation of our approach conducted on walmart e-commerce, and a synthetic data.

## 2    Background and Related Work

**Related Work.** The general approach of estimating seasonality is by decomposing the time-series into mean, trend and seasonal components, see [6] for an example. The seasonal component can be modeled as a cyclic/periodic component in the form of a triginometric series. However, because of leap years, seasonality in our setting is not the same as periodicity. However, it can still be computed by a regression of observation with the week number– but as we already pointed out this approach does not work for short series. One could still use hierarchical regressions [7] commonly used in panel data– in this approach we would use item catalog to specify a hierarchy of items, however this approach does not scale well to large datasets we use. This is inherent to the method itself because of computations involving large covariance matrices. Also, generally some sort of clustering is needed before applying these methods as described in [8].

The application of forecasting to settings like ours is relatively new, but the idea of seasonal factors is not new and has been investigated by others as well. For e.g., [9,10], explore Non-Negative Matrix Factorization (NMF), and Principal Component Analysis (PCA) respectively. In this paper, we are focused not on the particulars of whether factors be non-negative, or sparse or smooth; instead we are proposing that they have disjoint support which is an orthogonal idea that can be used along with each of these approaches. We build upon PCA since it is the most common way of estimating factors.

**Notation.** Given a vector $v$, we denote coordinate $i$ by $v_i$. For a matrix $M$, we denote row $i$ and column $j$ with $M[i,j]$; column $i$ with $m_i$. We say vectors $u$, $v$ have disjoint support iff $\forall i, u_i v_i = 0$. The reason we are interested in them is because they ensure orthogonality of arbitrary segments of $u$, $v$. For a natural number $n$, $[n]$ denotes the set $\{1, 2, \ldots, n\}$.

**Forecasting with Seasonal Basis.** Since this paper is about computing seasonality, we won't delve into the forecasting models, but we do want to illustrate how seasonality is incorporated in forecasting to motivate the problem. The following is a simple univariate local-level model which has a *mean* component $\mu$ and a seasonality component that is expressed by seasonal basis that form rows of $H$.

$$
\begin{aligned}
y_t &= \mu_t + h_{s(t)}^T \alpha_t + \epsilon_t & \epsilon_t &\sim N(0, \sigma^2), \\
\mu_t &= \mu_{t-1} + \eta_t & \eta_t &\sim N(0, \lambda_\mu \sigma^2), \\
\alpha_t &= \alpha_{t-1} + \omega_t & \omega_t &\sim N(0, \lambda_\omega \sigma^2 \mathbf{I}_k)
\end{aligned}
\tag{1}
$$

where $s(t)$ denotes the season at time $t$. For e.g., if we are making weekly forecasts it will be between 1 to 52, for daily forecasts it would vary from 1 to 365. One could add more components to the model like trend, and include price and calendar effects. Furthermore, this could be generalized to a multivariate model.

**Problem Statement.** Let $Y$ be an $n \times p$ sales matrix, where $p$ is the number of seasons, for e.g. $p = 52$ in Fig. 1. We will assume that rows of $Y$ are centered to take out the effect of mean. Also, note that $Y$ can have missing values. Our goal in this paper is to express $Y$ as $WH$, where $W$ is an $n \times k$ matrix of basis coefficients, and $H$ is a $k \times p$ matrix whose rows have disjoint support and $HH^T = I$, that minimize $\|Y - WH\|_F^2$, which is same as maximizing $Tr\left(HY^TYH^T\right)$. Note that the constraint $HH^T = I$ is just for uniqueness; we could also enforce $W^TW = I$ instead– depending on the algorithm one constraint is preferred over the other. We will extend the problem further by adding some temporal constraints on the rows of $H$ in Sect. 4. $k$ is typically a small number, and hence would be assumed to be a bounded constant when stating complexity results throughout this paper.

**Input:** $Y_{n \times p} = U_{n \times r} V_{r \times p}$, #factors
  $k$
**Output:** $W, H$
$Z_{r \times k} \leftarrow$ variables from a $rk$-D space
$S \leftarrow \bigcup_{1 \le i \le p} \bigcup_{1 \le j_1 < j_2 \le k} \{ v_i^T z_{j_1} \pm v_i^T z_{j_2} = 0 \}$
$r(S) \leftarrow$ regions of arrangement $S$
$opt \leftarrow 0$
**for** *regions* $\nabla \in r(S)$ **do**
  $Support(i) \leftarrow \mathtt{argmax}_j \left( v_i^T z_j \right)^2,$
  $\forall i \in [p]$
  $M_j \leftarrow$ columns of $Y$ with
  support $j, \forall j \in [k]$
  $currOpt \leftarrow \sum_{i=1}^k \sigma_1^2(M_i)$
  **if** $currOpt > opt$ **then**
    $opt \leftarrow currOpt$
    $H \leftarrow$ right singular vectors
    of $M_i, i = 1..k$
  **end**
**end**
$W \leftarrow Y H^T$
**return** $W, H$
**Algorithm 1.** Computing $DS$-basis
for a low rank matrix

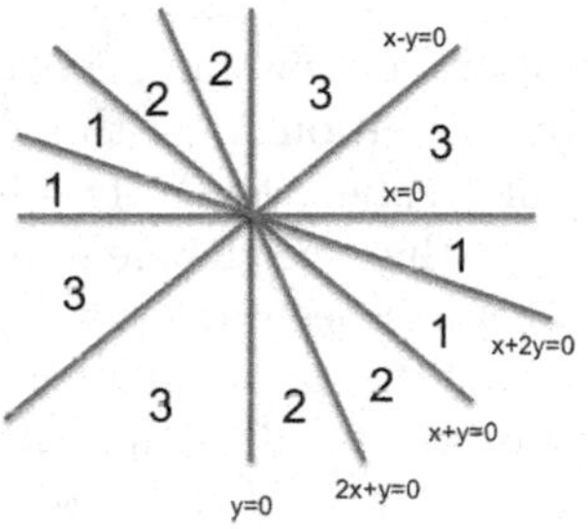

**Fig. 2.** Consider functions $1.x^2$, $2.y^2$, $3.$ $(x+y)^2$. The above arrangement of lines partitions 2D-space into regions which are annotated with $1/2/3$ according to which function is maximum in that region. The lines are just $f \pm g$ for each pair of functions $f^2$, $g^2$.

## 3 Computing $DS$-Basis

We first discuss the low rank case and propose a polynomial time algorithm. The algorithm can be used in general too by applying it on a low rank projection. We then discuss the results for general matrices including NP-hardness and approximation results. Finally, we show how these results can be extended if basis need to be sparse.

### 3.1 $DS$-Basis for Low Rank Matrices

We first reformulate the problem from Sect. 2 into a form that depends only on $W$, and not on the basis $H$. W.l.o.g, we assume $\|w_i\|_2 = 1, \forall i \in [k]$. Now, note that if the $i^{th}$-support is basis $j$, then $H_{j,i} = w_j^T y_i$. Hence, it follows that the optimal $W$ can be found by maximizing:

$$\sum_{i=1}^p \max \left( \left( w_1^T y_i \right)^2, \ldots, \left( w_k^T y_i \right)^2 \right)$$

s.t. $\|w_i\|_2 = 1, i \in [k]$. Now we use the fact that that $Y$ is low rank to express it as $Y = UV$, and replace $W$, $nk$ variables, with variables $Z = W^T U$, only $kr$ variables. We can then reformulate the objective as:

$$\sum_{i=1}^{p} \max\left( \left(z_1^T v_i\right)^2, \ldots, \left(z_k^T v_i\right)^2 \right) \tag{2}$$

maximization of a low-rank convex function over the unit sphere. A Polynomial Time Approximation Algorithm (PTAS) for this problem is possible by iterating over the unit sphere for $Z$, by discretizing it into grids of small size, and using the property that the change in objective is bounded by $\epsilon^2$, for a perturbation in $Z$ of $\epsilon$. This has been already discovered in [11], and similar approach has also been used in [12] to maximize a class of quasi-convex functions. But in neither of these cases, is an algorithm with polynomial running time independent of the error $\epsilon$ still known. In this paper, we present such an approach, that can also extend the result in [11] from PTAS to PTIME, and we hope can be extended to more low rank convex maximization problems as in [12].

Algorithm 1 and Fig. 2 present the algorithm for computing $DS$-basis of a low rank matrix. We restate that $k$ and the rank of $Y$ are assumed to be small constants in this section. Formally:

**Theorem 1 (Computing $DS$-basis of a low rank matrix is in PTIME).** *Given a matrix $Y$ of rank $r$, we can compute DS-basis $H$ of $Y$ in time $O\left(p^{rk+4}\right)$*

### 3.2   $DS$-Basis for Arbitrary Matrices

Once we venture beyond low-rank matrices to arbitrary matrices, the problem becomes NP-hard as we prove below.

**Theorem 2 (Computing a $DS$-basis of even constant size is NP-hard).** *Finding a DS-basis $H$ that maximizes $Tr\left(HY^TYH^T\right)$ s.t. $HH^T = I_k$ is NP-hard for any fixed $k \geq 3$.*

In general, it would be good to know how close one could approximately solve the problem for a general matrix. We give an incomplete answer by proving both lower and upper bounds on the optimal hardness of approximation. It would be great if the two matched so we knew how close an approximation is possible in polynomial time, but we leave that as an open problem.

**Theorem 3 (Approximating a $DS$-basis of size $k$).** *Let $max_H Tr(HY^T YH^T)$ be opt\*, where $H$ is a DS-basis and $HH^T = I_k$. Then opt\*, can be approximated to a ratio $1/k$ in PTIME. Furthermore, unless $P = NP$, it cannot be approximated to a ratio better than $1 - 1/p$ in PTIME.*

*Implications for Sparse PCA.* Algorithm 1 is very general in its scope, in that it first shows that there are only a polynomial and not exponential number of possible supports one needs to consider when looking for disjoint support of a low rank matrix. Of course, given the support one still needs to find the basis, which reduces to finding the dominant eigenvector of a low rank matrix. The framework extends to solving for principal components with a particular constraint as well, so long as the second stage is still tractable. In case of sparse pca, for instance, one can find the principal component of a low rank matrix with either $l_0/l_1$ constraint in polynomial time [13,14]. Hence the tractability results extend to these cases as well.

## 4    Adding Temporal Regularization

As can be seen from Fig. 1, *DS*-basis sometimes look counterintuitive, when the non-zero component switches between consecutive weeks of year for a short period of time. In general, we expect the year to be divided into contiguous segments of weeks that form the support for a basis, and non-contiguous supports are implausible. But in a noisy real-world dataset like ours, it can be hard to get to the optimal solution due to the presence of multiple outliers and other noise. To counter the noise, we enforce this domain knowledge via a prior/regularizer over the simple gaussian factor analysis approach as follows. We model the support using a Hidden Markov Model (HMM) that encourages consecutive time-periods to have similar support. However, we also have to account for the fact that our data is a time-series. This means we expect our basis curves to be smooth and not change too much from one time point to another.

$$y_t = w_{x_t} H_{x_t,t} + \epsilon_t \qquad \epsilon_t \sim N\left(0, \sigma^2 \mathbf{I}\right)$$
$$\mathbf{Pr}\left(x_t | x_{t-1}\right) = \rho \mathbf{1}_{x_t = x_{t-1}} + {}^{(1-\rho)}/_{(k-1)} \mathbf{1}_{x_t \neq x_{t-1}}$$
$$h_t = h_{t-1} + \eta_t \qquad \eta_t \sim N\left(0, \lambda \sigma^2 \mathbf{I}\right) \qquad (3)$$

These two regularization work against each other. Regularization on the support indicator $x_t$ tries to put consecutive seasons in the same support– this is tuned with the parameter $\rho$, if it is $1/k$, supports can change arbitrarily between time points, while if $\rho = 1$, consecutive weeks must have the same support leading to only one non-zero basis. There is also a penalty on the difference in basis $h$ between consecutive weeks, controlled by parameter $\lambda$– higher $\lambda$ means lower penalty, while $\lambda = 0$ forces $h$ to be constant. The seasonality and segmentation achieved with these regularizations look more natural, and as we will show in Sect. 5, lead to better forecast accuracy as well. However, because the consecutive supports are correlated, an approach like Algorithm 1 is no longer applicable.

Equation 3 is a special case of *Switching State Space Models* (SSSM) which combine ideas from HMM and State Space Models (SSM) to allow for both discrete and continuous hidden states. Unlike SSM, computing the distribution of $H$ given $W$ has been recognized as intractable in SSSM [15]. The hardness of computing posterior state distribution stems from the fact that at each time point,

**Input:** Sales Matrix $Y(n \times p)$,
        initial $W = W^0$, parameters
        $k, \sigma, \rho, \lambda$
**Output:** $W$, $H$
$W \leftarrow W^0$
**while** $H$ *has not converged* **do**
    // Compute $H, x$ given $W$
    Compute states $x_t, h_t$ with
    GPB(1) smoothing [5, 17]
    Normalize each row of $H$ to
    norm 1
    // Compute $W$ given $H$
    $W \leftarrow YH^T$
**end**
**return** $W$,$H$
**Algorithm  2.** AM-GPB(1)  algorithm to compute $DS$-basis

**Input:** Sales Matrix $Y(n \times p)$,
        initial $W = W^0$,parameters
        $k, \sigma, \rho, \lambda$
**Output:** $W$, $H$
$W \leftarrow W^0$
**while** $H$ *has not converged* **do**
    // Compute $x$ given $W$
    Compute $x_t$ using Viterbi
    algorithm
    // Compute $H$ given $x$
    **for** $i \in [k]$ **do**
        $s \leftarrow \{j \mid x_j = i\}$ // columns
        with support $i$

        $Y_s \leftarrow$ matrix with columns
        $y_j \forall j \in s$

$$L \leftarrow \begin{bmatrix} 1 & -1 & 0 & 0 & \dots & 0 \\ -1 & 2 & -1 & \ddots & \ddots & \vdots \\ 0 & -1 & 2 & -1 & \ddots & 0 \\ 0 & \ddots & \ddots & \ddots & \ddots & 0 \\ \vdots & \ddots & \ddots & -1 & 2 & -1 \\ 0 & \dots & 0 & 0 & -1 & 1 \end{bmatrix}$$

        $h_i \leftarrow$ first eigenvector of
        $Y_s^T Y + L/\lambda$
    **end**
    // Compute $W$ given $H$
    $W \leftarrow YH^T$
**end**
**return** $W$,$H$
**Algorithm 3.** AM-HMM algorithm to compute $DS$-basis

we have $k$ possibilities corresponding to the values of $x_t$. The final posterior thus is a mixture of $k^p$ gaussians. Various approximations have been used in the literature to deal with this intractability. The most common is to modify the kalman filter by merging the $k$ gaussians into 1 gaussian at each step [5,16,17]: the resulting filter is called GPB(1). This leads to a natural alternating minimization scheme, which we call AM-GPB(1), summarized in Algorithm 2: compute $H$ given $W$ using GPB(1), and $W$ given $H$ using regression. Time complexity per iteration can be shown to be $O\left(npk^4\right)$, dominated by the time for GPB(1).

However, GPB(1)-smoothing is expensive and the execution cost builds up because of the repeated calls involved with the alternating minimization involved. We also pursue an alternative way in which we put more emphasis on finding states $x$ instead. Observe that given support $x$, we can find basis $i$ as the first eigenvector of $Y^{(i)T}Y^{(i)} + L/\lambda$, where $Y^{(i)}$ is the matrix with columns of support $i$ from $Y$, and $L$ is a tridiagonal matrix with 2 on diagonal, except the first and last, and $-1$ off-diagonal. Also, once we know $H$, $W$ is just $YH^T$. Now, we use $W$ to find $x$ using the forward backward algorithm for state estimation in HMM. Note that this step completely ignores $H$, and just finds optimal $x$ for the given $W$. In other words, while GPB(1) smoothing focuses more on estimating $H$, this approach puts more emphasis on $x$. The time complexity per iteration now is $O\left(npk + p^3\right)$. We call it AM-HMM, summarized in Algorithm 3, and it also leads to a faster execution time as we will demonstrate empirically.

## 5  Empirical Evaluation

In this section, we will look at the impact of $DS$-basis on forecast accuracy in a real-world dataset, and also explore the robustness and performance of the algorithms proposed in this paper on a synthetic dataset. Our implementation is in C++ and R, and experiments are conducted on a MacBook Pro with 16 GB memory and 2.5 GHz Intel i7 processor.

### 5.1  E-Commerce Data

In this section, we use sales data from Walmart E-Commerce. 20 groups of items from different sections of the catalog are selected, with sizes varying from about 2 K to 10 K, for a total of around 50 K items. We should point out that these groups were not manually selected; they are actual groups of items assigned to a particular category in the catalogue. In that sense, the items they contain are representative of an e-commerce assortment. We will compare the forecasts from local level model in (1) with $k = 3$, $\lambda_\mu = \lambda_\omega = 0.1$. For forecasts, we choose six different weeks of year distributed throughout year. For each week, we forecasted six weeks ahead. Our benchmark for comparison are seasonal factors generated using PCA. To compare how a new forecast $f$ compares to benchmark $g$, we look at the metric of percentage improvement offered by $f$ over $g$: $|f-s|-|g-s|/|f-s|$, where $s$ is the sales. We compute $DS$-basis for each group using Algorithm 1[1], and $DS$-basis(temporal) with temporal regularization is computed using AM-HMM.

Figure 3 shows that there is a stark difference in comparison when it comes to items with less than a year of history and items with long history, with median improvement of 20–30% possible with $DS$-basis. This is in accordance with the argument made in Sect. 1 that having orthogonal basis is not sufficient when the time-series involved are short, since it can be hard to disambiguate between

---

[1] We don't explore the full search space but use randomization to run within a time budget.

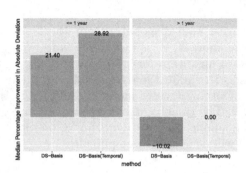

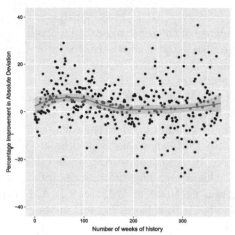

**Fig. 3.** Median Percent Improvement in error for items with less than or more than one year of sales history by using *DS*-basis over principal components. Improvement is $|f-s|-|g-s|/|f-s|$, where $s$ is sales, $f, g$ are forecasts using seasonality from PCA and *DS*-basis. *DS*-basis was computed using Algorithm 1, and *DS*-basis(temporal) with temporal regularization is computed using Algorithm 3

**Fig. 4.** Average Percent Improvement in error for items with a certain week of sales data; the shaded region shows the 95% confidence interval. This shows significant improvements for items with short time-series. Improvement is $|f-s|-|g-s|/|f-s|$, where $s$ is sales, $f, g$ are forecasts using seasonality from PCA and *DS*-basis computed by AM-HMM respectively.

different factors in a short time-span. But not only do we see improvements for short time-series, we don't experience any penalty for long time-series when using *DS*-basis(temporal) which is encouraging since it means the approach can be deployed for all items and not restricted to short series.

Figure 4 describes in detail how the improvement offered by *DS*-basis (temporal) varies with length of history a time-series has. We only plot the average improvement for items with given weeks of history to minimize the clutter of the graph resulting from too many points. Figure 4 shows, if we ignore the beginning, till say 10 weeks, there is a clear and marked improvement for items with less than 60 weeks of sales, often about 10–25%. For items with less than 10 weeks of history, initialization is the dominating factor, and performance is very volatile. From 50 to 150, most of the times improvement is positive, but after 150 weeks, there is no significant improvement.

## 5.2   Synthetic Data

In this section, we will evaluate our algorithms for computing *DS*-basis, and see if they are effective in finding the underlying basis and observation in the presence of noise and outliers, assuming that the underlying basis does have

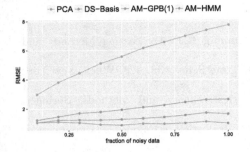

**Fig. 5.** RMSE in recovering true data using various decomposition methods as the fraction of noisy outliers in the data is increased.

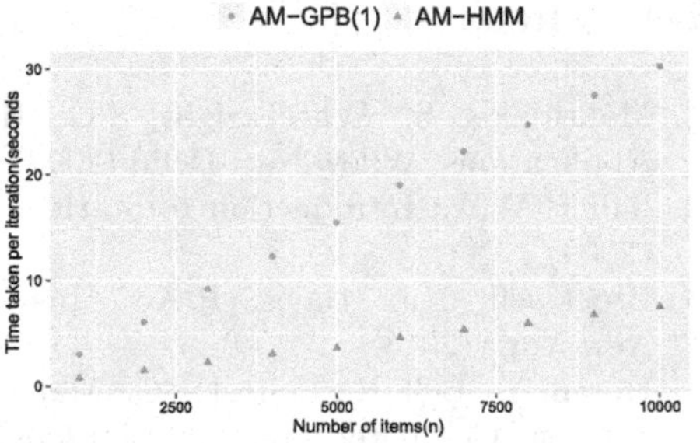

**Fig. 6.** Time taken per iteration for the two methods of computing smooth DS-basis

disjoint support. For this, given $0 \leq f \leq 1$, we generate a matrix $M$ of dimension $1000 \times 52$ as $M = WH + \epsilon + \mu$, where $W$ is $1000 \times 3$ matrix of $\mathcal{N}(0,1)$, and $H$ is a $3 \times 52$ smooth disjoint support factor where factors vary from one time-point to another by $\mathcal{N}(0,0.1)$. $\epsilon$ is $\mathcal{N}(0,1)$ error and $\mu$ is outlier noise: with probability $f$ it is $\mathcal{N}(0,10)$, else it is zero. Now to simulate the missing data, we divide rows of $M$ into 50 groups and from each group remove the first $0, 1, \ldots, 49$ entries. Note that $M$ is then about 50% sparse, but in a stair-case fashion since we assume the data is time-series and hence the missing data is at the beginning and not at random. We want to see now if one can recover true data: $WH$. We will look the Root Mean Square Error (RMSE); because of the construction of $M$, an algorithm that can recover the true $H$ can achieve an rmse of 1 from $WH$ on average, because of $\epsilon$. But that requires being able to work through missing data and outlier noise $\mu$.

Figure 5 shows the rmse achieved by different methods as the fraction of outliers $f$ is varied. We see that AM-HMM is remarkably robust to noise and can recover the true basis even with many outliers. AM-GPB(1) is also close but as $f$ is increased, it does slightly worse in recovering the basis. PCA does not work well at all in this scenario, and computing DS-basis without temporal regularization performs much worse as $f$ increases. This illustrates why in real-world data with many outliers having temporal regularization is crucial when we know the underlying basis is smooth.

Figure 6 compares the execution time of AM-HMM and AM-GPB(1) as the rows of $M$ are varied from 1 K to 10 K. Even though asymptotically, the two approaches have linear running time, in our experience AM-HMM is the only one that scales well for large groups, and we can see this in the rapidly increasing difference as we approach 10 K items in the plot.

# References

1. Makridakis, S., Wheelwright, S.C., Hyndman, R.J.: Forecasting Methods and Applications. Wiley, New Delhi (2008)
2. Fuller, W.A.: Introduction to Statistical Time Series, vol. 428. Wiley, New York (2009)
3. Brockwell, P.J., Davis, R.A.: Time Series: Theory and Methods. Springer, New York (2013)
4. Koren, Y., Bell, R., Volinsky, C.: Matrix factorization techniques for recommender systems. Computer **42**(8), 30–37 (2009)
5. Bar-Shalom, Y., Li, X.-R.: Estimation and Tracking- Principles, Techniques, and Software. Artech House Inc., Norwood (1993)
6. Cleveland, R.B., Cleveland, W.S., McRae, J.E., Terpenning, I.: STL: a seasonal-trend decomposition procedure based on loess. J. Official Stat. **6**(1), 3–73 (1990)
7. Gelman, A., Hill, J.: Data Analysis Using Regression and Multilevel/Hierarchical Models. Cambridge University Press, New York (2006)
8. Jha, A., Ray, S., Seaman, B., Dhillon, I.S.: Clustering to forecast sparse time-series data. In: 2015 IEEE 31st International Conference on Data Engineering (ICDE), pp. 1388–1399. IEEE (2015)
9. Sun, W., Malioutov, D.: Time series forecasting with shared seasonality patterns using non-negative matrix factorization. In: NIPS Time Series Workshop (2015)
10. Taylor, J.W., De Menezes, L.M., McSharry, P.E.: A comparison of univariate methods for forecasting electricity demand up to a day ahead. Int. J. Forecast. **22**(1), 1–16 (2006)
11. Asteris, M., Papailiopoulos, D., Kyrillidis, A., Dimakis, A.G.: Sparse PCA via bipartite matchings. In: Advances in Neural Information Processing Systems, pp. 766–774 (2015)
12. Goyal, V., Ravi, R.: An FPTAS for minimizing a class of low-rank quasi-concave functions over a convex set. Oper. Res. Lett. **41**(2), 191–196 (2013)
13. Asteris, M., Papailiopoulos, D.S., Karystinos, G.N.: The sparse principal component of a constant-rank matrix. IEEE Trans. Inf. Theor. **60**(4), 2281–2290 (2014)
14. Karystinos, G.N.: Optimal algorithms for binary, sparse, and $L_1$-norm principal component analysis. In: Pardalos, P., Rassias, T. (eds.) Mathematics Without Boundaries, pp. 339–382. Springer, New York (2014). https://doi.org/10.1007/978-1-4939-1124-0_11
15. Ghahramani, Z., Hinton, G.E.: Variational learning for switching state-space models. Neural Comput. **12**(4), 831–864 (2000)
16. Kim, C.-J.: Dynamic linear models with markov-switching. J. Econometrics **60**(1–2), 1–22 (1994)
17. Murphy, K.P.: Switching Kalman filters, technical report, Citeseer (1998)

# Event Detection and Summarization
# Using Phrase Network

Sara Melvin[1], Wenchao Yu[1(✉)], Peng Ju[1], Sean Young[2], and Wei Wang[1(✉)]

[1] Department of Computer Science, UCLA, Los Angeles, USA
yuwenchao@ucla.edu, weiwang@cs.ucla.edu
[2] University of California Institute for Prediction Technology,
UCLA, Los Angeles, USA

**Abstract.** Identifying events in real-time data streams such as Twitter is crucial for many occupations to make timely, actionable decisions. It is however extremely challenging because of the subtle difference between "events" and trending topics, the definitive rarity of these events, and the complexity of modern Internet's text data. Existing approaches often utilize topic modeling technique and keywords frequency to detect events on Twitter, which have three main limitations: (1) supervised and semi-supervised methods run the risk of missing important, breaking news events; (2) existing topic/event detection models are base on words, while the correlations among phrases are ignored; (3) many previous methods identify trending topics as events. To address these limitations, we propose the model, PhraseNet, an algorithm to detect and summarize events from tweets. To begin, all topics are defined as a clustering of high-frequency phrases extracted from text. All trending topics are then identified based on temporal spikes of the phrase cluster frequencies. PhraseNet thus filters out high-confidence events from other trending topics using number of peaks and variance of peak intensity. We evaluate PhraseNet on a three month duration of Twitter data and show the both the efficiency and the effectiveness of our approach.

**Keywords:** Event detection · Phrase network · Event summarization

## 1 Introduction

It has been of interest for many years to have an automated tool to alert and summarize newsworthy events in real-time. Identifying events in real-time is crucial for many occupations to make timely, actionable decisions. It is shown to be extremely challenging to identify these events because of the subtle difference between "events" and trending topics, the definitive rarity of these events, and the complexity of modern Internet's text data. Existing approaches often utilize topic modeling technique and keywords frequency to detect events on Twitter, which have three main limitations:

© Springer International Publishing AG 2017
Y. Altun et al. (Eds.): ECML PKDD 2017, Part III, LNAI 10536, pp. 89–101, 2017.
https://doi.org/10.1007/978-3-319-71273-4_8

1. Supervised and semi-supervised methods run the risk of missing important, breaking news events [3,5,10,12–14]. These methods share one common weakness, they rely on the seeding of keywords for their tool or human labeling of tweets to train their models. This approach runs the risk of missing some events since their model is scoped to identify only events that fall under their static list of keywords.
2. Many previous methods mistakenly identify trending topics as events [8,11], however the description of an "event" is a unique sub-component to all "topics". Figure 1 shows the difference between event distribution (Paris terrorist attack) and topic distribution (discussion of social media photos).
3. Existing methods [1,19] summarize their results with a small grouping of keywords that do not convey enough information for a user to know in real-time what occurred. These models are also base on unigram words, while the correlations among phrases are ignored.

To address the above limitations, we propose PhraseNet, a model for event detection using phrase network. Our method begins by extracting the high-frequency phrases from tweets. Each frequent phrase and relationship between phrases are then represented in a phrase network. A community detection algorithm is applied to the phrase network to identify a grouping of phases which we define as event candidates. Finally, the high-confidence events can be identified by three criteria extracted from the event candidate distributions over time: (1) number of peaks in distribution, (2) intensity of peaks and (3) variance of the distribution.

Defining the unique features of an event is key in designing an event detection model. Consider an event such as the Paris terrorist attack on the offices of Charlie Hebdo. As you can see in Fig. 1, the words to describe the event spike in a collective frequency on the day of the attack with only a couple of peaks post event. In contrast, words used in the discussion of the non-event topic of

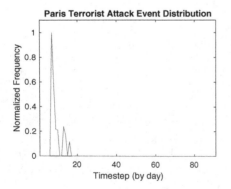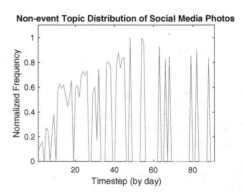

**Fig. 1.** A comparison between the distributions of an event and a topic. This figure shows the normalized frequency distribution between a non-event topic discussion of social media photos (right) and the event distribution describing the Paris Terrorist Attack (left) at the offices of Charlie Hebdo.

social media photo opinions spike in frequency during several different time steps throughout the data. Therefore, the characteristic of an event's distribution is defined to have very few peaks because an event description is usually unique; not normally shared by many other events.

In addition, non-event topics are discussed by the masses rises and falls with similar frequency throughout time because of the common interest in such topics stays fairly consistent. However, events are discussed during the occurrence and post-event to discuss opinions about the event or, if the events are planned, events can discussed prior in anticipation. These event peaks that occur prior- and post-event will be small in frequency compared to the moment the event occurs, therefore the standard deviation of an event's peak intensity will be larger than a non-event topic because of the varied interest in discussing the event. As you can see in Fig. 1, the Paris attack was not planned, there will be no peaks prior to the event occurrence.

Finally, our method, PhraseNet, leverages phrases and graph clustering to group correlated phrases together and help give more context to the identified event. You will see in Sect. 4.3 how PhraseNet summarizes compared to Twevent.

In summary, our contributions in this paper are:

1. *Event detection using phrase network*: We proposed the PhraseNet model to detect and summarize events on Twitter stream which includes three steps: (1) building phrase network using high-frequency phrases extracted from tweets, (2) detecting event candidates using community detection algorithm on phrase network, (3) identifying high-confidence events from candidate set using criteria such as number of peaks and variance of peak intensity in the event candidate distributions.
2. *Event summarization with phrases*: The proposed model summarizes events with phrases to give an interested user a short description and time duration of the detected event.
3. *Empirical improvements over Twevent*: We evaluate the PhraseNet model on a three month duration of Twitter data, and show that PhraseNet outperforms the baseline Twevent [9] by a large margin, which demonstrates the effectiveness of our model.

## 2    Problem Definition

In this section, we formally define a *phrase* as a sequence of contiguous tokens [6]:

$$p_m = \{w_{d,i}, \ldots, w_{d,i+n}\}, i + n \leq \mathcal{N}_d \tag{1}$$

where $w_{d,i}$ is a word (a.k.a. token) in the $i$-th place of the document $d$; $n \geq 0$. The $d$-th document is a sequence of $\mathcal{N}_d$ tokens. A *topic* consists of a set of phrases $\mathcal{P} = \{p_1, \ldots p_k\}$ where $p_m$ is a phrase and $k$ is the total number of phrases in the set ($m \in [1, k]$).

A *sliding window*, $T$, consists of $\tau$ amount of *time steps*, $t$. As the sliding window moves along, a sliding window *mean*, $\mu_T$, and the sliding window *standard deviation,* $\sigma_T$ are calculated as follows:

$$\mu_T = \frac{1}{\tau} \sum_{t=1}^{\tau} \left( \sum_{m=1}^{k} \mathcal{F}(p_m^{(t)}) \right) \tag{2}$$

$$\sigma_T = \frac{1}{\tau} \sum_{t=1}^{\tau} \left( \sum_{m=1}^{k} \mathcal{F}(p_m^{(t)}) - \mu_T \right)^2 \tag{3}$$

where $\tau$ is the number of time steps within the sliding window and $\mathcal{F}(p_m^{(t)})$ is the frequency of phrase $p_m$ at time step $t$ in the sliding window $T$.

A *trending topic,* or an *event candidate,* is identified by a peak in topic phrase frequency above a certain standard deviations from the topic's mean. Therefore, the peak is defined as:

$$\frac{\sum_{m=1}^{k} (\mathcal{F}(p_m^{(t)})) - \mu_T}{\sigma_T} > \theta \tag{4}$$

where $\theta$ is user-specified threshold. Therefore, an *event* is an unique subset of *trending topics,* or *event candidates,* that is formally defined in this method as a phrase cluster with very few peaks ($\leq \alpha$), a high frequency intensity of a peak ($\geq \beta$), and the largest standard deviation in peak height ($\geq \chi$).

# 3    Approach

## 3.1    Creating the Phrase Network

As mentioned in Sect. 2, to identify these phrases, the ToPMine algorithm [6] was used to identify the frequent phrases for a certain unit of time (e.g. an hour) $t$ and to partition each tweet into a combination of frequent. ToPMine algorithm includes two phases: (1) parse all the words into text segments; (2) create a hashmap of phrases and recursively merge if phrases appear frequently enough together.

The second phase is a bottom-up process that results in a partition on the original document that, when completed, creates a "bag-of-phrases." For example, the following tweet: *american sniper wins for putting bradley in that body #oscars2015.* Would be partitioned with the following phrases with a minimum support of 50: *american sniper, bradley.*

Now each frequent phrase found is considered a node in a graph. The edges between each frequent phrase reflect the co-occurrence of the phrases in the same tweet. The weight to the edge, $w_e$, is the Jaccard coefficient defined as $w_e = \frac{\mathcal{F}(p_a \wedge p_b)}{\mathcal{F}(p_a) + \mathcal{F}(p_b)}$, where the edge connects the phrases $p_a$ and $p_b$.

To calculate the most frequent co-occurring phrase pairs efficiently, the FP-Growth algorithm [7] was used. In this research, brute force scanning and tallying up co-occurrences became a bottleneck in PhraseNet, however, the FP-Growth exhibited the speed necessary to keep PhraseNet a real-time algorithm.

## 3.2   Phrases Clustering

After the graph is constructed, it is clustered into communities of phrases using the Louvain community detection method [2], which maximizes the modularity. The clusters identified by this method are event candidates. Hence, output for this stage is the set of event candidates $\Xi = \{\mathcal{P}_1, \ldots, \mathcal{P}_c\}$ where $c$ is number of event candidates in all time steps. The details are shown in Algorithm 1.

---

**ALGORITHM 1.** Phrase network construction and event candidate detection

---

    **Data**: Frequent patterns of phrases $P = \{p_i, \mathcal{F}(p_i)\}$
    **Result**: List of event candidates, $\Xi = \{\mathcal{P}_1, \ldots, \mathcal{P}_c\}$
1  Graph G=(V,E)
2  **for** $p_i, \mathcal{F}(p_i)$ *in* $P$ **do**
3     **for** $p_a, p_b$ *in* $p_i$ *where* $a \neq b$ **do**
4       **if** $p_a \notin V$ **then**
5         $V = V \cup p_a$
6       **if** $p_b \notin V$ **then**
7         $V = V \cup p_b$
8       $e = (p_a, p_b)$
9       e.weight $= \mathcal{F}(p_a \wedge p_b)/(\mathcal{F}(p_a) + \mathcal{F}(p_b))$
10      $E = E \cup e$
11 $\Xi = $ LouvainClustering(G)
12 **return** $\Xi$

---

## 3.3   Merging Event Candidates Across Time Steps

Since events could potentially carry on beyond the set time interval, each event candidate $\mathcal{P}_i$ is measured against the other event candidates of the next time step to measure whether the two event candidates should merge. The criteria used to determine the merge is the similarity score defined by Eq. (5). If the two event candidates with the highest score have a score greater than a threshold (we set 0.5 in this paper), then the event candidates will merge.

$$\text{similarity} = \max \left( \frac{\sum\limits_{p_s \in (\mathcal{P}_{i,t} \cap \mathcal{P}_{i,t+1})} w_s}{\sum\limits_{p_r \in \mathcal{P}_{i,t}} w_r}, \frac{\sum\limits_{p_s \in (\mathcal{P}_{i,t} \cap \mathcal{P}_{i,t+1})} w_s}{\sum\limits_{p_j \in \mathcal{P}_{i,t+1}} w_j} \right) \qquad (5)$$

For each time interval there is a set of phrase, $\mathcal{P}$ at time step $t$. Each phrase, $p_m$ has a weight, $w_m$ associated with it that will be normalized by the total number of phrases in the time interval $t$, denoted as $n$ in the equation below.

$$w_m = \frac{\mathcal{F}(p_m)}{\sum_{i=1}^{n} \mathcal{F}(p_i)} \qquad (6)$$

On completion of merging there remains a set of unique event candidates are maintained through all time steps. The event candidate distribution over time is created by defining the frequency of the phrase cluster over each time step. The frequency of a phrase cluster $\mathcal{P}$ will be denoted as $\mathcal{F}(\mathcal{P})$. Therefore, $\mathcal{F}(\mathcal{P}) = \sum_{m=1}^{k} w_m$ which is the sum of all phrase weights contained in the phrase cluster that make up $\mathcal{P}$.

### 3.4  Peak Detection

PhraseNet identifies potential events by first identifying the trending topics. Trending topics are discussions on a subject that becomes, all of a sudden, popular. To define "all of a sudden," the z-score was used to calculate the phrase cluster frequency, $\mathcal{F}(\mathcal{P})$, is $\theta$ standard deviations above the sliding window mean, $\mu_t$. The z-score was used to better identify peaks in a noisy environment. For example, a planned event may be discussed in advance thus showing a $\mathcal{F}(\mathcal{P}) > \mu_t$, however, these discussions are only small bumps compared to the height of the phrase community on the day of the planned event. To clarify the day and the duration of the event, whether planned or not planned, z-score helps filter the larger spikes in frequency compared to the small bumps.

Some events last longer than a time step, therefore, the sliding window average is updated as it slides, however a damping coefficient, $w_t$, is used to weight the phrase communities' peak. Therefore, the sliding window average shown in Eq. (2) is updated as follows:

$$\mu_T = \frac{1}{\tau} \sum_{t=1}^{\tau} w_t \left( \sum_{m=1}^{k} \mathcal{F}(p_m^{(t)}) \right) \tag{7}$$

where $w_t$ is zero for non-peak topic time steps and during peak time intervals of a topic the coefficient is $0 \leq w_t \leq 1$ where $w_t \in \mathbb{R}$. The exact definition of $w_t$ is a parameter for the user to define.

Finally, to focus on event candidate peaks, all time steps where the phrase community did not show a peak, their phrase community frequency is lowered to zero, however, all peak identified time steps maintain the phrase community frequency, $\sum_{m=1}^{k} \mathcal{F}(p_m^{(t)})$. This filtering is shown in Fig. 1.

Lastly, all event candidates are held to a certain threshold of key features and then sorted: the least number of peaks ($\alpha_i > \alpha_j$ where $i \neq j$), the largest standard deviation of peak heights ($\beta_i < \beta_j$ where $i \neq j$), and the highest peak intensity ($\chi_i < \chi_j$ where $i \neq j$). The last feature ($\chi$) is used to merely sort between the most popular phrase groups to aid in identifying the most urgent events. The first $\gamma$ of the event candidates are considered **events**. Each event that has a peak on the same day as another event are joined together for a total summary of the time step occurrences.

## 4    Results

It will be shown in this section how accurate and quick PhraseNet identifies events in comparison with Twevent.

## 4.1  Data and Parameters

Data was collected using Twitter's REST API[1] for the time period of January 1, 2015 to March 31, 2015. The sliding window for each time step was set for 24 h, from midnight to midnight. The experiment dataset only used English tweets thus using a total of 2,747,808 tweets. Each tweet was preprocessed to expand all contractions, all non-English characters were removed, and all stop words were removed.

The ToPMine algorithm uses the minimum support of 40 to find all frequent phrases and phrases were given a limit to search no more than 5-gram. In addition, the FP-Growth algorithm used a minimum support of 8. The $\theta$ value was placed at a 3, which means all event candidate peaks are identified as more than 3 standard deviations above the sliding window mean. The dampening coefficient, $\omega_t$, weight was defined as 0.1 and the allowed window of time for a true positive event peak to occur consisted of the true event date $\pm 5$ days. Lastly, the event key feature thresholds are the following: $\alpha = 10$, $\beta = .05$, and $\chi = .5$.

## 4.2  Experiment and Evaluation

Since ground truth was not available for this dataset, ground truth was defined from the "On This Day" website[2] and by various other reliable news sources. From the "On This Day" website, all events were filtered to only include English speaking country events (i.e. United States, England, Australia, Canada, and New Zealand) and terrorist attacks. In addition, all national holidays celebrated by the United States, U.K., Australia, Canada, and New Zealand identified in Wikipedia were added to the ground truth. Lastly, all sports related events were found via ESPN, BBC Sport, or NFL websites. Under this definition of ground truth, there are 102 events in total.

A sampling of true positives found by PhraseNet are listed in the table found in Table 1. This table exhibits the correlation of sub-events identified by peaks within the same time step. For example, the Grammy Awards are described by PhaseNet with some of the winners' names and included the word "Kanye" and "Beyonce" to note the fact that Kanye, again, interrupted a Grammy winner's speech to stick up for his friend Beyonce.

Considering the ground truth for identifying and labeling all true positives, false positives, and false negatives, it is impossible to determine every event that occurred within the data time frame, therefore this research uses the metrics of precision and recall. To show the trade off between precision and recall, the $F_1$ score is also provided for comparison. Precision is defined as the number of event candidates that correlate to known events divided by total number of event candidates. Recall is defined as the number of unique events detected divided by the total number of events possible listed in the ground truth. The final performance of PhraseNet is shown in Fig. 2 and detailed further in Table 2. It is

---

[1] https://dev.twitter.com/rest/public.
[2] http://www.onthisday.com/events/date/2015.

**Table 1.** A sampling of events identified and summarized by PhraseNet.

| Date of event detected | Description of event | PhraseNet phrase set |
| --- | --- | --- |
| January 1, 2015 | Steven Gerrard announced he will be leaving the Liverpool soccer team at the end of the season | gerrard, steven |
| January 5, 2015 | ESPN longtime host, Scott Stuart, died at the age of 49 | espn, sportscenter, stuart scott, rip |
| January 7, 2015 | Terrorist attack at a newspaper office, Charlie Hebdo, in Paris, France | charlie, hebdo, paris, attack, jesuischarlie, charliehebdo |
| January 11, 2015 | 72nd Golden Globes where George Clooney won a lifetime achievement award | clooney, george |
| January 15, 2015 | Oscar Nominations are Announced | oscar, nominations |
| January 15–19, 2015 | Pope Francis visits the Philippines for the first time in 20 years | francis, philippines, pope |
| January 21, 2015 | Barack Obama gives the State of the Union (sotu) speech | union, state, sotu, address, president, barackobama, obama |
| January 24, 2015 | Golden State Warrior scores the most NBA points and the most 3-pointers in a quarter | quarter, point |
| January 24, 2015 | FA Cup in the 4th Round | cup, fa |
| January 25, 2015 | WWE Royal Rumble | rumble, royal, royalrumble, wwe |
| January 31, 2015 | Anderson Silva vs. Nick Diaz UFC 183 Fight | silva, diaz |
| February 1, 2015 | 103rd Men's Australian Open where Novak Dokovic defeats Andy Murray | murray, andy |
| February 8, 2015 | Grammy Awards ceremony where "Stay With Me" by Sam Smith won best song, Beck was given Album of the Year, and Kanye West almost interrupts Beck's speech to argue that Beck's award should go to Beyonce | grammys, give, win, year, brits, kanye, west, beyonce, congrats, show, live, performance, watch, beck, awards, ago, pharrell, sam smith, night, samsmithworld, artist, nominated, tonight, shit, won, enter, album, connorfranta |
| February 14, 2015 | Valentine's Day | ago, gift, house, valentine, birthday, card, year, cards, red, blue, blackhawks, art, gift, violets, valentine day, roses, special, tomorrow, red, carpet |
| February 25, 2015 | BRIT Awards | brits, awards, brit, awards |
| **March 5, 2015** | Harrison Ford crash lands his plane | ford, harrison |
| March 10, 2015 | The family of Marvin Gaye win a record $7.3 million lawsuit for music copyright infringement (song: "Blurred Lines") | lines, blurred |
| March 12, 2015 | Sir Terence "Terry" Pratchett dies | terry, pratchett |
| March 16, 2015 | Two police officers were shot in Ferguson | ferguson, shot, police |
| March 24, 2015 | Co-pilot commits suicide by crashing Germanwings flight in the French Alps | germanwings, cockpit, pilot, locked, crash, plane french, alps, crash, plane |
| March 26, 2016 | U.S.A. Indiana Religious Freedom Act Protest | indiana, law, religious, freedom |

seen in the table that the best trade off between precision and recall is when $\gamma$ is 480 giving an $F_1$ score of .54.

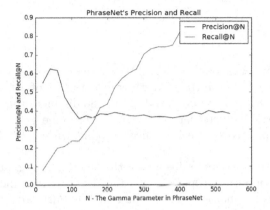

**Fig. 2.** This figure portrays the Precision@N and Recall@N where the N refers to the PhraseNet parameter $\gamma$. As you can see from this graph, as more event candidates are considered as events, the recall increases to almost 100%, however, with the increase in recall the precision of PhraseNet begins to slightly decrease.

**Table 2.** This table shows the Precision@N, Recall@N, and $F_1$ Score of PhraseNet. As you can see, the best precision occurs when $\gamma$ is set to 40, however, the recall becomes the best when $\gamma$ is set to 520. To determine the best trade off between precision and recall, is shown by the $F_1$ of .54 when $\gamma$ is 480.

|  | $\gamma = 20$ | $\gamma = 40$ | $\gamma = 80$ | $\gamma = 260$ | $\gamma = 400$ | $\gamma = 480$ | $\gamma = 500$ | $\gamma = 520$ |
|---|---|---|---|---|---|---|---|---|
| Precision@N | 55% | **63%** | 48% | 36% | 36% | 40% | 39% | 39% |
| Recall@N | 8% | 14% | 21% | 60% | 78% | 84% | 85% | **86%** |
| $F_1$ Score | .14 | .23 | .29 | .46 | .491 | **.542** | .537 | .541 |

For comparison, Twevent was used since it is the most similar state-of-the-art phrase event detection method. Twevent's source code was provided by the authors without the segmentation source code, therefore, the PhraseNet ToP-Mine output was used to create the necessary segments. In addition, the authors of Twevent specified to set the prior probability of segments to 0.01 based upon their previous calculations from Wikipedia and Microsoft N-Gram Web, however, it was found that the prior probability that gave the best $F_1$ score was .001, therefore, it was used for the comparison.

As you can see in Table 3, PhraseNet shows a distinct strength in discovering events compared to Twevent. In total Twevent identified 694 potential events for

**Table 3.** Precision and Recall for the best $F_1$ Score of both PhraseNet and Twevent.

|           | Precision | Recall | $F_1$ Score |
|-----------|-----------|--------|-------------|
| PhraseNet | 40%       | 84%    | .54         |
| Twevent   | 2%        | 15%    | .04         |

the three months of data, however, only 22 of those were confirmed true positives. In addition, Twevent identified 11 distinct events out of 102. In comparison when PhraseNet returned 480 potential events, 86 distinct events were correctly identified. These results were determined with the same ground truth list and with the all true positives were identified if found within ±5 days of the true event date.

Figure 1 showed an example displaying the key differences between an event distribution and a non-event topic distribution. Twevent identified the non-event topic of social media photos as an event and the Paris attack was not even identified, however, both of these cases were identified correctly by PhraseNet.

One reason for Twevent's performance is the mistake of identifying a non-event topic as an event. This is due to the mechanism that determines a "bursty" segment. Some words are frequent, however, their popularity in usage tends to rise and fall in its frequency throughout time. PhraseNet can find these groups of phrase segments and recognizes these multiple rises and falls as a characteristic of a non-event topic.

There was one common weakness made by Twevent and PhraseNet. They both mistakenly identified some non-event topics as events because these particular non-event topics showed event-like characteristics. For example, some artists have an army of users spreading a marketing campaign across social media to pre-order their new album. These types of discussions do not continue after the initial push from the artist's publicist, therefore, there shows a single high frequency peak on the day of the marketing campaign, yet no other frequency throughout the rest of the data.

### 4.3    Event Summarization: A Case Study

PhraseNet gives a more holistic picture about an event by leveraging phrases and graph clustering than other phrase focused event detection methods. For example, the Super Bowl event detected by PhraseNet consists of the following set of phrases: *superbowl, super bowl, pats, watch, year, vote, superbowlxlix, seattle, end, patriotswin, patriots, fans, call, katy perry, music, play, hase, commercial, f\*\*k, s\*\*t, depressing, game, seahawks, win, ago, nfl, chance, team, sb, halftime show, win sb, mousetrapspellingbee, video, youtube, kianlawley.* This description, correlated, aggregated, and produced by PhraseNet, explains that the Seattle Seahawks and the Patriots played in the NFL Super Bowl XLIL and, from the "patriotswin" hashtag, the Super Bowl was won by the Patriots. In addition, PhraseNet unveils that the Super Bowl half time show starred Katy Perry.

However, Twevent [9] gives a description of the same event with the following keywords: *rt, superbowl, ve, super bowl, ll, commercial, watch, game, seahawks, time, patriots.* This description of the Super Bowl leaves out the half time show description and who eventually won the game.

## 4.4 Scalability and Efficiency

PhraseNet can be implemented in real-time. PhraseNet has a complexity of $O(\tau n)$ where $\tau$ is the number of intervals of the sliding window (i.e. number of documents) and $n$ is the number of phrases within each sliding window, therefore, it scales to be a suitable algorithm for real-time. Under the experiment setting described in Sect. 4.1, the running time of PhraseNet is 8.12 s per time step where the experiment was run on a Macbook Pro 2.2 GHz Intel Core i7 with 16 GB of memory. It takes Twevent 45.95 s under the same setting.

## 5 Related Work

Twitter opens up doors to a faster way to gain information and to connect. People became a form of social "sensors" [16]. Many event detection algorithms have been proposed, both supervised and unsupervised, based on this platform.

**Supervised Methods.** Supervised methods focus on a certain set of seed keywords or hashtags which causes the method to miss events that have never been seen before or other important, unique, and rare events. This limits the ability of the system to rapidly evolving with its users and the evolving environment the users interact and live [3,5,10,12,13]. Thelwall et al. [18] showed evidence that strong negative or positive sentiment about a subject would separate out the events. However, the sentiment was found of a specific set of seeded keywords and hashtags used for tweet correlation which biases the detections to past data and recurring events.

**Unsupervised Methods.** Some event detection papers, such as Twevent, [9], consider trending (aka "bursty") topics as synonymous to events, however, not all topics are events [8,11,20]. Other methods are more semi-supervised methods since they need seeded events to learn from to identify events in the midst of other topics. FRED [14] use training data labeled as "newsworthy" to aid in seeding the model. In addition, GDTM [4] explores a graphical model approach which relies on keywords to seed their unsupervised topic modeling. Ritter et al. [15] developed a semi-supervised method which makes use of text annotation, however, in the midst of an informal environment such as Twitter, annotations could easily be mistaken. HIML [21] and EMBERS [17] methods required an already established taxonomy to find complex events. The taxonomy focuses on location information given in the text, which is hardly ever the case for Twitter data. TopicSketch [19] identifies "bursty topics" in real time where topics are defined as a word used more frequently at a rate greater than a threshold and does so uniquely. Agarwal et al. [1] similarly use keywords that occur together in the same tweet appearing in a short sliding window ("burstiness" of a keyword) to identify potential events. In addition, this method uses a greedy clique clustering method to incrementally find small, dense clusters which limits the final description of the event.

# 6    Conclusion

PhraseNet has exhibited to be an unsupervised, real-time Twitter event detection algorithm that summarizes events with a grouping of phrases. PhraseNet showed to have no bias towards certain types of events by being unsupervised, PhraseNet distinguished out non-event topics from events, and gave a short description of the events with a short keyword description. For potential future work, we want to identify dependencies between events and calculate the probability of influence unsupervised.

**Acknowledgement.** The work is partially supported by NIH U01HG008488, NIH R01GM115833, NIH U54GM114833, and NSF IIS-1313606. We thank the anonymous reviewers for their careful reading and insightful comments on our manuscript.

# References

1. Agarwal, M.K., Ramamritham, K., Bhide, M.: Real time discovery of dense clusters in highly dynamic graphs: identifying real world events in highly dynamic environments. VLDB **5**(10), 980–991 (2012)
2. Blondel, V.D., Guillaume, J.-L., Lambiotte, R., Lefebvre, E.: Fast unfolding of communities in large networks. J. Stat. Mech. Theor. Exp. **2008**(10), P10008 (2008)
3. Bollen, J., Mao, H., Zeng, X.: Twitter mood predicts the stock market. J. Comput. Sci. **2**(1), 1–8 (2011)
4. Chua, F.C.T., Asur, S.: Automatic summarization of events from social media. In: ICWSM (2013)
5. Du, N., Dai, H., Trivedi, R., Upadhyay, U., Gomez-Rodriguez, M., Song, L.: Recurrent marked temporal point processes: embedding event history to vector. In: KDD, pp. 1555–1564. ACM (2016)
6. El-Kishky, A., Song, Y., Wang, C., Voss, C.R., Han, J.: Scalable topical phrase mining from text corpora. VLDB **8**(3), 305–316 (2014)
7. Han, J., Pei, J., Yin, Y.: Mining frequent patterns without candidate generation. In: SIGMOD, vol. 29, pp. 1–12. ACM (2000)
8. Kwak, H., Lee, C., Park, H., Moon, S.: What is Twitter, a social network or a news media? In: WWW, pp. 591–600. ACM (2010)
9. Li, C., Sun, A., Datta, A.: Twevent: segment-based event detection from tweets. In: CIKM, pp. 155–164. ACM (2012)
10. Lin, C.X., Zhao, B., Mei, Q., Han, J.: PET: a statistical model for popular events tracking in social communities. In: KDD, pp. 929–938. ACM (2010)
11. Mathioudakis, M., Koudas, N.: TwitterMonitor: trend detection over the Twitter stream. In: SIGMOD, pp. 1155–1158. ACM (2010)
12. Popescu, A.-M., Pennacchiotti, M.: Detecting controversial events from Twitter. In: CIKM, pp. 1873–1876. ACM (2010)

13. Popescu, A.-M., Pennacchiotti, M., Paranjpe, D.: Extracting events and event descriptions from Twitter. In: WWW, pp. 105–106. ACM (2011)
14. Qin, Y., Zhang, Y., Zhang, M., Zheng, D.: Feature-rich segment-based news event detection on Twitter. In: IJCNLP, pp. 302–310 (2013)
15. Ritter, A., Etzioni, O., Clark, S., et al.: Open domain event extraction from Twitter. In: KDD, pp. 1104–1112. ACM (2012)
16. Sakaki, T., Okazaki, M., Matsuo, Y.: Earthquake shakes Twitter users: real-time event detection by social sensors. In: WWW, pp. 851–860. ACM (2010)
17. Saraf, P., Ramakrishnan, N.: EMBERS AutoGSR: automated coding of civil unrest events. In: KDD, pp. 599–608. ACM (2016)
18. Thelwall, M., Buckley, K., Paltoglou, G.: Sentiment in Twitter events. J. Assoc. Inf. Sci. Technol. **62**(2), 406–418 (2011)
19. Xie, W., Zhu, F., Jiang, J., Lim, E.-P., Wang, K.: TopicSketch: real-time bursty topic detection from Twitter. TKDE **28**(8), 2216–2229 (2016)
20. Yu, W., Aggarwal, C.C., Wang, W.: Temporally factorized network modeling for evolutionary network analysis. In: WSDM, pp. 455–464. ACM (2017)
21. Zhao, L., Ye, J., Chen, F., Lu, C.-T., Ramakrishnan, N.: Hierarchical incomplete multi-source feature learning for spatiotemporal event forecasting. In: KDD, pp. 2085–2094. ACM (2016)

# Generalising Random Forest Parameter Optimisation to Include Stability and Cost

C. H. Bryan Liu[1(✉)], Benjamin Paul Chamberlain[2], Duncan A. Little[1], and Ângelo Cardoso[1]

[1] ASOS.com, London, UK
bryan.liu@asos.com
[2] Department of Computing, Imperial College London, London, UK

**Abstract.** Random forests are among the most popular classification and regression methods used in industrial applications. To be effective, the parameters of random forests must be carefully tuned. This is usually done by choosing values that minimize the prediction error on a held out dataset. We argue that error reduction is only one of several metrics that must be considered when optimizing random forest parameters for commercial applications. We propose a novel metric that captures the stability of random forest predictions, which we argue is key for scenarios that require successive predictions. We motivate the need for multi-criteria optimization by showing that in practical applications, simply choosing the parameters that lead to the lowest error can introduce unnecessary costs and produce predictions that are not stable across independent runs. To optimize this multi-criteria trade-off, we present a new framework that efficiently finds a principled balance between these three considerations using Bayesian optimisation. The pitfalls of optimising forest parameters purely for error reduction are demonstrated using two publicly available real world datasets. We show that our framework leads to parameter settings that are markedly different from the values discovered by error reduction metrics alone.

**Keywords:** Bayesian optimisation · Parameter tuning
Random forest · Machine learning application · Model stability

## 1 Introduction

Random forests are ensembles of decision trees that can be used to solve classification and regression problems. They are very popular for practical applications because they can be trained in parallel, easily consume heterogeneous data types and achieve state of the art predictive performance for many tasks [6,14,15].

Forests have a large number of parameters (see [4]) and to be effective their values must be carefully selected [8]. This is normally done by running an optimisation procedure that selects parameters that minimize a measure of prediction error. A large number of error metrics are used depending on the problem specifics. These include prediction accuracy and area under the receiver

© Springer International Publishing AG 2017
Y. Altun et al. (Eds.): ECML PKDD 2017, Part III, LNAI 10536, pp. 102–113, 2017.
https://doi.org/10.1007/978-3-319-71273-4_9

operating characteristic curve (AUC) for classification, and mean absolute error (MAE) and root mean squared error (RMSE) for regression problems. Parameters of random forests (and other machine learning methods) are optimized exclusively to minimize error metrics. We make the case to consider monetary cost in practical scenarios and introduce a novel metric which measures the stability of the model.

Unlike many other machine learning methods (SVMs, linear regression, decision trees), predictions made by random forests are not deterministic. While a deterministic training method has no variability when trained on the same training set, it exhibits randomness from sampling the training set. We call the variability in predictions due solely to the training procedure (including training data sampling) the **endogenous variability**. It has been known for many years that instability plays an important role in evaluating the performance of machine learning models. The notion of instability for bagging models (like random forests) was originally developed by Breiman [1,2], and extended explicitly by Elisseeff et al. [5] to randomised learning algorithms, albeit focusing on generalisation/leave-one-out error (as is common in computational learning theory) rather than the instability of the predictions themselves.

It is often the case that changes in successive prediction values are more important than the absolute values. Examples include predicting disease risk [9] and changes in customer lifetime value [3]. In these cases we wish to measure a change in the external environment. We call the variability in predictions due solely to changes in the external environment **exogenous variability**. Figure 1 illustrates prediction changes with and without endogenous changes on top of exogenous change. Ideally we would like to measure only exogenous change, which is challenging if the endogenous effects are on a similar or larger scale.

Besides stability and error our framework also accounts for the cost of running the model. The emergence of computing as a service (Amazon elastic cloud, MS Azure etc.) makes the cost of running machine learning algorithms transparent and, for a given set of resources, proportional to runtime.

It is not possible to find parameter configurations that simultaneously optimise cost, stability and error. For example, increasing the number of trees in a random forest will improve the stability of predictions, reduce the error, but increase the cost (due to longer runtimes). We propose a principled approach to this problem using a multi-criteria objective function.

We use Bayesian optimisation to search the parameter space of the multi-criteria objective function. Bayesian optimisation was originally developed by Kushner [10] and improved by Močkus [12]. It is a non-linear optimisation framework that has recently become popular in machine learning as it can find optimal parameter settings faster than competing methods such as random/grid search or gradient descent [13]. The key idea is to perform a search over possible parameters that balances exploration (trying new regions of parameter space we know little about) with exploitation (choosing parts of the parameter space that are likely to lead to good objectives). This is achieved by placing a prior distribution on the mapping from parameters to the loss. An acquisition function

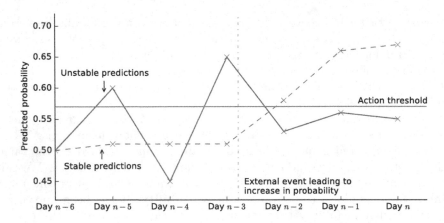

**Fig. 1.** Illustration of the change in predicted probability on successive days, in a scenario where action is taken when the prediction is over a certain threshold (red horizontal line), and some external event leading to increase in probability occurred sometime between days $n-3$ and $n-2$ (indicated by the dot-dashed grey vertical line). The solid line (blue) and dashed line (green) shows the change in the predicted probability if the model does or does not produce a fluctuation in successive predictions respectively. (Color figure online)

then queries successive parameter settings by balancing high variance regions of the prior (good for exploration) with low mean regions (good for exploitation). The optimal parameter setting is then obtained as the setting with the lowest posterior mean after a predefined number of query iterations.

We demonstrate the success of our approach on two large, public commercial datasets. Our work makes the following contributions:

1. A novel metric for the stability of the predictions of a model over different runs and its relationship with the variance and covariance of the predictions.
2. A framework to optimise model hyperparameters and training parameters against the joint effect of prediction error, prediction stability and training cost, utilising constrained optimisation and Bayesian optimisation.
3. A case study on the effects of changing hyperparameters of a random forest and training parameters on the model error, prediction stability and training cost, as applied on two publicly available datasets.

The rest of the paper is organized as follows: in Sect. 2 we propose a novel metric to assess the stability of random forest predictions, in Sect. 3 we propose a random forest parameter tuning framework using a set of metrics, in Sect. 4 we discuss the effects of the hyper-parameters on the metrics and illustrate the usefulness of the proposed optimization framework to explore the trade-offs in the parameter space in Sect. 5.

## 2  Prediction Stability

Here we formalise the notion of random forest stability in terms of repeated model runs using the same parameter settings and dataset (i.e. all variability is endogenous). The expected squared difference between the predictions over two runs is given by

$$\frac{1}{N}\sum_{i=1}^{N}\left[\left(\hat{y}_i^{(j)} - \hat{y}_i^{(k)}\right)^2\right],\tag{1}$$

where $\hat{y}_i^{(j)} \in [0,1]$ is the probability from the $j^{th}$ run that the $i^{th}$ data point is of the positive class in binary classification problems (note this can be extended to multiclass classification and regression problems). We average over $R \gg 1$ runs to give the Mean Squared Prediction Delta (MSPD):

$$\mathrm{MSPD}(f) = \frac{2}{R(R-1)}\sum_{j=1}^{R}\sum_{k=1}^{j-1}\left[\frac{1}{N}\sum_{i=1}^{N}\left[\left(\hat{y}_i^{(j)} - \hat{y}_i^{(k)}\right)^2\right]\right]\tag{2}$$

$$= \frac{2}{N}\sum_{i=1}^{N}\left[\frac{1}{R-1}\sum_{l=1}^{R}\left(\hat{y}_i^{(l)} - \mathbb{E}(\hat{y}_i^{(\cdot)})\right)^2\right.$$

$$\left. - \frac{1}{R(R-1)}\sum_{j=1}^{R}\sum_{k=1}^{R}\left(\hat{y}_i^{(j)} - \mathbb{E}(\hat{y}_i^{(\cdot)})\right)\left(\hat{y}_i^{(k)} - \mathbb{E}(\hat{y}_i^{(\cdot)})\right)\right]$$

$$= 2\mathbb{E}_{x_i}[\mathrm{Var}(f(x_i)) - \mathrm{Cov}(f_j(x_i), f_k(x_i))],\tag{3}$$

where $\mathbb{E}_{x_i}$ is the expectation over all validation data, $f$ is a mapping from a sample $x_i$ to a label $y_i$ on a given run, $\mathrm{Var}(f(x_i))$ is the variance of the predictions of a single data point over model runs, and $\mathrm{Cov}(f_j(x_i), f_k(x_i))$ is the covariance of predictions of a single data point over two model runs.[1]

The covariance, the variance and hence the model instability are closely related to the forest parameter settings, which we discuss in Sect. 4. It is convenient to measure stability on the same scale as the forest predictions and so in the experiments we report the RMSPD $= \sqrt{\mathrm{MSPD}}$.

## 3  Parameter Optimisation Framework

In industrial applications, where ultimately machine learning is a tool for profit maximisation, optimising parameter settings based solely on error metrics is inadequate. Here we develop a generalised loss function that incorporates our stability metric in addition to prediction error and running costs. We use this loss with Bayesian optimisation to select parameter values.

---

[1] A full derivation is available at our GitHub repository https://github.com/liuchbryan/generalised_forest_tuning.

## 3.1  Metrics

Before composing the loss function we define the three components:

*Stability.* We incorporate stability (defined in Sect. 2) in to the optimization framework with the use of the RMSPD.

*Error reduction.* Many different error metrics are used with random forests. These include F1-score, accuracy, precision, recall and Area Under the receiver operating characteristics Curve (AUC) and all such metrics fit within our framework. In the remainder of the paper we use the AUC because for binary classification, most other metrics require the specification of a threshold probability. As random forests are not inherently calibrated, a threshold of 0.5 may not be appropriate and so using AUC simplifies the exposition [3].

*Cost reduction.* It is increasingly common for machine learning models to be run on the cloud with computing resources paid for by the hour (e.g. Amazon Web Services). Due to the exponential growth in data availability, the cost to run a model can be comparable with the financial benefit it produces. We use the training time (in seconds) as a proxy of the training cost.

## 3.2  Loss-Function

We choose a loss function that is linear in cost, stability and AUC that allows the relative importance of these three considerations to be balanced:

$$L = \beta \, \text{RMSPD}(N_t, d, p) + \gamma \, \text{Runtime}(N_t, d, p) - \alpha \, \text{AUC}(N_t, d, p), \qquad (4)$$

where $N_t$ is the number of trees in the trained random forest, $d$ is the maximum depth of the trees, and $p$ is the proportion of data points used in training; $\alpha, \beta, \gamma$ are weight parameters. We restrict our analysis to three parameters of the random forest, but it can be easily extended to include additional parameters (e.g. number of features bootstrapped in each tree).

The weight parameters $\alpha, \beta$ and $\gamma$ are specified according to business/research needs. We recognise the diverse needs across different organisations and thus refrain from specifying what constitutes a "good" weight parameter set. Nonetheless, a way to obtain the weight parameters is to quantify the gain in AUC, the loss in RMSPD, and the time saved all in monetary units. For example, if calculations reveal 1% gain in AUC equates to £50 potential business profit, 1% loss in RMSPD equates to £10 reduction in lost business revenue, and a second of computation costs £0.01, then $\alpha, \beta$ and $\gamma$ can be set as 5,000, 1,000 and 0.01 respectively.

## 3.3  Bayesian Optimisation

The loss function is minimized using Bayesian optimisation. The use of Bayesian optimisation is motivated by the expensive, black-box nature of the objective function: each evaluation involves training multiple random forests, a complex process with internal workings that are usually masked from users. This rules out gradient ascent methods due to unavailability of derivatives. Exhaustive search strategies, such as grid search or random search, have prohibitive runtimes due to the large random forest parameter space.

A high-level overview on Bayesian Optimisation is provided in Sect. 1. Many different prior functions can be chosen and we use the Student-t process implemented in pybo [7,11].

## 4 Parameter Sensitivity

Here we describe three important random forest parameters and evaluate the sensitivity of our loss function to them.

### 4.1 Sampling Training Data

Sampling of training data – drawing a random sample from the pool of available training data for model training – is commonly employed to keep the training cost low. A reduction in the size of training data leads to shorter training times and thus reduces costs. However, reducing the amount of training data reduces the generalisability of the model as the estimator sees less training examples, leading to a reduction in AUC. Decreasing the training sample size also decreases the stability of the prediction. This can be understood by considering the form of the stability measure of $f$, the RMSPD (Eq. 2). The second term in this equation is the expected covariance of the predictions over multiple training runs. Increasing the size of the random sample drawn as training data increases the probability that the same input datum will be selected for multiple training runs and thus the covariance of the predictions increases. An increase in covariance leads to a reduction in the RMSPD (see Eq. 3).

### 4.2 Number of Trees in a Random Forest

Increasing the number of trees in a random forest will decrease the RMSPD (and hence improve stability) due to the Central Limit Theorem (CLT). Consider a tree in a random forest with training data bootstrapped. Its prediction can be seen as a random sample from a distribution with finite mean and variance $\sigma^2$.[2] By averaging the trees' predictions, the random forest is computing the sample mean of the distribution. By the CLT, the sample mean will converge to a Gaussian distribution with variance $\frac{\sigma^2}{N_t}$, where $N_t$ is the number of trees in the random forest.

---

[2] This could be any distribution as long as its first two moments are finite, which is usually the case in practice as predictions are normally bounded.

To link the variance to the MSPD, recall from Eq. 2 that MSPD captures the interaction between the variance of the model and covariance of predictions between different runs:

$$\text{MSPD}(f) = 2\mathbb{E}_{x_i}[\text{Var}(f(x_i)) - \text{Cov}(f_j(x_i), f_k(x_i))].$$

The covariance is bounded below by the negative square root of the variance of its two elements, which is in turn bounded below by the negative square root of the larger variance squared:

$$\text{Cov}(f_j(x_i), f_k(x_i)) \geq -\sqrt{\text{Var}(f_j(x_i))\text{Var}(f_k(x_i))}$$
$$\geq -\sqrt{(\max\{\text{Var}(f_j(x_i)), \text{Var}(f_k(x_i))\})^2}. \qquad (5)$$

Given $f_j$ and $f_k$ have the same variance as $f$ (being the models with the same training proportion across different runs), the inequality 5 can be simplified as:

$$\text{Cov}(f_j(x_i), f_k(x_i)) \geq -\sqrt{(\max\{\text{Var}(f(x_i)), \text{Var}(f(x_i))\})^2} = -\text{Var}(f(x_i)). \quad (6)$$

MSPD is then bounded above by a multiple of the expected variance of $f$:

$$\text{MSPD}(f) \leq 2\mathbb{E}_{x_i}[\text{Var}(f(x_i)) - (-\text{Var}(f(x_i)))] = 4\mathbb{E}_{x_i}[\text{Var}(f(x_i))], \qquad (7)$$

which decreases as $N_t$ increases, leading to a lower RMSPD estimate.

While increasing the number of trees in a random forest reduces error and improves stability in predictions, it increases the training time and hence monetary cost. In general, the runtime complexity for training a random forest grows linearly with the number of trees in the forest.

### 4.3    Maximum Depth of a Tree

The maximum tree depth controls the complexity of each decision tree and the computational cost (running time) increases exponentially with tree depth. The optimal depth for error reduction depends on the other forest paramaters and the data. Too much depth causes overfitting. Additionally, as the depth increases the prediction stability will decrease as each model tends towards memorizing the training data. The highest stability will be attained using shallow trees, however if the forest is too shallow the model will underfit resulting in low AUC.

## 5    Experiments

We evaluate our methodology by performing experiments on two public datasets: (1) the Orange small dataset from the 2009 KDD Cup and (2) the Criteo display advertising challenge Kaggle competition from 2014. Both datasets have a mixture of numerical and categorical features and binary target labels (Orange: 190 numerical, 40 categorical, Criteo: 12 numerical, 25 categorical).

We report the results of two sets of experiments: (1) Evaluating the effect of changing random forest parameters on the stability and loss functions (2) Bayesian optimisation with different weight parameters.

We train random forests to predict the upselling label for the Orange dataset and the click-through rate for the Criteo dataset. Basic pre-processing steps were performed on both datasets to standardise the numerical data and transform categoricals into binary indicator variables. We split the datasets into two halves: the first as training data (which may be further sampled at each training run), and the later as validation data. All data and code required to replicate our experiments is available from our GitHub repository.[3]

## 5.1    Parameter Sensitivity

In the first set of experiments we evaluate the effect of varying random forest parameters on the components of our loss function.

Figure 2 visualises the change in the RMSPD with relation to the number of trees in the random forest. The plots show distributions of prediction deltas for the Orange dataset. Increasing the number of trees (going from the left to the right plot) leads to a more concentrated prediction delta distribution, a quality also reflected by a reduction in the RMSPD.

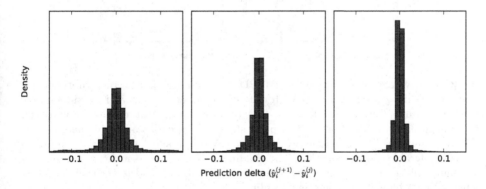

**Fig. 2.** The distribution of prediction deltas (difference between two predictions on the same validation datum) for successive runs of random forests with (from left to right) 8, 32, and 128 trees, repeated ten times. The RMSPD for these three random forests are 0.046, 0.025, and 0.012 respectively. Training and prediction are done on the Orange small dataset with upselling labels. The dataset is split into two halves: the first 25 k rows are used for training the random forests, and the latter 25 k rows for making predictions. Each run re-trains on all 25 k training data, with trees limited to a maximum depth of 10.

Figure 3 shows the AUC, runtime, RMSPD and loss functions averaged over multiple runs of the forest for different settings of number of trees and maximum

---

[3] https://github.com/liuchbryan/generalised_forest_tuning.

tree depth. It shows that the AUC plateaus for a wide range of combinations of number of trees and maximum depth. The RMSPD is optimal for large numbers of shallow trees while runtime is optimised by few shallow trees. When we form a linear combination of the three metrics, the optimal solutions are markedly different from those discovered by optimising any single metric in isolation. We show this for $\alpha = 1, \beta = 1, \gamma = 0.01$ and $\alpha = 2, \beta = 1, \gamma = 0.005$.

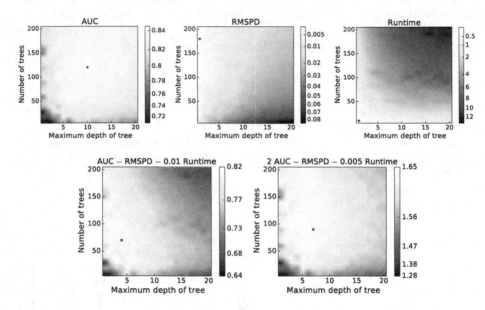

**Fig. 3.** The average AUC (top left), RMSPD (top middle), and average runtime (top right) attained by random forests with different number of trees and maximum tree depth (training proportion is fixed at 0.5) over five train/test runs, as applied on the Orange dataset. The bottom two plots shows the value attained in the specified objective functions by the random forests above. A lighter spot on the maps represents a more preferable parametrization. The shading is scaled between the minimum and maximum values in each chart. The optimal configuration found under each metric is indicated by a blue star. (Color figure online)

## 5.2   Bayesian Optimization of the Trilemma

We also report the results of using the framework to choose the parameters. The aim of these experiments is to show that (1) Bayesian optimisation provides a set of parameters that achieve good AUC, RMSPD and runtime, and (2) by varying the weight parameters in the Bayesian optimisation a user is able to prioritise one or two of the three respective items.

Table 1 summarises the trilemma we are facing – all three parameter tuning strategies improves two of the three practical considerations with the expense of the consideration(s) left.

**Table 1.** Effect of the common hyperparameter tuning strategies on the three practical considerations. Plus sign(s) means a positive effect to the measure (and hence more preferred), and minus sign(s) means a negative effect to the measure (and hence not preferred). The more plus/minus sign within the entry, the more prominent the effect of the corresponding strategy.

| Hyperparameter tuning strategy | AUC gain | RMSPD reduction | Cost savings |
|---|---|---|---|
| Increase training proportion | + | + | − |
| Increase number of trees | + | + | − − |
| Reduce maximum depth of trees | − | + | + + |

The results of our experiments on Bayesian optimisation of the trilemma are shown in Tables 2 and 3. The first row in both tables shows the results for a vanilla random forest with no optimisation of the hyper-parameters discussed in the previous section: 10 trees, no limit on the maximum depth of the tree, and using the entire training data set (no sampling). The Bayesian optimisation for each set of weight parameters was run for 20 iterations, with the RMSPD calculated over three training runs in each iteration.

The first observation from both sets of results is that Bayesian optimisation is suitable for providing a user with a framework that can simultaneously improve AUC, RMSPD and runtime as compared to the baseline. Secondly, it is clear that by varying the weight parameters, Bayesian optimisation is also capable of prioritising specifically AUC, RMSPD or runtime. Take for example the third and fourth rows of Table 2; setting $\beta = 5$ we see a significant reduction in the RMSPD in comparison to the second row where $\beta = 1$. Similarly, comparing the fourth row to the second row, increasing $\alpha$ from 1 to 5 gives a 1% increase in AUC. In the final row we see that optimising for a short runtime keeps the RMSPD low in comparison to the non-optimal results on the first row and sacrifices the AUC instead.

**Table 2.** Results of Bayesian optimisation for the Orange dataset at various settings of $\alpha$, $\beta$ and $\gamma$, the weight parameters for the AUC, RMSPD and runtime respectively. The Bayesian optimiser has the ability to tune three random forest hyper-parameters: the number of trees, $N_t^*$, the maximum tree depth, $d^*$, and size of the training sample, $p^*$. Key results are emboldened and discussed further in the text.

| $\alpha$ | $\beta$ | $\gamma$ | $N_t^*$ | $d^*$ | $p^*$ | AUC | RMSPD | Runtime |
|---|---|---|---|---|---|---|---|---|
| No optimisation: | | | | | | 0.760 | 0.112 | 1.572 |
| 1 | 1 | 0.01 | 166 | 6 | 0.100 | 0.829 | 0.011 | 1.142 |
| 1 | 5 | 0.01 | 174 | 1 | 0.538 | 0.829 | **0.002** | 1.452 |
| 5 | 1 | 0.01 | 144 | 12 | 0.583 | **0.839** | 0.013 | 5.292 |
| 1 | 1 | **0.05** | 158 | 4 | 0.100 | 0.8315 | 0.0082 | **1.029** |

For the Criteo dataset (Table 3) we see on the second and third row that again increasing the $\beta$ parameter leads to a large reduction in the RMSPD. For this dataset the Bayesian optimiser is more reluctant to use a larger number of estimators to increase AUC because the Criteo dataset is significantly larger (around 100 times) than the Orange dataset and so using more trees increases the runtime more severely. To force the optimiser to use more estimators we reduce the priority of the runtime by a factor of ten as can be seen in the final two rows. We see in the final row that doubling the importance of the AUC ($\alpha$) leads to a significant increase in AUC (4.5%) when compared to the non-optimal results.

**Table 3.** Results of Bayesian optimisation for the Criteo dataset. The table shows the results of the Bayesian optimisation by varying $\alpha$, $\beta$ and $\gamma$ which control the importance of the AUC, RMSPD and runtime respectively. The Bayesian optimiser has the ability to tune three hyper-parameters of the random forest: the number of trees, $N_t^*$, the maximum depth of the tree, $d^*$, and size of the training sample, $p^*$. Key results are emboldened and discussed further in the text.

| $\alpha$ | $\beta$ | $\gamma$ | $N_t^*$ | $d^*$ | $p^*$ | AUC | RMSPD | Runtime |
|---|---|---|---|---|---|---|---|---|
| No optimisation: | | | | | | 0.685 | 0.1814 | 56.196 |
| 1 | 1 | 0.01 | 6 | 8 | 0.1 | 0.7076 | **0.04673** | 1.897 |
| 1 | 5 | 0.01 | 63 | 3 | 0.1 | 0.6936 | **0.01081** | 4.495 |
| 1 | 1 | 0.05 | 5 | 5 | 0.1 | 0.688 | 0.045 | 1.136 |
| 2 | 1 | 0.05 | 9 | 9 | 0.1 | 0.7145 | 0.03843 | 2.551 |
| 1 | 1 | 0.001 | 120 | 2 | 0.1 | 0.6897 | 0.007481 | 7.153 |
| **2** | 1 | 0.001 | 66 | 15 | 0.1 | **0.7300** | 0.02059 | 11.633 |

# 6    Conclusion

We proposed a novel metric to capture the stability of random forest predictions, which is key for applications where random forest models are continuously updated. We show how this metric, calculated on a sample, is related to the variance and covariance of the predictions over different runs. While we focused on random forests in this text, the proposed stability metric is generic and can be applied to other non-deterministic models (e.g. gradient boosted trees, deep neural networks) as well as deterministic training methods when training is done with a subset of the available data.

We also propose a framework for multi-criteria optimisation, using the proposed metric in addition to metrics measuring error and cost. We validate this approach using two public datasets and show how optimising a model solely for error can lead to poorly specified parameters.

# References

1. Breiman, L.: Bagging predictors. Mach. Learn. **24**(2), 123–140 (1996)
2. Breiman, L.: Heuristics of instability in model selection. Ann. Stat. **24**(6), 2350–2383 (1996)
3. Chamberlain, B.P., Cardoso, A., Liu, C.H.B., Pagliari, R., Deisenroth, M.P.: Customer lifetime value prediction using embeddings. In: Proceedings of the 23rd ACM SIGKDD International Conference on Knowledge Discovery and Data Mining, pp. 1753–1762 (2017)
4. Criminisi, A.: Decision forests: a unified framework for classification, regression, density estimation, manifold learning and semi-supervised learning. Found. Trends® Comput. Graph. Vis. **7**(2–3), 81–227 (2012)
5. Elisseeff, A., Evgeniou, T., Pontil, M.: Stability of randomized learning algorithms. J. Mach. Learn. Res. **6**(1), 55–79 (2005)
6. Fernández-Delgado, M., Cernadas, E., Barro, S., Amorim, D., Amorim Fernández-Delgado, D.: Do we need hundreds of classifiers to solve real world classification problems? J. Mach. Learn. Res. **15**, 3133–3181 (2014)
7. Hoffman, M.W., Shahriari, R.: Modular mechanisms for Bayesian optimization. In: NIPS Workshop on Bayesian Optimization (2014)
8. Huang, B.F.F., Boutros, P.C.: The parameter sensitivity of random forests. BMC Bioinform. **17**(1), 331 (2016)
9. Khalilia, M., Chakraborty, S., Popescu, M.: Predicting disease risks from highly imbalanced data using random forest. BMC Med. Inf. Dec. Making **11**(1), 51 (2011)
10. Kushner, H.J.: A new method of locating the maximum point of an arbitrary multipeak curve in the presence of noise. J. Basic Eng. **86**(1), 97–106 (1964)
11. Martinez-Cantin, R.: BayesOpt: a bayesian optimization library for nonlinear optimization, experimental design and bandits. J. Mach. Learn. Res. **15**, 3735–3739 (2014)
12. Močkus, J.: On bayesian methods for seeking the extremum. In: Marchuk, G.I. (ed.) Optimization Techniques 1974. LNCS, vol. 27, pp. 400–404. Springer, Heidelberg (1975). https://doi.org/10.1007/3-540-07165-2_55
13. Snoek, J., Larochelle, H., Adams, R.: Practical bayesian optimization of machine learning algorithms. In: Advances in Neural Information Processing Systems, pp. 2951–2959 (2012)
14. Tamaddoni, A., Stakhovych, S., Ewing, M.: Comparing churn prediction techniques and assessing their performance: a contingent perspective. J. Serv. Res. **19**(2), 123–141 (2016)
15. Vanderveld, A., Pandey, A., Han, A., Parekh, R.: An engagement-based customer lifetime value system for e-commerce. In: Proceedings of the 22nd ACM SIGKDD International Conference on Knowledge Discovery and Data Mining, pp. 293–302 (2016)

# Have It Both Ways—From A/B Testing to A&B Testing with Exceptional Model Mining

Wouter Duivesteijn[1](✉), Tara Farzami[2], Thijs Putman[2], Evertjan Peer[1],
Hilde J. P. Weerts[1], Jasper N. Adegeest[1], Gerson Foks[1],
and Mykola Pechenizkiy[1]

[1] Technische Universiteit Eindhoven, Eindhoven, the Netherlands
{w.duivesteijn,m.pechenizkiy}@tue.nl,
{e.peer,h.j.p.weerts,j.n.adegeest,g.foks}@student.tue.nl
[2] StudyPortals B.V., Eindhoven, the Netherlands
{tara,thijs}@studyportals.com

**Abstract.** In traditional A/B testing, we have two variants of the same product, a pool of test subjects, and a measure of success. In a randomized experiment, each test subject is presented with one of the two variants, and the measure of success is aggregated per variant. The variant of the product associated with the most success is retained, while the other variant is discarded. This, however, presumes that the company producing the products only has enough capacity to maintain one of the two product variants. If more capacity is available, then advanced data science techniques can extract more profit for the company from the A/B testing results. Exceptional Model Mining is one such advanced data science technique, which specializes in identifying subgroups that behave differently from the overall population. Using the association model class for EMM, we can find subpopulations that prefer variant A where the general population prefers variant B, and vice versa. This data science technique is applied on data from StudyPortals, a global study choice platform that ran an A/B test on the design of aspects of their website.

**Keywords:** A/B testing · Exceptional Model Mining · Association
Online controlled experiments · E-commerce · Website optimization

## 1 Introduction

A/B testing [20] is a form of statistical hypothesis testing involving two versions of a product, A and B. Typically, A is the control version of a product and B represents a new variation version, considered to replace A if it proves to be more successful. An A/B test requires two further elements: a pool of test subjects, and a measure of success. Each test subject in the pool is presented with a randomized choice between A and B. The degree to which this product version is successful with this test subject is measured. Having collected results over the full pool of test subjects, the success degree is aggregated per version. Subsequently, a decision is made whether the new variation version B is a

Y. Altun et al. (Eds.): ECML PKDD 2017, Part III, LNAI 10536, pp. 114–126, 2017.
https://doi.org/10.1007/978-3-319-71273-4_10

(substantial) improvement over the control version A. For making this decision, a vast statistical toolbox is available [6, 7].

Since the rise of the internet, A/B tests have become ubiquitous. It is a simple, cheap, and reliable manner to assess the efficacy of the redesign of a web page. Running two versions of a web page side by side is not too intrusive to your online business, and standard web analytics suites will tell you all you need to know on which of the versions deliver the desired results. In fact, through proper web analytics tools, we can obtain substantially more information on the factors that influence the success of versions A and B.

Having performed an A/B test, the standard operating procedure is the following. An assessment is made whether the new variation version B performs (substantially) better than the current control version A. From that assessment, a hard, binary decision is made: either version A or version B is the winner. The loser is discarded, and the winner becomes the standard version of the web page that is rolled out and presented to all visitors from this moment onwards. There is beauty in the simplicity, and this 'exclusive or' procedure inspires the slash in the name of the A/B test.

For large companies, making such a coarse decision leaves potential unused. If you own a high-traffic website, then even a small increase in click-through rate gets multiplied by a large volume of visitors, which results in a vast increase in income. It makes sense to use the traditional conclusion of an A/B test to determine the default page that should be displayed to a visitor of which we know nothing. But it is not uncommon to have some meta-information on the visitors to your website: which language setting does their browser have, which OS do they use, in which country are they located, etcetera. If we can identify subpopulations of the dataset at hand, defined in terms of such metadata, for which the A/B test reaches the opposite conclusion from the general population, then we can generate more revenue with a more sophisticated strategy: we maintain both versions of the web page, and present a visitor with either A or B depending on whether they belong to specific subgroups. Rather than choosing either A or B, we can instead choose to have it both ways: this paper turns the A/B test into an A&B test.

## 2   Related Work

First, we provide a brief summary on the current state of the art in mining of A/B testing results. Thus we explain how our problem formulation is different from existing body of work. Then we overview relevant research in the areas of local pattern mining and exceptional model mining that motivate our approach for the chosen problem formulation.

### 2.1   Utility of A/B Testing

In a marketing context, A/B testing has been studied extensively [20]. Analysis of the results from an A/B test has made it to the Encyclopedia of Machine

Learning and Data Mining [6], and an extensive survey on experiment design choices and results analysis is available [7]. This last paper encompasses a discussion of accompanying A/B tests with A/A tests to establish a proper baseline, extending the test to the multivariate case (more than two product versions), result confidence intervals, randomization methods to divide the test subjects fairly over the versions, sample size effects, overlapping experiments, and the effect of bots on the process. Regardless of the setting of all of these facets, the goal of A/B testing always remains to make a crisp decision at the end, selecting either A or B and discarding the alternative(-s).

If the main business goal is to increase the average performance with respect to e.g. a click through rate (CTR) rather than really find our whether A or B is statistically significantly better, then the Contextual Multi-Armed Bandits (cMAB) is the commonly considered alternative optimization approach to A/B testing. cMABs help to address an exploration-exploitation trade-off: using, i.e. exploring effectiveness, of A and B provides feedback about its effectiveness (exploration), but collecting that feedback on both A and B is an opportunity cost of exploitation, i.e. using one of the variants we already know is effective. To balance exploration with exploitation lots of policy learning bandit algorithms were considered, particularly in web analytics, e.g. [22,23].

In data mining for user modeling and convergence prediction two related problem formulations have been studied – predictive user modeling with actionable attributes [26] and uplift prediction [18]. While in traditional predictive modeling, the goal is to learn a model for predicting accurately the class label for unseen instances, in targeting applications, a decision maker is interested not only to generate accurate predictions, but to maximize the probability of the desired outcome, e.g. user clicking. Assuming that possibly neither of marketing actions A and B is always best, the problem can be formulated as learning to choose the best marketing action at instance level (rather than globally).

The paper that you are currently reading does not have a mission to promote either A/B testing or cMABs or uplift prediction; we merely observe that A/B tests are performed anyway, and strive to help companies performing such tests to learn more actionable insight from their data that would allow to domain experts to decide whether to stay with A, or switch to B or use both A and B, each for a particular context or customer segment.

## 2.2    Local Pattern Mining

The subfield of Data Mining with which this paper is concerned is Local Pattern Mining [4,17]: describing only part of the dataset at hand, while disregarding the coherence of the reminder. The Local Pattern Mining subtask that is particularly relevant here, is Theory Mining [15], where subsets of the dataset are sought that are *interesting* in some sense. Typically, not just any subset is sought. Instead, the focus is on subsets that are easy to interpret. A canonical choice to enforce that is to restrict the search to subsets that can be described as a conjunction of a few conditions on single attributes of the dataset. Hence, if the dataset concerns people, we would find subsets of the form

"Age $\geq 30$ ∧ Smokes = yes ⇒ (interesting)". Such subsets are referred to as *subgroups*. Limiting the search to subgroups ensures that the results can be interpreted in terms of the domain of the dataset at hand; the resulting subgroups represent pieces of information on which a domain expert can act.

Many choices can be made to define 'interesting'. One such choice is to make this a supervised concept: we set apart one attribute of the dataset as the *target*, and seek subsets that feature an unusual distribution of that target. This is known as Subgroup Discovery (SD) [9,11,25]. In the running example of a dataset concerning people, if the target would be whether the person develops lung cancer or not, SD would find results such as "Smokes = yes ⇒ Lung cancer = yes". This of course does not mean that all smokers fall in the 'yes' category; it merely implies a skew in the target distribution.

### 2.3 Exceptional Model Mining

Exceptional Model Mining (EMM) can be seen as a generalized form of SD. Instead of singling out one attribute of the data as the target, in EMM one typically selects several target attributes. The exceptionality of a subgroup is no longer evaluated in terms of an unusual distribution of the single target, but instead in terms of an unusual interaction between the multiple targets. This interaction is captured by some kind of modeling, which inspired the name of EMM. Exceptional Model Mining was first introduced in 2008 [13]. An extensive overview of the *model classes* (types of interaction) that have been investigated can be found in [3]; as examples, one can think of an unusual correlation between two targets [13], an unusual slope of a regression vector on any number of targets [2], or unusual preference relations [19].

Algorithms for EMM include a form of beam search [3] that works for all model classes, a fast sampling-based algorithm for a few dedicated model classes [16], an FP-Growth-inspired tree-based exhaustive algorithm that works for almost all model classes [14], a tree-constrained gradient ascent algorithm for linear models using sofy subgroup membership [10], and a compression-based method that improves the resulting models at the cost of interpretability [12].

## 3   The StudyPortals A/B Test Setting

Since the Bologna process contributed to harmonizing higher-education qualifications throughout Europe, locating (part of) one's study programme in another country than one's own has become streamlined. This offers opportunities for students to acquire international experience while still studying, which is something from which both the students and the higher education institutions can benefit. The harmonization of how higher education is structured enables a fair comparison of programmes across country boundaries.

Such a comparison being possible does not necessarily imply that it is also easy. In 2007, three (former) students identified that there was a hole in the information market, and they filled that hole with a hobby project that eventually resulted in StudyPortals [21].

## 3.1   StudyPortals

In 2007, two alumni from the Technische Universiteit Eindhoven and one from the Kungliga Tekniska Högskolan created MastersPortal: a central database for European Master's programmes. The goal was to become the primary destination for students wanting to study in Europe. In April 2008, the website presented 2 700 studies at 200 universities from 30 countries, and attracted 80 000 visits per month. Since then, the scope of the website has expanded. The subject ranges beyond Master's programmes, also encompassing Bachelor's and PhD programmes, short courses, scholarships, distance learning, language learning, and preparation courses. The website is no longer restricted to Europe, but expanded globally. In September 2016, MastersPortal presented 56 000 studies at 2 000 universities from 100 countries, and attracted 1.4 million unique sessions per month. The overarching company StudyPortals logged 14.5 million unique visitors in the first nine months of 2016, with 7 page views per second during the busiest hour of the year. This growth allows the company to employ 150 team members in five offices on three continents.

StudyPortals generates revenue from the visitors to their websites through the universities, who pay for activity on the pages presenting their programmes. A study programme's web page generates revenue in three streams: (1) Cost Per Mille (thousand page views); (2) Cost Per Lead; (3) Cost Per Click. The first revenue stream depends on the attractiveness of links towards the programme's web page. The second revenue stream depends on whether the person viewing the programme's web page fill their personal information in the university lead form. The design of a programme's web page has a low impact on these two revenue streams. The third revenue stream is the one that StudyPortals can influence directly through appropriate web page design.

## 3.2   The Third Revenue Stream and the A/B Test

Figure 1 displays the mobile version of a university's web page on the MastersPortal website. The orange button at the bottom left of the page links through to the website of the university itself. When a user clicks on that button, Study-Portals receives revenue in the Cost Per Click revenue stream. With the volume of web traffic StudyPortals experiences, a small increase in the click-through rate represents a substantial increase in income.

The advance of smartphones and tablets has vastly increased the importance of the mobile version of websites. These versions come with their own UI requirements and quirks. Figure 1a displays the page design that was in use in September 2016; having an orange rectangle that is clickable is one of those UI design elements that is typical of mobile websites as opposed to desktop versions. However, the website visitors, being human beings, are creatures of habit. They might prefer clickable elements of websites to resemble traditional buttons, as they remember from their desktop dwelling times. To test this hypothesis, Study-Portals designed an alternative version of their mobile website (cf. Fig. 1b). These variants become the subject of our A/B test: the rectangular version is the control version A, and the more buttony version is the variation B.

(a) Control version                    (b) Buttony variation

**Fig. 1.** The A and B variants of the A/B test at hand: two versions of buttons on university profile pages of the mobile version of the MastersPortal website.

## 3.3   The Data at Hand

StudyPortals collected raw data on the A/B test results for a period of time. From this raw, anonymized data, a traditional flat-table dataset was generated through data cleaning and feature engineering. The full process is beyond the scope of this paper; it involved removing redundant information, removing the users that have seen both versions of the web page (as is customary in A/B testing), aggregating location information (available on city level) to country level, merging various versions of the distinct OSs (e.g., eight distinct versions of iOS were observed; these sub-OSs were flattened into one OS), etcetera. In the end, the columns in the dataset include device characteristics, location information, language data, and scrolling characteristics. The dataset spans 3 065 records.

Finally, we are particularly interested in two columns: the one holds the information with which version of the web page (A/B) the visitor was presented, and the other holds whether the visitor merely viewed or also clicked. The goal of traditional A/B testing is to find out whether version A or B leads to more clicks; the main contribution of this paper is to identify subpopulations where these two columns display an unusual interaction: can we find subgroups where the click rate interacts exceptionally with the web page version?

# 4    Data Science to Be Applied

Finding subsets of the dataset at hand where several columns of special interest interact in an unusual manner is the core task of Exceptional Model Mining (EMM). This interaction can be gauged in many ways. This section discusses the EMM framework and its specific instantiation for the problem at hand.

## 4.1    The Exceptional Model Mining Framework

EMM [3,13] assumes a flat-table dataset $\Omega$, which is a bag of $N$ records of the form $r = \{a_1, \ldots, a_k, t_1, \ldots, t_m\}$. We call the attributes $a_1, \ldots, a_k$ the *descriptors* of the dataset. These are the attributes in terms of which subgroups will be *defined*; the ones on the left-hand side of the $\Rightarrow$ sign in the examples of Sect. 2.2. The other attributes, $t_1, \ldots, t_m$, are the *targets* of the dataset. These are the attributes in terms of which subgroups will be *evaluated*; the most exceptional target interaction indicates the most interesting subgroup.

Subgroups are defined in terms of conditions on descriptors. These induce a subset of the dataset: all records satisfying the conditions. For notational purposes, we identify a subgroup with that subset, so that we write $S \subseteq \Omega$, and denote $|S|$ for the number of records in a subgroup. We also denote $S^C$ for the complement of subgroup $S$ in dataset $\Omega$, i.e.: $S^C = \Omega \backslash S$.

To instantiate the EMM framework, we need to define two things: a model class, and a quality measure for that model class. The model class specifies what type of interaction we are interested in. This can sometimes be fixed by a single word, such as 'correlation'; it can also be a more convoluted concept. The choice of model class may put restrictions on the number and type of target columns that are allowed: if one chooses the regression model class [2], one can accommodate as many targets as one wishes, but if one chooses the correlation model class [13], this fixes the number of targets $m = 2$ and demands both those targets to be numeric. Once a model class has been fixed, we need to define a quality measure (QM), which quantifies exactly what in the selected type of interaction we find interesting. For instance, in the correlation model class, maximizing $\rho$ as QM would find those subgroups featuring perfect positive target correlation, minimizing $|\rho$ would find those subgroups featuring uncorrelated targets, and maximizing $|\rho_S - \rho_{S^C}|$) would find those subgroups $S$ for which the target correlation deviates from the target correlation on the subgroup complement $S^C$.

## 4.2    Instantiating the Framework: The Association Model Class

As alluded to in Sect. 3.3, the StudyPortals dataset comes naturally equipped with $m = 2$ nominal targets: $t_1$ is the binary column representing whether the page visitor merely viewed or also clicked, and $t_2$ is the binary column representing whether the visitor was presented with web page version A or B. Therefore, the natural choice of EMM instance would be the association model class [3, Sect. 5.2]. Essentially, this is the nominal-target equivalent of the correlation model class [13, Sect. 3.1]: we strive to find subgroups for which the association between view/click and A/B is exceptional.

**Table 1.** Target cross table

|     | View  | Click |
|-----|-------|-------|
| A   | $n_1$ | $n_2$ |
| B   | $n_3$ | $n_4$ |

## 4.3   Instantiating the Framework: Yule's Quality Measure

Having fixed the model class, we need to define an appropriate quality measure. As has been observed repeatedly [3, 13, 19], one can easily achieve huge deviations in target behavior for very small subgroups. To ensure the discovery of subgroups that represent substantial effects within the datasets, a common approach is to craft a quality measure by multiplying two components: one reflecting target deviation, and one reflecting subgroup size.

**The Target Deviation Component.** For the quality measure component representing the target deviation, we build on the cells of the target contingency table, depicted in Table 1. Given a subgroup $S \subseteq \Omega$, we can assign each record in $S$ to the appropriate cell of this contingency table, which leads to count values for each of the $n_i$ such that $n_1 + n_2 + n_3 + n_4 = |S|$. From such an instantiated contingency table, we can compute Yule's Q [1], which is a special case of Goodman and Kruskal's Gamma for $2 \times 2$ tables. Yule's Q is defined as $Q = (n_1 \cdot n_4 - n_2 \cdot n_3)/(n_1 \cdot n_4 + n_2 \cdot n_3)$. A positive value for $Q$ implies a positive association between the two targets, i.e. high values on the diagonal of the contingency table and low values on the antidiagonal. Hence, a positive value for $Q$ indicates that people presented with web page variant $B$ click the button more often than people presented with web page variant $A$. We denote by $Q_S$ the value for $Q$ instantiated by the subgroup $S$.

Analogous to the component developed for Pearson's $\rho$ in the correlation model class [13, Sect. 3.1], we contrast Yule's Q instantiated by a subgroup with Yule's Q instantiated by that subgroup's complement: $\varphi_Q(S) = |Q_S - Q_{S^C}|$. Hence, this component detects schisms in target interaction: subgroups whose view/click-A/B association is markedly different from the rest of the dataset.

**The Subgroup Size Component.** To represent subgroup size, we take the entropy function $\varphi_{ef}$ as described in [13, Sect. 3.1] (denoted $H(p)$ there). The components rewards 50/50 splits between subgroup and complement, while punishing subgroups that either are tiny or cover the vast majority of the dataset.

**Combining the Components: Yule's Quality Measure.** Combining the components into an association model class quality measure is straightforward:

$$\varphi_{\text{Yule}}(S) = \varphi_Q(S) \cdot \varphi_{ef}(S)$$

Multiplication is chosen to ensure subgroups score well on both components.

## 5 Experiments

On the entire dataset, Yule's Q has a value of $\varphi_Q(\Omega) = -0.031$. Hence, the results of the traditional A/B test would be a resounding victory for variant A: the less buttony control version of Figure 1a generates more clicks than the more buttony variation of Fig. 1b. Whether the difference is significant is another question, but the new variation is clearly not significantly better than the already-in-place control version. In traditional A/B testing, that would be the end of the analysis: the new variant B does not outperform the current variant A, so we keep variant A and discard variant B. The main contribution of this paper is that with EMM, we can draw more sophisticated conclusions.

### 5.1 Experimental Setup

For empirical evaluation, we select the beam search algorithm for EMM whose pseudocode is given in [3, Algorithm 1], parametrized with $w = 10$ and $d = 2$. We have also trialed more generous values for the beam width $w$, which did not affect the results much. The search depth $d$ is deliberately kept modest: this parameter controls the number of conjuncts allowed in a subgroup description, hence modest settings guarantee good subgroup interpretability.

The beam search algorithm, the association model class, and Yule's quality measure have been implemented in Python as part of a Bachelor's project in a course on Web Analytics. The code will be made available upon request. In the following section, we report the top-five subgroups found with the thusly parametrized and implemented EMM algorithm.

### 5.2 Found Subgroups

The top-five subgroups found are presented in Table 2, in order of descending quality. Subgroup definitions are provided along with the values for the compound quality measure $\varphi_{\text{Yule}}$, the value of the Yule's Q component on both the

**Table 2.** Top-five subgroups found with the association model class for Exceptional Model Mining. The subgroup definitions are listed along with their values for Yule's quality measure, the within-subgroup value for Yule's Q, the outside-subgroup value for Yule's Q, and the subgroup size.

| Subgroup definition | $\varphi_{\text{Yule}}(S)$ | $\varphi_Q(S)$ | $\varphi_Q(S^C)$ | $|S|$ |
|---|---|---|---|---|
| Browser_lang = EN-GB | 0.1540 | 0.1287 | −0.1172 | 979 |
| Browser_lang = EN-GB ∧ Viewheight = small | 0.1300 | 0.2852 | −0.0722 | 363 |
| Browser_lang = TR | 0.0859 | −1.0000 | −0.0164 | 53 |
| Browser_lang = EN-GB ∧ OS_name = iOS | 0.0797 | 0.2661 | −0.0599 | 204 |
| Country = NG | 0.0783 | 0.2000 | −0.0554 | 281 |

subgroup and its complement, and the subgroup size. Recall that the total number of records in the dataset is 3 065, and the value for Yule's Q on the whole dataset is $\varphi_Q(\Omega) = -0.031$.

The best subgroup found, $S_1$, is defined by people having British English set as their browser language. More extreme values for Yule's Q itself can be found elsewhere in the table; $S_1$ has other distinctive qualities. What sets it apart, is that there is a clear dichotomy in Q-values between subgroup and complement: the Q-value on $S_1$ is substantially (though not spectacularly) elevated from the behavior on the whole dataset, and *at the same time*, the Q-value on the $S_1^C$ is substantially *depressed* from the behavior on the whole dataset. This means that people using British English as their browser language generate markedly more revenue when presented with version B of the web page, whereas people using any other browser language generate markedly more revenue when presented with version A of the web page. Moreover, $S_1$ has a substantial size. These two factors make $S_1$ the subgroup for which business action is most apposite: we have clearly distinctive behavior between two sizeable groups of website visitors, and presenting each group with the version of the web page appropriate for that group stands to substantially increase overall revenue.

The second- ($S_2$) and fourth-ranked ($S_4$) subgroups are specializations of $S_1$. $S_2$ specifies visitors that view the website using a relatively small mobile browser screen; they strongly prefer version B. Small screens can be found in relatively old smartphones, so this population contains people that are relatively slow in adopting new technology. It stands to reason that this population would also prefer a more traditionally-shaped button. $S_4$ specifies visitors that run the iOS operating system. They too strongly prefer version B, which is remarkable, since the buttons of version B do not conform to Apple's design standards. Perhaps the unusual button design draws more attention.

The third-ranked subgroup are those people that have set their browser language to Turkish. This subgroup may be too small to deliver actionable results, covering less than 2% of the dataset. However, the Q-value measured on this subgroup is strong: this subgroup displays a crystal clear preference for version A. This is a marked departure from the previously presented subgroups.

The final subgroup presented in Table 2, ranked fifth, concerns people from Nigeria. Yule's Q indicates that these people prefer version B. Given that the official language of Nigeria is English, the version preference is unsurprising: this subgroup overlaps substantially with $S_1$.

## 6   Conclusions

Having performed an A/B test—where a pool of test subjects are randomly presented with either version A or version B of the same product, a measure of success is aggregated by version, and the experimenter is presented with the results—the typical subsequent action is to make a crisp decision to either maintain the control version A, or replace it with the new variation version B, while the losing alternative is discarded. In this paper, we argue that that action can be overly coarse. Instead, we present an alternative approach: A&B testing.

The procedure of the A&B test is the exact same as that of a traditional A/B test, but the subsequent action is much more sophisticated. We analyze the results of the traditional A/B test with Exceptional Model Mining, to find coherent subgroups of the overall population that display an unusual response to the A/B test: the resulting subgroups feature an unusual association between the A/B decision and the measure of success at hand. Hence, while the general population might generate more revenue when presented with the one version, the resulting subgroups might generate more revenue when presented with the other version. If the company performing the A/B test can afford the upkeep of both versions, then knowledge of these subgroups can be invaluable.

As proof of concept, we roll out the A&B test on data generated by Study-Portals, an online information platform for higher education. From the results of the A/B test (cf. Fig. 1), we derive several subgroups displaying unusual behavior (cf. Table 2). The largest schism lies between people using British English as browser language ($\sim 1/3$ of the population, preferring version B), and people using any other browser language ($\sim 2/3$ of the population, preferring version A). In other words, the results suggest that British prefer buttony buttons.

A natural next step would be to verify empirically whether identified subgroups lead to effective personalization serving either A or B version to corresponding web portal visitors. Since it is common for StudyPortals and other companies to run a number of A/B testing experiments, and there is a motivation to provide personalized content and personalized layout, it is interesting to develop a framework for automation of website personalization based on findings of EMM. It would also make sense to extend this paper by refining the employed quality measure, incorporating the economics of the underlying decision problem directly [8].

While the main application within this paper lies in the context of web analytics, it is important to notice that the methodology of A&B testing is applicable on any controlled experiment. Hence, A&B testing is relevant in diverse fields such as medical research [5], education [24], etcetera. In future work, we plan to roll out A&B testing in clinical trials near you.

# References

1. Adeyemi, O.: Measures of association for research in educational planning and administration. Res. J. Math. Stat. **3**(3), 82–90 (2010)
2. Duivesteijn, W., Feelders, A., Knobbe, A.J.: Different slopes for different folks – mining for exceptional regression models with cook's distance. In: Proceedings of KDD, pp. 868–876 (2012)
3. Duivesteijn, W., Feelders, A.J., Knobbe, A.: Exceptional model mining – supervised descriptive local pattern mining with complex target concepts. Data Min. Knowl. Disc. **30**(1), 47–98 (2016)
4. Hand, D.J.: Pattern detection and discovery. In: Hand, D.J., Adams, N.M., Bolton, R.J. (eds.) Pattern Detection and Discovery. LNCS (LNAI), vol. 2447, pp. 1–12. Springer, Heidelberg (2002). https://doi.org/10.1007/3-540-45728-3_1

5.  Jakowski, M., Jaroszewicz, S.: Uplift modeling for clinical trial data. In: Proceedings of ICML 2012 Workshop on Machine Learning for Clinical Data Analysis (2012)

6.  Kohavi, R., Longbotham, R.: Online controlled experiments and A/B tests. In: Sammut, C., Webb, G.I. (eds.) Encyclopedia of Machine Learning and Data Mining, pp. 1–8. Springer, New York (2016). https://doi.org/10.1007/978-1-4899-7502-7_891-1

7.  Kohavi, R., Longbotham, R., Sommerfield, D., Henne, R.M.: Controlled experiments on the web: survey and practical guide. Data Min. Knowl. Discov. **18**(1), 140–181 (2009)

8.  Kleinberg, J., Papadimitrou, C., Raghavan, P.: A microeconomic view of data mining. Data Min. Knowl. Disc. **2**(4), 311–324 (1998)

9.  Klösgen, W.: Explora: a multipattern and multistrategy discovery assistant. In: Advances in Knowledge Discovery and Data Mining, pp. 249–271 (1996)

10.  Krak, T.E., Feelders, A.: Exceptional model mining with tree-constrained gradient ascent. In: Proceedings of SDM, pp. 487–495 (2015)

11.  Lavrač, N., Kavšek, B., Flach, P.A., Todorovski, L.: Subgroup discovery with CN2-SD. J. Mach. Learn. Res. **5**, 153–188 (2004)

12.  van Leeuwen, M.: Maximal exceptions with minimal descriptions. Data Min. Knowl. Discov. **21**(2), 259–276 (2010)

13.  Leman, D., Feelders, A., Knobbe, A.: Exceptional model mining. In: Daelemans, W., Goethals, B., Morik, K. (eds.) ECML PKDD 2008. LNCS (LNAI), vol. 5212, pp. 1–16. Springer, Heidelberg (2008). https://doi.org/10.1007/978-3-540-87481-2_1

14.  Lemmerich, F., Becker, M., Atzmueller, M.: Generic pattern trees for exhaustive exceptional model mining. In: Flach, P.A., De Bie, T., Cristianini, N. (eds.) ECML PKDD 2012. LNCS (LNAI), vol. 7524, pp. 277–292. Springer, Heidelberg (2012). https://doi.org/10.1007/978-3-642-33486-3_18

15.  Mannila, H., Toivonen, H.: Levelwise search and borders of theories in knowledge discovery. Data Min. Knowl. Discov. **1**(3), 241–258 (1997)

16.  Moens, S., Boley, M.: Instant exceptional model mining using weighted controlled pattern sampling. In: Blockeel, H., van Leeuwen, M., Vinciotti, V. (eds.) IDA 2014. LNCS, vol. 8819, pp. 203–214. Springer, Cham (2014). https://doi.org/10.1007/978-3-319-12571-8_18

17.  Morik, K., Boulicaut, J.-F., Siebes, A. (eds.): Local Pattern Detection. Springer, Heidelberg (2005). https://doi.org/10.1007/b137601

18.  Rzepakowski, P., Jaroszewicz, S.: Decision trees for uplift modeling with single and multiple treatments. Knowl. Inf. Syst. **32**(2), 303–327 (2012)

19.  Rebelo de Sá, C., Duivesteijn, W., Soares, C., Knobbe, A.: Exceptional preferences mining. In: Calders, T., Ceci, M., Malerba, D. (eds.) DS 2016. LNCS (LNAI), vol. 9956, pp. 3–18. Springer, Cham (2016). https://doi.org/10.1007/978-3-319-46307-0_1

20.  Siroker, D., Koomen, P.: A/B Testing: The Most Powerful Way to Turn Clicks Into Customers. Wiley, Hoboken (2013)

21.  StudyPortals. www.studyportals.com

22.  Tang, L., Jiang, Y., Li, L., Li, T.: Ensemble contextual bandits for personalized recommendation. In: Proceedings of RecSys, pp. 73–80 (2014)

23.  Tang, L., Rosales, R., Singh, A.P., Agarwal, D.: Automatic ad format selection via contextual bandits. In: Proceedings of CIKM, pp. 1587–1594 (2013)

24. Williams, J.J., Li, N., Kim, J., Whitehill, J., Maldonado, S., Pechenizkiy, M., Chu, L., Heffernan, N.: MOOClets: A Framework for Improving Online Education through Experimental Comparison and Personalization of Modules. Working Paper No. 2523265 (2014). http://tiny.cc/moocletpdf
25. Wrobel, S.: An algorithm for multi-relational discovery of subgroups. In: Komorowski, J., Zytkow, J. (eds.) PKDD 1997. LNCS, vol. 1263, pp. 78–87. Springer, Heidelberg (1997). https://doi.org/10.1007/3-540-63223-9_108
26. Žliobaitė, I., Pechenizkiy, M.: Learning with actionable attributes: attention - boundary cases! In: Proceedings of ICDM Workshops, pp. 1021–1028 (2010)

# Koopman Spectral Kernels for Comparing Complex Dynamics: Application to Multiagent Sport Plays

Keisuke Fujii[1]([✉]), Yuki Inaba[2], and Yoshinobu Kawahara[1,3]

[1] Center for Advanced Intelligence Project, RIKEN, Osaka, Japan
keisuke.fujii.zh@riken.jp
[2] Japanese Institute of Sports Sciences, Tokyo, Japan
yuki.inaba@jpnsport.go.jp
[3] The Institute of Scientific and Industrial Research, Osaka University, Osaka, Japan
ykawahara@sanken.osaka-u.ac.jp

**Abstract.** Understanding the complex dynamics in the real-world such as in multi-agent behaviors is a challenge in numerous engineering and scientific fields. Spectral analysis using Koopman operators has been attracting attention as a way of obtaining a global modal description of a nonlinear dynamical system, without requiring explicit prior knowledge. However, when applying this to the comparison or classification of complex dynamics, it is necessary to incorporate the Koopman spectra of the dynamics into an appropriate metric. One way of implementing this is to design a kernel that reflects the dynamics via the spectra. In this paper, we introduced Koopman spectral kernels to compare the complex dynamics by generalizing the Binet-Cauchy kernel to nonlinear dynamical systems without specifying an underlying model. We applied this to strategic multiagent sport plays wherein the dynamics can be classified, e.g., by the success or failure of the shot. We mapped the latent dynamic characteristics of multiple attacker-defender distances to the feature space using our kernels and then evaluated the scorability of the play by using the features in different classification models.

## 1 Introduction

Groups of organisms competing and cooperating in nature are assumed to behave as complex and nonlinear dynamical systems, which currently elude formulation [7,9]. Understanding the complex dynamics of living organisms or artificial agents (and the component parts) is a challenging research area in biology [5], physics [7], and machine learning. In the field of physics, decomposition or spectral methods that factorize the dynamics into modes from the data are used such as proper orthogonal decomposition (POD) [1,25] or dynamic mode decomposition (DMD) [23,24]. The problem of learning dynamical systems in machine learning has been discussed such as in terms of Bayesian approaches [10] and

© Springer International Publishing AG 2017
Y. Altun et al. (Eds.): ECML PKDD 2017, Part III, LNAI 10536, pp. 127–139, 2017.
https://doi.org/10.1007/978-3-319-71273-4_11

predictive state representation [19]. This topic is closely related to the decomposition technique in physics, aiming to estimate a prediction model by examining the obtained modes.

In this paper, we consider the following discrete-time nonlinear dynamical system:

$$\boldsymbol{x}_{t+1} = \boldsymbol{f}\left(\boldsymbol{x}_t\right) \tag{1}$$

where $\boldsymbol{x}_i$ is a state vector on the state space $\mathcal{M}$ (i.e., $x \in \mathcal{M} \subset \mathbb{R}^d$) and $\boldsymbol{f}$ is a state transition function that assumes the dynamical system to be nonlinear. A recent development is the use of Koopman spectral analysis with reproducing kernels (called kernel DMD). This defines a mode that can yield direct information about the nonlinear latent dynamics [16]. However, to compare or classify these complex dynamics, it is necessary to incorporate their Koopman spectrum into a metric appropriate for representing the similarity between the nonlinear dynamical systems.

Several works have applied approximation with a low-dimensional linear subspace to represent this similarity [12,30,33]. One approach has used the Binet-Cauchy (Riemannian) distance with a variety of kernels on a Grassman manifold [12], such as the kernel principal angle [33], and the trace and determinant kernel [30], which were designed for application in face recognition [33] and movie clustering [30]. The algorithm essentially calculates the Binet-Cauchy distance between two subspaces in the feature space, defined by the product of the canonical correlations. However, the main applications assumed a linear dynamical model [12,30,33] and thus generalization to nonlinear dynamics without specifying an underlying model remains to be addressed. In this paper, we map the latent dynamics to the feature space using the kernels, allowing binary classification to be applied to real-world complex dynamical systems.

Organized human group tasks such as navigation [13] or ballgame teams [8] provide excellent examples of complex dynamics and pose challenges in machine learning because of their switching and overlapping hierarchical subsystems [8], characterized by recursive shared intentionality [28]. Measurement systems have been developed that capture information regarding the position of a player in a ballgame, allowing analysis of particular shots [11]; however, plays involving collaboration between several teammates have not yet been addressed. In games such as basketball or football, coaches analyze team formations and players repeatedly practice moves that increase the probability of scoring ("scorability"). However, the selection of tactics is an ill-posed problem, and thus basically requires the implicit experience-based knowledge of the coach. An algorithm is needed that clarifies scorable moves involving multiple players in the team.

Previous research has classified team moves on a global scale by directly applying machine learning methods derived mainly from natural language processing. These include recursive neural networks (RNN) using optical flow images of the trajectories of all players [31] or the application of latent Dirichlet allocation (LDA) to the arrangement of individual trajectories [22]. However, the contribution of team movement to the success of a play remains unclear.

Previously, we reported that three maximum attacker-defender distances separately explained scorability [8], but the study addressed only the outcome of a play, rather than its time evolution and the interactions that it comprised. An algorithm is required that uses mapping to feature space to discriminate between successful and unsuccessful moves while accounting for these complex factors. In this paper, we map the latent dynamic characteristics of multiple attacker-defender distances [8] to the feature space using our kernels acquired by kernel DMD and then evaluated scorability.

The rest of the paper is organized as follows. Section 2 briefly reviews the background of Koopman spectral kernels, while Sect. 3 discusses methods for computing them. We then extended this to empirical example of actual human locomotion in Sect. 4. For application to multiple sporting agents, Sect. 5 reports our findings using the data on actual basketball games. Our approach proved capable of capturing complex team moves. Finally, Sect. 6 presents our discussion and conclusions.

## 2 Background

### 2.1 Koopman Spectral Analysis and Dynamic Mode Decomposition

Spectral analysis (or decomposition) for analyzing dynamical systems is a popular approach aimed at extracting low-dimensional dynamics from the data. Common techniques include global eigenmodes for linearized dynamics, discrete Fourier transforms, and POD for nonlinear dynamics [25], as well as multiple variants of these techniques. DMD has recently attracted particular attention in areas of physics such as fluid mechanics [23] and several engineering fields [2,26] because of its ability to define a mode that can yield direct information even when applied to time series with nonlinear latent dynamics [23,24]. However, the original DMD has numerical disadvantages, related to the accuracy of the approximate expressions of the Koopman eigenfunctions derived from the data. A number of variants have been proposed to address this shortcoming, including exact DMD [29], optimized DMD [4], and baysian DMD [27]. Sparsity-promoting DMD [14] provides a framework for the approximation of the Koopman eigenfunctions with fewer bases. Extended DMD [32], which works on predetermined kernel basis functions, has also been proposed. These Koopman spectral analyses have been generalized to a reproducing kernel Hilbert space (RKHS) [16], an approach which is called *kernel DMD*.

In Koopman spectral analysis, the Koopman operator $\mathcal{K}$ [18] is an infinite dimensional linear operator acting on the scalar function $g_i : \mathcal{M} \to \mathbb{C}$. That is, it maps $g_i$ to the new function $\mathcal{K}g_i$ as follows:

$$(\mathcal{K}g_i)(\boldsymbol{x}) = (g_i \circ \boldsymbol{f})(\boldsymbol{x}), \tag{2}$$

where $\mathcal{K}$ denotes the composition of $g_i$ with $\boldsymbol{f}$. We can see that $\mathcal{K}$ acts linearly on the function $g_i$. The dynamics defined by $\boldsymbol{f}$ may be nonlinear. Since $\mathcal{K}$ is a linear operator, it can generally perform eigenvalue decomposition:

$$\mathcal{K}\varphi_j\left(\boldsymbol{x}\right) = \lambda_j\varphi_j\left(\boldsymbol{x}\right), \tag{3}$$

where $\lambda_j \in \mathbb{C}$ is the $j$th eigenvalue (called the *Koopman eigenvalue*) and $\varphi_j$ is the corresponding eigenfunction (called the *Koopman eigenfunction*). We denote the concatenation of $g_j$ to $\boldsymbol{g} := [g_1, \ldots, g_p]^{\mathrm{T}}$. If each $g_j$ lies within the space spanned by the eigenfunction $\varphi_j$, we can expand the vector-valued $\boldsymbol{g}$ in terms of these eigenfunctions as $\boldsymbol{g}(\boldsymbol{x}) = \sum_{j=1}^{\infty} \varphi_j(\boldsymbol{x})\boldsymbol{\psi}_j$, where $\boldsymbol{\psi}_j$ is a set of vector coefficients called *Koopman modes*. By iterative application of Eqs. (2) and (3), the following equation is obtained:

$$\left(\boldsymbol{g} \circ \boldsymbol{f}\right)\left(\boldsymbol{x}\right) = \sum_{j=1}^{\infty} \lambda_j \varphi_j\left(\boldsymbol{x}\right)\boldsymbol{\psi}_j. \tag{4}$$

Therefore, $\lambda_j$ characterizes the time evolution of the corresponding Koopman mode $\boldsymbol{\psi}_j$, i.e., the phase of $\lambda_j$ determines its frequency and the magnitude determines the growth rate of its dynamics.

DMD is a popular approach for estimating the approximations of $\lambda_j$ and $\boldsymbol{\psi}_j$ from a finite length observation data sequence $y_0, y_1, \ldots, y_\tau$ ($\in \mathbb{R}^p$), where $y_t := \boldsymbol{g}(\boldsymbol{x}_t)$. Let $\boldsymbol{A} = [y_0, y_1, \ldots, y_{\tau-1}]$ and $\boldsymbol{B} = [y_1, y_2, \ldots, y_\tau]$. Then, DMD basically approximates those by calculating the eigendecomposition of the least-squares solution to

$$\min_{P' \in \mathbb{R}^{p \times p}} (1/\tau) \sum_{t=0}^{\tau} \|y_{t+1} - \boldsymbol{P}'y_t\|^2, \tag{5}$$

i.e., $\boldsymbol{B}\boldsymbol{A}^\dagger (:= \boldsymbol{P})$ ($\bullet^\dagger$ is the pseudo-inverse of $\bullet$). Let the $j$-th right and left eigenvector of $\boldsymbol{P}$ be $\boldsymbol{\psi}_j$ and $\boldsymbol{\kappa}_j$, respectively, and assume that these are normalized so that $\boldsymbol{\kappa}_i^*\boldsymbol{\psi}_j = \delta_{ij}$ ($\delta_{ij}$ is the Kronecker's delta). Then, since any vector $\boldsymbol{b} \in \mathbb{C}^p$ can be written as $\boldsymbol{b} = \sum_{j=1}^{p} (\boldsymbol{\kappa}_i^*\boldsymbol{b})\boldsymbol{\psi}_j$, we have $\boldsymbol{g}(\boldsymbol{x}) = \sum_{j=1}^{p} \varphi_j(\boldsymbol{x})\boldsymbol{\psi}_j$ by applying it to $\boldsymbol{g}(\boldsymbol{x})$. Therefore, by applying $\mathcal{K}$ to both sides, we have

$$\left(\boldsymbol{g} \circ \boldsymbol{f}\right)\left(\boldsymbol{x}\right) = \sum_{j=1}^{p} \lambda_j\varphi_j\left(\boldsymbol{x}\right)\boldsymbol{\psi}_j, \tag{6}$$

indicating a modal representation corresponding to Eq. (4) for the finite sum.

## 2.2   Kernels for Comparing Nonlinear Dynamical Systems

Selection of an appropriate representation of the data is a fundamental issue in pattern recognition. The important point is to design the features (i.e., kernels) that reflect structure of the data. Time series data is challenging to design the features because of the difficulty in reflecting the data structure (including time length). Researchers have developed alternative kernel methods, including the use of graphs [15,17], subspaces [12,33] or trajectories [30]. In this paper, a kernel design applicable to dynamical systems was required. Several methods were proposed, based on the subspace angle with kernel methods such as for an auto-regressive moving average (ARMA) model [30]. These methodologies were previously reviewed [12], from the viewpoint of the Riemannian distance (or metric) on the Grassman manifold.

The Grassmann manifold $\mathcal{G}(m, D)$ is the set of $m$-dimensional linear subspaces of $\mathbb{R}^D$. Formally, the Riemannian distance between two subspaces is the geodesic distance on the Grassmann manifold. However, a more intuitive and computationally efficient way of defining the distances uses the principal angles [20]. A previous review [12] categorized the various Riemannian distances into the projection and Binet-Cauchy distance. The former has been used in applications such as face recognition [3,12], and the latter has been applied in video clustering [30] and face recognition [33], and has been generalized to (specific nonlinear) dynamical systems [30]. We then adopted the Binet-Cauchy distance when comparing complex systems.

The Binet-Cauchy distances were basically obtained with the product of canonical correlations using a variety of kernels [30]. However, the main applications assumed linear dynamical model [12,30,33] such as ARMA model. Thus, it is necessary to generalize to nonlinear dynamics without any specific underlying model, into which the Koopman spectrum of dynamics is incorporated. We called the kernels *Koopman spectral kernels*.

## 3   Design of Koopman Spectral Kernels

### 3.1   DMD with Reproducing Kernel

Conceptually, DMD can be considered as producing a local approximation of the Koopman eigenfunctions using a set of linear monomials of the observables as the basis functions. In practice, however, this is certainly not applicable to all systems (in particular, beyond the region of validity for local linearization). Then, DMD with reproducing kernels [16] approximates the Koopman eigenfunctions with richer basis functions.

Let $\mathcal{H}$ be the RKHS embedded with the dot product determined by a positive definite kernel $k$. Additionally, let $\phi : \mathcal{M} \to \mathcal{H}$ be a feature map, and an instance of $\phi$ with respect to $\boldsymbol{x}$ is denoted by $\phi_{\boldsymbol{x}}$ (i.e., $\phi_{\boldsymbol{x}} := \phi(\boldsymbol{x})$). Then, we define the Koopman operator $\mathcal{K}_{\mathcal{H}} : \mathcal{H} \to \mathcal{H}$ in the RKHS by

$$\mathcal{K}_{\mathcal{H}}\phi_{\boldsymbol{x}} = \phi_{\boldsymbol{x}} \circ \boldsymbol{f}. \tag{7}$$

Note that almost of the theoretical claims in this study do not necessarily require $\phi$ to be in the RKHS (it is sufficient to consider that $\phi$ stays within a Hilbert space), but this assumption should perform the calculation in practice.

In this paper, we robustify the kernel DMD by projecting data onto the direction of POD [4,16,29]. First, a centered Gram matrix is defined by $\bar{G} = \mathbf{H}G\mathbf{H}$, where $G$ is a Gram matrix, $\mathbf{H} = \mathbf{I} - 1\tau$, $\mathbf{I}$ is a unit matrix, and $1\tau$ is a $\tau$-by-$\tau$ matrix, for which each element takes the value $1/\tau$. The Gram matrix $G_{xx}$ of the kernel $k(\boldsymbol{y}_i, \boldsymbol{y}_j)$ is defined at $\boldsymbol{y}_i$ and $\boldsymbol{y}_j$ ($i$ and $j$ dimensions) of the observation data matrix $\boldsymbol{A}$. Similarly, the Gram matrix $G_{xy}$ of the kernel between $\boldsymbol{A}$ and $\boldsymbol{B}$ can be calculated. At this time, $G_{xx} = \mathcal{M}_\tau^* \mathcal{M}_\tau$ and $G_{xy} = \mathcal{M}_\tau^* \mathcal{M}_+$, where $\mathcal{M}_\tau^*$ indicates the Hermitian transpose of $\mathcal{M}_\tau$. Also, $\mathcal{M}_\tau := [\phi_{\boldsymbol{x}_0}, .., \phi_{\boldsymbol{x}_{\tau-1}}]$ and $\mathcal{M}_+ := [\phi_{\boldsymbol{x}_1}, .., \phi_{\boldsymbol{x}_\tau}]$, where $\phi_{\boldsymbol{x}_i}$ is considered as a feature map of $\boldsymbol{x}_i$ from the state space $\mathcal{M}$ to the RKHS $\mathcal{H}$.

Here, suppose that the eigenvalues and eigenvectors can be truncated based on eigenvalue magnitude. In other words, $\bar{G} \approx \bar{B}\bar{G}\bar{B}^*$ where $p$ ($\leq \tau$) eigenvalues are adopted. Then, a principal orthogonal direction in the feature space is given by

$$\nu_j = \mathcal{M}_\tau \mathbf{H} \bar{S}_{jj}^{-1/2} \beta_j, \tag{8}$$

where $\beta_j$ is the $j$th row of $\bar{B}$. Let $\mathcal{U} = [\nu_1, \ldots, \nu_j] = \mathcal{M}_\tau \mathbf{H} \bar{B} \bar{S}^{-1/2}$. Since $\mathcal{M}_+ = \mathcal{K}_\mathcal{H} \mathcal{M}_\tau$, the projection of $\mathcal{K}_\mathcal{H}$ onto the space spanned by $\nu_j$ is given as follows:

$$\hat{F} = \mathcal{U} \mathcal{K}_\mathcal{H} \mathcal{U} = \bar{S}^{-1/2} \bar{B}^* \mathbf{H} (\mathcal{M}_\tau \mathcal{M}_+) \mathbf{H} \bar{B} \bar{S}^{-1/2}. \tag{9}$$

Note that $G_{xy} = \mathcal{M}_\tau^* \mathcal{M}_+$. Then, if we let $\hat{F} = \hat{T}^{-1} \hat{\Lambda} \hat{T}$ be the eigendecomposition of $\hat{F}$, we obtain the centered DMD mode $\bar{\varphi}_j = \mathcal{U} b_j = \mathcal{M}_\tau \mathbf{H} \bar{B} \bar{S}^{-1/2} \mathbf{b}_j$, where $b_j$ is the $j$th row of $\hat{T}^{-1}$. The diagonal matrix $\hat{\Lambda}$ comprising the eigenvalues represents the temporal evolution of the mode.

## 3.2   Koopman Spectral Kernels

For calculating the similarity between the dynamical systems $DS_i$ and $DS_j$, we compute Koopman spectral kernels based on the idea of Binet-Cauchy kernels. The Binet-Cauchy kernels are basically calculated from the traces of compound matrices [30] defined as follows. Let $M$ be a matrix in $\mathbb{R}^{m \times n}$. For $q \leq min(m,n)$, define $I_q^n = \{\boldsymbol{i} = i_1, \cdots, i_q : 1 \leq i_1 < ... < i_q \leq n, i_i \in \mathbb{N}\}$, and likewise $I_q^m$. We denote by $C_q(M)$ the $q$th compound matrix, that is, the $\binom{m}{q} \times \binom{n}{q}$ matrix whose elements are the minors $det((M_{k,l})_{k \neq i, l \neq j})$, where $\boldsymbol{i} \in I_q^n$ and $\boldsymbol{j} \in I_q^m$ are assumed to be arranged in lexicographical order. In the unifying viewpoint [30], Binet-Cauchy kernels is a general representation including various kernels [6,15, 17,21], divided into two strategies. One is the trace kernel obtained by setting $q = 1$ (i.e., $C_1(M) = M$), which directly reflects the property of temporal evolution of the dynamical systems, including diffusion kernel [17] and graph kernel [15]. Second is the determinant kernel obtained by setting order $q$ to be equal to the order of the dynamical systems $n$ (i.e., $C_n(M) = det(M)$), which extracts coefficients of dynamical systems, including the Martin distance [21] and the distance based on the subspace angle [6].

We expand the kernels to applying Koopman spectral analysis, which are called the *Koopman trace kernel* and *Koopman determinant kernel*, respectively. Both kernels reflect the Koopman eigenvalue, the eigenfunction, and the mode (i.e., system trajectory including the initial condition). However, richer information of system trajectory does not necessarily increase expressiveness such as in classification with real-world data. Therefore, we also expanded the kernel of principal angle [33] to applying Koopman spectral analysis, which is called *Koopman kernel of principal angle*. The kernel principal angle is theoretically a simple case of the trace kernel [30], which is defined as the inner product of linear subspaces in this feature space. In this paper, for a simple comparison, we compute the kernel with the inner product of the Koopman modes (i.e. not the trajectory and independent of initial condition).

**Koopman Trace Kernel and Determinant Kernel.** First, for the trace kernel, we generalize the kernel assmuing the ARMA model [30], to nonlinear dynamical systems without specifying an underlying model. The trace kernel of $DS_i$ and $DS_j$ can be theoretically defined as follows:

$$k\left(DS_i, DS_j\right) := \sum_{t=0}^{\infty}\left(e^{-\kappa t}g_i\left(x_{i,t}\right)^{\mathrm{T}}Wg_j\left(x_{j,t}\right)\right), \tag{10}$$

where $g_i$ and $g_j$ is the observation function and $W$ is an arbitrary semidefinite matrix (here, $W = 1$). Moreover, for converging the above equation, we suppose the exponential discount $\mu(t) = e^{-\kappa t}(\kappa > 0)$. In this paper, noises in observation and latent dynamics are not considered. Koopman trace kernel can be computed using the modal representation given by the kernel DMD as follows:

$$k\left(DS_i, DS_j\right) = \varphi_i\left(x_{i,0}\right)^{\mathrm{T}}\sum_{t=0}^{\infty}\left(e^{-\kappa t}\Lambda_i^t\left(\Psi_i^{\mathrm{T}}W\Psi_j\right)\Lambda_j^t\right)\varphi_j\left(x_{j,0}\right), \tag{11}$$

where, $\Lambda_i$ is a diagonal matrix consisting of Koopman eigenvalues, $\Psi_i$ is the Koopman mode, and $\varphi_i$ is the Koopman eigenfunction (also for $j$). Although the equation includes an infinite sum, we can efficiently compute the matrix $M := \sum_{t=0}^{\infty}(e^{-\kappa t}\Lambda_i^t(\Psi_i^{\mathrm{T}}W\Psi_j)\Lambda_j^t)$ using the following Sylvester equation $M = e^{-\kappa}\Lambda_i^{\mathrm{T}}M\Lambda_j + \Psi_i^{\mathrm{T}}W\Psi_j$, where the Koopman mode $\Psi = \mathcal{U}^*H\mathcal{M}_\tau H\mathcal{U}\hat{T}^{-1}$ for $i$ and $j$. For creating a trace kernel independent of the initial conditions [30], we take expectation over $x_{i,0}$ and $x_{j,0}$ in the trace kernel, yielding

$$k\left(DS_i, DS_j\right) = tr\left(\Sigma_{\varphi_i(x_{i,0}),\varphi_j(x_{j,0})}M\right), \tag{12}$$

where the initial Koopman eigenvalue $\varphi(x_0) = a^*(\mathcal{M}_\tau H\mathcal{U})^*\mathcal{M}_{\tau,0}$ for $i$ and $j$ [16]. Here, $a$ is the left eigenvector of $\hat{F}$ and $\mathcal{M}_{\tau,0}$ is a vector indicating the first single column of $\mathcal{M}_\tau$. $\Sigma_{\varphi_i(x_{i,0}),\varphi_j(x_{j,0})} \in \mathbb{C}^{p\times p}$ is the covariance of all initial values $\varphi_n\left(x_0\right) \in \mathbb{C}^{p\times n}$ of $DS_i$ for each index 1, ... $p$ of eigenvalues ($p$ was fixed for all $i$). Similarly, the determinant kernel using the representation given by kernel DMD can be computed:

$$k\left(DS_i, DS_j\right) = det\left(\Psi_i M\Psi_j^{\mathrm{T}}\right), \tag{13}$$

where $M = e^{-\kappa}\Lambda_i^{\mathrm{T}}M\Lambda_j + \varphi_i(x_{i,0})\varphi_j(x_{j,0})^{\mathrm{T}}$. Determinant kernels independent of the initial condition can only be computed for a single output system [30].

**Koopman Kernel of Principal Angle.** The kernel of principal angle can be computed using the Koopman modes given by kernel DMD. With respect to $DS_i$, we define the kernel of principal angles as the inner product of the Koopman modes in the feature space: $A^*A = \hat{T}_i^{-1}\mathcal{U}_i^*\mathbf{H}G_{xxi}\mathbf{H}\mathcal{U}_i\hat{T}_i$. If the rank of $\hat{F}$ is $r_i$, $A^*A$ is a $r_i$-order square matrix. Also for $DS_j$, we create a similar matrix $B^*B$. Furthermore, we define the inner product of the linear subspaces between $DS_i$ and $DS_j$ as $A^*B = \hat{T}_i^{-1}\mathcal{U}_i^*\mathbf{H}G_{xxij}\mathbf{H}\mathcal{U}_j\hat{T}_j$. $G_{xxij}$ is a $n_i \times n_j$ matrix obtained by picking up the upper-right part of the centered Gram matrix obtained by connecting $A_i$ and $A_j$ in series ($n_i$ and $n_j$ are the lengths of the time series). Then, using these matrices, we solve the following generalized eigenvalue problem:

$$\begin{pmatrix} 0 & (A^*B)^* \\ A^*B & 0 \end{pmatrix} V = \lambda_{ij} \begin{pmatrix} B^*B & 0 \\ 0 & A^*A \end{pmatrix} V, \tag{14}$$

where the size of $\lambda_{ij}$ is finally adjusted to $r_{ij} = \min(r_i, r_j)$ in descending order, and $V$ is a generalized eigenvector. The eigenvalue $\lambda_{ij}$ is the kernel of principal angle.

## 4    Embedding and Classification of Dynamics

A direct but important application of this analysis is the embedding and classification of dynamics using extracted features. A set of Koopman spectra estimated from the analysis can be used as the basis for a low-dimensional subspace representing the dynamics. The classification of dynamics can be performed using feature vectors determined by the Koopman spectral kernels. We used the Gaussian kernel, with the kernel width set as the median of the distances from a data matrix.

Before applying our approach to multiagent sports data, an experiment was conducted using open-source real-world data. In this case, human locomotion data were taken from the CMU Graphics Lab Motion Capture Database (available at http://mocap.cs.cmu.edu). To verify the classification performance, we computed the trace kernel of an auto-regressive (AR) model, representing a conventional linear dynamical model. For embedding of the distance matrix with our kernels, components of the distance matrix between $DS_i$ and $DS_j$ in the feature space were obtained using $dist(DS_i, DS_j) = k(A_i, A_i) + k(A_j, A_j) - 2k(A_i, A_j)$. Figure 1a–c shows the embedding of the sequences using multidimensional scaling (MDS) with the distance matrix, computed with the Koopman kernel principal angle, Koopman determinant kernel, and trace kernel of the AR model, respectively. Classification of performances into jumping, running, and walking was computed using the k-nearest neighbor algorithm. Error rates of the test data were small in this order: the Koopman kernel of principal angle (0.261), Koopman determinant kernel (0.348), trace kernel of the AR model (0.522), and Koopman trace kernel (0.601). Two Koopman spectral kernels performed better in classification than the kernel of the linear dynamical model.

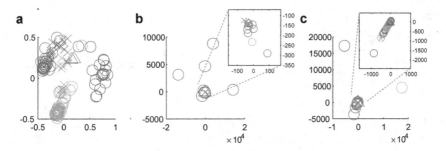

**Fig. 1.** MDS embedding of (a) Koopman kernel of principal angle, (b) Koopman determinant kernel, and (c) trace kernel of AR model. Blue, red, and green indicate jump, run, and walk, respectively (x and triangle show the movements with turn and stop, respectively). (Color figure online)

## 5  Application to Multiagent Sport Plays

We used player-tracking data from two international basketball games in 2015 collected by the STATS SportVU system. The total playing time was 80 min, and the total score of the two teams was 276. Positional data comprised the xy position of every player and the ball on the court, recorded at 25 frames per second. We eliminated transitions in attack to automatically extract the time periods to be analyzed (called an *attack-segment*). We defined an attack-segment as the period from all players on the attacking side court entry to 1 s before a shot was made. We analyzed a total of 192 attack-segments, 77 of which ended in a successful shot.

Next, we calculated effective attacker-defender distances to predict the success or failure of the shot (details were given by [8]), which were temporally and spatially corrected (Fig. 2a). Although all of the distances were 25 dimensions (five attackers and defenders), we previously reduced to four dimensions [8]: (1) ball-mark distance, (2) ball-help distance, (3) pass-mark distance, and (4) pass-help distance (Fig. 2b–c). These distances were used to create seven input vector series: (i) a one-dimensional distance (1), (ii) a two-dimensional distance comprising (1) and (2), and (iii–iv) three- and four-dimensional (1–3, 1–4) important distances, respectively. For verification, (v) total 25 distances and (vi) 25-dimensional Euclidean distances without spatiotemporal correction were calculated. We also used (vii) the xy position (total 20 dimensions) of all the ten players.

When predicting the outcome of a team-attack movement, it is preferable to compute the posterior probability rather than the outcome identification of the shot accuracy itself. We used a naive Bayes classifier and a related vector machine (RVM) for classification. Figure 3a shows the result of applying the naive Bayes classifier. The horizontal axis shows the seven input vector series and the vertical axis the classification error. The Koopman kernel principal angles derived by inputting four important distances demonstrated minimum error of 35.9%. The result of applying the RVM is shown in Fig. 3b, using the same axes.

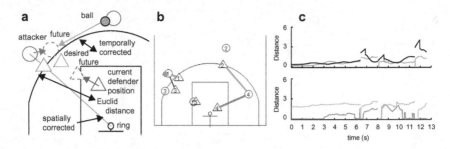

**Fig. 2.** Diagrams and examples of attacker-defender distance. (a) Diagram of attacker-defender distance with spatiotemporal correction. (b) Examples of four important distances. Orange, black, pink and light blue indicate the ball-mark, ball-help, pass-mark, and pass-help distance, respectively. (c) Example of time series in the same four important attacker-defender distances. (Color figure online)

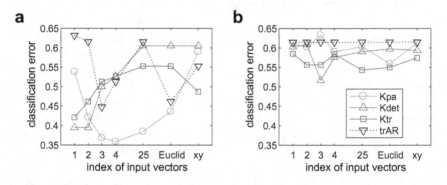

**Fig. 3.** Results from applying (a) the naive Bayes classifier and (b) the relevant vector machine. Kpa, Kdet, Ktr, and trAR are Koopman kernel of principal angle, Koopman determinant kernel and trace kernel, and trace kernel with AR model, respectively.

The performance of the naive Bayes classifier was superior to that of the RVM. In both cases, the Koopman spectral kernels produced better classification than the kernel of the linear dynamical model.

Figure 4a–c show embedding via MDS with the distance matrix of the Koopman kernel of principal angle countered by frequencies of success and failure of the shot. For example, the best case of the four important attacker-defender distances (Fig. 4a) showed the expressiveness in scorability due to wide distribution across the plot. In contrast, they were less widely distributed when only single distance (Fig. 4b) or the xy coordinates of all players (Fig. 4c) were used.

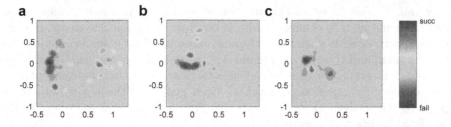

**Fig. 4.** MDS embedding of Koopman kernel of principal angle with three input vector series. The series consisted of (a) four important distances, (b) single important distance, and (c) xy coordinates of all players. Red and blue indicates success and failure of the shot, respectively. (Color figure online)

## 6   Discussion and Conclusion

The results of the two empirical examples showed that the best performances of the Koopman spectral kernels (Koopman determinant kernel and kernel of principal angle) are superior to that of the AR model assuming a linear dynamical model. Our proposed kernels can be computed in a closed form; but practically, the values of the Koopman determinant kernel were too large and the performance of the Koopman trace kernel was no better than that of the others. In contrast, the Koopman kernel of principal angle showed effective expressiveness only using Koopman modes.

When applied to multiagent sports data, the highest performance was provided by the classifier using the four important distances. This vector series reflects four characteristics: the scorability of a player in the current and future (i) shot, (ii) dribble, and (iii) pass, and (iv) the scorability of a dribbler after the pass. The proposed kernel reflected the time series of all interactions between players and was more effective for the classification than the kernel based on the information only on the shot itself. Well-trained teams aim to create scoring opportunities by continuously selecting tactical passes and dribbles or by improvising when no shooting opportunity is available.

However, even the best classification was not high (64.1% accuracy) when applied to real multiagent sports data. Two factors may have been neglected by our framework. The first is the existence of local interactions between players, such as local competitive and cooperative play by the attackers and defenders [8] when seen in higher spatial resolution than was available in this study. The approach needs to reflect the hierarchical characteristics of global dynamics and local dynamics. The second is the limitation of the input vector series to the attacker-defender distances. To achieve more accurate classifiers, not only the most important factor (i.e., distance) but also further hand-made time-series input vector series (e.g., Cartesian coordinates or specific movement parameters) should be used.

Overall, we developed Koopman spectral kernels that can be computed in closed form and used to compare multiple nonlinear dynamical systems. In competitive sports, coaches spend considerable amounts of time analyzing videos of their own team and the opposing team. Application of a system such as the one presented here may save time and create tactical plans that can currently be generated only by experienced coaches. More generally, the algorithm can be applied to the analysis of the complex dynamics of groups of living organisms or artificial agents, which currently elude formulation.

**Acknowledgements.** We would like to thank Charlie Rohlf and the STATS team for their help and support for this work. This work was supported by JSPS KAKENHI Grant Numbers 16H01548.

# References

1. Bonnet, J., Cole, D., Delville, J., Glauser, M., Ukeiley, L.: Stochastic estimation and proper orthogonal decomposition: complementary techniques for identifying structure. Exp. Fluids **17**(5), 307–314 (1994)
2. Brunton, B.W., Johnson, L.A., Ojemann, J.G., Kutz, J.N.: Extracting spatial-temporal coherent patterns in large-scale neural recordings using dynamic mode decomposition. J. Neurosci. Methods **258**, 1–15 (2016)
3. Chang, J.M., Beveridge, J.R., Draper, B.A., Kirby, M., Kley, H., Peterson, C.: Illumination face spaces are idiosyncratic. In: Proceedings of International Conference on Image Processing, Computer Vision, & Pattern Recognition, vol. 2, pp. 390–396 (2006)
4. Chen, K.K., Tu, J.H., Rowley, C.W.: Variants of dynamic mode decomposition: boundary condition, Koopman, and Fourier analyses. J. Nonlinear Sci. **22**(6), 887–915 (2012)
5. Couzin, I.D., Krause, J., Franks, N.R., Levin, S.A.: Effective leadership and decision-making in animal groups on the move. Nature **433**(7025), 513–516 (2005)
6. De Cock, K., De Moor, B.: Subspace angles between ARMA models. Syst. Control Lett. **46**(4), 265–270 (2002)
7. Fodor, E., Nardini, C., Cates, M.E., Tailleur, J., Visco, P., van Wijland, F.: How far from equilibrium is active matter? Phys. Rev. Lett. **117**(3), 038103 (2016)
8. Fujii, K., Yokoyama, K., Koyama, T., Rikukawa, A., Yamada, H., Yamamoto, Y.: Resilient help to switch and overlap hierarchical subsystems in a small human group. Scientific Reports 6 (2016)
9. Fujii, K., Isaka, T., Kouzaki, M., Yamamoto, Y.: Mutual and asynchronous anticipation and action in sports as globally competitive and locally coordinative dynamics. Scientific Reports 5 (2015)
10. Ghahramani, Z., Roweis, S.T.: Learning nonlinear dynamical systems using an EM algorithm. In: Advances in Neural Information Processing Systems, pp. 431–437 (1999)
11. Goldman, M., Rao, J.M.: Live by the three, die by the three? The price of risk in the NBA. In: Proceedings of MIT Sloan Sports Analytics Conference (2013)
12. Hamm, J., Lee, D.D.: Grassmann discriminant analysis: a unifying view on subspace-based learning. In: Proceedings of International Conference on Machine Learning, pp. 376–383 (2008)

13. Hutchins, E.: The technology of team navigation. In: Intellectual Teamwork: Social and Technological Foundations of Cooperative Work, vol. 1, pp. 191–220 (1990)
14. Jovanović, M.R., Schmid, P.J., Nichols, J.W.: Sparsity-promoting dynamic mode decomposition. Phys. Fluids **26**(2), 024103 (2014)
15. Kashima, H., Tsuda, K., Inokuchi, A.: Kernels for graphs. Kernel Methods Comput. Biol. **39**(1), 101–113 (2004)
16. Kawahara, Y.: Dynamic mode decomposition with reproducing kernels for Koopman spectral analysis. In: Proceedings of Advances in Neural Information Processing Systems, pp. 911–919 (2016)
17. Kondor, R.I., Lafferty, J.: Diffusion kernels on graphs and other discrete input spaces. In: Proceedings of International Conference on Machine Learning, vol. 2, pp. 315–322 (2002)
18. Koopman, B.O.: Hamiltonian systems and transformation in Hilbert space. Proc. Natl. Acad. Sci. **17**(5), 315–318 (1931)
19. Kulesza, A., Jiang, N., Singh, S.P.: Spectral learning of predictive state representations with insufficient statistics. In: Proceedings of Association for the Advancement of Artificial Intelligence, pp. 2715–2721 (2015)
20. Loan, C.V., Golub, G.: Matrix Computations, 3rd edn. Johns Hopkins University Press, Baltimore (1996)
21. Martin, R.J.: A metric for ARMA processes. IEEE Trans. Signal Process. **48**(4), 1164–1170 (2000)
22. Miller, A.C., Bornn, L.: Possession sketches: mapping NBA strategies (2017)
23. Rowley, C.W., Mezić, I., Bagheri, S., Schlatter, P., Henningson, D.S.: Spectral analysis of nonlinear flows. J. Fluid Mech. **641**, 115–127 (2009)
24. Schmid, P.J.: Dynamic mode decomposition of numerical and experimental data. J. Fluid Mech. **656**, 5–28 (2010)
25. Sirovich, L.: Turbulence and the dynamics of coherent structures. I. Coherent structures. Q. Appl. Math. **45**(3), 561–571 (1987)
26. Susuki, Y., Mezić, I.: Nonlinear koopman modes and power system stability assessment without models. IEEE Trans. Power Syst. **29**(2), 899–907 (2014)
27. Takeishi, N., Kawahara, Y., Tabei, Y., Yairi, T.: Bayesian dynamic mode decomposition. In: Proceedings of the International Joint Conference on Artificial Intelligence (2017)
28. Tomasello, M., Carpenter, M.: Shared intentionality. Dev. Sci. **10**(1), 121–125 (2007)
29. Tu, J.H., Rowley, C.W., Luchtenburg, D.M., Brunton, S.L., Kutz, J.N.: On dynamic mode decomposition: theory and applications. J. Comput. Dyn. **1**(2), 391–421 (2014)
30. Vishwanathan, S., Smola, A.J., Vidal, R.: Binet-Cauchy kernels on dynamical systems and its application to the analysis of dynamic scenes. Int. J. Comput. Vis. **73**(1), 95–119 (2007)
31. Wang, K.C., Zemel, R.: Classifying NBA offensive plays using neural networks. In: Proceedings of MIT Sloan Sports Analytics Conference (2016)
32. Williams, M.O., Kevrekidis, I.G., Rowley, C.W.: A data-driven approximation of the Koopman operator: extending dynamic mode decomposition. J. Nonlinear Sci. **25**(6), 1307–1346 (2015)
33. Wolf, L., Shashua, A.: Learning over sets using kernel principal angles. J. Mach. Learn. Res. **4**, 913–931 (2003)

# Modeling the Temporal Nature of Human Behavior for Demographics Prediction

Bjarke Felbo[1,3], Pål Sundsøy[2], Alex 'Sandy' Pentland[1], Sune Lehmann[3],
and Yves-Alexandre de Montjoye[1,4(✉)]

[1] MIT Media Lab, Massachusetts Institute of Technology, Cambridge, USA
[2] Telenor Research, Oslo, Norway
[3] DTU Compute, Technical University of Denmark, Kgs. Lyngby, Denmark
[4] Department of Computing and Data Science Institute,
Imperial College London, London, UK
demontjoye@imperial.ac.uk

**Abstract.** Mobile phone metadata is increasingly used for humanitarian purposes in developing countries as traditional data is scarce. Basic demographic information is however often absent from mobile phone datasets, limiting the operational impact of the datasets. For these reasons, there has been a growing interest in predicting demographic information from mobile phone metadata. Previous work focused on creating increasingly advanced features to be modeled with standard machine learning algorithms. We here instead model the raw mobile phone metadata directly using deep learning, exploiting the temporal nature of the patterns in the data. From high-level assumptions we design a data representation and convolutional network architecture for modeling patterns within a week. We then examine three strategies for aggregating patterns across weeks and show that our method reaches state-of-the-art accuracy on both age and gender prediction using only the temporal modality in mobile metadata. We finally validate our method on low activity users and evaluate the modeling assumptions.

**Keywords:** Call Detail Records · Mobile phone metadata
Temporal patterns · User modeling · Demographics prediction

## 1 Introduction

For the first time last year, there were more active mobile phones in the world than humans [17]. Every time one of these phones is being used to text or call, it generates mobile phone metadata or CDR (Call Detail Records). Collected at large scale this metadata – records of who calls or texts whom, for how long, and from where – provide a unique lens into the behavior of humans and societies. For instance, mobile phone metadata have been used to plan disaster response and inform public health policy [2,24]. The potential of mobile phone metadata is particularly high in developing countries where basic statistics such as population density or mobility are often either missing or suffer from severe biases [21].

© Springer International Publishing AG 2017
Y. Altun et al. (Eds.): ECML PKDD 2017, Part III, LNAI 10536, pp. 140–152, 2017.
https://doi.org/10.1007/978-3-319-71273-4_12

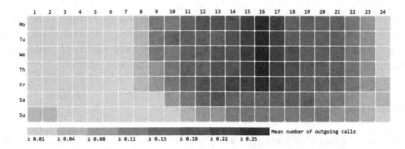

**Fig. 1.** The mean number of outgoing calls averaged across the population. Differences between workdays and weekends are clearly visible as well as different times of the day.

Last year, an expert advisory group to the United Nations emphasized the importance of mobile phone data in measuring and ultimately achieving the Sustainable Development Goals [23].

The potential of mobile phone data in developing countries has, however, been hindered by the absence of demographic information, such as age or gender, associated with the data. This issue has caused a growing interest in predicting demographic information from mobile phone metadata. While previous work has focused on developing increasingly complicated features, we here propose a novel way of modeling mobile phone metadata using deep learning. From high-level assumptions regarding the nature of temporal patterns, we design a data representation and convolutional network (ConvNet) architecture that reach state-of-the-art accuracy inferring both age and gender using only the temporal modality.

## 2   Related Work

Previous work has relied heavily on hand-engineered features to predict demographics and other information from mobile phone metadata. Sarraute et al. [19] and Herrera-Yagüe et al. [8] both combined hand-engineered features with various machine learning algorithms to predict gender from mobile phone metadata while de Montjoye et al. used them to predict personality traits [15]. Martinez et al. used an support vector machine (SVM) and random forest (RF) on similar features as well as a custom algorithm based on k-means to predict gender [6]. Finally, Dong et al. used a double-dependent factor graph model to predict demographic information in a mobile phone social graph [5]. While promising, the graph-based approach requires demographic information about a large fraction of the population to be known a priori, making it impractical in most countries where training data is not available at scale and must be collected through surveys.

The current state of the art in predicting demographics from mobile phone data is a recent paper by Jahani et al. [10] which relies on a large number of hand-engineered features (1440) provided by the open-source bandicoot toolbox [16] and a carefully tuned SVM with a radial basis function kernel. The features used are divided into two categories (individual, spatial) and based on carefully engineered definitions such as how to group together calls and text messages into conversations or compute the churn rate of common locations.

# 3    Data and Assumptions

A mobile phone produces a record every time it sends or receives a text message or makes or receives a phone call. These records (called mobile phone metadata, or CDRs) are generated by the carrier's infrastructure and are highly standardized. CDRs contain the type of interaction (text/call), direction (in/out), timestamp (date and time), recipient ID, call duration (if call) and cell tower to which the phone was connected to. The dataset we work with, provided by an anonymous carrier, contains more than 250 million anonymized mobile phone records for 150.000 people in a Western European country covering a period of 14 weeks.

We state the following three assumptions about the nature of the temporal patterns in mobile phone metadata:

1. **The day of the week and time of day of an observed pattern holds predictive power**
   Previous work showed that increasing the temporal granularity of the hand-engineered features in the bandicoot toolbox by differentiating between day-time and nighttime activity yields a substantial accuracy boost [10]. For instance, the percentage of initiated calls at night during the weekend was one of the most useful features to predict gender. Consequently, we assume that information on the specific time of the week that a pattern occurred contains useful information to predict demographic attributes.

2. **Temporal patterns are similar across days of the week**
   While the time of day matters (e.g. night vs. day), we furthermore assume that such temporal patterns have similarities across days of the week which could help predict demographic attributes. For instance, one could imagine that a relevant temporal pattern on Friday night may help model a similar pattern on Saturday night.

3. **Local temporal patterns can be combined into predictive global features**
   The current state-of-the-art approach relies on complex hand-engineered (and non-linear) features such as the response rate within conversations, churn between antennas, and entropy of contacts [10]. We assume that the convolutional network (ConvNet) will be able to combine local temporal patterns on the scale of hours to find global features (i.e. on the scale of days/weeks), thereby removing the need for such high-level hand-engineered features. ConvNets have similarly been used in previous work to learn a hierarchy of features directly from raw visual data [13].

# 4 Representation, Architecture and Aggregation

## 4.1 Week-Matrix Representation

Assumptions 1 and 2 from Sect. 3 are used to derive our data representation for a week of mobile phone metadata. We represent the data as eight matrices summarizing mobile phone usage on a given week with hours of the day on the x-axis and the weekdays on the y-axis (see Fig. 1). These eight matrices are the number of unique contacts, calls, texts and the total duration of calls for incoming and outgoing interactions respectively. Every cell in the matrices represents the amount of activity for a given variable of interest in that hour interval (e.g. between 2 and 3 pm). In this way, we effectively bin any number of interactions during the week. These eight matrices are combined into a 3-dimensional matrix with a separate 'channel' for each of the 8 variables of interest. This 3-dimensional matrix is named a 'week-matrix'.

The week-matrix representation is a logical result of our Assumptions 1 and 2. Our first assumption focuses on the importance of high temporal granularity, which is why our data representation summarizes mobile phone usage for each hour, thereby splitting local patterns into separate bins such that they may be captured by a suitable classification algorithm. Our second assumption focuses on the similarity of temporal patterns across weekdays, making it logical to design the week-matrix to have the weekdays on the y-axis such that similar patterns are located in neighboring cells in the matrix (see Fig. 1 for clear temporal patterns in mobile phone usage across weekdays). We shift the time in the matrices by 4 h such that it is easier to capture mobile phone usage occurring across midnight (Fig. 1 shows that there is especially a lot of activity occurring the night between Saturday and Sunday). Each row in the matrix thus contains data from 4 am–4 am instead of from midnight to midnight. This shift also moves the low-activity (and potentially less informative) areas to the borders of the matrix.

## 4.2 ConvNet Architecture

We use our assumptions (see Sect. 3) to develop the ConvNet architecture used to model a single week of mobile metadata. The choice of architecture is crucial to finding predictive patterns and has been equated to a choice of prior [1].

Assumption 2 emphasizes the similarity of temporal patterns across weekdays. We therefore design an architecture consisting of five horizontal conv. layers followed by a vertical conv. filter and a dense layer (see Table 1 and Fig. 2). The horizontal conv. layers learn to capture patterns within a single day, reusing the same parameters across different times of day and across the different weekdays. For a 1D conv. filter with filter size four (as illustrated in Fig. 2) the value of a single neuron at the position $k$ in the next layer is:

$$o_k = \sigma \left( b + \sum_{l=0}^{3} w_l i_{k+l} \right), \tag{1}$$

where $w_l$ is position $l$ in the weight matrix for that filter and $b$ is the bias [18]. The input is defined as $i_k$ for position $k$ in the previous layer. $\sigma$ is a non-linear activation function, which in this case is the leaky ReLU [14]. A single conv. layer consists of multiple filters with the specified size, allowing the conv. layer to capture many different patterns across the entire input using only a few parameters.

The intraday patterns captured by the horizontal conv. layers are then combined using the vertical conv. layer across the different weekdays to find global features. Lastly, the dense layer and the softmax layer combine these global features to predict the demographic attribute (see Fig. 2).

Assumptions 1 and 3 emphasizes the importance of capturing information about local temporal patterns. Consequently, we design an architecture that does not use pooling layers, which would throw away information about the location of the patterns in the week-matrix. Similarly, we make sure of a small conv. filter size for the first four conv. filter to focus on capturing local patterns.

There are many different parameters that can be tuned when choosing the architecture and the optimization procedure for training the ConvNet. Bayesian optimization is used for tuning seven of these as proposed in [20], covering e.g. the learning rate, L2 regularization, and the number of filters in the horizontal conv. layers. The vertical conv. layer has a fixed number of 400 filters. The dense layer has 400 neurons, whereas the softmax layer has as many neurons as the number of classes (two for gender and three for age).

### 4.3 Aggregation of Patterns Across Weeks

The ConvNet architecture described models only a single week of data at a time, whereas each user has multiple weeks of data that should all be utilized when predicting a demographic attribute. Based on our three assumptions (see Sect. 3) it makes sense to design the ConvNet architecture to model a single week at a time, making it possible to reuse the same convolutional filters across multiple weeks. There are several ways to aggregate the features captured by the ConvNet for individual weeks, making our method utilize the data for multiple weeks. We examine three different approaches: averaging the predictions, adding a long short-term memory (LSTM) module to the ConvNet and modeling the features captured by the ConvNet with an SVM.

The most basic approach for modeling multiple weeks of data is to pass each week-matrix through the ConvNet architecture and then average the probabilities from the softmax layer. In this way, an overall prediction can be found across all weeks of data for a given user. An issue with this averaging approach is that it limits the contribution of a given week to the final prediction.

Another way of modeling multiple weeks of data is by modifying the ConvNet architecture to include a long short-term memory (LSTM) module [9]. The LSTM is a specialized variant of the recurrent neural network (RNN), which uses recurrent connections between the neurons to capture patterns in sequences of inputs. We design a ConvNet-LSTM such that it has the same architecture for finding patterns as our ConvNet architecture, but without the final softmax layer for classification (i.e. $conv_1$–$dense_7$ as seen in Fig. 2). This architecture is then

**Table 1.** Architecture for the convolutional network. The filter size describes the number of neurons in the previous layer that each neuron in the current conv. layer is connected to. A filter with size M × 1 takes as input M neurons located side-by-side horizontally, whereas a 1 × N filter uses N neurons located side-by-side vertically.

| Layer name | Conv. filter size |
|------------|-------------------|
| *input*    | –                 |
| *conv*$_1$ | $4 \times 1$      |
| *conv*$_2$ | $4 \times 1$      |
| *conv*$_3$ | $4 \times 1$      |
| *conv*$_4$ | $4 \times 1$      |
| *conv*$_5$ | $12 \times 1$     |
| *conv*$_6$ | $1 \times 7$      |
| *dense*$_7$| –                 |
| *softmax*$_8$ | –              |

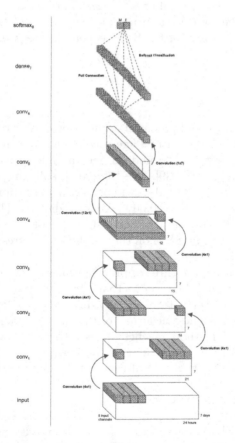

**Fig. 2.** Illustration of the convolutional network architecture. The depth of a conv. layer equals the number of filters in that layer. Dimensions are not to scale.

connected to a 2-layer LSTM module with 128 hidden units in each layer. In this way, the week-matrices can be modeled with an end-to-end architecture that can utilize convolutional layers to find patterns within a week and recurrent layers to find patterns across weeks. It is trained using the default settings of the Adam optimization method [12]. L2 regularization of $10^{-4}$ and recurrent dropout [7] of 0.5 is used to avoid overfitting. The ConvNet-LSTM is implemented using Keras [3] and Theano [22].

Lastly, we use an SVM with a radial basis function kernel to design a 2-step model (ConvNet-SVM). The ConvNet is used to transform the raw data into learned high-level features for each week with the SVM then modeling patterns across weeks. Using ConvNets to find good representations of raw data for modeling with SVMs has previously been done for generic visual recognition [4], but to our knowledge this is the first time it is done for combining patterns across individual observations in the dataset (i.e. weeks in this case). We extract the feature activations for $dense_7$ and $softmax_8$ (see Fig. 2. For each user we compute the mean and standard deviation for these extracted feature activations across the different weeks. A total of $800 + 2n_c$ features are extracted this way, where $n_c$ is the number of classes in the problem at hand (2 for gender, 3 for age). The number of features for the SVM is constant regardless of the number of weeks for a given user.

## 5    Results

In line with previous work and potential applications, we demonstrate the effectiveness of our method on gender and age prediction. We consider a binary gender variable (largest class: 56.3%) and an age variable discretized by the data provider into three groups: [18–39], [40–49], [50+], splitting the dataset almost equally (largest class: 35.7%). Our dataset contains data of approximately 150.000 people. We split it into training (100.000 people), validation (10.000 people), and test set (40.000 people). We compare our results to a state-of-the-art approach, Bandicoot-SVM [10], using an SVM on the bandicoot features trained and tested on the same data as our method.

We report results using the three approaches for aggregating patterns across weeks described in Sect. 4. Table 2 shows that our 2-step model (ConvNet-SVM), which extracts the high-level features found using the ConvNet and models them with an SVM yields the highest accuracy of the three approaches.

Our ConvNet-SVM method reaches state-of-the-art accuracy and slightly outperforms it on both age and gender prediction ($p < 10^{-5}$ with a one-tailed t-test). Our method reaches the state-of-the-art using only the temporal modality in mobile metadata, whereas the current state-of-the-art approach also exploits patterns related to mobility (see Sect. 7).

Mobile phone usage in developing countries is still fairly low [17] making it important for our method to perform well on low-activity users (see Fig. 4 for the distribution of interactions per user). To test the performance of our method, we train and evaluate it on low-activity users (users with fewer interactions than the median) and show that our model reaches state-of-the-art and even slightly

**Table 2.** Accuracy of classifiers on the test set when predicting age and gender.

|  | Age | Gender |
| --- | --- | --- |
| Random | 35.7% | 56.3% |
| Bandicoot-SVM | 61.6% | 78.2% |
| ConvNet (averaging) | 60.7% | 78.3% |
| ConvNet-LSTM | 61.3% | 78.4% |
| ConvNet-SVM | **63.1%** | **79.7%** |

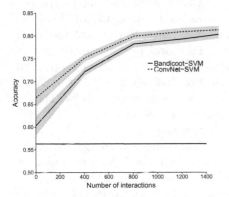

**Fig. 3.** Accuracy on gender prediction as a function of the number of interactions (across all 14 weeks) visualized using generalized additive model (GAM) smoothing. The x-axis is constrained to contain roughly 50% of the users. The black solid line is the baseline accuracy when predicting everyone as part of the majority class.

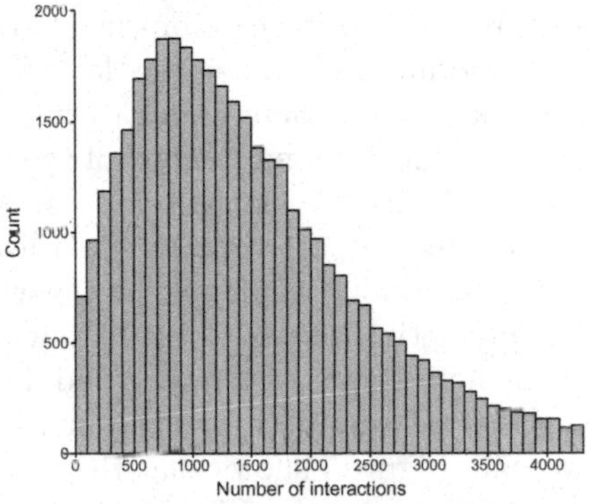

**Fig. 4.** Histogram of the distribution of the number of interactions. The top 5% users in terms of number of interactions are not included.

**Table 3.** Accuracy on the original and the temporally randomized week-matrices.

|  | Age | Gender |
| --- | --- | --- |
| Original | 60.7% | 78.3% |
| Permuted | 54.0% | 70.4% |
| Change | −11.0% | −10.1% |

outperforms it ($p < 0.01$ with a one-tailed t-test) with an accuracy of 76.9% vs. 75.7% for the Bandicoot-SVM. Figure 3 shows the accuracy of our method and the Bandicoot-SVM as a function of the number of interactions (calls + texts) when trained on all users showing that we perform particularly well on users with few interactions.

# 6    Evaluating Assumptions

Designing a ConvNet architecture for a particular modeling task involves many choices regarding filter sizes, layer types, etc. We derived many of our choices from the three assumptions stated in Sect. 3. In this section we evaluate these assumptions to qualify our choices.

**Evaluating Assumption 1:** The first assumption states that the weekday and time of day of an observed pattern holds predictive power. One way we can evaluate this assumption is by comparing the performance of a ConvNet on the original data with the performance of a ConvNet using the same hyperparameters and architecture but using data that has been temporally randomized. We temporally randomize the dataset by assigning values to cells at random in the week-matrix, thereby destroying potential temporal patterns in the week-matrices while keeping the rest of the information intact (total activity, etc.). To quantify the impact of the temporal randomization independently of the SVM, we evaluate the performance when averaging predictions across weeks. Table 3 shows temporally randomizing the week-matrices decreases accuracy by 11% when predicting age and by 10.1% when predicting gender.

The importance of the time and day of the interactions is indicated by examining the week-matrices which our model is most confident belong to a man or a woman. Figure 5 shows that the top "men" week-matrix has a higher number of outgoing contacts during the hours from 7 am to 4 pm on workdays while the top "female" week-matrix's outgoing contacts are spread across the day.

**Evaluating Assumption 2:** The second assumption states that temporal patterns are similar across weekdays. To evaluate our assumption, we examine the performance of ConvNet architectures on a 1-dimensional representation of the data. While this 1D representation contains the same information as the week-matrix, the hours of the weekdays are arranged next to each other horizontally instead of vertically ($168 \times 1$ instead of $24 \times 7$, see Fig. 1) therefore preventing the ConvNet to exploit similarity in patterns across days of the week. We test multiple ConvNet architectures (examples in Table 4) that have the same number of conv. layers as our ConvNet architecture and a comparable number of parameters and show that all of these architectures yield a lower accuracy than our ConvNet and the current state-of-the-art approach.

**Evaluating Assumption 3:** The third assumption states that local temporal patterns captured by convolutional filters (see Eq. 1) can be combined into predictive global features, thereby eliminating the need for hand-engineered features. To evaluate this assumption, we examine the global features learned with our deep learning method by comparing the patterns captured by the neurons of our ConvNet[1] with the bandicoot features. We only consider the individual bandicoot features as our ConvNet does not capture location and movement information used for the mobility features.

---

[1] For this comparison we use the mean activation of neurons in the $FC_7$ layer.

**Table 4.** Examples of 1-dimensional ConvNet architectures that we have tested. These contain convolutional, dense, max-pool and softmax layers as denoted by the prefix. The filter size is shown in the suffix. The mark $(s)$ means that the conv. layer has a stride of 2. Padding is used such that only pooling and a stride of 2 decreases the dimensions.

| ConvNet 1 | ConvNet 2 |
|-----------|-----------|
| *input* | |
| *conv*5 | *conv*13 |
| *conv*5 | *conv*13 |
| *pool*2 | *conv*13$(s)$ |
| *conv*5 | *conv*13 |
| *conv*5 | *conv*13 |
| *pool*2 | *conv*13$(s)$ |
| *conv*5 | |
| *conv*5 | |
| *dense* | |
| *softmax* | |

**Table 5.** Top 5 bandicoot features captured by the neurons.

| Features | $|r|$ |
|----------|-------|
| Interevent time (call) | 0.786 |
| Number of contacts (text) | 0.782 |
| Interevent time (text) | 0.769 |
| Entropy of contacts (call) | 0.764 |
| Number of interactions (text) | 0.761 |

Table 5 shows that the ConvNet captures information very similar to the one encoded in high-level hand-engineered features such as interevent time and entropy of contacts, suggesting that our deep learning model combines local temporal patterns into global features.

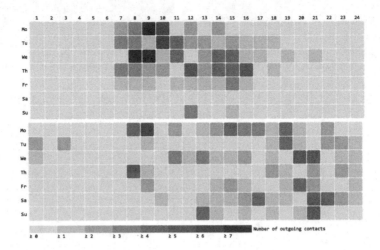

**Fig. 5.** Visualization of a single channel, the number of unique outgoing contacts, in the week-matrix most predictive of the male gender (top) and of female gender (bottom). The week-matrix most predictive of male gender has a higher number of outgoing contacts during the hours from 7 am to 4 pm on workdays while the "female" week-matrix's outgoing contacts are spread across the day.

## 7   Discussion

Our results (Table 2) show that the ConvNet-SVM outperforms the ConvNet-LSTM despite the ConvNet-SVM not capturing the ordering of the week-matrices. While an in-depth study is outside the scope of this paper, these results suggest that there are no strong inter-week patterns that are crucial for predicting demographic attributes.

The state-of-the-art approach found that two mobility features (percent interactions at home and entropy of antennas) were among the top 5 most predictive features for one of their two benchmark datasets [10]. In contrast, our ConvNet-SVM method reached state-of-the-art accuracy despite not using mobility information at all. In future work, we would like to use deep learning methods for modeling the other modalities in mobile phone metadata as well, thereby likely increasing the prediction accuracy.

Our weekmatrix representation have been added to bandicoot[2] and our trained ConvNets for Caffe [11] are available[3].

---

[2] Version $\geq$ 0.4 at http://bandicoot.mit.edu under `bc.special.punchcard`.
[3] https://github.com/yvesalexandre/convnet-metadata/.

# References

1. Bengio, Y., Courville, A., Vincent, P.: Representation learning: a review and new perspectives. TPAMI **35**(8), 1798–1828 (2013)
2. Bengtsson, L., Lu, X., Thorson, A., Garfield, R., Von Schreeb, J.: Improved response to disasters and outbreaks by tracking population movements with mobile phone network data: a post-earthquake geospatial study in Haiti. PLoS Med. **8**(8), e1001083 (2011)
3. Chollet, F.: keras (2015). https://github.com/fchollet/keras
4. Donahue, J., Jia, Y., Vinyals, O., Hoffman, J., Zhang, N., Tzeng, E., Darrell, T.: Decaf: a deep convolutional activation feature for generic visual recognition. In: PMLR. arXiv arXiv:1310.1531 (2013)
5. Dong, Y., Yang, Y., Tang, J., Yang, Y., Chawla, N.V.: Inferring user demographics and social strategies in mobile social networks. In: Proceedings of the 20th ACM SIGKDD International Conference on Knowledge Discovery and Data Mining, pp. 15–24. ACM (2014)
6. Frias-Martinez, V., Frias-Martinez, E., Oliver, N.: A gender-centric analysis of calling behavior in a developing economy using call detail records. In: AAAI Spring Symposium: Artificial Intelligence for Development (2010)
7. Gal, Y.: A theoretically grounded application of dropout in recurrent neural networks. In: NIPS. arXiv arXiv:1512.05287 (2016)
8. Herrera-Yagüe, C., Zufiria, P.J.: Prediction of telephone user attributes based on network neighborhood information. In: Perner, P. (ed.) MLDM 2012. LNCS (LNAI), vol. 7376, pp. 645–659. Springer, Heidelberg (2012). https://doi.org/10.1007/978-3-642-31537-4_50
9. Hochreiter, S., Schmidhuber, J.: Long short-term memory. Neural Comput. **9**(8), 1735–1780 (1997)
10. Jahani, E., Sundsøy, P., Bjelland, J., Bengtsson, L., de Montjoye, Y.A., et al.: Improving official statistics in emerging markets using machine learning and mobile phone data. EPJ Data Sci. **6**(1), 3 (2017)
11. Jia, Y., Shelhamer, E., Donahue, J., Karayev, S., Long, J., Girshick, R., Guadarrama, S., Darrell, T.: Caffe: Convolutional architecture for fast feature embedding. arXiv arXiv:1408.5093 (2014)
12. Kingma, D., Ba, J.: Adam: a method for stochastic optimization. In: ICLR. arXiv arXiv:1412.6980 (2015)
13. LeCun, Y., Bengio, Y., Hinton, G.: Deep learning. Nature **521**(7553), 436–444 (2015)
14. Maas, A.L., Hannun, A.Y., Ng, A.Y.: Rectifier nonlinearities improve neural network acoustic models. In: Proceedings of ICML (2013)
15. de Montjoye, Y.-A., Quoidbach, J., Robic, F., Pentland, A.S.: Predicting personality using novel mobile phone-based metrics. In: Greenberg, A.M., Kennedy, W.G., Bos, N.D. (eds.) SBP 2013. LNCS, vol. 7812, pp. 48–55. Springer, Heidelberg (2013). https://doi.org/10.1007/978-3-642-37210-0_6
16. de Montjoye, Y.A., Rocher, L., Pentland, A.S.: bandicoot: a Python toolbox for mobile phone metadata. J. Mach. Learn. Res. **17**(175), 1–5 (2016). http://jmlr.org/papers/v17/15-593.html
17. News, I.: Mobile subscriptions near the 7 billion mark - does almost everyone have a phone? (2013). Accessed 5 Jan 2016. http://itunews.itu.int/en/3741-Mobile-subscriptions-near-the-78209billion-markbrDoes-almost-everyone-have-a-phone.note.aspx

18. Nielsen, M.A.: Neural Networks and Deep Learning. Determination Press (2015)
19. Sarraute, C., Blanc, P., Burroni, J.: A study of age and gender seen through mobile phone usage patterns in Mexico. In: ASONAM (2014)
20. Snoek, J., Larochelle, H., Adams, R.P.: Practical Bayesian optimization of machine learning algorithms. In: NIPS (2012)
21. Stuart, E., Samman, E., Avis, W., Berliner, T.: The data revolution: finding the missing millions. Overseas Development Institute (2015)
22. Theano Development Team: Theano: a Python framework for fast computation of mathematical expressions. arXiv arXiv:1605.02688 (2016)
23. United Nations: A world that counts - mobilising the data revolution for sustainable development (2014). UN Independent Expert Advisory Group on a Data Revolution for Sustainable Development
24. Wesolowski, A., Qureshi, T., Boni, M.F., Sundsøy, P.R., Johansson, M.A., Rasheed, S.B., Engø-Monsen, K., Buckee, C.O.: Impact of human mobility on the emergence of dengue epidemics in Pakistan. PNAS **112**(38), 11887–11892 (2015)

# MRNet-Product2Vec: A Multi-task Recurrent Neural Network for Product Embeddings

Arijit Biswas$^{(\boxtimes)}$, Mukul Bhutani, and Subhajit Sanyal

Core Machine Learning, Amazon, Bangalore, India
{barijit,mbhutani,subhajs}@amazon.com

**Abstract.** E-commerce websites such as Amazon, Alibaba, Flipkart, and Walmart sell billions of products. Machine learning (ML) algorithms involving products are often used to improve the customer experience and increase revenue, e.g., product similarity, recommendation, and price estimation. The products are required to be represented as features before training an ML algorithm. In this paper, we propose an approach called MRNet-Product2Vec for creating generic embeddings of products within an e-commerce ecosystem. We learn a dense and low-dimensional embedding where a diverse set of signals related to a product are explicitly injected into its representation. We train a Discriminative Multi-task Bidirectional Recurrent Neural Network (RNN), where the input is a product title fed through a Bidirectional RNN and at the output, product labels corresponding to fifteen different tasks are predicted. The task set includes several intrinsic characteristics about a product such as price, weight, size, color, popularity, and material. We evaluate the proposed embedding quantitatively and qualitatively. We demonstrate that they are almost as good as sparse and extremely high-dimensional TF-IDF representation in spite of having less than 3% of the TF-IDF dimension. We also use a multimodal autoencoder for comparing products from different language-regions and show preliminary yet promising qualitative results.

## 1 Introduction

Large e-commerce companies such as Amazon, Alibaba, Flipkart, and Walmart sell billions of products through their websites. Data scientists across these companies try to solve hundreds of machine learning (ML) problems everyday that involve products, e.g., duplicate product detection, product recommendation, safety classification, and price estimation. The first step towards training any ML model usually involves creating feature representation of the relevant entities, i.e., products in this scenario. However, searching through hundreds of data resources within a company, identifying the relevant information, processing and transforming a product related data to a feature vector is a tedious and time consuming process. Furthermore, teams of data scientists performing such tasks on a regular basis makes the overall process inefficient and wasteful.

© Springer International Publishing AG 2017
Y. Altun et al. (Eds.): ECML PKDD 2017, Part III, LNAI 10536, pp. 153–165, 2017.
https://doi.org/10.1007/978-3-319-71273-4_13

For typical ML tasks such as classification, regression, and similarity retrieval, a product can be represented in several ways. One of the most common approaches to represent a product is using an order-independent bag-of-words approach leveraging the textual metadata associated with the product. In this approach, one constructs a TF-IDF vector representation based on the title, description, and bullet points of a product. Although these representations are effective as features in a wide variety of classification tasks, they are usually high-dimensional and sparse, e.g., a TF-IDF representation with only 300 K product titles and a minimum document frequency of 5 represents each product using more than 20 K dimensions, where typically only 0.05% of the features are non-zero. Using these high-dimensional features creates several problems in practice[1]: (a) overfitting, i.e., does not generalize to novel test data, (b) training ML algorithms using these high-dimensional features is usually computational and storage inefficient, (c) computing semantically meaningful nearest neighbors is not straightforward and (d) they cannot be directly used in downstream ML algorithms such as Deep Neural Networks (DNN) as that increases the number of parameters significantly. On the other hand, using dense and low-dimensional features could alleviate these issues. In this paper, our goal is to create a generic, low-dimensional and dense product representation which can work almost as effectively as the high-dimensional TF-IDF representation.

We propose a novel discriminative Multi-task Learning Framework where we inject different kinds of signals pertaining to a product into its embedding. These signals capture different static aspects, such as color, material, weight, size, subcategory, target-gender, and dynamic aspects such as price, popularity, and views of a product. Each signal is captured by formulating a classification/regression or decoding task depending on the type of the corresponding label. The proposed architecture contains a Bidirectional Recurrent Neural Network (RNN) with LSTM cells as the input layer which takes the sequence of words in a product title as input and creates a hidden representation, which we refer to as "product embeddings". During training phase, the embeddings are fed into multiple classification/regression/decoding units corresponding to the training tasks. The full multi-task network is trained jointly in an end-to-end manner. We refer to the proposed approach as MRNet (Discriminative Multi-task Recurrent Neural Network) and the embeddings created using this method are referred to as MRNet-Product2Vec. Section 3 elaborates more on this.

Products sold on e-commerce websites usually belong to multiple Product Groups (PG) such as furniture, jewelry, clothes, books, home, and sports items. Some of the signals which we inject within products are PG-specific. For example, the weights of home items have a very different distribution than the weights of jewelry. Similarly, sizes of clothes (L, XL, XXL etc.) could be quite different from the sizes of furniture (king, queen etc.). We believe that a common embedding for all products across all PGs will not be able to capture the intra-PG variations. Hence, we initially learn the embeddings in a PG-specific manner and then use a sparse autoencoder to project the PG-specific embeddings to a PG-agnostic space.

---

[1] Curse of dimensionality [5].

This ensures that MRNet-Product2Vec can also be used when the train or test data for an ML model belong to multiple PGs. Section 3.2 provides more details on this.

We encode different signals about products in the embeddings such that the embeddings are as generic as possible. However, creating embeddings that will work well for every product related ML task without further feature processing is not easy and perhaps impossible. So, we create these embeddings keeping two particular e-commerce use-cases in mind: (a) Anyone building an ML model with products can use these embeddings to build a good baseline model with little effort. (b) Someone who has a set of task specific features can use these embeddings as a means to augment with the generic signals captured in these representations. Our end-goal is to provide a generic feature representation for each product in an e-commerce system, such that data scientists don't have to spend days or months to build their first prototype.

We evaluate MRNet-Product2Vec in both quantitative and qualitative ways. MRNet-Product2Vec is applied to five different classification tasks: (i) plugs, (ii) Ship In Its Own Container (SIOC), (iii) browse category, (iv) ingestible and (v) SIOC with unseen population (Sect. 4.1). We compare MRNet-Product2Vec with a TF-IDF bag-of-words (sparse high-dimensional) on title words and show that in spite of having a much lower dimension than TF-IDF, MRNet-Product2Vec is comparable to TF-IDF representation. It performs almost as good as TF-IDF in two of these tasks, better than TF-IDF in two of these tasks and worse than TF-IDF in the remaining task. In Sects. 4.2 and 4.3, we provide the qualitative analysis of MRNet-Prod2Vec. In Sect. 5, we use a variant of multimodal autoencoder [8] that can be used to compare products sold in different language-regions/countries. Preliminary qualitative results using this approach are also provided.

## 2   Prior Work

There have been several prior works on entity embeddings using deep neural networks. Perhaps the most famous work on entity embeddings is the word2vec method [6], where continuous and distributed vector representations of words are learned based on their co-occurrences in a large text corpus. There are also a few prior research works for creating product embeddings for recommendation. Prod2Vec [2] uses a word2vec-like approach that learns vector representations of products from email receipt logs by using a notion of a purchase sequence as a "sentence" and products within the sequence as "words". The product representations are used for recommendation. The authors in [9] propose Meta-Prod2Vec, which extends the Prod2Vec [2] loss by including additional interaction terms involving products' meta-data. However, these embeddings are specifically fine-tuned for a predefined end-task, i.e., recommendation and may not perform well on a wide variety of product related ML tasks.

Traditionally multi-task learning has been used when one or all of the individual tasks have smaller training datasets and the tasks are somehow correlated [1].

The training data from other correlated tasks should improve the learning of a particular task. However, we do not have any paucity of data and the tasks which are used to train MRNet-Product2Vec are largely uncorrelated. We have used "unrelated" multi-task learning such that the learned representations are generic. To the best of our knowledge, this is the first work, that performs multi-task learning in an RNN to explicitly encode different kinds of static and dynamic signals for a generic entity embedding.

## 3   Proposed Approach

In this section, we describe the proposed embedding MRNet-Product2Vec. In MRNet-Product2Vec, we feed the vector representation of each word in a product title to a Bidirectional RNN. We use word2vec [6] to create a dense and compact representation of all words in the product catalog. A large corpus of text is created comprising the titles and descriptions of 143 million randomly selected products from the catalog. We use Gensim[2] to learn a 128 dimensional word2vec representation of all the words in the corpus which occur at least 10 times.

### 3.1   MRNet-Product2Vec

The proposed embeddings MRNet-Product2Vec are created by explicitly introducing different kinds of static and dynamic signals into the embeddings using a Discriminative Multi-task Bidirectional RNN. The goal of injecting different signals is to create embeddings which are as generic as possible. We believe that the learned embeddings will be effective in any ML task, which is correlated with one or more of the tasks for which we train our embeddings (see Sect. 4.3).

We describe fifteen different tasks which are used to learn our product embeddings. These tasks were selected primarily because we thought that the corresponding signals are intrinsic and should be included in a generic product embedding. However, these set of tasks are not exhaustive and may not capture all possible information about the products. Future research could incorporate more tasks during training and also study if dense product embeddings of a small fixed dimension (say, 128 or 256) can capture more signals effectively.

The set of present tasks can be grouped in several ways. Some of these capture static information that are unlikely to change over the lifetime of a product, e.g., size, weight, and material. Some are also dynamic which are likely to change every week or month, e.g., price or number of views. Some of these tasks are classification problems, where some others are regression or decoding. We summarize all the tasks in Table 1 and omit the details due to lack of space. Color, size, material, subcategory, and item type are formulated as multi-class classification problems where the most frequent ones are treated as individual classes and the remaining ones are grouped as one class. Rest of the classification tasks are binary. As mentioned earlier, this list of tasks may not be exhaustive. However, they capture a

---

[2] https://radimrehurek.com/gensim/.

wide variety of aspects regarding a product and effective encoding of these signals should create embeddings that are generic enough to address a wide class of ML problems pertaining to products.

**Table 1.** Tasks used to train MRNet-Product2Vec.

|                | Static                                                                                             | Dynamic            |
| -------------- | -------------------------------------------------------------------------------------------------- | ------------------ |
| Classification | Color, Size, Material, Subcategory, Item type, Hazardous, Batteries, High value, Target gender      | Offer, Review      |
| Regression     | Weight                                                                                             | Price, View Count  |
| Decoding       | TF-IDF representation (5000 dim.)                                                                   |                    |

The block diagram of MRNet Product2Vec is shown in Fig. 1a. The word2vec representation of each word in a product title is fed through a Bidirectional RNN layer containing LSTM cells. The hidden layer representation from the forward and backward RNNs are concatenated to create the product embedding which is used to predict multiple task labels as described above. The network is trained jointly with all of these tasks.

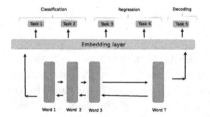 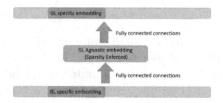

(a) MRNet-Product2Vec for PG-specific embeddings.  (b) Sparse Autoencoder for PG-agnostic embeddings.

**Fig. 1.** Architecture of different components of MRNet-Product2Vec.

Let us assume that the word2vec representation of words in a product title with T words are denoted as $\{\mathbf{x}_1, \mathbf{x}_2,..., \mathbf{x}_T\}$. We use a Bidirectional RNN, which has a forward RNN and a backward RNN. Let us assume that $\mathbf{h}_t^f$ and $\mathbf{h}_t^b$ denote the hidden states of the forward and backward RNN respectively at time $t$. The recursive equations for the forward and the backward RNN are given by:

$$\mathbf{h}_t^f = \phi(W^f\mathbf{x}_t + U^f\mathbf{h}_{t-1}^f) \tag{1}$$

$$\mathbf{h}_t^b = \phi(W^b\mathbf{x}_t + U^b\mathbf{h}_{t-1}^b) \tag{2}$$

Where $W^f$ and $W^b$ are the feedforward weight matrices for the forward and backward RNNs respectively. $U^f$ and $U^b$ are the recursive weight matrices for the forward and backward RNNs respectively. $\phi$ is usually a nonlinearity such as tanh or RELU. We use $\mathbf{h}_T = [\mathbf{h}_T^f, \mathbf{h}_T^b]$, as the final hidden representation of a product after all words in a product title have been fed through both the forward and backward RNNs. RNNs are trained using Backpropagation Through Time (BPTT) [10]. Although RNNs are designed to model sequential data, it has been found that simple RNNs are unable to model long sequences because of the vanishing gradient problem [3]. Long Short Term Memory units [4] are designed to tackle this issue where along with the standard recursive and feed-forward weight matrices there are input, forget, and output gates, which control the flow of information and can remember arbitrarily long sequences. In practice, it has been observed that RNNs with LSTM units are better than traditional RNNs (Eqs. 1 and 2). Hence, we use LSTM units in the forward and backward RNNs. We skip the details of LSTM units and suggest interested readers to look at this article[3] for an intuitive explanation on LSTMs.

Suppose, we want to train our network with N different tasks. Out of the N tasks, $N^c$ are classification, $N^r$ are regression and $N^d$ are decoding, i.e., $N^c + N^r + N^d = N$. $\mathbf{l}_m^c$ denotes the loss of the m-th classification task, $\mathbf{l}_p^r$ denotes the loss corresponding to the p-th regression task and $\mathbf{l}_q^d$ denotes the loss of the q-th decoding task. The losses corresponding to all the tasks are normalized such that one task with a higher loss cannot dominate the other tasks. While training we optimize the following loss which is the sum of the losses of all N tasks.

$$L = \sum_{m=1}^{m=N^c} \mathbf{l}_m^c + \sum_{p=1}^{p=N^r} \mathbf{l}_p^r + \sum_{q=1}^{q=N^d} \mathbf{l}_q^d \tag{3}$$

The hidden representation $\mathbf{h}_T$ corresponding to a product is projected to multiple output vectors ($\mathbf{o}_n$ for $n$-th task) using task specific weights and biases (Eq. 4). The loss is computed as a function of the output vector and the target vector according to the type of a task. For example, if the task is a five-way classification, $\mathbf{h}_T$ is projected to a five dimensional output, followed by a softmax layer and eventually a cross-entropy loss is computed between the softmax output and the true target labels. For the regression task, $\mathbf{h}_T$ is projected to a scalar and a squared loss is computed with respect to the true score. Similarly, in the decoding task, $\mathbf{h}_T$ is projected to a 5000 dimensional vector $\mathbf{o}_n$ and a squared loss is computed between the projected representation and the target 5000 dimensional TF-IDF representation.

$$\mathbf{o}_n = W_n \mathbf{h}_T + \mathbf{b}_n \tag{4}$$

**Optimization in Deep Multi-task Neural Network:** The cost function in Eq. (3) can be optimized in two different ways:

– **Joint Optimization:** At each iteration of training, update all weights of the network using gradients computed with respect to the total loss as defined in Eq. (3). However, if each training example does not have labels corresponding to all the tasks, training in this way may not be possible.

---

[3] http://colah.github.io/posts/2015-08-Understanding-LSTMs/.

- **Alternating Optimization:** At each iteration of training, randomly choose one of the tasks and optimize the network with respect to the loss corresponding to that task only. In this case, only the weights which correspond to that particular task and the weights of task-invariant layers (the Bidirectional RNN in our case) are updated. This style of training is useful when we do not have all task labels for a product. However, the training might be biased towards a specific task if the number of training examples corresponding to that task is significantly higher than other tasks.

While we were training MRNet, it was difficult to obtain all task labels for each product. Hence, alternating optimization was an obvious choice for us. We sample training batches from each task uniformly to avoid biasing towards any specific task.

## 3.2 Product Group (PG) Agnostic Embeddings

While training the proposed network, we trained a separate model for each PG because the label distribution could be very different across PGs. For example, the median price and the price range of jewelry is very different from that of books. Similarly, the materials used in clothes (cotton, polyester etc.) are different from that of kitchen items which are usually made of aluminium, metal or glass. If we train one model across all PGs, the embeddings are unlikely to capture the finer intra-PG variations in their representations. Hence, we build one model for each PG. In this paper, we train 23 different models for 23 different PGs.

The PG-specific embeddings can be used for any ML problem which is either PG-specific (all train and test data are from a particular PG) or when there are a large number of training examples from each PG such that separate PG-specific models can be trained. However, in many practical situations, none of the above might be true. Hence, it is also important to have product embeddings which are PG-agnostic such that ML models can be trained with products spanning multiple PGs. We handle this problem by training a sparse autoencoder [7] that projects the PG-specific embeddings to a PG-agnostic space (Fig. 1b).

Let us assume that each PG-specific embedding has a dimension $d$ and there are total $G$ (23 in our case) PG-specific embeddings. First, we represent the embeddings from PG $g$, using a $Gd$ dimensional vector, where the PG-specific embedding is placed at the index range $(g-1)d+1 : gd$ and the rest are filled with zeros. This vector is called $\mathbf{x}_{ga}$. We train a sparse autoencoder where we reconstruct $\mathbf{x}_{ga}$ through a fully connected network containing a hidden layer of dimension $2d$. The hidden layer representation is used as the PG-agnostic embedding. We enforce sparsity such that the autoencoder can learn interesting structures from the data and does not end up learning an identity function.

## 4 Experimental Results

In this section, we evaluate MRNet-Product2Vec in various ways. In Sect. 4.1, we discuss the quantitative results while qualitative studies are discussed in Sects. 4.2 and 4.3.

**Architecture and Framework Details:** In MRNet-Product2Vec, there is one layer of Bidirectional RNN containing LSTM nodes followed by multiple classification/regression/decoding units. We train each PG-specific model with maximum one million training samples from a PG corresponding to each training task. It took around 30 min to train each epoch using one Grid K520 GPU. While training the PG-agnostic embeddings, we used 500 K randomly chosen products from each PG (total 11.5 million for 23 PGs). The sparse autoencoder took around 20 min per epoch while training.

### 4.1 Quantitative Analysis

Product embeddings can be created in many possible ways. They can capture different kinds of signals about products and can have varying performance in different end-tasks. To get a sense of the efficacy of MRNet-Product2Vec, we consider five different classification tasks. These tasks are different from the tasks that are used to train MRNet-Product2Vec.

1. **Plugs:** In this binary classification problem, the goal is to predict if a product has an electrical plug or not. This dataset has 205,535 labeled products. Here we perform five-fold cross validation and report the average AUC.
2. **SIOC:** This classification problem tries to predict if a product can ship in its own container (SIOC) provided by the seller or if the e-commerce company needs to provide an additional container for this product. This is also a binary classification problem. This dataset has 296,961 labeled examples. Five-fold cross validation is performed and the average AUC over five-folds is reported.
3. **Browse Categories:** This is a multi-class classification problem where products from the PG toy are classified into 75 different website browse categories (e.g.: baby toys, puzzles, and outdoor toys). There are a total of 150,197 samples in this dataset. We perform five-fold cross validation and report the average accuracy.
4. **Ingestible Classification:** We apply MRNet-Product2Vec on a product classification problem which predicts if a product is ingestible or not. However, only 1500 training samples are available for learning a classifier. We perform five-fold cross validation and report the average AUC over five-folds.
5. **SIOC (unseen population):** We believe that if the test data distribution is significantly different from the train data distribution, dense embeddings such as MRNet-Product2Vec should perform better than sparse/high-dimensional TF-IDF representations. We simulate that by modifying the SIOC dataset using the following steps. First, we split the full dataset into fixed training and test parts. We filter out each test product for which the maximum intersection of it's title with any training product title is larger than a threshold $t_h$. All the remaining products are used as the test data set. The lower the threshold, the difference between the test and the train data distribution increases. We fixed a training dataset of 150 K examples and used $t_h = 0.2$ with 271 test examples (106 +ve and 165 −ve). We report the AUC.

We compare MRNet-Product2Vec with the sparse and high-dimensional TF-IDF representation of product titles for all the five classification tasks. The sparse TF-IDF representation for each classification task was created using only the training examples corresponding to that task. We use the PG-agnostic version of MRNet-Product2Vec (dimensionality: 256) for SIOC, PLUGs, ingestible and SIOC (unseen population) as the data spanned multiple PGs. We use the PG-specific version (dimensionality: 128) of MRNet-Product2Vec for browse category classification as all the samples were from the same PG, i.e., toy. We use MRNet-Product2Vec and TF-IDF in two different classifiers, Logistic Regression and Random Forest, and report the evaluation metric for both of them in Table 2. We observe that on Plugs and SIOC, MRNet-Product2Vec is almost as good as sparse and high-dimensional TF-IDF in spite of having a much lower dimension than that of TF-IDF. However, on Browse Categories, MRNet-Product2Vec performs much worse than TF-IDF. This happens because out of the 15 tasks, only the subcategory classification task is somewhat related to browse categories. Hence, the browse category related information that is encoded in MRNet-Product2Vec is not sufficient enough for this "hard" 75-class classification task. MRNet-Product2Vec performs better than sparse and high-dimensional TF-IDF on ingestible and SIOC (unseen population). Since the dense embeddings are semantically more meaningful, i.e., it knows that a chair and a sofa are similar objects, they should be able to learn classifiers even from smaller training datasets (such as ingestible) and generalize well for unseen test population (SIOC with unseen population). However, sparse and high-dimensional TF-IDF is not as effective in these scenarios. Overall, we observe that MRNet-Product2Vec is mostly comparable to TF-IDF in spite of having less than 3% of the TF-IDF dimension.

**Table 2.** Results on five classification tasks. RF: Random Forest, LR: Logistic Regression. TF-IDF dim.: >10K, MRNet-Product2Vec dim.: 256 and 128. All numbers are relative w.r.t TF-IDF-LR.

| Task | MRNet-RF | MRNet-LR | TF-IDF-RF |
|---|---|---|---|
| Plugs | −2.8% | −9.72% | −2.8% |
| SIOC | −5.81% | −18.60% | −9.3% |
| Browse categories | −16.67% | −26.38% | −25.0% |
| Ingestible | 0% | **+2.15%** | −11.8% |
| SIOC (unseen) | **+10%** | 0% | −3.33% |

## 4.2  Nearest Neighbor Analysis

We study the characteristics of MRNet-Product2Vec by analyzing the nearest neighbors (NN) of several products. Since computing meaningful NNs is not straightforward using the sparse TF-IDF features, we do not show the NNs using this method. We created a universe of 220 K products from the PG furniture and computed the NNs of several randomly chosen products based on the Euclidean distance. In Fig. 2, we show the first nine NNs of four hand-picked products. In (a), MRNet-Product2Vec finds several **grey** colored tables as NNs. In (b), several **full-sized** beds are obtained as NNs. In (c), MRNet-Product2Vec fetches four **blue-colored** tables and two "drum barrel" tables as NNs. In (d), MRNet-Product2Vec produces several **tools/tool-boxes** as NNs. Overall, we can see that MRNet-Product2Vec has learned several intrinsic characteristics of products such as size, color, type etc., which were used to train MRNet-Product2Vec.

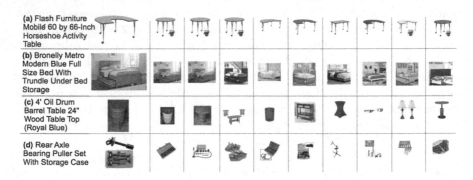

**Fig. 2.** Nearest neighbors computed using MRNet-Product2Vec for each query product (first column) (best viewed in electronic copy).

### 4.3  MRNet-Product2Vec Feature Interpretation

MRNet-Product2Vec is trained with multiple tasks to incorporate different product related signals. We performed some preliminary analysis on the PG-agnostic embeddings (256 dimensional) to detect if a subset of features encode a particular signal (such as size, weight, electrical properties etc.). First, MRNet-Product2Vec is used for the battery classification task (one of our training tasks) and multiple Random Forest (RF) models with randomly chosen subsets of the training data are built. We found out that there are 29 features that always appear in the top quartile (64) of all the features with respect to RF feature importance. That indicates that these 29 features in MRNet-Product2Vec are indicative of product's electrical properties and some of these should play a role in plugs classification (evaluation task). Indeed, we find that there are 28 features which are important in the context of plugs classification and 8 of these features were also important in the battery classification task. Likewise, we find that 13 important features for the weight classification training task are also important for the SIOC evaluation task. This demonstrates that MRNet-Product2Vec encodes different product characteristics which play a crucial role in the final evaluation tasks.

# 5   Language Agnostic MRNet-Product2Vec

E-commerce companies usually sell products across multiple countries and language-regions. There are many scenarios when it is important to compare products that have details such as title, description, and bullet-points in different languages. When a seller lists a product in a country, the e-commerce company would like to know if that product is already listed in other countries for accurate stock-accounting and price estimation. A customer from UK might like to know if a product which she liked in the France website is available for purchase from the UK website or not. Often it is also required to apply a trained ML model from a particular language-region to a different language-region product because labeled data in that language may not be available. For each of these use-cases, it is important to learn cross-language transformations, such that products from different countries/language-regions can be compared seamlessly. To address this issue, we propose to use a variant of multimodal autoencoder [8] that can project MRNet-Product2Vec trained in different languages to a common space for comparison.

Let us assume that we have $P$ paired product titles from two different countries UK and France, i.e., for each product title in French, we know the corresponding UK title and vice versa. We separately train MRNet-Product2Vec for UK-english and French and refer them as MRNet-Product2Vec-UK and MRNet-Product2Vec-FR respectively. MRNet-Product2Vec-UK and MRNet-Product2Vec-FR are used to obtain the embeddings for $P$ products. The corresponding embeddings for the p-th product are defined as $\mathbf{x}_p^{UK}$ (dim. 256) and $\mathbf{x}_p^{FR}$ (dim. 256) respectively. Now, we train an autoencoder (Fig. 3a) which has input $\mathbf{x}_p$ (dim. 512), output $\mathbf{y}_p$ (dim. 512) and a hidden layer of dimension 256. Let us assume that $\mathbf{0}$ denotes a zero vector of 256 dimension. The training data for this network consists of three parts: (1) $\mathbf{x}_p = [\mathbf{x}_p^{UK}, \mathbf{0}]$ and $\mathbf{y}_p = [\mathbf{0}, \mathbf{x}_p^{FR}]$, (2) $\mathbf{x}_p = [\mathbf{0}, \mathbf{x}_p^{FR}]$ and $\mathbf{y}_p = [\mathbf{x}_p^{UK}, \mathbf{0}]$ and (3) $\mathbf{x}_p = [\mathbf{x}_p^{UK}, \mathbf{x}_p^{FR}]$ and $\mathbf{y}_p = [\mathbf{x}_p^{UK}, \mathbf{x}_p^{FR}]$. We train the autoencoder with batches of size 256, where each batch is randomly selected from the full training data. The trained network is used to project a product's MRNet-Product2Vec-FR to the corresponding MRNet-Product2Vec-UK space. The projected MRNet-Product2Vec-UK representation is used to find the nearest UK products corresponding to a French product. We demonstrate a few French products and their UK nearest neighbors in Fig. 3b. We could have also projected the UK products to the French embedding space or project both of them to the common shared space for comparison. Although the results are preliminary, this demonstrates that we can use a multimodal autoencoder to effectively compare embeddings from different language-regions.

# 6   Discussion and Future Work

In this paper, we propose a novel variant of e-commerce product embeddings called MRNet-Product2Vec, where different product related signals are explicitly injected into their embeddings by training a Discriminative Multi-task Bidirectional RNN. Initially, PG-specific embeddings are learned and then a PG-agnostic

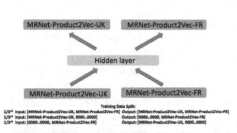

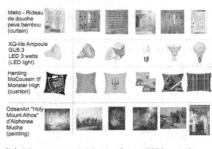

(b) Nearest neighbors from UK products w.r.to a French product.

(a) Architecture of Multimodal Autoencoder

**Fig. 3.** Language Agnostic MRNet-Product2Vec (best viewed in electronic copy).

version is learned using a sparse autoencoder. We evaluate the proposed embeddings qualitatively and quantitatively and establish it's effectiveness. We also propose a multimodal autoencoder for comparing products across different countries (i.e., languages) and provide initial results using that. MRNet-Product2Vec has been applied to generate product embeddings of around 2 billion products and have been made available internally within our company for product related ML model building. We periodically retrain MRNet to keep the model updated with the dynamic signals and also update the resulting embeddings of all products. We note that MRNet-ProductVec is suitable for cold-start scenarios as the embeddings can be created using only product titles, which are available as part of the catalog data. Although MRNet-Product2Vec has been trained with the proposed set of 15 different tasks, the framework provides the flexibility to learn embeddings with any other set of tasks or fine-tune the already learnt embeddings with additional tasks.

# References

1. Caruana, R.: Multitask learning. In: Thrun, S., Pratt, L. (eds.) Learning to Learn. Springer, Boston (1998). https://doi.org/10.1007/978-1-4615-5529-2_5
2. Grbovic, M., Radosavljevic, V., Djuric, N., Bhamidipati, N., Savla, J., Bhagwan, V., Sharp, D.: E-commerce in your inbox: product recommendations at scale. In: Proceedings of the 21th ACM SIGKDD. ACM (2015)
3. Hochreiter, S.: The vanishing gradient problem during learning recurrent neural nets and problem solutions. Int. J. Uncert. Fuzz. Knowl.-Based Syst. **6**(02), 107–116 (1998)
4. Hochreiter, S., Schmidhuber, J.: Long short-term memory. Neural Comput. **9**(8), 1735–1780 (1997)
5. Keogh, E., Mueen, A.: Curse of dimensionality. In: Sammut, C., Webb, G.I. (eds.) Encyclopedia of Machine Learning, pp. 257–258. Springer, New York (2011). https://doi.org/10.1007/978-0-387-30164-8
6. Mikolov, T., Sutskever, I., Chen, K., Corrado, G.S., Dean, J.: Distributed representations of words and phrases and their compositionality. In: NIPS (2013)

7.  Ng, A.: Sparse autoencoder. CS294A Lecture Notes **72**, 1–19 (2011)
8.  Ngiam, J., Khosla, A., Kim, M., Nam, J., Lee, H., Ng, A.Y.: Multimodal deep learning. In: ICML (2011)
9.  Vasile, F., Smirnova, E., Conneau, A.: Meta-prod2vec: product embeddings using side-information for recommendation. In: Proceedings of the 10th ACM Conference on Recommender Systems. ACM (2016)
10. Werbos, P.J.: Backpropagation through time: what it does and how to do it. Proc. IEEE **78**(10), 1550–1560 (1990)

# Optimal Client Recommendation for Market Makers in Illiquid Financial Products

Dieter Hendricks[(✉)] and Stephen J. Roberts

Machine Learning Research Group, Oxford-Man Institute of Quantitative Finance,
Department of Engineering Science, University of Oxford, Oxford, UK
dieter.hendricks@eng.ox.ac.uk

**Abstract.** The process of liquidity provision in financial markets can result in prolonged exposure to illiquid instruments for market makers. In this case, where a proprietary position is not desired, pro-actively targeting the *right* client who is likely to be interested can be an effective means to offset this position, rather than relying on commensurate interest arising through natural demand. In this paper, we consider the inference of a client profile for the purpose of corporate bond recommendation, based on typical recorded information available to the market maker. Given a historical record of corporate bond transactions and bond meta-data, we use a topic-modelling analogy to develop a probabilistic technique for compiling a curated list of client recommendations for a particular bond that needs to be traded, ranked by probability of interest. We show that a model based on Latent Dirichlet Allocation offers promising performance to deliver relevant recommendations for sales traders.

## 1 Introduction

The exchange of financial products primarily relies on the principle of matching willing counter-parties who have opposing interest in the underlying product, resulting in a demand-driven natural transaction at an agreed price. There are, however, cases where there is insufficient commensurate demand on one side at the desired price level, resulting in one of the parties needing to either wait for willing counter-parties or adjust their price. Where transaction immediacy is required, the client may approach a *market maker* (such as a bank or broker) who will facilitate the required trade by guaranteeing the other side of the transaction and charging a fee (the *spread*) for this service. This process of facilitating client transactions is termed *liquidity provision*, as the client can pay a fee to trade an otherwise illiquid asset immediately.

From the market maker's perspective, providing this liquidity of course results in taking a proprietary position in the underlying product, affecting their inventory and/or cash on hand. The management of this inventory and how it relates to quoted spread to account for associated risks is widely studied (see [2, 8, 11, 12] as examples), however is beyond the scope of this paper. We are interested in the particular case where a market maker has provided liquidity in a product and is not

Y. Altun et al. (Eds.): ECML PKDD 2017, Part III, LNAI 10536, pp. 166–178, 2017.
https://doi.org/10.1007/978-3-319-71273-4_14

interested in a long-term proprietary position, viz. they would like to mitigate or eliminate this exposure by targeting interested clients to offset the position. Finding suitable clients is the task of *sales traders*, who use their knowledge of potential clients' interests to find a match for the required trade, however understanding the nuanced preferences of all the clients is an arduous task. This paper seeks to create a system which will automate client profile inference and assist the sales traders by providing them with a curated list of clients to contact, who are most likely to be interested in the product. A successful system will expedite the liquidation of the market maker's product exposure, assisting with regulatory [9,16] and inventory management [1] concerns.

The products we consider are *corporate bonds*, which are fixed-term financial instruments issued by companies as a means of raising capital for operations. An investor who owns a corporate bond is usually entitled to interest payments from the issuer in the form of regular *coupons*, and redemption of the *face value* of the bond at *maturity*. The *yield* (interest rate) associated with a corporate bond is typically higher than a comparable government-issued bond. This yield differential is commensurate with the perceived credit-worthiness of the underlying company, the nature of the issue (senior/subordinated, secured/unsecured, callable/non-callable, etc.), the liquidity of the market place and the contractual provisions for contingencies in the event of issuer default [10,17]. From an investor's perspective, corporate bonds offer a relatively stable investment compared to, say, buying stocks in the company, since the instrument does not participate in the underlying profits of the company and bondholders are preferential creditors in the case of company bankruptcy. Following the initial issuance, corporate bonds are traded between investors in the *secondary market* until maturity, where market makers facilitate transactions by providing liquidity when required, leading to product exposures which need to be offset, as discussed above.

We will use a topic modelling analogy to frame the problem and develop a client profile inference technique. In the Natural Language Processing (NLP) community, many authors have focused on probabilistic generative models for text corpora, which infer latent statistical structure in groups of documents to reveal likely topic attributions for words [5,6,19,24]. One such model is Latent Dirichlet Allocation (LDA) [6], which is a three-level hierarchical Bayesian model under which *documents* are modelled as a finite mixture of *topics*, and topics in turn are modelled as a finite mixture over *words* in the vocabulary. Learning the relevant topic mixture and word mixture probabilities provides an explicit statistical representation for each document in the corpus. If one considers *documents* as *products* and *words* as *clients*, this has a natural analogy to the client recommendation problem we seek to solve. By observing product-client (document-word) transactions, we can infer a posterior probability of trade over *relevant* clients (topic with highest probability mass) for a particular product. These ideas are made more concrete in Sect. 2. Sampling from this posterior probability distribution provides us with a mechanism for client recommendation (most likely matches), coupled with a probability of trade, which will assist sales trades to gauge recommendation confidence.

This paper proceeds as follows: Sect. 2 discusses the analogy between topic modelling and bond recommendation. Section 3 introduces LDA as a candidate technique for client profile inference. Section 4 discusses some baseline models for comparison. Section 5 introduces some custom metrics to quantify recommendation efficacy, in the context of bond recommendation. Section 6 discusses the data and results, and Sect. 7 provides some concluding remarks.

## 2    A Topic Modelling Approach: Terminology and Analogies

We will frame the problem using the exposition in Blei et al. [6] as a guide, making appropriate modifications to reflect the bond recommendation use-case.

The *word* ($w$) represents the basic observable unit of discrete data, where each word belongs to a finite vocabulary set indexed by $\{1, ..., W\}$. Where appropriate, we may use the convention of a superscript ($w^i$) to indicate location in a sequence (such as in a document or topic), and subscript ($w_t$) to indicate a word observed at a particular time. Words are typically represented using unit-basis $W$-length vectors, with a 1 coinciding with the associated vocabulary index and zeros elsewhere. In our context, words represent *clients*, viz. $w = i$ is a unit vector associated with client $i$. We have used the term *client interest*, as we may abstract the actual trade status of our recorded data (traded, not traded, indication of interest, traded away, passed) to an indicator representing *interest* or *no interest*. In each case, the client was interested in the underlying bond and requested a price, regardless of whether they actually traded with us, another bank or changed their mind. This is the behaviour we would like to predict and has the added benefit of reducing the sparsity of our dataset. In future work, we may consider relaxing this assumption to determine if certain trade statuses contain more relevant information for likely client interest.

A *document* ($d$) is a sequence of $N$ words $d = \{w^1, w^2, ..., w^N\}$, where $w^n$ is the $n^{th}$ word in the sequence. In our context, a document relates to a specific *product*, where, like a *document* is a collection of *words*, a *product* represents a collection of *clients* who have expressed interest to trade.

A *topic* ($z$) is a collection of $M$ words $z = \{w^1, w^2, ..., w^M\}$ which are related in some way, representing an abstraction of words which can act as a basic building block of documents. In our context, a topic refers to a *client group*, viz. a set of clients that are regarded as *similar* based on the products they are interested in.

A *corpus* (**w**) is a collection of $D$ documents, $\mathbf{w} = \{d^1, d^2, ..., d^D\}$. In our context, the corpus represents the set of *products* which the market maker is interested in trading with its clients.

### 2.1    The Product-Client Term-Frequency Matrix

In the topic modelling analogy, a corpus can be summarised by a *document-word matrix*, which is essentially a 2-d matrix where, for each document (row), we count the frequency of each possible word in the vocabulary (columns) in the document. This summary is justified by the *exchangeability* assumption typical in topic modelling, where temporal and spatial ordering of documents and words are ignored to ensure tractable inference.

For our application, we can compute an analogous *product-client matrix* where, for each product (row), we count the number of times a client (column) has expressed interest in the product. While we suspect the temporal property of client interest is an important property (clients trade bonds in response to particular market conditions, to renew exposure close to maturity or as part of a regular portfolio rebalancing scheme), we will ignore these effects in this study and revisit these properties in future work. We will, however, ensure only *active* bonds are used to populate the product-client matrix, i.e. bonds which have a start date before the training period start and maturity date after the chosen testing day.

The product-client summary of records we use in this study results in a highly sparse matrix, with relatively few clients dominating trading activity. Since equal weight is placed on zero and non-zero counts, this will make inference for clients who trade less frequently more difficult. One remedy used in the topic modelling literature is to convert the raw document-word matrix into a Term Frequency-Inverse Document Frequency (TF-IDF) matrix [21,23]. Under this scheme, for our application, the weighting of a client associated with a product increases proportionally with the number of times they have traded the product, but this is offset by the number of times the product is traded among all clients. We will use the standard formulation,

$$tf\text{-}idf(w, d, \mathbf{w}) = tf(w, d) \cdot idf(w, \mathbf{w}), \tag{1}$$

where

$$tf(w, d) = 0.5 + 0.5 \cdot \frac{f_{w,d}}{\max\{f_{w^*,d} : w^* \in d\}}$$

and

$$idf(w, \mathbf{w}) = \log \frac{D}{|\{d \in \mathbf{w} : w \in d\}|}.$$

Here, $f_{w,d}$ is the raw count of the number of times client $w$ was interested in product $d$, $D$ is the total number of products and $\mathbf{w}$ is the set of all products.

## 3    Latent Dirichlet Allocation

Latent Dirichlet Allocation (LDA) [6] is a probabilistic generative model typically used in Natural Language Processing (NLP) to infer latent topics present in sets of documents. Documents are modelled as a mixture of topics sampled from a Dirichlet prior distribution, where each topic, in turn, corresponds to a multinomial distribution over words in the vocabulary [13]. The learned document-topic and topic-word distributions can then be used to identify the best topics which describe the document, as well as the best words which describe the associated topics [7].

As discussed in Sect. 2, we will consider *documents* as *products* and *words* as *clients*, allowing us to infer a posterior probability of trade (or at least client interest) over *relevant* clients (topic with highest probability mass) for a particular product.

LDA is traditionally a *bag-of-words* model, assuming document and word *exchangeability*. This means an entire corpus is used to infer document-topic and topic-word distributions, ignoring potential effects of spatial and temporal ordering. Given the particular problem of corporate bond recommendation, certain spatial and temporal features may be useful for more accurate recommendations. For example, the maturity date and frequency of coupon payment associated with a particular bond may influence the client's probability of trading. The duration and convexity characteristics of a bond and it's impact on the client's overall exposures may affect their willingness to trade. In this paper, we will ignore the effects of bond characteristics and temporal ordering of transactions, using only the bond *issue* and *maturity* dates to ensure they are *active* for the training and testing periods.

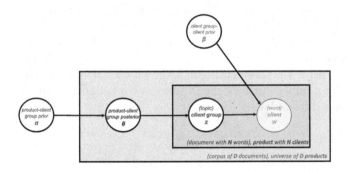

**Fig. 1.** Graphical representation of LDA in plate notation, indicating interpretation of *words*, *topics* and *documents* as *clients*, *client groups* and *products*.

To formalise ideas, we will reproduce the key aspects of the mathematical exposition of LDA (we follow conventions and notation set out in Wallach [24]), modified to reflect the product recommendation use-case. This is complemented by the *plate notation* representation of LDA in Fig. 1.

Client generation is defined by the conditional distribution $P(w_t = i | z_t = k)$, described by $T(W-1)$ free parameters, where $T$ is the *number of client groups* and $W$ is the *total number of clients*. These parameters are denoted by $\mathbf{\Phi}$, with $P(w_t = i | z_t = k) \equiv \phi_{i|k}$. The $k^{th}$ row of $\mathbf{\Phi}$ ($\phi_k$) thus contains the distribution over *clients* for *client group* $k$.

Client group generation is defined by the conditional distribution $P(z_t = k | d_t = d)$, described by $D(T-1)$ free parameters, where $D$ is the *total number of products traded by the market maker*. These parameters are denoted by $\mathbf{\Theta}$, with $P(z_t = k | d_t = d) \equiv \theta_{k|d}$. The $d^{th}$ row of $\mathbf{\Theta}$ ($\theta_d$) thus contains the distribution over *client groups* for *product* $d$.

The joint probability of a *set of products* **w** and a set of associated latent *groups of interested clients* **z** is

$$P(\mathbf{w}, \mathbf{z} | \mathbf{\Phi}, \mathbf{\Theta}) = \prod_i \prod_k \prod_d \phi_{i|k}^{N_{i|k}} \theta_{k|d}^{N_{k|d}}, \tag{2}$$

where $N_{i|k}$ is the number of times *client $i$* has been generated by *client group $k$*, and $N_{k|d}$ is the number of times *client group $k$* has been interested in *product $d$*. As in Blei et al. [6], we assume a Dirichlet prior over $\mathbf{\Phi}$ and $\mathbf{\Theta}$, i.e.

$$P(\mathbf{\Phi} | \beta \mathbf{m}) = \prod_k \text{Dirichlet}(\phi_k | \beta \mathbf{m}) \tag{3}$$

and

$$P(\mathbf{\Theta} | \alpha \mathbf{n}) = \prod_d \text{Dirichlet}(\theta_d | \alpha \mathbf{n}). \tag{4}$$

Combining these priors with Eq. 2 and integrating over $\mathbf{\Phi}$ and $\mathbf{\Theta}$ yields the probability of the *set of products* given hyperparameters $\alpha \mathbf{n}$ and $\beta \mathbf{m}$:

$$P(\mathbf{w} | \alpha \mathbf{n}, \beta \mathbf{m}) = \sum_{\mathbf{z}} \Big( \prod_k \frac{\prod_i \Gamma(N_{i|k} + \beta m_i)}{\Gamma(N_k) + \beta} \frac{\Gamma(\beta)}{\prod_i \Gamma(\beta m_i)}$$
$$\prod_d \frac{\prod_k \Gamma(N_{k|d} + \alpha n_k)}{\Gamma(N_d + \alpha)} \frac{\Gamma(\alpha)}{\prod_k \Gamma(\alpha n_k)} \Big). \tag{5}$$

In Eq. 5, $N_k$ is the total number of times *client group $k$* occurs in **z** and $N_d$ is the number of *clients* interested in *product $d$*. This posterior is intractable for exact inference, but a number of approximation schemes have been developed, notably Markov Chain Monte Carlo (MCMC) [15] and variational approximation [13,14].

For our study, we made use of the *scikit-learn* [20] open-source Python library, which includes an implementation of the online variational Bayes algorithm for LDA, described in Hoffman et al. [13,14]. They make use of a simpler, tractable distribution to approximate Eq. 5, optimising the associated variational parameters to maximise the Evidence Lower Bound (ELBO), and hence minimising the Kullback-Leibler (KL) divergence between the approximating distribution and the true posterior.

## 4   Baseline Models for Comparison

1. *Empirical Term-Frequency (ETF)*: We can use the normalised product-client term-frequency matrix discussed in Sect. 2.1 to construct an empirical probability distribution over clients for each product. This encodes the historical intensities of client interest, without exploiting any latent structure.
2. *Non-negative Matrix Factorisation (NMF)*: NMF aims to discover latent structure in a given non-negative matrix by using the product of two low-rank non-negative matrices as an approximation to the original, and minimising the distance of the reconstruction to the original, measured by the

Frobenius norm [18]. Applied to our problem, for a specified number of client groups, NMF can be used to reveal an *unnormalised* probability distribution over client groups for each product, and distribution over clients for each client group, from a given term-frequency matrix. These probabilities can be normalised for comparison with other models.

# 5   Evaluating Recommendation Efficacy

Recommender systems are usually evaluated in terms of their predictive accuracy, but the appropriate metrics should be chosen to reflect success in the specific application [22]. The data we have for inference and testing purposes is framed in terms of *positive interest*, viz. the presence of a record indicates a client was interest in the associated product, and the absence of a record indicates no interest. In addition, we are interested in capturing the accuracy of a "top $N$" client list, as opposed to a binary classifier. In terms of the standard confusion matrix metrics, we will thus focus on true and false positive results, however we have implemented a nuanced interpretation based on our application:

- *Cumulative True Positives (CTP)*: A client recommendation for a particular product is classified as a True Positive (TP) if the recommended client matches the actual client for that product on the testing day. The total number of TPs for a testing day is thus the total number of correctly matched recommendations. Given our use-case, where the $N$ *best* (ranked) recommendations are sampled, we compute the *cumulative* TPs as the number of TPs captured within the first $x$ recommendations, $x = 1, ..., N$. More formally, the CTP for product $j$ captured within the first $x$ recommendations is given by

$$\text{CTP}_j^x = \sum_{i=1}^{x} \mathbb{I}_{(w_j^i = w_j^*)}, \tag{6}$$

where $w_j^i$ is the $i^{th}$ recommended client for product $j$ and $w_j^*$ is the actual client who traded product $j$.

- *Relevant False Positives (RFP)*: A client recommendation is classified as a Relevant False Positive (RFP) if is does not match the actual client for that product on that day, but the recommended client traded another product instead. The rationale here is that the model captures the property of general client trading interest, so may be useful for the sales traders to discuss possibilities with the client, even though the model has matched the client to the incorrect product. These are measured at the first recommendation level ($x = 1$). For product $j$,

$$\text{RFP}_j = \mathbb{I}_{\left( (w_j^1 \neq w_j^*) \cap (w_j^1 \in \{w_k^*\}_{k \neq j}) \right)}. \tag{7}$$

- *Irrelevant False Positives (IFP)*: A client recommendation is classified as an Irrelevant False Positive (IFP) if is does not match the actual client for that product on that day, and the recommended client did not trade another product. This captures the *wasted resources* property of a false positive, as the sales trader could have spent that time targeting the right client. These are measured at the first recommendation level ($x = 1$). For product $j$,

$$\text{IFP}_j = \mathbb{I}_{\left((w_j^1 \neq w_j^*) \cap (w_j^1 \notin \{w_k^*\}_{k \neq j})\right)}. \tag{8}$$

## 6   Data and Results

*Data:* BNP Paribas (BNPP) provided daily recorded transactions with clients for various corporate bond products over the period 5 January 2015 to 10 February 2017, including records where clients did not end up trading with the bank. To maintain privacy, the Client and Product ID's were anonymised in the provided dataset. The data includes the following fields:

- *TradeDateKey*: Date of the transaction (*yyyymmdd*)
- *VoiceElec*: Whether the transaction was performed over the phone (*VOICE*) or electronically (*ELEC, ELECDONE*)
- *BuySell*: The trade direction of the transaction
- *NotionalEUR*: The notional of the bond transaction, in EUR
- *Seniority*: The seniority of the bond
- *Currency*: The currency of the actual transaction
- *TradeStatus*: Indicates whether the bond was actually traded with the bank (*Done*), price requested but traded with another LP (*TradedAway*), bank decided to pass on the trade (*Passed*), client requested price without immediate intention to trade (*IOI*) or client did not end up trading (*TimeOut, NotTraded*). This field also refers to entries which are aggregate bond positions based on quarterly reports (*IPREO*). Some entries also indicate an *UNKNOWN* trade status.
- *IsinIdx*: The unique product ID associated with the bond.
- *ClientIdx*: The unique client ID.

Some metadata was also provided, related to properties of the traded bonds:

- *Currency*: Currency of the bond
- *Seniority*: Seniority of the bond
- *ActualMaturityDateKey*: Maturity date of bond (*yyyymmdd*)
- *IssueDateKey*: Issue date of bond (*yyyymmdd*)
- *Mat*: Maturity as number of days since "00 Jan 2000" (*00000100*)
- *IssueMat*: Issue date as number of days since "00 Jan 2000" (*00000100*)
- *IsinIdx*: Unique product ID associated with bond
- *TickerIdx*: Bond type index

This data was parsed by: (1) Removing *TradeStatus = IPREO* or *UNKNOWN*, (2) Collapsing the *TradeStatus* column into a single *client interest* indicator, (3) Isolating either *Buys* or *Sells* for inference related to a particular trade direction, (4) Ensuring bonds are "active" for the relevant period, i.e. issued before start of training and matures after testing date, and finally, (5) construct a product-client term frequency matrix as described in Sect. 2.1.

*Results:* Due to space constraints, we will only show results for the SELL trade direction, however results for BUYS were quite similar. Figure 2 shows $CTP^x$ for $x = 1, ..., 100$ for a number of candidate models, with parameter inference from a single training period (5 Jan 2015 to 30 Nov 2016) and model testing on a single period (1 Dec 2016 to 10 Feb 2017). A crude baseline which all models beat is random client sampling (without replacement), indicated by the solid black line, suggesting that there is useful information in the historical transaction record for the purpose of client recommendation. The ETF model does surprisingly well, capturing 40% TP matches within the first 20 recommendations. We find that the LDA models offer superior accuracy beyond 10 recommendations, indicating that the latent structure is useful for the purpose of refining posterior probability of trade. These results do, however, aggregate results over the entire testing period, whereas the intended use-case will be on a daily basis, using the previous day's transactions to refine recommendations.

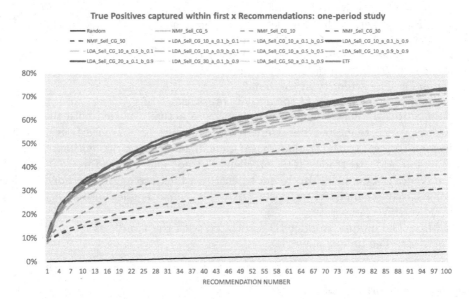

**Fig. 2.** Comparison of candidate models for *single period* training (5 Jan 2015 to 30 Nov 2016) and testing (1 Dec 2016 to 10 Feb 2017), evaluating cumulative true positives captured within first $x$ recommendations. Client SELL interest.

Table 1 shows the results from a *through-time* study, where a specified *window size* (WS) (number of days) was used for parameter inference, test metrics calculated for the day after, and the study moved forward one day. Results shown are averaged over all the testing days in the data set. Here, it is clear that, while the ETF model offers comparable CTP results to other models, it offers poor RFP and IFP results. For the highlighted LDA model, on average, 79% of the "incorrectly" recommended clients still traded on that day, albeit a different product. For a sales trader, making contact with these clients could start the conversation about their interests and be converted into a trade. Although it may not solve

**Table 1.** Summarised results for *through-time* study, varying estimation windows and hyperparameter values. Averaged over testing days in period 05 Jan 2015 to 10 Feb 2017. *WS* = Inference Window Size and *CG* = Client Groups. Client SELL interest.

| Model | WS | CG | $\alpha$ | $\beta$ | $CTP^1$ | $CTP^2$ | $CTP^3$ | $CTP^4$ | $CTP^5$ | $CTP^6$ | $CTP^7$ | $CTP^8$ | $CTP^9$ | $CTP^{10}$ | $\sigma(CTP^{10})$ | RFP | IFP |
|---|---|---|---|---|---|---|---|---|---|---|---|---|---|---|---|---|---|
| ETF | 100 | | | | 0.11 | 0.17 | 0.21 | 0.24 | 0.27 | 0.29 | 0.31 | 0.33 | 0.34 | 0.36 | 0.06 | 0.49 | 0.40 |
| NMF | 100 | 5 | | | 0.10 | 0.14 | 0.16 | 0.19 | 0.21 | 0.23 | 0.24 | 0.26 | 0.27 | 0.28 | 0.07 | 0.83 | 0.07 |
| NMF | 100 | 10 | | | 0.10 | 0.13 | 0.14 | 0.16 | 0.17 | 0.19 | 0.20 | 0.21 | 0.22 | 0.23 | 0.06 | 0.77 | 0.13 |
| NMF | 100 | 20 | | | 0.11 | 0.13 | 0.14 | 0.15 | 0.16 | 0.17 | 0.18 | 0.19 | 0.20 | 0.21 | 0.05 | 0.72 | 0.17 |
| NMF | 100 | 50 | | | 0.10 | 0.12 | 0.13 | 0.14 | 0.14 | 0.15 | 0.15 | 0.16 | 0.16 | 0.17 | 0.04 | 0.63 | 0.27 |
| LDA | 100 | 5 | 0.1 | 0.9 | 0.10 | 0.14 | 0.17 | 0.20 | 0.23 | 0.25 | 0.27 | 0.29 | 0.31 | 0.32 | 0.08 | 0.84 | 0.06 |
| LDA | 100 | 10 | 0.1 | 0.9 | 0.10 | 0.14 | 0.18 | 0.20 | 0.23 | 0.25 | 0.27 | 0.29 | 0.30 | 0.32 | 0.08 | 0.81 | 0.09 |
| LDA | 100 | 20 | 0.1 | 0.9 | 0.10 | 0.15 | 0.18 | 0.21 | 0.23 | 0.25 | 0.27 | 0.29 | 0.31 | 0.32 | 0.08 | 0.79 | 0.11 |
| LDA | 100 | 50 | 0.1 | 0.9 | 0.11 | 0.15 | 0.18 | 0.21 | 0.23 | 0.25 | 0.27 | 0.28 | 0.30 | 0.31 | 0.07 | 0.77 | 0.12 |
| ETF | 500 | | | | 0.11 | 0.17 | 0.22 | 0.25 | 0.27 | 0.30 | 0.32 | 0.34 | 0.36 | 0.38 | 0.07 | 0.54 | 0.35 |
| NMF | 500 | 5 | | | 0.11 | 0.16 | 0.18 | 0.21 | 0.23 | 0.25 | 0.27 | 0.28 | 0.30 | 0.31 | 0.09 | 0.80 | 0.09 |
| NMF | 500 | 10 | | | 0.10 | 0.13 | 0.14 | 0.16 | 0.16 | 0.18 | 0.19 | 0.20 | 0.21 | 0.22 | 0.06 | 0.78 | 0.12 |
| NMF | 500 | 20 | | | 0.11 | 0.12 | 0.14 | 0.15 | 0.16 | 0.17 | 0.18 | 0.19 | 0.20 | 0.21 | 0.05 | 0.73 | 0.17 |
| NMF | 500 | 50 | | | 0.10 | 0.11 | 0.12 | 0.13 | 0.14 | 0.15 | 0.15 | 0.16 | 0.17 | 0.17 | 0.06 | 0.65 | 0.23 |
| LDA | 500 | 5 | 0.1 | 0.9 | 0.11 | 0.16 | 0.21 | 0.22 | 0.24 | 0.26 | 0.28 | 0.30 | 0.33 | 0.34 | 0.10 | 0.82 | 0.06 |
| LDA | 500 | 10 | 0.1 | 0.9 | 0.11 | 0.16 | 0.20 | 0.23 | 0.24 | 0.27 | 0.30 | 0.31 | 0.33 | 0.34 | 0.09 | 0.81 | 0.09 |
| LDA | 500 | 20 | 0.1 | 0.9 | 0.11 | 0.16 | 0.21 | 0.22 | 0.25 | 0.27 | 0.29 | 0.30 | 0.32 | 0.35 | 0.09 | 0.79 | 0.10 |
| LDA | 500 | 50 | 0.1 | 0.9 | 0.12 | 0.17 | 0.20 | 0.23 | 0.25 | 0.27 | 0.29 | 0.30 | 0.32 | 0.34 | 0.09 | 0.79 | 0.11 |

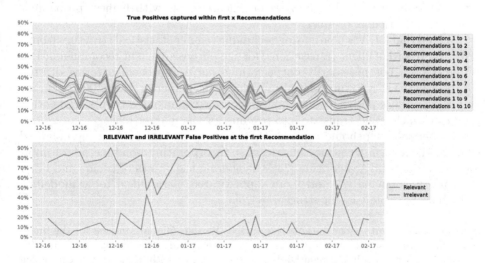

**Fig. 3.** True Positives, Relevant and Irrelevant False Positives. LDA with $CG = 20$, $\alpha = 0.1$, $\beta = 0.9$, 500-day rolling training window, 05 Jan 2015 to 10 Feb 2017. Client SELL interest.

the direct problem of offsetting a particular position, it could still translate into revenue for the market maker. We found that increasing the WS to 500 days alleviates the sparse data problem somewhat and offers marginal improvements in performance, however more sophisticated data balancing techniques [3] should be explored to ensure accurate inference for clients who trade less frequently.

Figure 3 shows $CTP^x$ for $x = 1, ..., 10$, RFP and IFP for each testing day, using the highlighted *through-time* LDA model in Table 1 (WS = 500, CG = 20, $\alpha = 0.1$, $\beta = 0.9$). We see that this model offers relatively consistent recommendation performance. There is a significant increase in CTP accuracy around the end of December 2016, but this is largely due to relatively few "typical" clients trading. These clients would have traded frequently in the past, thus are more likely to be recommended in the first instance. There is also a decrease in performance around the beginning of February 2017. This could be due to a change in client preferences due the expiry of a certain class of bonds. This does suggest that simple moving inference windows may insufficient to capture temporal trends, and a more sophisticated modelling approach may be required.

## 7   Conclusion

We proposed a novel perspective for framing financial product recommendation using a topic modelling analogy. By considering *documents* as *products* and *words* as *clients*, we can use classical NLP techniques to develop a probabilistic generative model to infer an explicit statistical representation for each *product* as a mixture of *client groups* (topics), where each *client group* is a mixture of *clients*. By observing product-client (document-word) transactions, we can infer a posterior probability of trade over *relevant* clients (topic with highest probability mass) for a particular product.

We find that LDA is a promising technique to infer statistical structure from a historical record of client transactions, for the purpose of client recommendation. While it does not necessarily outperform a naïve approach in terms of "top 10" true positive recommendations, it does offer superior "top 100" accuracy and *relevant* false positive performance, where recommended clients trade other products which could translate into revenue for the market maker.

Further research should consider the advantages of inference using balanced product-client term frequency matrices [3], incorporating bond metadata information into the LDA algorithm [25], considering the effects of trends and other temporal phenomena [7], and more sophisticated hierarchical topic modelling techniques to exploit latent structure [4,5].

**Acknowledgements.** The authors thank BNP Paribas Global Markets for the financial support and provision of data necessary for this study. The discussions with Joe Bonnaud, Laurent Carlier, Julien Dinh, Steven Butlin and Philippe Amzelek provided meaningful context and intuition for the problem.

# References

1. Amihud, Y., Mendelson, H.: Asset pricing and the bid-ask spread. J. Financ. Econ. **17**(2), 223–249 (1986)
2. Avellaneda, M., Stoikov, S.: High-frequency trading in a limit order book. Quant. Finan. **8**(3), 217–224 (2016)
3. Batista, G.E., Prati, R.C., Monard, M.C.: A study of the behavior of several methods for balancing machine learning training data. ACM SIGKDD Explor. Newslett. **6**(1), 20–29 (2004)
4. Blei, D.M., Griffiths, T.L., Jordan, M.I.: The nested Chinese Restaurant Process and Bayesian nonparametric inference of topic hierarchies. J. ACM (JACM) **57**(2), 7 (2010)
5. Blei, D.M., Griffiths, T.L., Jordan, M.I., Tenenbaum, J.B.: Hierarchical topic models and the nested Chinese Restaurant Process. In: Advances in Neural Information Processing (2004)
6. Blei, D.M., Ng, A.Y., Jordan, M.I.: Latent dirichlet allocation. J. Mach. Learn. Res. **3**, 993–1022 (2003)
7. Bolelli, L., Ertekin, Ş., Giles, C.L.: Topic and trend detection in text collections using latent dirichlet allocation. In: Boughanem, M., Berrut, C., Mothe, J., Soule-Dupuy, C. (eds.) ECIR 2009. LNCS, vol. 5478, pp. 776–780. Springer, Heidelberg (2009). https://doi.org/10.1007/978-3-642-00958-7_84
8. Das, S., Magdon-Ismail, M.: Adapting to a market shock: optimal sequential market-making. In: Advances in Neural Information Processing Systems (2009)
9. Duffie, D.: Market making under the proposed volcker rule. Rock Center for Corporate Governance at Stanford University Working Paper No. 106 (2012)
10. Elton, E.J., Gruber, M.J., Agrawal, D., Mann, C.: Explaining the rate spread on corporate bonds. J. Finan. **56**(1), 247–277 (2001)
11. Ghoshal, S., Roberts, S.: Optimal FX market making under inventory risk and adverse selection constraints. Working paper (2016)
12. Guéant, O.: Optimal market making (2017). arXiv:1605.01862 [q-fin.TR]
13. Hoffman, M., Bach, F.R., Blei, D.M.: Online learning for Latent Dirichlet Allocation. In: Advances in Neural Information Processing (2010)
14. Hoffman, M., Blei, D.M., Wang, C., Paisley, J.W.: Stochastic variational inference. J. Mach. Learn. Res. **14**(1), 1303–1347 (2013)
15. Jordan, M.I.: Learning in Graphical Models, vol. 89. Springer, Berlin (1998)
16. Kaal, W.A.: Global Encyclopedia of Public Administration, Public Policy, and Governance: Dodd-Frank Act. Springer (2016)
17. Kim, I., Ramaswamy, K., Sundaresan, S.: The Valuation of Corporate Fixed Income Securities. Columbia University, Manuscript (1988)
18. Lee, D.D., Seung, H.S.: Learning the parts of objects by non-negative matrix factorization. Nature **401**, 788–791 (1999)
19. MacKay, D.J.C., Bauman Peto, L.C.: A hierarchical Dirichlet language model. Nat. Lang. Eng. **1**(3), 289–308 (1995)
20. Pedregosa, F., Varoquaux, G., Gramfort, A., Michel, V., Thirion, B., Grisel, O., Blondel, M., Prettenhofer, P., Weiss, R., Dubourg, V., et al.: Scikit-learn: machine learning in python. J. Mach. Learn. Res. **12**, 2825–2830 (2011)
21. Robertson, S.: Understanding inverse document frequency: on theoretical arguments for IDF. J. Documentation **60**(5), 503–520 (2004)
22. Shani, G., Gunawardana, A.: Evaluating Recommendation Systems. Recommender Systems Handbook. Springer, US (2011)

23. Jones, K.S.: A statistical interpretation of term specificity and its application in retrieval. J. Documentation **28**, 11–21 (1972)
24. Wallach, H.M.: Topic modeling: beyond bag-of-words. In: Proceedings of the 23rd International Conference on Machine Learning, pp. 977–984 (2006)
25. Wang, X., Grimson, E.: Spatial latent dirichlet allocation. In: Advances in Neural Information Processing Systems, pp. 1577–1584 (2008)

# Predicting Self-reported Customer Satisfaction of Interactions with a Corporate Call Center

Joseph Bockhorst[✉], Shi Yu, Luisa Polania, and Glenn Fung

Machine Learning Unit, Strategic Data & Analytics, American Family Insurance,
6000 American Parkway, Madison, WI 53783, USA
jbockhor@amfam.com

**Abstract.** Timely identification of dissatisfied customers would provide corporations and other customer serving enterprises the opportunity to take meaningful interventions. This work describes a fully operational system we have developed at a large US insurance company for predicting customer satisfaction following all incoming phone calls at our call center. To capture call relevant information, we integrate signals from multiple heterogeneous data sources including: speech-to-text transcriptions of calls, call metadata (duration, waiting time, etc.), customer profiles and insurance policy information. Because of its ordinal, subjective, and often highly-skewed nature, self-reported survey scores presents several modeling challenges. To deal with these issues we introduce a novel modeling workflow: First, a ranking model is trained on the customer call data fusion. Then, a convolutional fitting function is optimized to map the ranking scores to actual survey satisfaction scores. This approach produces more accurate predictions than standard regression and classification approaches that directly fit the survey scores with call data, and can be easily generalized to other customer satisfaction prediction problems. Source code and data are available at https://github.com/cyberyu/ecml2017.

## 1 Introduction

In a competitive customer-driven landscape where businesses are constantly competing to attract and retain customers; customer satisfaction is one of the top differentiators. While digitization and other forces continue to increase consumer choice, understanding and improving customer satisfaction are often core elements of the business strategy of modern companies. It enables service providers to unveil timely opportunities to take meaningful interventions to improve customer experience and to train customer representatives (CR) in an optimal way.

In order to measure the effectiveness of a CR during a phone interaction with a customer, generally a customer survey is taken shortly after the call takes place. However, due to survey expense, typically only a small percentage of calls are measured. When CR performance is calculated from a small sample of surveys performance scores have high variability and there is potential misrepresentation of CR performance.

© Springer International Publishing AG 2017
Y. Altun et al. (Eds.): ECML PKDD 2017, Part III, LNAI 10536, pp. 179–190, 2017.
https://doi.org/10.1007/978-3-319-71273-4_15

The focus of this work is to describe the design and implementation of a deployed machine-learning-based system used to automatically predict customer satisfaction following phone calls. Our discovery and system design process can be divided into four stages:

1. **Extraction, processing and linking of raw data:** Raw data is collected and linked from four primary sources: call logs, historical survey scores, customer and policy databases, and call transcription and related content derived from audio recordings.
2. **Feature engineering:** Call data is processed to create informative features.
3. **Model design and creation:** In this stage we focus on the design and creating of the customer satisfaction predictive models.
4. **Aggregation of model predictions to the group level:** At the last stage, we aggregate individual model predictions to the group level (by call queue, by CR, in a given period of time, etc.). We also provide estimated bounds for the group average predictions.

## 2    Related Work

Research studies on emotion recognition using human-human real-life corpus extracted from call center calls are limited. In [15], a system for emotion recognition in the call center domain, using lexical and paralinguistic cues, is proposed. The goal was to classify parts of dialogs into three emotional states. Training and testing was performed on a corpus of 18 h of real dialogs between agent and customer, collected in a service of complaints. A similar work [2], also proposes to classify call center calls between three emotional states, namely, anger, positive and neutral. The authors used classical descriptors, such as zero crossing rate and Mel-frequency cepstral coefficients, and support vector machines as the classifier. They used service complaints and medical emergency conversations from call centers, and adopted a cross-corpus methodology for the experiments, meaning that they use one corpus as training set and another corpus as test set. They attained a classification accuracy between 40% to 50% for all the experiments.

Park and Gates [10] developed a method to automatically measure customer satisfaction by analyzing call transcripts in near real-time. They identified several linguistics and prosodic features that are highly correlated with behavioral aspects of the speakers and built machine learning models that predict the degree of customer satisfaction in a scale from 1 to 5 with an accuracy of 66%. Sun et al. [13] adopted a different approach, based on fusion techniques, to predict the user emotional state from dialogs extracted from a Chinese Mobile call center corpus. They implemented a statistical model fusion to alleviate the data imbalance problem and combined n-gram features, sentiment word features and domain-specific words features for classification.

Recently, convolutional neural networks have been used on raw audio signals to automatically learn meaningful features that lead to successful prediction of self-reported customer satisfaction from call center conversations in Spanish [12]. This approach starts by pretraining a network on debates from French TV shows with the goal of detecting salient information in raw speech that correlates with emotion. Then, the last layers of the network are finetuned with more than 18000 conversations from several call centers. The CNN-based system achieved comparable performance to the systems based on traditional hand-designed features.

There are many machine learning problems, referred to as ordinal ranking problems, where the goal is to classify patterns using a categorical scale which shows a natural order between labels, but not a meaningful numeric difference between them. For example, emotion recognition in the call center domain usually involves rating based on an ordinal scale. Indeed, psychometric studies show that human ratings of emotion do not follow an absolute scale [8,9]. Ordinal ranking is fundamentally different from nominal classification techniques in that order is relevant and the labels are not treated as independent output categories. The ordinal ranking problems may not be optimally addressed by the standard regression either since the absolute difference of output values is nearly meaningless and only their relative order matters [3].

There are several algorithms which specifically benefit from the ordering information and yield better performance than nominal classification and regression approaches. For example, Herbrich et al. [5] proposed a support vector machines approach based on comparing training examples in a pairwise manner. A constraint classification approach that works with binary classifiers and is based on the pairwise comparison framework was proposed by Har-Peled et al. [4]. Crammer and Singer [1] developed an ordinal ranking algorithm based on the online perceptron algorithm with multiple thresholds.

Some areas where ordinal ranking problems are found include medical research [11], brain computer interface [17], credit rating [7], facial beauty assessment [16], image classification [14], and more. All these works are examples of applications of ordinal ranking models, where exploiting ordering information improves their performance with respect to their nominal counterparts.

## 3   Overview of the Proposed System

Our main goal is to develop a model to predict satisfaction scores for all incoming customer calls in order to (i) take meaningful timely interventions to improve customer experience and (ii) obtain a robust understanding on how care center performance and training can be enhanced, ultimately for our customer's benefit.

Our company recently adopted a system which automatically transcribes phone calls to text. The transcriptions generated by this system are key for our deployed system. The company customer care center monitors customer satisfaction by offering surveys conducted by a third party vendor to 10% of incoming calls. Each care center CR has around five surveys completed per month, which is only about 0.5–1% of all assigned calls. There are four topics measured by the survey: (a) If the customer felt "valued" during the call; (b) If the issue was resolved; (c) How polite the CR was, and (d) How clearly the CR communicated during the call. Scores range form 1 to 10 (1 lowest, 10 highest) and the four scores are averaged into an additional variable called RSI (Representative Satisfaction Index). In this paper we focus on predicting the RSI.

Several difficulties, in terms of modeling, are discovered after a quick initial inspection of the training data:

- The customer satisfaction scores (RSI) are highly biased towards the highest score (10), while calls with scores lower than 8 are less than 4%. This highly skewed distribution makes building a predictive model more complex.
- Survey scores are customer responses, thus are subjective, qualitative states heavily impacted by personal preferences.
- The measurement scale of survey scores is ordinal; one cannot say, for example, that a score of 10 indicates double satisfaction as a score of 5. Most, if not all, standard regression techniques implicitly assume an interval or ratio scale.

Figure 1 displays an overview of the deployed system. The system workflow can be summarized by the following steps:

1. After a call ends, a transcript of the call is automatically produced by a speech-to-text system developed by Voci (vocitec.com).
2. Calls are partitioned into temporal segments and non-text features are engineered. The rationale of temporal segmentation is that certain events are more relevant depending of when they occur in the call. For example: detecting negative sentiment trends in the first quarter of the call but positive at the end may lead to a higher satisfaction score than when the opposite is true.
3. Textual features are constructed and merged with non-text features. The fused feature vectors are used as input features for the models described in the next step.
4. Ranking model scoring. The ranking model is trained by sampling ordered pairs based on satisfaction scores.

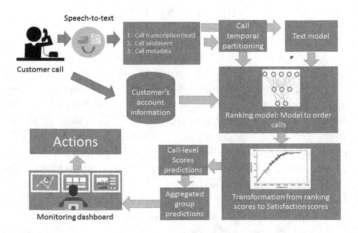

**Fig. 1.** Overview of the deployed system

5. Mapping from raking scores to satisfaction scores using an isotonic model. Individual (per call) satisfaction predictions are generated.
6. Aggregation of calls at the group level are stored in a database. Example groups include: per CR, per queue and per time period.
7. Aggregations are used for real-time reporting though a monitoring dashboard.

## 4    Representation

This section describes the pipeline of extracting features from various types of input data sources related to a phone call which are passed on to the models. Available input data sources are call transcriptions, call logs, and other customer and policy data. Figure 2 displays our input data model.

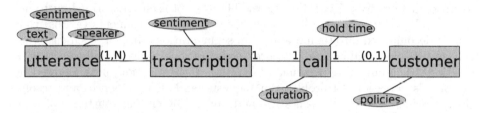

**Fig. 2.** Partial entity-relationship data model of input data. Numbers indicate cardinality ratios between entities. Not all attributes are shown.

Calls are transcribed to sequences of non-overlapping *utterances*, chunks of semi-continuous speech by a single speaker flanked on either side by either a change of speaker or a break in speech. Each utterance contains the transcribed text along with related attributes including the predicted speaker, either customer or company representative, start and end times, and predicted sentiment. Concatenating the text of all utterances gives us the full transcribed text of a call. In addition to the call transcription, we generate features from the telephony system logs. Examples of log level attributes are assigned call-center queue, waiting time and transfer indicators. For calls that are linked to specific customers we use additional customer and policy data.

### 4.1    Feature Engineering

Our feature engineering process takes linked input data for a call and produces a fixed-length feature vector.

**Temporal Segment Features.** Each temporal segment feature represents an aspect of the call in a certain temporal range, for example, the minimum sentiment score of any customer utterance in the last quarter of the call. A temporal

**Table 1.** A temporal segment feature is created for each of the 300 combinations ($5 \times 3 \times 4 \times 5$) of component values.

| Component | Possible values |
|---|---|
| Utterance function | negSent(), negCount(), duration(), consNeg(), sentScore() |
| Speaker | representative, customer, either |
| Aggregate function | min(), max(), mean(), std() |
| Temporal range | [0.0, 0.25), [0.25, 0.5), [0.5, 0.75), [0.75, 1.0], [0.9, 1.0] |

segment feature is defined by (i) a numerical utterance function[1], (ii) a speaker, (iii) an aggregate function and, (iv) a temporal range (see Table 1).

**Temporal Segment Text-Features.** The text of each transcribed customer call can also be viewed as a linear sequence of temporal elements (words) thus can be decomposed into temporal textual segmentations. In fact, each customer call consists of several natural temporal segmentations, which usually starts with greetings, then customer personal information authentication, next followed by customer's narrations of problems or requests, and then responses and resolutions provided by the representative, and finishes by ending courtesies of both parties. To predict customer satisfaction, we assume that late segmentations of a call (i.e., problem explanations, resolutions) are more informative than early parts (i.e., greetings, authentication), therefore we create separate textual models by decomposing the transcribed text of a call into different temporal segments.

We denote $\mathcal{D}$ as the corpus of transcribed text of all calls, where $d_i \in \mathcal{D}$, $i = 1...N$ is a document of transcribed text of the $i$-th call. Each $d_i$ is composed of a sequence of words $w_{i,j}$, $j = 1...M_i$ where $M_i$ is the total number of words in $d_i$. And we further decompose all the words in a document into four sub-documents $q_{i1}, q_{i2}, q_{i3}, q_{i4}$, where

$$q_{i1} = \{w_{i,1}, ..., w_{i,s_1}\},$$
$$q_{i2} = \{w_{i,s_1+1}, ..., w_{i,s_2}\},$$
$$q_{i3} = \{w_{i,s_2+1}, ..., w_{i,s_3}\},$$
$$q_{i4} = \{w_{i,s_3+1}, ..., w_{i,M_i}\}.$$

Since each call has different lengths, and we haven't applied any method to automatically segment a call according to the content, we simply set $s_1, s_2, s_3$ respectively to the rounded integers of $\frac{M_i}{4}, \frac{2M_i}{4}, \frac{3M_i}{4}$, thus gives us four even

---

[1] negSent() is an indicator that is 1 if the utterance sentiment label is *Negative*, negCount() is the number of *Negative* or *Mostly Negative* sentiment phrases in the utterance, duration() is the length of the utterance in seconds, consNeg() is an indicator that is 1 if the current and previous utterance have negative sentiment, and sentScore() maps utterance sentiment labels (*Negative, Mostly Negative, Neutral, Mostly Positive, Positive*) to numerical scores $(-1, -0.5, 0, 0.5, 1)$.

temporal segments, where each segment contains words appeared in a quarter part, from beginning to end, of a call and we call them *quarter documents*.

Analogously, using the same $s_1, s_2, s_3$ chosen before, we define four sets of *tail documents*

$$t_{i1} = \{w_{i,1}, ..., w_{i,M_i}\},$$
$$t_{i2} = \{w_{s_1+1,1}, ..., w_{i,M_i}\},$$
$$t_{i3} = \{w_{s_2+1,1}, ..., w_{i,M_i}\},$$
$$t_{i4} = \{w_{s_3+1,1}, ..., w_{i,M_i}\},$$

as segmented documents of various lengths. Notice that $t_{i1}$ is equivalent to $d_i$, and $t_{i2}, t_{i3}, t_{i4}$ are respectively the remaining $75\%, 50\%, 25\%$ part of a call.

Thus, we obtain eight corpora of call text (four quarter documents and four tail documents) and each corpus represents a temporally segmented snapshot of the textual content. Next, we construct standard TF-IDF profiles on each individual corpus, where a row represents a call, and benchmark the best corpora using a held-out training and validation set. We find that the corpus composed by $t_{i3}$, represented by 5000 TFIDF weights, gives the best performance and we select that for modeling.

**Other Features.** Additional features are created from telephony logs, such as duration of call, queue, in-queue waiting time, and policy count information such as the number of auto policies, number of property policies, *etc.* held by the customer's household. Our system has a total of 5,340 natural features, and following one-hot-encoding of categorical features the final model ready dataset contains 5,501 features.

## 5    Models

Here we describe our approach to learning a predictive model of ordinal satisfaction ratings, such as RSI. The modeling task is to learn a function $f(\mathbf{x}) = \hat{y}$ mapping feature vector $\mathbf{x}$ to predicted RSI $\hat{y}$ such that on average the difference between the predicted score and actual score $y$ is small. Our approach involves two models: a linear ranking model $r(\mathbf{x})$ that maps examples to rank scores and a non-decreasing, non-linear model $s(\hat{r})$ mapping rank scores to RSI. We form $f()$ through composition: $f(\mathbf{x}) = s(r(\mathbf{x}))$. We term this approach RS + IR for "rank score + isotonic regression".

Unlike standard linear and non-linear regression methods that directly model $y$, the RS + IR approach is consistent with the ordinal scale of the satisfaction score. A second advantage of RS + IR is that since the rank score model is learned from pairs of examples (see below), a larger pool of training examples are available and the class labels of the training set can be balanced, which is especially important for data sets like those considered here that are strongly skewed towards the high end of the satisfaction scale.

**Rank Score Model.** We learn a model to rank examples by RSI using the pairwise transform [6]. The pairwise transform induces a rank score function $r(\mathbf{x})$ by learning a linear binary classifier from an auxiliary training set of examples $(\mathbf{u}, v)$ that are formed from pairs of examples $(\mathbf{x_i}, y_i), (\mathbf{x_j}, y_j)$ in the original ordinal training set that have different satisfaction scores[2].

The features of the auxiliary examples are the component-wise difference between the original examples, $\mathbf{u}_{ij} = \mathbf{x_i} - \mathbf{x_j}$. The binary class value $v_{ij}$ indicates whether or not example $i$ has a higher satisfaction than example $j$: $v_{ij}$ is $+1$ if $y_i > y_j$ and $-1$ if $y_i < y_j$. The linear binary classifier $r(\mathbf{u})$, which is learned from the auxiliary training set to predict which of two examples has a higher satisfaction score, is subsequently used as a rank score function $r(\mathbf{x})$. That $r()$ can be used as a ranking function follows from its linearity $r(\mathbf{x_i} - \mathbf{x_j}) = r(\mathbf{x_i}) - r(\mathbf{x_j})$ and by noticing that $r(\mathbf{x_i}) > r(\mathbf{x_j})$ is consistent with the prediction that $y_i$ is larger than $y_j$.

**Rank Score to Satisfaction.** The second sub-model $s(r)$ is one-dimensional, non-decreasing function mapping rank scores to satisfaction scores. After learning the rank score model $r()$ we calculate the rank score of all examples in the original training set, order examples by their rank scores and smooth the resulting sequence of satisfaction scores. We then fit an isotonic regression model using training examples sampled uniformly from the smoothed function.

## 6    Results

In this section we describe the results of experiments conducted on a data set of 8,726 incoming phone calls from between March 23, 2015 and Dec 29, 2015 for which we have customer satisfaction survey results. We randomly selected 75% (6,108) for the training set and the remainder served as our test set.

### 6.1    Individual Predictions

To assess the value of our "rank score + isotonic regression" (RS + IR) approach to predicting phone call representative satisfaction index (RSI) scores we compared it with three standard regression methods (ridge regression, Lasso and random forest regression) and one classification method, linear support vector machine[3]. Ridge regression and Lasso are both penalized linear regression methods, but use different loss functions: L2 for ridge and L1 for lasso. Random forest regression is a non-linear approach that trains different ensembles of

---

[2] All the auxiliary examples may not be needed. We have found that while there are over 10 million auxiliary examples that can be formed from our training set, the rank score model is well converged when trained with 10,000 examples. We experimented with various techniques for sampling the auxiliary examples (biased for large RSI difference, small RSI difference, *etc.*), and found that simple uniform sampling works best.

[3] All comparison models trained using the scikit-learn Python package.

least-squares linear models for non-overlapping partitions of the input space. We use cross-validation on the training set to set hyperparameters ($\alpha$ for ridge and lasso, max_depth and min_samples_per_split for random forest, and $C$ for SVM).

| | Pe. | Sp. | MAE |
|---|---|---|---|
| Ridge | 0.300 | 0.231 | 0.811 |
| Lasso | 0.303 | 0.227 | 0.815 |
| Random forest | 0.149 | 0.150 | 0.835 |
| Rank Score | 0.255 | **0.239** | * |
| RS+IR | **0.312** | **0.239** | **0.784** |

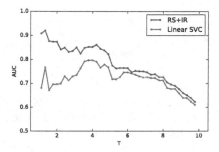

**Fig. 3.** (Left) Regression results (Pe: Pearson correlation, Sp: Spearman correlation, MAE: mean absolute error). (Right) classification results

Figure 3(left) shows test set results. The RS + IR model outperforms the other models in terms of Pearson correlation, Spearman correlation and mean absolute error (MAE). Also, RS + IR has better Pearson correlation than the rank score alone, showing the value of the non-linear mapping from rank score to prediction. If actions are taken in response to model predictions, for example reaching out to potentially dissatisfied customers, when predicted RSI falls below a given threshold $T$ classification models are more appropriate than regression. The right panel shows the area under an ROC curve as $T$ varies for our approach and linear SVM. Even though we trained a different SVM model for each value of $T$ and only a single RS + IR model, the AUC of the RS + RI model dominates the SVM over the whole range of $T$, especially for smaller thresholds.

## 6.2  Group Predictions

Since users of the productionalized system view reports on mean predicted satisfaction scores for various collections of calls, for example by department, call-center queue, and hour-of-day, we have investigated our system's accuracy for call groups. We use two kinds of groupings: random and by *topic*. We formed random groups of a given size by sampling calls with replacement from the test set. For the topic groups we used hand-crafted text-based predicates, which were created by another business unit for tagging calls related to various products and services and aspects of the customer journey. Each topic predicate is a Boolean function that takes a single sentence as input. A call belongs to a topic $T$ if $T(s)$ is true for any sentence $s$ in the call. Thus, a given call may belong to zero, one or many topics. There are a total of 107 topics groups with group sizes ranging from 1 to 1,560.

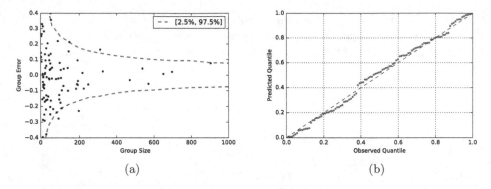

(a)                                          (b)

**Fig. 4.** (a) Dashed lines indicate 95% confidence band for randomly selected groups. Points are observed group errors of topics groups containing more than 50 calls. (b) Quantile/quantile plot of group errors for the topics groups.

We define the group error to be the difference between the mean of the predicted scores for all calls in the group and the mean of the actual satisfaction scores. We form random groups with between 10 and 1,000 calls and for each group size we formed 5,000 replicate random groups. The dashed blue lines in Fig. 4(a) show 95% confidence bands for the group error of the random groups. That is, for a given group size the group error of 95% of groups of that size in our simulation fell between the bands. We can see from this figure that group error decreases with group size.

We use the bands of Fig. 4(a) to determine tolerance levels for deciding when to raise alarms due to differences between predicted and actual satisfaction scores. The points represent the errors of the topics groups. The errors for 45 of the 48 topics groups (93.75%) with more than 50 calls fall between the 95% confidence bands. This provides evidence that the topic groups have similar error profiles to natural groupings by topic. Figure 4(b) shows the quantile/quantile plot for the group error of all 107 topics groups using the errors of random groups of similar size to compute the observed percentile. As the points lie close to the ideal diagonal line, we conclude that the error profiles of random groups and topics group are similar.

Figure 5(a) shows the predicted and actual mean satisfaction for topics groups with at least 50 calls. The area of each bubble is proportional to the number of calls in the group, which ranges from a minimum of 50 to a maximum of 1,560. There is a general agreement (Pearson correlation = 0.73) between the predicted and actual group means. And in general, as with the random groups, larger groups have smaller within group errors. Figure 5(b) shows the predicted group mean with 95% confidence interval (dependent on the group size) and the actual group mean for these same 48 groups. Again, as this is a different view of the same data represented by the points of Fig. 4(a), we see that the confidence bounds determined by random group errors do an excellent job of describing the distribution of errors in the topics groups.

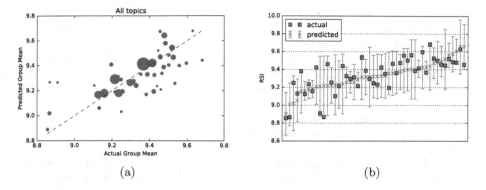

**Fig. 5.** (a) Mean predicted RSI vs. mean satisfaction RSI for the topics groups. Bubble area is proportional to group size. Group sizes range from 50 to 1,560. (b) Mean predicted RSI for the topics groups with 95% confidence intervals.

## 7  Conclusions and Lessons Learned

This paper presents an efficient and accurate method for predicting self-reported satisfaction scores of customer phone calls. Our approach has been implemented into a production system that is currently predicting caller satisfaction of approximately 30,000 incoming calls each business day and generating frequent reports read by call-center mangers and decision makers in our company.

We described several techniques that we suspect will generalize to related tasks. (1) Rather than applying regression models directly on the ordinal data, we use a linear ranking sub-model along with a non-linear isotonic regression sub-model for predicting satisfaction. We presented empirical evaluation that shows this approach yields more accurate satisfaction predictions than standard regression models. (2) Temporally segmented features constructed from call meta-information and transcribed text are shown to be useful to capture informative signals relevant to customer satisfaction. (3) The average satisfaction prediction for groups of calls, instead of by only individual calls, agrees very strongly with actual satisfaction scores, especially for large groups. (4) We provided methods for determining system tolerance levels based on the deviation between predicted and actual group predictions that we use to verify that the production system is performing as expected.

## References

1. Crammer, K., Singer, Y.: Online ranking by projecting. Neural Comput. **17**(1), 145–175 (2005)
2. Devillers, L., Vaudable, C., Chastagnol, C.: Real-life emotion-related states detection in call centers: a cross-corpora study. In: INTERSPEECH 2010, pp. 2350–2353 (2010)

3. Gutiérrez, P., Perez-Ortiz, M., Sanchez-Monedero, J., Fernández-Navarro, F., Hervas-Martinez, C.: Ordinal regression methods: survey and experimental study. IEEE Trans. Knowl. Data Eng. **28**(1), 127–146 (2016)

4. Har-Peled, S., Roth, D., Zimak, D.: Constraint classification: a new approach to multiclass classification. In: Cesa-Bianchi, N., Numao, M., Reischuk, R. (eds.) ALT 2002. LNCS (LNAI), vol. 2533, pp. 365–379. Springer, Heidelberg (2002). https://doi.org/10.1007/3-540-36169-3_29

5. Herbrich, R., Graepel, T., Obermayer, K.: Large margin rank boundaries for ordinal regression. In: Advances in Neural Information Processing Systems, pp. 115–132 (1999)

6. Herbrich, R., Graepel, T., Obermayer, K.: Support vector learning for ordinal regression. In: International Conference on Artificial Neural Networks, pp. 97–102 (1999)

7. Kim, K., Ahn, H.: A corporate credit rating model using multiclass support vector machines with an ordinal pairwise partitioning approach. Comput. Oper. Res. **39**(8), 1800–1811 (2012)

8. Metallinou, A., Narayanan, S.: Annotation and processing of continuous emotional attributes: challenges and opportunities. In: IEEE International Conference and Workshops on Automatic Face and Gesture Recognition, pp. 1–8 (2013)

9. Ovadia, S.: Ratings and rankings: reconsidering the structure of values and their measurement. Int. J. Soc. Res. Methodol. **7**(5), 403–414 (2004)

10. Park, Y., Gates, S.: Towards real-time measurement of customer satisfaction using automatically generated call transcripts. In: Proceedings of the 18th ACM Conference on Information and Knowledge Management, pp. 1387–1396. ACM (2009)

11. Pérez-Ortiz, M., Cruz-Ramírez, M., Ayllón-Terán, M., Heaton, N., Ciria, R., Hervás-Martínez, C.: An organ allocation system for liver transplantation based on ordinal regression. Appl. Soft Comput. **14**, 88–98 (2014)

12. Segura, C., Balcells, D., Umbert, M., Arias, J., Luque, J.: Automatic speech feature learning for continuous prediction of customer satisfaction in contact center phone calls. In: Abad, A., Ortega, A., Teixeira, A., García Mateo, C., Martínez Hinarejos, C.D., Perdigão, F., Batista, F., Mamede, N. (eds.) IberSPEECH 2016. LNCS (LNAI), vol. 10077, pp. 255–265. Springer, Cham (2016). https://doi.org/10.1007/978-3-319-49169-1_25

13. Sun, J., Xu, W., Yan, Y., Wang, C., Ren, Z., Cong, P., Wang, H., Feng, J.: Information fusion in automatic user satisfaction analysis in call center. In: International Conference on Intelligent Human-Machine Systems and Cybernetics (IHMSC), vol. 1, pp. 425–428 (2016)

14. Tian, Q., Chen, S., Tan, X.: Comparative study among three strategies of incorporating spatial structures to ordinal image regression. Neurocomputing **136**, 152–161 (2014)

15. Vaudable, C., Devillers, L.: Negative emotions detection as an indicator of dialogs quality in call centers. In: IEEE International Conference on Acoustics, Speech and Signal Processing (ICASSP), pp. 5109–5112 (2012)

16. Yan, H.: Cost-sensitive ordinal regression for fully automatic facial beauty assessment. Neurocomputing **129**, 334–342 (2014)

17. Yoon, J., Roberts, S., Dyson, M., Gan, J.: Bayesian inference for an adaptive ordered probit model: an application to brain computer interfacing. Neural Netw. **24**(7), 726–734 (2011)

# Probabilistic Inference of Twitter Users' Age Based on What They Follow

Benjamin Paul Chamberlain[1](✉), Clive Humby[2], and Marc Peter Deisenroth[1]

[1] Department of Computing, Imperial College London, London, UK
`b.chamberlain14@ic.ac.uk`
[2] Starcount Insights, 2 Riding House Street, London, UK

**Abstract.** Twitter provides an open and rich source of data for studying human behaviour at scale and is widely used in social and network sciences. However, a major criticism of Twitter data is that demographic information is largely absent. Enhancing Twitter data with user ages would advance our ability to study social network structures, information flows and the spread of contagions. Approaches toward age detection of Twitter users typically focus on specific properties of tweets, e.g., linguistic features, which are language dependent. In this paper, we devise a language-independent methodology for determining the age of Twitter users from data that is native to the Twitter ecosystem. The key idea is to use a Bayesian framework to generalise ground-truth age information from a few Twitter users to the entire network based on what/whom they follow. Our approach scales to inferring the age of 700 million Twitter accounts with high accuracy.

## 1 Introduction

Digital social networks (DSNs) produce data that is of great scientific value. They have allowed researchers to study the flow of information, the structure of society and major political events (e.g., the Arab Spring) quantitatively at scale.

Owing to its simplicity, size and openness, Twitter is the most popular DSN used for scientific research. Twitter allows users to generate data by *tweeting* a stream of 140 character (or less) messages. To consume content users *follow* each other. Following is a one-way interaction, and for this reason Twitter is regarded as an *interest network* (Gupta 2013). By default, Twitter is entirely public, and there are no requirements for users to enter personal information.

The lack of reliable (or usually any) demographic data is a major criticism of the usefulness of Twitter data. Enriching Twitter accounts with demographic information (e.g., age) would be valuable for scientific, industrial and governmental applications. Explicit examples include opinion polling, product evaluations and market research.

We assume that people who are close in age have similar interests as a result of age-related life events (e.g., education, child birth, marriage, employment, retirement, wealth changes). This is an example of the well-known homophily

© Springer International Publishing AG 2017
Y. Altun et al. (Eds.): ECML PKDD 2017, Part III, LNAI 10536, pp. 191–203, 2017.
https://doi.org/10.1007/978-3-319-71273-4_16

**Fig. 1.** Twitter profile for @williamockam that we created to illustrate our method. The profile contains the name, Twitter handle, number of tweets, number of followers, number of people following and a free-text description field with age information.

principle, which states that people with related attributes form similar ties (McPherson 2001). For age inference in Twitter, we exploit that most Follows[1] are indicative of a user's interests. Putting things together, we arrive at our central hypothesis that (a) somebody follows what is interesting to them, (b) their interests are indicative of their age. Hence, we propose to infer somebody's age based on what/whom they Follow. We created the artificial @williamockam account shown in Fig. 1 to use as a running example of our method.

The contribution of this paper is a probabilistic model that is massively scalable and infers every Twitter user's age based on what/whom they Follow without being restricted by national/linguistic boundaries or requiring data that few users provide (e.g. photos or large numbers of tweets). Our model handles the high levels of noise in the data in a principled way. We infer the age of 700 million Twitter accounts with high accuracy. In addition we supply a new public dataset to the community.

## 2   Related Work

There is a large body of excellent research on enhancing social data with demographic attributes. This includes work on gender (Burger 2011), political affiliation (Pennacchiotti 2011), location (Cheng 2010) and ethnicity (Mislove 2011; Pennacchiotti 2011). Also of note is the work of Fang (2015) who focus on modelling the correlations between various demographic attributes.

Following the seminal work of Schler (2006), the majority of research on age detection of Twitter users has focused on linguistic models of tweets (Al Zamal 2012; Nguyen 2011; Rao 2010). Notably, (Nguyen 2013) developed a linguistic model for Dutch tweets that allows them to predict the age category (using logistic regression) of Twitter users who have tweeted more than ten times in Dutch. They performed a lexical analysis of Dutch language tweets and obtained ground truth through a labour intensive manual tagging process. The principal features were unigrams, assuming that older people use more positive language, fewer pronouns and longer sentences. They concluded that age prediction works well for young people, but that above the age of 30, language tends to homogenise.

---

[1] we use capitalisation to indicate the Twitter specific usage of this word.

Additionally, tweet-based methods struggle to make predictions for Twitter users with low tweet counts. In practice, this is a major problem since we calculated that the median number of tweets for the 700 m Twitter users in our data set is only 4 (the *tweets* field shown in Fig. 1 is available as account metadata for all accounts).

The user name has also been considered as a source of demographic information. This was first done by Liu (2013) to detect gender and later by Oktay (2014) to estimate the age of Twitter users from the first name supplied in the free-text *account name* field (e.g. William in Fig. 1). In their research, they use US social security data to generate probability distributions of birth years given the name. They show that for some names, age distributions are sharply peaked. A potential issue with this approach is that methods based on the "user name" field rely on knowledge of the user's true first name and their country of birth (Oktay 2014). In practice, this assumption is problematic since Twitter users often do not use their real names, and their country of birth is generally unknown.

**Table 1.** Ground-truth data set: Age categories and counts. "features" gives the average number of feature accounts followed

| Idx | Age | Count | Freq. | Features |
|---|---|---|---|---|
| 0 | Under 12 | 7,753 | 5.9% | 23.7 |
| 1 | 12–13 | 20,851 | 15.8% | 27.9 |
| 2 | 14–15 | 30,570 | 23.1% | 30.8 |
| 3 | 16–17 | 23,982 | 18.1% | 28.7 |
| 4 | 18–24 | 33,331 | 25.2% | 26.0 |
| 5 | 25–34 | 9,286 | 7.0% | 23.1 |
| 6 | 35–44 | 3,046 | 2.3% | 22.6 |
| 7 | 45–54 | 1,838 | 1.0% | 16.0 |
| 8 | 55–64 | 962 | 0.7% | 11.4 |
| 9 | Over 65 | 596 | 0.5% | 11.2 |

Approaches to combine lexical and network features include Al Zamal (2012); Pennacchiotti (2011), who show that using the graph structure can improve performance at the expense of scalability. Kosinski (2013) used Facebook-Likes to predict a broad range of user attributes mined from 58,466 survey correspondents in the US. Their approach of solely using Facebook Likes as features for learning has the benefit of generalising readily to different locales. Culotta (2015) have applied a similar Follower based approach to Twitter to predict demographic attributes, however their approach of using aggregate distributions of website visitors as ground-truth is restricted to predicting the aggregate age of groups of users. Our work is inspired by the generality of the approaches of Kosinski (2013) and Culotta (2015), however our setting differs in two ways. We use data native to the Twitter ecosystem to generalise from a few examples to make individual predictions for the entire Twitter population. Secondly we do not make the assumption that our sample is an unbiased estimate of the Twitter population and we explicitly account for this bias to make good population predictions. For these reasons it is hard to get ground truth and careful probabilistic modelling is required to infer the age of arbitrary Twitter users.

# 3    Probabilistic Age Inference in Twitter

Our age inference method uses ground-truth labels (users who specify their age), which are then generalised to 700 m accounts based on the shared interests, which we derive from Following patterns.

To extract ground-truth labels we crawl the Twitter graph and download user descriptions. To do this we implemented a distributed Web crawler using Twitter access tokens mined through several consumer apps. To maximize data throughput while remaining within Twitter's rate limits we built an asynchronous data mining system connected to an access token server using Python's Twisted library Wysocki (2011).

Our crawl downloaded 700 m user descriptions. Fig. 1 shows the profile with associated metadata fields for the fictitious @williamockam account, which we use to illustrate our approach. We index the free-text description fields using Apache SOLR (Grainger 2014) and search the index for REGular EXpression (REGEX) patterns that are indicative of age (e.g., the phrase: "I am a 22 year old" in Fig. 1) across Twitter's four major languages (English, Spanish, French, Portuguese). For repeatability we include our REGEX code in the git repository. Twitter is ten years old and contains many out-of-date descriptions. To tackle the stale data problem we restricted the ground-truth to active accounts, defined to be accounts that had tweeted or Followed in the last three months (we do not have access to Twitter's logs). This process discovered 133,000 active users who disclosed their age (i.e., 0.02% of the 700 m indexed accounts), which we

**Table 2.** Public dataset labels: age categories and counts.

| Idx | Age range | Count |
|-----|-----------|-------|
| 1   | 10–19     | 4486  |
| 2   | 20–29     | 4485  |
| 3   | 30–39     | 4487  |
| 4   | 40–49     | 4485  |
| 5   | 50–59     | 4484  |
| 6   | 60–69     | 4481  |
| 7   | 70–79     | 4481  |

use as "ground-truth" labels. For each of these we download every account that they Followed. Figure 1 shows that @williamockam Follows 73 accounts and we downloaded each of their user IDs. We use ten age categories with a higher resolution in younger ages where there is more labelled data. For our ground-truth data set, the age categories, number of accounts, relative frequency and average number of features per category are shown in Table 1.

Applying REGEX matches to free-text fields inevitably leads to some false positives due to unanticipated character combinations when working with large data sets. In addition, many Twitter accounts, while correctly labelled, may not represent the interests of human beings. This can occur when accounts are controlled by machines (bots), accounts are set up to look authentic to distribute spam (spam accounts) or account passwords are hacked in order to sell authentic looking Followers. To reduce the impact of spurious accounts on the model we note that (1) incorrectly labelled accounts can have a large effect on the model as they are distant in feature space from other members of the class/label (2) incorrectly labelled accounts that have a small effect on the model (e.g. because they only follow one popular feature) do not matter much by definition. To measure the effect of each labelled account on the model we compute the Kullback-Leibler divergence $KL(P||P_{\backslash i})$ between the full model and a model evaluated with one data point missing. Here, $P$ is the likelihood of the full, labelled data set, and $P_{\backslash i}$ is the likelihood of the model using the labelled data set minus the $i^{th}$ data point. This methodology identifies any accounts that have a particularly large impact on our predictive distribution. We flagged any training examples that were more than three median absolute deviations from the median score for manual inspection. This process excluded 246 accounts from our training data and examples are shown in Table 3. We also randomly sampled 100 data points from across the full ground-truth set and manually verified them by inspecting the descriptions, tweets and who/what they Follow.

**Table 3.** Spurious data points identified by taking the Median Absolute Deviation of the leave-one-out KL-Divergence.

| Handle | Twitter description | REGEX age | Reason to exclude |
|---|---|---|---|
| RIAMOpera | Opera at the Royal Irish... Presenting: Ormindo Jan 11 | 11 | An Irish Opera |
| TiaKeough13 | My name Tia I'm 13 years old | 13 | Hacked account |
| 39yearoldvirgin | I'm 39 years old... if you're a woman, I want to meet you | 39 | Probably not 39 |
| 50Plushealths | Retired insurance Agent After 40 years of services | Retired | Using reciprocation software |
| MrKRudd | Former PM of Australia... Proud granddad of Josie & McLean | Grandparent | Outlier. Former AUS PM |

For reproducibility we make an anonymised sample of the data and our code publicly available[2]. The data is in two parts: (1) A sparse bipartite adjacency matrix; (2) a vector of age category labels. This dataset was collected and cleaned according to the methodology described above and then down-sampled to give approximately equal numbers of labels in each of seven classes detailed in Table 2. It includes only accounts that explicitly state an age (i.e. no grandparents or retirees). The adjacency matrix is in the format of a standard (sparse) design matrix and includes only features that are Followed by at least 10 examples. The high level statistics of this network are described in Table 4.

---

[2] https://github.com/melifluos/bayesian-age-detection.

## 3.1  Age Inference Based on Follows

Given a set of 133,000 labelled data points (ground-truth, i.e., Twitter users who reveal their age) we wish to infer the age of the remaining 700 m Twitter users. For this purpose, we define a set of features that can be extracted automatically. The features are based on the Following patterns of Twitter users. Once the features are defined, we propose a scalable probabilistic model for age inference.

Our age inference exploits the hypothesis that someone's interests are indicative of their age, and uses Twitter Follows as a proxy for interests. Therefore, the features of our model are the 103,722 Twitter accounts that are Followed by more than ten labelled accounts, which can be found automatically. Of the 73 accounts Followed by @williamockam, 8 had sufficient support to be included in our model. These were: Lord_Voldemort7, WaltDisneyWorld, Applebees, UniStudios, UniversalORL, HorrorNightsORL, HorrorNights and OlanRogers.

Table 5 shows the number of labelled accounts Following each feature for @williamockam. The support is the number of *labelled* Followers summed over all age categories, while Followers gives the total number of Followers (labelled and unlabelled). A general trend across all features (not only the ones relevant to @williamockam) is that the age distribution is peaked towards "younger" ages as not many older people reveal their age (we show this for the accounts with the highest support in our data set in the appendix on our git repo). To improve the predictive performance of the model in higher age categories we adapted our REGEX to search for grandparents and retirees. This augmented our training data with 176,748

**Table 4.** Public dataset adjacency matrix statistics. Subscript 1 describes labelled acounts and 2 describes features. $V$ denotes vertices, $E$ edges and $D$ degree.

| Attribute | Value |
|-----------|-------|
| $|V_1|$ | 31,389 |
| $|V_2|$ | 50,190 |
| $|E|$ | 1,810,569 |
| avg $D_1$ | 57.7 |
| max $D_1$ | 2049 |
| std $D_1$ | 95.2 |
| avg $D_2$ | 36.1 |
| max $D_2$ | 4405 |
| std $D_2$ | 96.2 |

**Table 5.** Follower counts for the eight @williamockam features. The support gives their total number of Followers in our labelled data set and Followers is their total number on Twitter. Fractional counts are from assigning a distribution to grandparents.

| Twitter handle | Support | <12 | 12–13 | 14–15 | 16–17 | 18–24 | 25–34 | 35–44 | 45–54 | 55–64 | ≥65 | Followers |
|----------------|---------|-----|-------|-------|-------|-------|-------|-------|-------|-------|-----|-----------|
| Lord_Voldemort7 | 273 | 5 | 35 | 75 | 55 | 87 | 13 | 0 | 1 | 1 | 1 | $2.0 \times 10^6$ |
| WaltDisneyWorld | 435 | 61 | 100 | 89 | 80 | 65 | 20 | 4 | 7 | 4 | 4 | $2.5 \times 10^6$ |
| Applebees | 191 | 18 | 43 | 38 | 30 | 37 | 9 | 8 | 2.33 | 2.33 | 3.33 | $0.57 \times 10^6$ |
| UniStudios | 60 | 7 | 7 | 14 | 14 | 13 | 5 | 0 | 0 | 0 | 0 | $0.27 \times 10^6$ |
| UniversalORL | 65 | 5 | 13 | 10 | 15 | 14 | 4 | 0 | 1.66 | 1.66 | 0.66 | $0.40 \times 10^6$ |
| HorrorNightsORL | 5 | 0 | 0 | 0 | 1 | 3 | 1 | 0 | 0 | 0 | 0 | $0.04 \times 10^6$ |
| HorrorNights | 18 | 1 | 3 | 1 | 4 | 6 | 0 | 1 | 0.66 | 0.66 | 0.66 | $0.08 \times 10^6$ |
| OlanRogers | 16 | 0 | 2 | 0 | 7 | 7 | 0 | 0 | 0 | 0 | 0 | $0.11 \times 10^6$ |

**Table 6.** Posterior distributions (4) for the eight features Followed by @williamockam. Probabilities are $\times 10^{-5}$

| Twitter handle | Support | <12 | 12–13 | 14–15 | 16–17 | 18–24 | 25–34 | 35–44 | 45–54 | 55–64 | ≥65 | Followers |
|---|---|---|---|---|---|---|---|---|---|---|---|---|
| Lord_Voldemort7 | 273 | 111.7 | 190.9 | 258.0 | 252.3 | 248.6 | 145.9 | 31.9 | 38.9 | 77.6 | 177.5 | $2.0 \times 10^6$ |
| WaltDisneyWorld | 435 | 725.0 | 538.2 | 441.2 | 377.6 | 267.3 | 233.2 | 194.2 | 270.7 | 254.5 | 224.4 | $2.5 \times 10^6$ |
| Applebees | 191 | 231.8 | 206.3 | 176.6 | 150.3 | 129.8 | 137.4 | 226.7 | 132.4 | 139.6 | 139.2 | $0.57 \times 10^6$ |
| UniStudios | 60 | 80.6 | 56.0 | 59.3 | 59.5 | 49.3 | 48.1 | 11.3 | 2.8 | 2.3 | 2.3 | $0.27 \times 10^6$ |
| UniversalORL | 65 | 67.4 | 63.0 | 56.6 | 60.5 | 50.7 | 42.0 | 21.1 | 62.7 | 86.4 | 40.6 | $0.40 \times 10^6$ |
| HorrorNightsORL | 5 | 0.3 | 0.7 | 1.5 | 4.0 | 8.3 | 9.4 | 2.0 | 0.3 | 0.1 | 0.1 | $0.04 \times 10^6$ |
| HorrorNights | 18 | 14.0 | 13.7 | 11.3 | 15.5 | 16.1 | 9.4 | 29.1 | 29.9 | 36.8 | 29.3 | $0.08 \times 10^6$ |
| OlanRogers | 16 | 4.3 | 9.1 | 10.6 | 21.9 | 19.8 | 5.0 | 1.6 | 1.3 | 1.3 | 1.3 | $0.11 \times 10^6$ |

people labelled as retired and 63,895 labelled as grandparents. In our ten-category model, retired people are added to the 65+ category. Grandparents are assigned a uniform distribution across the three oldest age categories, which roughly reflects the age distribution of grandparents in the US (UScensus 2014)[3], such that we ended up with approximately 374,000 labelled accounts in our ground-truth data.

**Probabilistic Model for Age Inference.** We adopt a Bayesian classification paradigm as this provides a consistent framework to model the many sources of uncertainty (noisy labels, noisy features, survey estimates) encountered in the problem of age inference.

Our goal is to predict the age label of an arbitrary Twitter user with feature vector $X$ given the set of feature vectors $\mathbf{X}$ and corresponding ground-truth age labels $\mathbf{A}$. Within a Bayesian framework, we are therefore interested in the posterior predictive distribution

$$P(A|X, \mathbf{X}, \mathbf{A}) \propto P(X|A, \mathbf{X}, \mathbf{A})P(A), \qquad (1)$$

where $P(A)$ is the prior age distribution and $P(X|A, \mathbf{X}, \mathbf{A})$ the likelihood.

The prior $P(A)$ is based on a survey of American internet users conducted by Duggan (2013). They sampled 1,802 over-18-year olds using random cold calling and recorded their demographic information and social media use. 288 of their respondents were Twitter users, yielding a small data set that we use for the prior distributions of over 18 s. For under 18 s we inferred the corresponding values of the prior using US census data (UScensus 2010), which leads to the categorical prior

$$P(A) = \mathrm{Cat}(\pi) = [1, 2, 2, 3, 14, 23, 23, 22, 6, 4] \times 10^{-2}. \qquad (2)$$

---

[3] This value was used as the US is the largest *Twitter country*.

The likelihood $P(X|A, \mathbf{X}, \mathbf{A})$ is obtained as follows: For scalability we make the Naive Bayes assumption that the decision to Follow an account is independent given the age of the user. This yields the likelihood

$$P(X|A, \mathbf{X}, \mathbf{A}) = \prod_{i=1}^{M} P(X_i|A, \mathbf{A}, \mathbf{X})^{X_i}, \tag{3}$$

where $X_i \in \{0, 1\}$ and $i$ indexes the features. $X_i = 1$ means "user $\chi$ Follows feature account $i$".[4]

We model the likelihood factors $P(X_i|A, \mathbf{A}, \mathbf{X})$ as Bernoulli distributions

$$P(X_i|A = a) = \text{Ber}(\mu_{ia}), \tag{4}$$

$i = 1, \ldots, M$, where $M$ is the number of features and there are 10 age categories indexed by $a = 1, \ldots, 10$. Since our labelled data is severely biased towards "younger" age categories we cannot simply learn multinomial distributions $P(A|X_i)$ for each feature based on the relative frequencies of their followers (see Table 1). To smooth out noisy observations of less popular accounts we use a hierarchical Bayesian model. Inference is simplified by using the Bernoulli's conjugate distribution, the beta distribution

$$\text{Beta}(\mu_{ia}|b_{ia}, c_a) \tag{5}$$

on the Bernoulli parameters $\mu_{ia}$. We seek hyper-parameters $b_{ia}, c_{ia}$ of the prior $\text{Beta}(\mu_{ia}|\mathbf{X}, \mathbf{A})$, which do not have a large effect when ample data is available, but produce sensible distributions when it is not. To achieve this we set $c_a$ to be constant across all features $X_i$ (hence dropping the $i$ subscript) and proportional to the total number of observations $n_a$ in each age category (the count column in Table 1). We then set $b_{ia} \propto \frac{n_a n_i}{K}$, where $K = 7 \times 10^8$ is the total number of Twitter users and $n_i$ is the number of Followers of feature $i$ (the Followers column of Table 5 for @williamockam's features). Then, the expected prior probability that user $\chi$ Follows account $i$ is $\mathbb{E}[\mu_{ia}|A = a] = \frac{b_{ia}}{b_{ia} + c_a} = \frac{n_i}{K + n_i}$, i.e., it is constant across age classes and varies in proportion to the number of Followers across features. The effect of this procedure is to reduce the model confidence for features where data is limited. Due to conjugacy, the posterior distribution on $\mu_{ia}$ is also Beta distributed. Integrating out $\mu_{ia}$ we obtain

$$P(X_i = 1|A = a, \mathbf{X}, \mathbf{A}) = \int_0^1 P(X_i = 1|\mu_i, A)P(\mu_i|\mathbf{X}, \mathbf{A}, A)d\mu_i \tag{6}$$

$$= \int_0^1 \mu_{ia}P(\mu_{ia}|\mathbf{X}, \mathbf{A})d\mu_{ia} = \mathbb{E}[\mu_{ia}|\mathbf{X}, \mathbf{A}] = \frac{n_{ia} + b_{ia}}{n_a + b_{ia} + c_a}, \tag{7}$$

---

[4] We only consider cases where $X_i = 1$ since the Twitter graph is sparse: In the full Twitter graph there are $7 \times 10^8$ nodes with $5 \times 10^{10}$ edges, which implies a density of $1.6 \times 10^{-7}$, i.e., the default is to follow nobody. Hence, not following an account does not contain enough information to justify the additional computational cost.

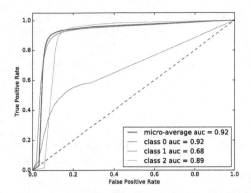

**Fig. 2.** Receiver operator characteristics for three class age detection (0 = under 18, 1 = 18–45, 2 = 45+). The dashed line indicates random performance.

where $n_{ia}$ is the number of labelled Twitter users in age category $a$ who Follow feature $X_i$, which are given in Table 5 for the @williamockam features and $n_a$ is the number of Twitter users in category $a$ in the ground-truth (See Table 1). Performing this calculation yields the likelihoods for the @williamockam features shown in Table 6. We are now able to compute the predictive distribution in (1) to infer the age of an arbitrary Twitter user. The predictive distribution for @williamockam is shown in Fig. 4 and is calculated by taking the product of the likelihoods from Table 6 with the prior in (2) and normalising.

The generative process in our model for the likelihood term in (1) is as follows.

1. Draw an age category $A \sim \mathrm{Cat}(\pi)$
2. For each feature $i$ draw $\mu_{ia} \sim \mathrm{Beta}(\mu_{ia}|b_{ia}, c_a)$
3. For each account draw the Follows: $X_i \sim \mathrm{Ber}(\mu_{ia})$

**Table 7.** The most discriminative features based on the posterior distribution over age in (6). Descriptions are taken from the $1^{st}$ line of their Wikipedia pages. See the git repo for a full table with probabilities and handles.

| <12 | 12–13 | 14–15 | 16–17 | 18–24 |
|---|---|---|---|---|
| Vlogger | Child presenter | Child singer | Singer | Metalcore band |
| Minecraft gamer | YouTuber | Child singer | Metalcore band | Rock band |
| Internet personality | Child actress | Child singer | Deathcore singer | Rapper |
| Vlogger | Child actress | Child singer | | |
| Gaming commentator | Girl band | Child singer | Electronic band | Rock band |
| 25–34 | 35–44 | 45–64[a] | 65+ | |
| Hip hop duo | Hip hop artist | Evangelist | Political journalist | |
| Boy band | Rapper | Evangelist | Retired cyclist | |
| Boy band | History channel | Evangelist | Golf channel | |
| Comedian | Record label | Faith group | Retired rugby player | |
| Adult actress | Boxer | Faith magazine | Boxer | |

[a] Both categories have the same features

In Table 7, we report the five features with the highest posterior age values of $P(A|X_i = 1)$ for each age category. The account descriptions are taken from the first line of the relevant Wikipedia page. The youngest Twitter users are characterised by an interest in internet celebrities and computer games players. Music genres are important in differentiating all age groups from 12–45. 25–34 year olds are in part marked by entities that saw greater prominence in the past. This group is also distinguished by an interest in pornographic actors. Age categories 45–54 and 55–64 have the same top five and are differentiated by their interest in religious topics. Users older than 65 are identifiable through an interest in certain sports and politics.

## 4    Experimental Evaluation

We demonstrate the viability of our model for age inference in huge social networks by applying it to 700 m Twitter accounts. We conducted three experiments: (1) We compare our approach with the language-based model by Nguyen (2013), which can be considered the state of the art for age inference. (2) We compare our age inference results with the survey by Duggan (2013).

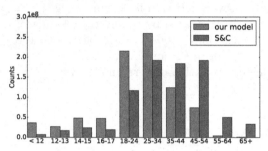

(3) We assess the quality of our age inference on a 10% hold-out set of ground-truth labels and compare it with results obtained from inference based solely on the prior derived from census and survey data in (2) for age prediction.

**Fig. 3.** Red bars show #accounts that our model allocated to each age class using the mode of the predictive posterior. Blue bars show #accounts that would have been allocated to each age class if ages were drawn from the Survey and Census (S&C) prior. (Color figure online)

### 4.1    Comparison with Dutch Language Model

For comparison with the state-of-the-art work of Nguyen (2013) based on linguistic features (Dutch tweets) we consider the performance of our model as a three-class classifier using age bands: under 18, 18–44 and 45+.

**Table 8.** Statistics for age prediction on a held-out test set.

|  |  | <12 | 12–13 | 14–15 | 16–17 | 18–24 | 25–34 | 35–44 | 45–54 | 55–64 | ≥65 |
|---|---|---|---|---|---|---|---|---|---|---|---|
|  | Test cases | 651 | 1,731 | 2,678 | 2,036 | 2,670 | 776 | 230 | 5,058 | 5,145 | 20,487 |
| Ours | Recall | **0.19** | **0.20** | **0.38** | **0.23** | **0.33** | **0.25** | 0.18 | **0.32** | **0.41** | **0.30** |
|  | Precision | **0.22** | **0.33** | **0.36** | **0.24** | **0.31** | **0.15** | 0.07 | **0.14** | **0.19** | **0.79** |
|  | Micro F1 | **0.31** |  |  |  |  |  |  |  |  |  |
| S&C | Recall | 0.01 | 0.02 | 0.02 | 0.03 | 0.14 | 0.23 | **0.23** | 0.22 | 0.06 | 0.04 |
|  | Precision | 0.02 | 0.04 | 0.06 | 0.05 | 0.06 | 0.02 | 0.01 | 0.12 | 0.12 | 0.49 |
|  | Micro F1 | 0.07 |  |  |  |  |  |  |  |  |  |

Table 9 lists the performance of our age inference algorithm on a 10% hold-out test set and the Dutch Language Model (DLM) proposed by Nguyen (2013). The corresponding performance statistics are shown in Table 9.

Both methods perform equally well with a Micro F1 score of 0.86. The precision and recall show that the DLM approach is efficient, extracting information from only a small training set (support). This is because significant engineering work went into labelling and feature design. In contrast, our feature generation process is automatic and scalable. While we do not achieve the same performance for the lower age categories, for the oldest age category, our approach performs substantially better than the method by Nguyen (2013), suggesting that a hybrid method could perform well. We leave this for future work.

The major advantages of our model to the state-of-the-art approach are twofold: First, we have applied our age inference to 700 m Twitter users, as opposed to being limited to a sample of Dutch Twitter users with a relatively high number of Tweets. Second, generating our training set is fully automatic and relies only on Twitter data[5], i.e., no manual labelling or verification is required.

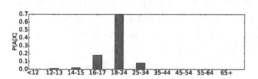

**Fig. 4.** Posterior age distribution for @williamockam.

Figure 2 shows the areas under the receiver-operator characteristics (ROC) curves for our three-class model. The curves are generated by measuring the true positive and false positive rates for each class over a range of classification thresholds. A perfect classifier has an area under the curve (AUC) equal to one, while a completely random classifier follows the dashed line with an $AUC = 0.5$. Performance is excellent for classes under 18 and over 45, but weaker for 18–45 where training data was limited, which we note as an area for improvement in future work.

**Table 9.** Performance for three-class age model.

|           | Our approach | | | DLM (Nguyen 2013) | | |
|-----------|------|-------|-----|------|-------|-----|
|           | <18  | 18–44 | ≥45 | ≤18  | 18–44 | ≥45 |
| Support   | 7,096 | 3,676 | 30,690 | 1,576 | 608 | 310 |
| Precision | 0.76 | 0.39 | **0.96** | **0.93** | **0.67** | 0.82 |
| Recall    | 0.68 | 0.50 | **0.95** | **0.98** | **0.75** | 0.45 |
| Micro F1  | **0.86** | | | **0.86** | | |

## 4.2   Comparison with Survey and Census Data

We report results on inferring the age of arbitrary Twitter users with the ten category model. Figure 3 shows aggregate classification results for 700m Twitter accounts compared with expected counts based on survey data (S&C) Duggan (2013). Our model predicts that over 50% of Twitter users are between 18 and 35,

---

[5] Nguyen (2013) used additional LinkedIn data for labelling.

i.e., the bias of the original training set has been removed due to the Bayesian treatment. It is likely that S&C under-represents young people as we did not factor in the increased rates of technology uptake amongst the younger people when converting census data.

### 4.3   Quality Assessment

In the following, we assess the quality of our age inference model (10 categories) on a 10% hold-out test data set.

Table 8 shows the performance statistics for this experiment. The majority of the test cases are in the younger age categories (due to the bias of young people revealing their age) and in older age categories (due to the inclusion of grandparents and retirees). Table 8 shows that the precision depends on the size of the data (e.g., predicting 25–44 year categories is hard) whereas the recall is fairly stable across all age categories.[6] Our model significantly outperforms an approach based only on the survey and census data (S&C), which we use as a prior. This highlights the ability of our model to adapt to the data.

## 5   Conclusion

We proposed a probabilistic model for age inference in Twitter. The model exploits generic properties of Twitter users, e.g., whom/what they follow, which is indicative of their interests and, therefore, their age. Our model performs as well as the current state of the art for inferring the age of Twitter users without being limited to specific linguistic or engineered features. We have successfully applied our model to infer the age of 700 million Twitter users demonstrating the scalability of our approach. The method can be applied to any attributes that can be extracted from user profiles.

**Acknowledgements.** This work was partly funded by an Industrial Fellowship from the Royal Commission for the Exhibition of 1851. The authors thank the anonymous reviewers for providing many improvements to the original manuscript.

## References

Al Zamal, F., Liu, W., Ruths, D.: Homophily and latent attribute inference: inferring latent attributes of twitter users from neighbors. In: ICWSM (2012)

Burger, J.D., Henderson, J., Kim, G., Zarrella, G.: Discriminating gender on Twitter. In: EMNLP (2011)

Cheng, Z., Caverlee, J., Lee, K.: You are where you tweet: a content-based approach to geo-locating Twitter users. In: CIKM (2010)

Culotta, A., Nirmal, R.K., Cutler, J.: Predicting the demographics of Twitter users from website traffic data. In: AAAI (2015)

---

[6] Without the inclusion of grandparents and retirees in the training set, the predictive performance would rapidly drop off for ages greater than 35.

Duggan, M., Brenner, J.: The demographics of social media Users–2012. http://tinyurl.com/jk3v9tu. Retrieved 12 Sep 2015

Fang, Q., Sang, J., Xu, C., Hossain, M.S.: Relational user attribute inference in social media. IEEE Trans. Multimedia **17**(7), 1031–1044 (2015)

Grainger, T., Potter, T.: Solr in Action. Manning Publications Co., Cherry Hill (2014)

Gupta, P., Goel, A., Lin, J., Sharma, A., Wang, D., Zadeh, R.: WTF: the who to follow service at Twitter. In: WWW (2013)

Kosinski, M., Stillwell, D., Graepel, T.: Private traits and attributes are predictable from digital records of human behavior. PNAS **110**(15), 5802–5805 (2013)

Liu, W., Ruths, D.: Whats in a name? using first names as features for gender inference in Twitter. In: AAAI Spring Symposium on Analyzing Microtext (2013)

McPherson, M., Smith-Lovin, L., Cook, J.M.: Birds of a feather: homophily in social networks. Ann. Rev. Sociol. **27**(1), 415–444 (2001)

Mislove, A., Lehmann, S., Ahn, Y.Y.: Understanding the demographics of Twitter users. In: ICWSM (2011)

Nguyen, D., Gravel, R., Trieschnigg, D., Meder, T.: How old do you think i am? a study of language and age in Twitter. In: ICWSM (2013)

Nguyen, D., Noah, A., Smith, A., Rose, C.P.: Author age prediction from text using linear regression. In: LaTeCH (2011)

Oktay, H., Firat, A., Ertem, Z.: Demographic breakdown of Twitter users: an analysis based on names. In: BIGDATA (2014)

Pennacchiotti, M., Popescu, A.M.: A machine learning approach to Twitter user classification. In: ICWSM (2011)

Rao, D., Yarowsky, D., Shreevats, A., Gupta, M.: Classifying latent user attributes in Twitter. In: SMUC (2010)

Schler, J., Koppel, M., Argamon, S., Pennebaker, J.W.: Effects of age and gender on blogging. In: AAAI-CAAW (2006)

Wysocki, R., Zabierowski, W.: Twisted framework on game server example. In: CADSM (2011)

U.S. Census Bureau, 2010 Census. Profile of General Population and Housing Characteristics: 2010. https://goo.gl/VAGMN1. Retrieved 12 Sep 2015

U.S. Census Bureau, American Community Survey, 2014 Grandparent Statistics. https://goo.gl/CqGXWI. Retrieved 15 Nov 2015

# Quantifying Heterogeneous Causal Treatment Effects in World Bank Development Finance Projects

Jianing Zhao[1]($\boxtimes$), Daniel M. Runfola[2], and Peter Kemper[1]

[1] College of William and Mary, Williamsburg, VA 23187-8795, USA
{jzhao,kemper}@cs.wm.edu
[2] AidData, 427 Scotland Street, Williamsburg, VA 23185, USA
drunfola@aiddata.org

**Abstract.** The World Bank provides billions of dollars in development finance to countries across the world every year. As many projects are related to the environment, we want to understand the World Bank projects impact to forest cover. However, the global extent of these projects results in substantial heterogeneity in impacts due to geographic, cultural, and other factors. Recent research by Athey and Imbens has illustrated the potential for hybrid machine learning and causal inferential techniques which may be able to capture such heterogeneity. We apply their approach using a geolocated dataset of World Bank projects, and augment this data with satellite-retrieved characteristics of their geographic context (including temperature, precipitation, slope, distance to urban areas, and many others). We use this information in conjunction with causal tree (CT) and causal forest (CF) approaches to contrast 'control' and 'treatment' geographic locations to estimate the impact of World Bank projects on vegetative cover.

## 1 Introduction

We frequently seek to test the effectiveness of targeted interventions - for example, a new website design or medical treatment. Here, we present a case study of using recent theoretical advances - specifically the use of tree-based analysis [3] - to estimate heterogeneous causal effects of global World Bank projects on forest cover over the last 30 years.

The World Bank is one of the largest contributors to development finance in the world, seeking to promote human well-being through a wide variety of programs and related institutions [1]. However, this goal is frequently at odds with environmental sustainability - building a road can necessitate the removal of trees; building a factory that supplies jobs can lead to the pollution of proximate forests. Multiple environmental safeguards have been put in place to offset these challenges, but relatively little is known about their efficacy across large scales.

We adopt the commonly applied approach of selecting "control" cases (i.e., areas where World Bank projects have very little funding) to contrast to

© Springer International Publishing AG 2017
Y. Altun et al. (Eds.): ECML PKDD 2017, Part III, LNAI 10536, pp. 204–215, 2017.
https://doi.org/10.1007/978-3-319-71273-4_17

"treated" cases (i.e., areas where World Bank projects have a large amount of funding). This is analogous to similar approaches in the medical literature, where humans are put into control and treatment groups, and individuals that are similar along all measurable attributes are contrasted to one another after a medicine is administered. This is necessary due to the generalized challenge of all observational studies: it is impossible observe the exact same unit of observation with and without a World Bank project simultaneously - in the same way it would be impossible to examine a patient that was and was not given medication at the same time. Further complicating the challenge presented in this paper is the scope of the World Bank - with tens of thousands of project locations worldwide, there is considerable variation in the aims of different projects, the project's size, location, socio-economic, environmental, and historical settings. This variation makes traditional, aggregate estimates of impact unhelpful, as such aggregates mask variation in where World Bank projects may be helping - or harming - the environment. Following this, we investigate the research question *What is the impact of world bank projects on forest cover?*

To examine this question, we first integrate information on the geographic location of World Bank projects with additional, satellite derived information on the geographic, environmental, and economic characteristics of each project. We apply four different models to this dataset, and contrast our findings to illustrate the various tradeoffs in these approaches. Specifically, we test Transformed Outcome Trees (TOTs), Causal Trees (CTs), Random Forest TOTs (RFTOTs), and Causal Forests (CFs). We follow the work of Athey and Imbens [3], who demonstrated how regression trees and random forests can be adjusted to estimate heterogenous causal effects. This work is based on the Rubin Causal Model (or potential outcome framework), where causal effects are estimated through comparisons between observed outcomes and the "counterfactual" outcomes one would have observed under the absence of an aid project [9]. While traditional tree-based approaches rely on training with data with known outcomes, Athey and Imbens illustrated that one can estimate the conditional average treatment effect on a subset with regressions trees after an appropriate data transformation process.

Many approaches to estimating heterogeneous effects have emerged over the last decade. LASSO [14] and support vector machines (SVM) [15] may serve as two popular examples. For this paper, we focus on very recent tree-based techniques that are very promising for causal inference. In [12], Su et al. proposed a statistical test as the criterion for node splitting. In [3], Athey and Imbens derived TOTs and CTs, an idea that is followed up on by Wagner and Athey [16] with CF (causal forest, random forests of CTs), and similarly Denil et al. in [6] who use different data for the structure of the tree and the estimated value within each node. Random forests naturally gave rise to the question of confidence intervals for the estimates they deliver. Following this, Meinshausen introduced quantile regression forests in [10] to estimate a distribution of results, and Wagner et al. [17] provided guidance for confidence intervals with random forests. Several authors, including Biau [4], recognize a gap between theoretical underpinnings and the practical applications of random forests.

The contribution of this paper is twofold: we evaluate and compare a number of proposed methods on simulated data where the ground truth is known and apply the most promising for the analysis of a real world data set. Practical experience results on tree-based causal inference methods are rare. To the best of our knowledge, this is the first investigation on the analysis of a spatial data set of world wide range with a large scale set of projects and dimensions. When it comes to applications for causal inference techniques, A/B testing for websites (such as eBay) is a more common [13]. A/B testing is conducted by diverting some percentage of traffic for a website A to a modified variant B of said website for evaluation purposes. This leads to a large amount of data with clearly defined treated and untreated groups where cases vary mainly by user activity. While the difference between A and B is precisely defined and typically small, the huge number of cases helps to recognize treatment effects. This is very different to the World Bank data which is both much more limited in size, and also spread all over the world (resulting in large diversity across projects). The rest of the paper is structured as follows. In Sect. 2, we present the basic methodology for the calculation of CT and CF. Section 3 introduces the data set, its characteristics, preprocessing steps and the calculation of propensity scores necessary for the estimation of each type of tree. In Sect. 4, we present the outcome of the analysis. We conclude in Sect. 5.

## 2    Methodology

Causal inference is to a vast part a missing data problem as we can not observe a unit at the same time receiving and not receiving treatment to compare the outcomes. We introduce some notation and recall common concepts to be able to address this problem in a more formal way.

***Causal Effects.*** Suppose we have a data set with $n$ independently and identically distributed (iid) units $U_i = (X_i, Y_i)$ with $i = 1, \cdots, n$. Each unit has an observed feature vector $X_i \in \mathbb{R}^d$, a response (i.e., the outcome of interest) $Y_i \in \mathbb{R}$ and binary treatment indicator $W_i \in \{0,1\}$. For a unit-level causal effect, the Rubin causal model considers the treatment effect on unit $i$ being $\tau(X_i) = Y_i(1) - Y_i(0)$, the difference between treated $Y_i(1)$ and untreated $Y_i(0)$ outcome. One can be interested in an overall average treatment effect across all units $U$ or investigate treatment effects of subsets that are characterized by their features $X$. The latter describes heterogeneous causal effects and is often of particular interest. In our case, it is interesting to identify characteristics of subsets of projects where the environment is affected strongly (positive or negative) by a World Bank project. The heterogenous causal effect is defined as $\hat{\tau}(x) = \mathbb{E}[Y_i(1) - Y_i(0) \mid X_i = x]$ following [8].

***Causal Tree.*** A regression tree defines a partition of a set of units $U_i = (X_i, Y_i)$ as each leaf node holds a subset of units satisfying conditions on $X$ expressed along the path from root node to leaf. This helps for the condition in $\hat{\tau}(x) = \mathbb{E}[Y_i(1) - Y_i(0) \mid X_i = x]$. In observational studies, a unit is either treated or not, so we know either $Y_i(1)$ or $Y_i(0)$, but not both. However, one can still estimate $\tau(x)$ if one assumes unconfoundedness: $W_i \perp\!\!\!\perp (Y_i(1), Y_i(0)) \mid X_i$. Athey and Imbens [3] showed that one can estimate the causal effect as:

$$\hat{\tau}(X_i) = \sum_{i \in T} Y_i \cdot \frac{W_i/\hat{e}(X_i)}{\sum_{j \in T} W_j/\hat{e}(X_j)} - \sum_{i \in C} Y_i \cdot \frac{(1 - W_i)/(1 - \hat{e}(X_i))}{\sum_{j \in C}(1 - W_j)/(1 - \hat{e}(X_j))} \quad (1)$$

where $e(X_i)$ is the propensity score of project $i$ which is calculated by logistic regression, T represents treatment units, and C control units. Hence one can adapt the calculation of a regression tree to support calculation of $\hat{\tau}(X_i)$ by (1) by adjusting the splitting rule in the tree generation process.

In a classic regression tree, mean square error (MSE) is often used as the criterion for node splitting, and the average value within the node is used as the estimator. Following Athey and Imbens [3], we use (1) as the estimator and the following equation as the new MSE for any given node $J$ in the causal tree.

$$MSE - \sum_{i \in J}(Y_i(1) - Y_i(0) - \hat{\tau}(X_i))^2 = \sum_{i \in J} \tau(X_i)^2 - \sum_{i \in J} \hat{\tau}(X_i)^2 \quad (2)$$

The right equation follows if one assumes that $\sum_{i \in J} \tau(X_i) = \sum_{i \in J} \hat{\tau}(X_i)$. The key observation is that $\sum_{i \in J} \tau(X_i)^2$ is constant and does not impact $\Delta MSE$. For a split, data in node $P$ is split into a left $L$ and right $R$ node, $\Delta MSE = MSE_P - MSE_L - MSE_R = \sum_{i \in P} \tau(X_i)^2 - \sum_{i \in L} \hat{\tau}(X_i)^2 - \sum_{i \in R} \hat{\tau}(X_i)^2$. The ground truth $\tau(X_i)$ cancels out in $\Delta MSE$ and we can grow the tree without knowledge of $\tau(X_i)$. However, there is one more constraint we need to add to the splitting rule aside from $MSE$. To use (1) for the calculation of $\hat{\tau}(X_i)$, neither set $T$ nor $C$ can be empty. Due to characteristics of the data in our applied study, we found that cases where only C or T units existed in children naturally emerged, so we added a corresponding additional stopping criterion to the splitting rule to prevent splits that would lead to situations where $T$ or $C$ had less than a fixed minimum cardinality.

**Causal Forest.** While a single causal tree allows us to estimate the causal effect, it leads to the problem of overfitting and subsequent challenges for pruning the tree. A common solution is to use an ensemble method such as bootstrap aggregating or bagging, namely a variant of Breiman's random forest [5]. If one applies the random forest approach to causal trees, the result is called a causal forest. Computation of a causal forest scales well as it can naturally be run in parallel. The same adjustments for generating a single CT apply to the generation of a random forest of CTs. We implemented a causal forest algorithm with the help of the scikit learn package. We can estimate the causal effect $\tau_{CF}(X_i)$ from a causal forest (a set CF of causal trees) for a unit $i$ as the average across the estimates obtained from its trees: $\hat{\tau}_{CF}(X_i) = \frac{1}{|CF|} \sum_{t \in CF} \hat{\tau}_t(X_i)$.

## 3    Data

*Data Pre-processing.* This analysis relies on three key types of data: satellite data to measure vegetation, data on the geospatial locations of World Bank projects, and covariate datasets[1]. Our primary variable of interest is the fluctuation of vegetation proximate to World Bank projects, which is derived from long-term satellite data [11]. There are many different approaches to using satellite data to approximate vegetation on a global scale, and satellites have been taking imagery that can be used for this purpose for over three decades. Of these approaches, the most frequently used is the Normalized Difference Vegetation Index (NDVI), which has the advantage of the longest continuous time record. NDVI measures the relative absorption and reflectance of red and near-infrared light from plants to quantify vegetation on a scale of $-1$ to $1$, with vegetated areas falling between 0.2 and 1 [7]. While the NDVI does have a number of challenges - including a propensity to saturate over densely vegetated regions, the potential for atmospheric noise (including clouds) to incorrectly offset values, and reflectances from bright soils providing misleading estimates - the popularity of this measurement has led to a number of improvements over time to offset many of these errors. This is especially true of measurements from longer-term satellite records, such as those used in this analysis, produced from the MODIS and AVHRR satellite platforms [11].

The second primary dataset used in this analysis measures where - geographically - World Bank projects were located. This dataset was produced by [2], relying on a double-blind coding system where two experts independently assign latitude and longitude coordinates, precision codes, and standardized place names to each geographic feature. Disagreements are then arbitrated by a third party.

In addition to the project name, the World Bank provided information on the amount of funding for each project and the year it was implemented, alongside a number of other ancillary variables. The database also provides information on the number of locations associated with each project - i.e., a single project may build multiple schools. These range from $n = 1$ to $n = 649$ project locations for a single project.

*Data Characteristics.* The temporal coverage of the covariates is variable across sources. For NDVI, precipitation, and temperature we have highly granular, yearly information on characteristics at each World Bank project location. From this information, we generate additional information regarding the trend (positive or negative) before and after project implementation, as well as simple averages in the pre- and post periods. Many variables only have a single measurement - population density, accessibility to urban areas, slope, and elevation are all measured circa 2000, while distances to roads and rivers are measured circa 2010. Figure 1(a) shows average annual NDVI values of all project locations for each year since 1982. The mean values are non-negative for all projects over all years, and typical values are around 0.2 which is a lower bound for areas with vegetation. Figure 1(b) shows the distributions of slope values for a time series of NDVI values that starts in 1982 and ends with the year before each project starts. Approximately 75% of all projects have an upward trend in NDVI values across this time period. Figure 1(c) shows that treated and control projects have

---

[1] For detailed information, check https://github.com/zjnsteven/appendix.

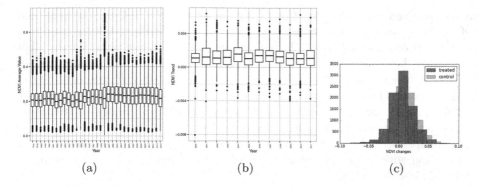

**Fig. 1.** Properties of NDVI values at World Bank project locations

very similar empirical distributions for changes in NDVI values when the pre- and post- averages are contrasted.

*Data Interpretation for the Context of Measuring Heterogenous Treatment Effects.* One key attribute of causal attribution is a dataset which distinguishes between treated and untreated cases. In the case of a clinical trial, human beings who receive treatment might be contrasted to a control group of other humans of similar characteristics who do not receive a treatment. Because World Bank projects either exist or not, here we attempt to replicate the treated and untreated conditions by contrasting World Bank projects that were funded at very low levels ("control") to those that were funded at high levels ("treated"). This is reflective of a hypothesis that the observed treatment effect should positively correlate with the amount of funding, i.e., huge amounts of funding are expected to have a bigger effect than small amounts of funding. Following this, we assign $W_i = 1$ if a project's funding is in the upper third of all funded projects. While an imperfect representation of an area at which no World Bank project exists, by leveraging locations where a World Bank project exists but at a very low intensity we mitigate potential confounding sources of bias associated with locations the World Bank chooses to site projects at. Further, we bias our results in the more conservative negative direction - i.e., we will tend to under-estimate the impact of World Bank projects relative to null cases. Future research will consider the difference between this and a true null case in which locations with no aid at them are contrasted.

As a single project typically takes place at several project locations, we consider each project location as an individual unit - i.e., a school may be effective in one location, but not another, even if they were implemented by the same funding mechanism. Further, to capture potential geographic heterogeneity this might introduce, for each unit's feature vector (i.e., selected covariates), we include the longitude and latitude of the project location. The total length of the feature vector is $d = 40$. All covariates are numerical and their values are not normalized. For our outcome measure (i.e., the variable we seek to estimate the impact on), we contrast the pre-treatment and post-treatment average NDVI values at a project's location. Let $ndvi_i(92, 03)$ denote the average of NDVI values observed for project location $i$ over the years from 1992 to 2003 (the year before the project is implemented, which varies across projects; 2003 is used here for illustration). Let $ndvi_i(05, 12)$ describe the corresponding value for the eight years after the project starts. The response $Y_i = ndvi_i(05, 12) - ndvi_i(92, 03)$ is thus the difference of the two averages. Figure 1(c) shows histograms of $Y_i$ values for treated and control projects. In order to calculate $Y^*$ for $Y$, we calculate the propensity score $e(x)$, which describes the expected likelihood of treatment $W_i$ for a given unit of observation. As described above, while there are many methods for estimating $e(x)$, here we use logistic regression to provide a better comparison with the econometric approaches commonly employed in the international aid community.

## 4    Experiments and Results

We follow a two stage procedure to examine the effectiveness of both the CT and CF algorithms, specifically considering our unique context of the effectiveness of World Bank projects. First, we test and evaluate which approach is most suited to our application using simulated synthetic data where we know the ground truth and where we can vary the size of sample data. Second, we apply these algorithms to examine the efficacy of World Bank projects based on satellite imagery. We implemented the CT and CF algorithms as well as Athey and Imbens transformed outcome tree (TOT) approach [3] and a random forest variant of TOT (RFTOT) using scikit-learn. The latter serve as a baseline for the performance of CT and CF algorithms.

*Experimental Results for Simulated Data.* First, we iteratively simulate synthetic datasets with known parameters to evaluate how the estimation of propensity score, dataset size, and degree of similarity between the control and treatment groups impact the accuracy of the result. To do this, we follow a bi-partite data generation process, in which two equations are used (one for treated cases and another for control cases).

We use each of the following two equations to produce one half of all data points. $Y_i^1$ gives the result for treated cases; $Y_i^0$ is for the control group. Here, from $x_1$ to $x_8$, $x_j \sim \mathcal{N}(0, 1)$ as well as $\varepsilon \sim \mathcal{N}(0, 1)$.

$$Y_i(1) = W_i^1 + \sum_{j=1}^{k} x_j * W_i^1 + \sum_{j=1}^{8} x_j + \varepsilon, \quad Y_i(0) = W_i^0 + \sum_{j=1}^{k} x_j * W_i^0 + \sum_{j=1}^{8} x_j + \varepsilon \quad (3)$$

As used in Table 1, k is defined as the number of covariates which contribute to heterogeneity in the causal effect. The true value of the causal effect is then $\tau(X_i) = Y_i(1) - Y_i(0) = 1 + \sum_{i=1}^{k} x_i$, with $W^1 = 1$ and $W^0 = 0$.

The first scenario we examine considers synthetic datasets with a randomized treatment assignment (each unit has the same probability to be treated, $e(x) = 0.5$). Figure 2 shows corresponding results for n = 2000, and includes both single tree and random forest implementations of Transformed Outcome

**Table 1.** Mean square error (forest has 1000 trees, feature ratio = 0.8)

| Sample size | CF | | CT | | TOT | | RFTOT | |
|---|---|---|---|---|---|---|---|---|
| | Mean | Std | Mean | Std | Mean | Std | Mean | Std |
| 1000 | 0.60 | 0.001 | 1.27 | 0.02 | 9.96 | 0.24 | 7.74 | 0.13 |
| 5000 | 0.58 | 0.001 | 0.99 | 0.02 | 7.95 | 0.03 | 5.61 | 0.05 |
| 10000 | 0.51 | 0.00001 | 0.86 | 0.005 | 7.45 | 0.02 | 5.14 | 0.02 |

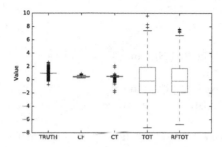

**Fig. 2.** Estimated treatment effects for randomized assignment, $e(x) = 0.5$

Trees (TOT; [3]) for comparison. The resultant distributions all encompass the true mean results, but with considerable difference in overall metrics of error. The Causal Forest approach is the most accurate across all simulations as well as the tightest overall distribution; this is in contrast to the TOT forest implementation. For single trees, the CT performs much better than the TOT and even outperforms the RFTOT.

The second scenario considers synthetic datasets with varying numbers of observations (n = 1000, 5000, and 10,000). We calculate the mean square error for CT, CF, TOT and RFTOT. The results in Table 1 show that - as expected - the error gets smaller as the number of observations increases. Of particular importance, we note that in the case of smaller datasets, the CF implementation strongly outperforms the single-tree CT implementation under all the scenarios we test.

We also test the convergence of each method as the size of data increases, as shown in Fig. 3. Figure 3a shows the MSE of each methods with increasing data size, while Fig. 3b shows a zoomed-in version of the MSE of the CF approach (due to the lower magnitude of MSE observed). At least for this specific data generation process, the CF and CT outperform other approaches, which is why we focus on them for the analysis of the World Bank data set where we can not measure accuracy.

*Results for World Bank Data.* Following the simulation results, we seek to identify and contrast the benefits and drawbacks associated with applying CT and CF approaches to a real-world scenario. In this case study, we identify the impact of international aid - specifically, World Bank projects - on forest cover. First,

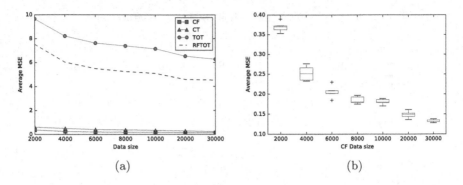

(a)                                            (b)

**Fig. 3.** MSE changes with data size

we use a single CT to estimate the causal effect $\hat{\tau}(X_i)$ of a single project $i$ with (Eq. 1) applied to the leaf where the project is located. Second, we implement a Causal Forest.

While our simulations, as well as the existing literature, suggest the Causal Tree has many drawbacks relative to a Causal Forest, it can enable practitioners to make inferences that are precluded by forest-based approaches. Most notably, the structure of single trees can provide insight into the explicit drivers of impacts - in this case, of World Bank projects. As an example, in the Causal Tree implementation here, we find that the year a project started was an important driver of effectiveness - specifically, projects starting before 2005 were more effective than those after 2005. This type of insight is particularly helpful, as it allows for analysis into the causes of impact heterogeneity. However, the lack of information on the robustness of findings in a single tree approach, coupled with the relative inaccuracy of CT as contrasts to CF, indicates that such findings should be approached with caution until better methods for identifying the robustness of CT tree shapes are derived.

The Causal Forest (CF) implementation represents a set of CTs and thus creates a distribution of values for each World Bank project $i$. These distributions are then aggregated to a single value to estimate $\hat{\tau}(X_i)$, or the distributions themselves are analyzed to examine the robustness of a given finding. In Fig. 4, we show the detailed distributions for selected example projects. These examples

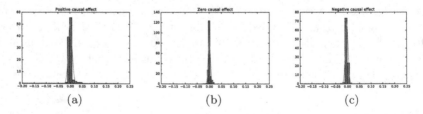

(a)                         (b)                         (c)

**Fig. 4.** CF calculated distributions of treatment effect estimates for specific projects: (a) Saint Lucia Hurricane Tomas Emergency Recovery Loan; (b) Sustainable Tourism Development Project; (c) Emergency Infrastructure Reconstruction Project.

provide an illustration of how applied CF results can provide indications not only of what projects are likely having a negative impact on the environment, but also the robustness of these estimates. By writing a second-stage algorithm which identified projects with distributions following certain characteristics (i.e., a mean centered around 0 with a Gaussian distribution; a negative-centered mean with a left-skewed distribution), it is possible to highlight the subset(s) of projects for which more robust findings exist. Figure 5(a) shows a histogram of CF calculated $\hat{\tau}(X_i)$ values for all world bank projects in our data set. Most of the projects have a slightly negative to no impact on the forest cover, which is in line with World Bank objectives to offset potential negative environmental outcomes. Figure 5(b) provides evidence that while the World Bank is generally successful in meeting it's goal of mitigating environmental impacts, the rate at which positive and negative deviation occurs is highly variable by geographic region. We can see that most outliers are in the positive direction, with Asia being a notable exception. The projects in Oceania are in a narrow range, however, projects in other continents have a wide range.

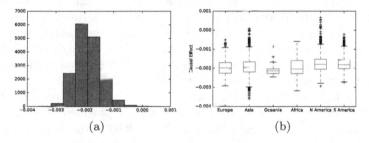

$$\text{(a)} \qquad\qquad\qquad\qquad \text{(b)}$$

**Fig. 5.** (a) Causal effect distribution of all World Bank projects combined and (b) separated by continents

While both the CT and CF approaches allow for the examination of the relative importance of factors in driving heterogeneity, the interpretation and robustness of these findings is highly variable. In the case of the CT, the position of a variable in the single tree can be interpreted as importance; i.e., splits higher in the tree are more influential on the results, and path-dependencies can be examined. However, the robustness of the shape of the CT approach is unknown, and both our simulations and existing literature suggest CT findings are likely to be less accurate than CF implementations. Conversely, in a CF each covariate can be ranked across all trees in terms of the purity improvements it can provide, giving a relative indication of importance across all trees (see Footnote 1). While these findings are more robust, they do not enable the interpretation of explicit thresholds (i.e., the year variable may be important, but the explicit year that is split on may change in the RF approach), and path dependencies are not made explicit. In our case study, we find that the first five variables in the CT and CF cases are stable between approaches, but we identify significant variance in deeper areas of the tree. For a practitioner, this allows an understanding of what the major drivers of aid effectiveness are; for example here, the purity metric highlights the dollars committed and environmental conditions as major drivers of forest cover loss, and also highlights a disparity between projects located at different latitudes; all factors which can enable a deeper understanding of what is causing success and failure in World Bank environmental initiatives. This is consistent with past findings which illustrate a stable set of covariates in the top-level of trees across a CF [13]. Further, we note that the 15 most highly ranked covariates in the CF approach are generally uncorrelated, providing an indication that the information they provide is not redundant (see Footnote 1). However, we leave the interpretation of the shape of the random forest, and the insights that can be gained from it, to future research.

## 5    Discussion and Conclusions

This paper sought to examine the research question *What is the impact of world bank projects on forest cover?* To examine this, we contrasted four different approaches all based on variations of regression trees and random forests of trees: Transformed Outcome Trees (TOTs), Causal Trees (CTs), Random Forest TOTs (RFTOTs), and Causal Forests (CF). We found that the method selected can have significant influence on the causal effect (or lack thereof) estimated, and provide evidence suggesting CF is more accurate than alternatives in our study context. By applying the CF approach to the case of World Bank projects, we were able to compute estimates for causal effects of individual projects; further, the prominent appearance of some covariates in trees provided us with guidance on which covariates were most important in mediating the impacts of World Bank projects. While - for most projects - the effect on forest cover is close to zero, we identified some notable exceptions, positive as well as negative ones. We also identified two key questions that have not yet been answered in the academic literature. The first of these is how to select proper limitations on the makeup of terminal nodes - i.e., if splits that result in nodes without both control and treatment cases should be prevented, omitted, or otherwise constrained. Even after propensity score adjustments, terminal nodes with no adequate comparison cases become difficult (if not impossible) to interpret. Second, there is little literature in the machine learning space regarding how to cope with spatial spillover between treated and control cases. The Stable Unit Treatment Value Assumption (SUTVA) is common practice, but in practice the effects of a project can not be expected to be purely local in nature when observations are geographically situated.

## References

1. World Bank Group. http://www.worldbank.org/en/about/what-we-do
2. AidData (2016). http://aiddata.org/subnational-geospatial-research-datasets
3. Athey, S., Imbens, G.: Recursive partitioning for heterogeneous causal effects (2015)

4. Biau, G.: Analysis of a random forests model. JMLR **13**(1), 1063–1095 (2012)
5. Breiman, L.: Random forests. Mach. Learn. **45**(1), 5–32 (2001)
6. Denil, M., Matheson, D., de Freitas, N.: Narrowing the gap: random forests in theory and in practice. In: ICML (2014)
7. Dunbar, B.: NDVI: satellites could help keep hungry populations fed as climate changes (2015). http://www.nasa.gov/topics/earth/features/obscure_data.html
8. Hirano, K., Imbens, G., Ridder, G.: Efficient estimation of average treatment effects using the estimated propensity score. Econometrica **71**(4), 1161–1189 (2003)
9. Imbens, G.W., Rubin, D.B.: Causal Inference for Statistics, Social, and Biomedical Sciences: An Introduction. Cambridge University Press, Cambridge (2015)
10. Meinshausen, N.: Quantile regression forests. JMLR **7**, 983–999 (2006)
11. NASA: The land long term data record (2015). http://ltdr.nascom.nasa.gov/cgi-bin/ltdr/ltdrPage.cgi
12. Su, X., Tsai, C.L., Wang, H., Nickerson, D.M., Li, B.: Subgroup analysis via recursive partitioning. J. Mach. Learn. Res. **10**, 141–158 (2009)
13. Taddy, M., Gardner, M., Chen, L., Draper, D.: A nonparametric Bayesian analysis of heterogeneous treatment effects in digital experimentation (2014)
14. Tibshirani, R.: Regression shrinkage and selection via the LASSO. J. R. Stat. Soc. Scr. B **58**, 267–288 (1994)
15. Vapnik, V.N.: Statistical Learning Theory. Wiley-Interscience, New York (1998)
16. Wager, S., Athey, S.: Estimation and inference of heterogeneous treatment effects using random forests (2015)
17. Wager, S., Hastie, T., Efron, B.: Confidence intervals for random forests: the jackknife and the infinitesimal jackknife. JMLR **15**(1), 1625–1651 (2014)

# RSSI-Based Supervised Learning
# for Uncooperative Direction-Finding

Tathagata Mukherjee[2]($\boxtimes$), Michael Duckett[1], Piyush Kumar[1],
Jared Devin Paquet[4], Daniel Rodriguez[1], Mallory Haulcomb[1], Kevin George[2],
and Eduardo Pasiliao[3]

[1] CompGeom Inc., 3748 Biltmore Ave., Tallahassee, FL 32311, USA
{michael,piyush,mallory}@compgeom.com
[2] Intelligent Robotics Inc., 3697 Longfellow Road, Tallahassee, FL 32311, USA
{tathagata,kevin}@intelligentrobotics.org
[3] Munitions Directorate, AFRL, 101 West Eglin Blvd, Eglin AFB, FL 32542, USA
eduardo.pasiliao@us.af.mil
[4] REEF, 1350 N. Poquito Rd, Shalimar, FL 32579, USA
jaredpaquet@ufl.edu

**Abstract.** This paper studies supervised learning algorithms for the problem of *uncooperative direction finding* of a radio emitter using the received signal strength indicator (RSSI) from a rotating and uncharacterized antenna. Radio Direction Finding (RDF) is the task of finding the direction of a radio frequency emitter from which the received signal was transmitted, using a single receiver. We study the accuracy of radio direction finding for the 2.4 GHz WiFi band, and restrict ourselves to applying supervised learning algorithms for RSSI information analysis. We designed and built a hardware prototype for data acquisition using off-the-shelf hardware. During the course of our experiments, we collected more than three million RSSI values. We show that we can reliably predict the bearing of the transmitter with an error bounded by 11°, in both indoor and outdoor environments. We do not explicitly model the multi-path, that inevitably arises in such situations and hence one of the major challenges that we faced in this work is that of automatically compensating for the multi-path and hence the associated noise in the acquired data.

**Keywords:** Data mining · Radio direction finding
Software defined radio · Regression · GNURadio · Feature engineering

## 1 Introduction

One of the primary problems in sensor networks is that of node localization [4,9,21]. For most systems, GPS is the primary means for localizing the network. But for systems where GPS is denied, another approach must be used. A way to achieve this is through the use of special nodes that can localize themselves without GPS, called anchors. Anchors can act as reference points through which

Y. Altun et al. (Eds.): ECML PKDD 2017, Part III, LNAI 10536, pp. 216–227, 2017.
https://doi.org/10.1007/978-3-319-71273-4_18

other nodes may be localized. One step in localizing a node is to find the direction of an anchor with respect to the node. This problem is known as Radio Direction Finding (RDF or DF). Besides sensor networks, RDF has applications in diverse areas such as emergency services, radio navigation, localization of illegal, secret or hostile transmitters, avalanche rescue, wildlife tracking, indoor position estimation, tracking tagged animals, reconnaissance and sports [1,13,15,16,23,25] and has been studied extensively both for military [8] and civilian [6] use.

Direction finding has also been studied extensively in academia. One of the most commonly used algorithms for radio direction finding using the signal received at an antenna array is called MUSIC [22]. MUSIC and related algorithms are based on the assumption that the signal of interest is Gaussian and hence they use second order statistics of the received signal for determining the direction of the emitters. Porat et al. [18] study this problem and propose the MMUSIC algorithm for radio direction finding. Their algorithm is based on the Eigen decomposition of a matrix of the fourth order cumulants. Another commonly used algorithm for determining the direction of several emitters is called the ESPRIT algorithm [19]. The algorithm is based on the idea of having *doublets* of sensors.

Recently, researchers have used both unsupervised and supervised learning algorithms for direction finding. Graefenstein et al. [10] used a robot with a custom built rotating directional antenna with fully characterized radiation pattern for collecting the RSSI values. These RSSI values were normalized and iteratively rotated and cross-correlated with the known radiation pattern for the antenna. The angle with the highest cross-correlation score was reported as the most probable angle of the transmitter. Zhuo et al. [27] used support vector machines with a known antenna model for classifying the directionality of the emitter at 3 GHz. Ito et al. [14] studied the related problem of estimating the orientation of a terminal based on its received signal strength. They measured the divergence of the signal strength distribution using Kulback-Leibler Divergence [17] and estimated the orientation of the receiver. In a related work, Satoh et al. [20] used directional antennas to sense the 2.4 GHz channel and applied Bayesian learning on the sensed data to localize the transmitters.

In this paper we use an uncalibrated receiver (directional) to sense the 2.4 GHz channel and record the resulting RSSI values from a directional as well as an omni-directional source. We then use feature engineering along with machine learning and data mining techniques (see Fig. 1) to learn the bearing information for the transmitter. Note that this is a basic ingredient of a DF system. Just replicating our single receiver with multiple receivers arranged in a known topology, can be used to determine the actual location of the transmitter. Hence pushing the boundary of this problem will help DF system designers incorporate our methods in their design. Moreover, such a system based only on learning algorithms would make them more accessible to people irrespective of their academic leaning. We describe the data acquisition system next.

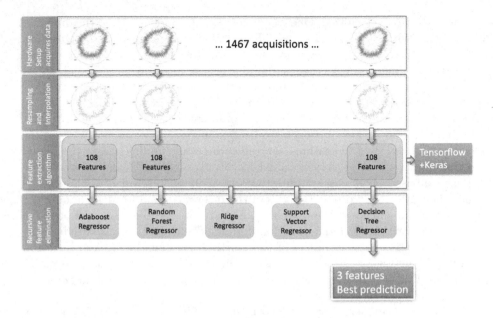

... 1467 acquisitions ...

108 Features | 108 Features | 108 Features | Tensorflow +Keras

Adaboost Regressor | Random Forest Regressor | Ridge Regressor | Support Vector Regressor | Decision Tree Regressor

3 features
Best prediction

**Fig. 1.** Data processing system for learning algorithm

## 2   Data Acquisition System

The data collection system is driven by an Intel NUC (Intel i7 powered) running Ubuntu 16.04. For sensing the medium, we use an uncharacterized COTS 2.4 GHz WiFi Yagi antenna. This antenna is used both as a transmitter as well as a receiver. For the receiver, the antenna is mounted on a tripod and attached to a motor so that it can be rotated as it scans the medium. We also use a TP-Link 2.4 GHz 15 dBi gain WiFi omni-directional antenna for transmission to ensure that the system is agnostic to the type of antenna being used for transmission. For both transmission and reception the antenna is connected to an Ettus USRP B210 software defined radio. To make the system portable and capable of being used anywhere we power the system with a 12V6Ah deep cycle LiFePO battery. A Nexus 5X smart-phone is used to acquire compass data from its on-board sensors and this data is used for calibrating the direction of the antenna at the start of each experiment.

There are two main components for our setup: the receiver and the transmitter. For our tests, we placed the receiver at the origin of the reference frame. The transmitter was positioned at various locations around the receiver. The transmitter was programmed to transmit at 2.4 GHz, and the receiver was used to sense the medium at that frequency as it rotated about its axis. Our experiments were conducted both indoors and outdoors.

For our analysis we consider one full rotation of the receiver as the smallest unit of data. Each full rotation is processed, normalized and considered as a unique data point that is associated with a given bearing to the transmitter.

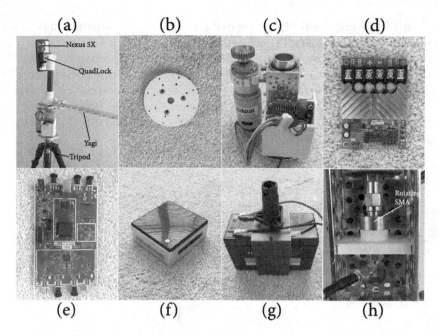

**Fig. 2.** (a): The full yagi setup, (b): plate adapter, (c): Pan Gear system composed of motor and motor controller mounted on standing bracket, (d): motor controller, (e): B210 Software Defined Radio, (f): NUC compact computer, (g): StarkPower lithium ion battery and holder, (h): chain of connectors from B210 to antenna including a rotating SMA adapter located in standing bracket

For each experiment we collected several rotations at a time, with the transmitter being fixed at a given bearing with respect to the receiver, by letting the acquisition system operate for a certain amount of time. We call each experiment a run and each run consists of several rotations.

There are two important aspects of the receiver that need to be controlled: the rotation of the yagi antenna and the sampling rate of the SDR. The rotation API has two important functions that define the phases of a run: first, *finding north* and aligning the antenna to this direction, so that every time the angles are recorded with respect to this direction; and second, the actual *rotation*, that makes the antenna move and at the same time uses the Ettus B210 for recording the spectrum. In the first phase the yagi is aligned to the magnetic north using the compass of the smart phone that we used in our system. In the second phase the yagi starts to rotate at a constant angular velocity. While rotating, the encoder readings are used to determine the angle of the antenna with respect to the magnetic north, and the RSSI values are recorded with respect to these angles. It should be noted that the angles from the compass are not used because the encoder readings are more accurate and frequent. The end of each rotation is determined based on the angles obtained from the encoder values.

In order to record the RSSI, we created a GNURadio Companion flow graph [2]. Our flow graph gets the I/Q values from the B210 (*UHD: USRP Source*) at a sample rate and center frequency of 2 MHz and 2.4 GHz respectively. We run the values through a high-pass filter to alleviate the DC bias. The data is then chunked into vectors of size 1024 (*Stream to Vector*), which is then passed through a Fast-Fourier Transform (*FFT*) and then flattened out (*Vector to Stream*). This converts the data from the time-domain to the frequency-domain. The details of the flow-graph is shown in Fig. 3.

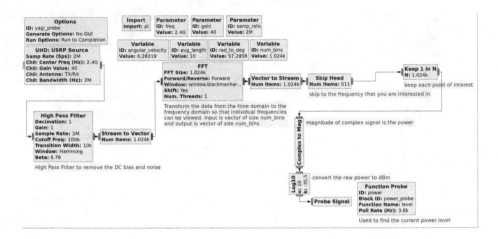

**Fig. 3.** GNURadio companion flow graph

# 3    Data Analysis

Now we are ready to describe the algorithms used for processing the data. Our approach has three phases: the feature engineering phase takes the raw data and maps it to a feature space; the learning phase uses this representation of the data in the feature space to learn a model for predicting the direction of the transmitter; and finally we use a cross validation/testing phase to test the learned model on previously unseen data. We start with feature engineering.

## 3.1    Feature Engineering

As mentioned before, our data consists of a series of *runs*. Each run consists of several rotations and each rotation is vector of (angle, power) tuples. The length of this vector is dependent on the total time of a rotation (fixed for each run) and speed at which the SDR samples the spectrum, which varies. Typically, each rotation has around 2200 tuples. In order to use this raw data for further analysis we transformed each rotation into a vector of fixed dimension, namely $k = 360$. We achieved this by simulating a continuous mapping from angles to powers based on the raw data for a single rotation and by reconstructing the vector using this mapping for $k$ evenly spaced points within the range 0 to $2\pi$. The new set of rotation vectors denoted by $R$ is a subset of $\mathbb{R}^k$. For our analysis, we let $k = 360$ because each run is representative of a sampling from a circle.

During the analysis of our data, we noticed a drift in one of the features (*moving average max value* which is defined below). This led us to believe that the encoder measurements were changing with time during a run (across rotations). Plotting each run separately revealed a linear trend with high correlation (Fig. 5). In order to correct the drift, we computed the least squares regression for the most prominent runs (runs which displayed very high correlation), averaged the slopes of the resulting lines, and used this value to negate the drift. The negation step was done on the raw data for each run because at the start of each run, the encoder is reset to zero. Once a run is corrected it can be split into rotations. Since each rotation vector can be viewed as a time-series, we use time series fea-

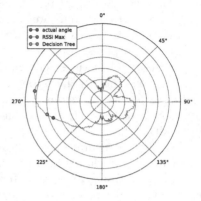

**Fig. 4.** Example rotation with markers for the actual angle and two predicted angles using the max RSSI and Decision Tree methods

ture extraction techniques to map the data into a high dimensional feature space. Feature extraction from time series data is a well studied problem, and we use the techniques described by Christ et al. [5] to map the data into the feature space. In all there were 86 features that were extracted using the algorithm. In addition to the features extracted using this method, we also added a few others based on the idea of a moving average [12].

More precisely, we use the *moving average max value*, which is the index (angle) in the rotation vector where the max power is observed after applying a moving average filter. The filter takes a parameter $d$, the size of the moving average which for a given angle is computed by summing the RSSI values corresponding to the preceding $d$ angles, the angle itself and the succeeding $d$ angles. Finally this sum is divided by the total number of points $(2d+1)$, which is always odd. We use the *moving average max value* with filter sizes ranging from 3 to 45, using every other integer. This gives an additional 22 features, which brings the total to 108 features.

## 3.2 Learning Algorithms

Note that we want to predict the bearing (direction) of the transmitter with respect to the receiver for each rotation. As the bearing is a continuous variable, we formulate this as a regression problem. We use several regressors for predicting the bearing: (1) *SVR: Support vector regression* is a type of regressor that uses an $\epsilon$-insensitive loss function to determine the error [17]. We used the RBF kernel and `GridSearchCV` for optimizing the parameters with cross validation

(2) *KRR*: *Kernel ridge regression* is similar to SVR but uses a squared-error loss function [17]. Again, we use the RBF kernel with `GridSearchCV` to get the optimal parameter (3) *Decision Tree* [3]: we used a max depth of 4 for our model and finally (4) *AdaBoost with Decision Tree* [7,26]: short for *adaptive boosting*, uses another learning algorithm as a "weak learner". We used a decision tree with max depth 4 as our "weak learner".

Although we have a total of 108 features, not all of them will be important for prediction purposes. As a result, we try two different approaches for selecting the most useful features: (1) the first one ranks each feature through a scoring function, and (2) the second prunes features at each iteration and is called *recursive feature extraction with cross-validation* [11]. For the first we use the function *SelectKBest*, and for the later we used *RFECV*, both implemented in ScikitLearn.

We also use neural networks for the prediction task. We used Keras for our experiments, which is a high-level neural networks API, written in Python and capable of running on top of TensorFlow. We used the Sequential model in Keras, which is a linear stack of layers. The results on our dataset are described in Sect. 4.

# 4    Experiments and Results

In this section we present the results of our experiments. In total, we collected 1467 *rotations* (after drift correction) at 76 unique angles (an example of a rotation reduced to 360 vertices can be seen in Fig. 4). After reducing each rotation to 360 power values, we ran the dataset through the feature extractor, which produced 108 total features. We tried out several regressors, namely, (SVR, KR, DT, and AB) and strategies:- (moving average max value without learning, moving average max value with learning, SelectKBest, RFECV, and neural networks). The objective for this set of tests was to find the predictor that yielded the lowest mean absolute error (MAE), which is the average of the absolute value of each error from that test.

## 4.1    Regressors

We used the data from the feature selection phase to test a few regressors. For each regressor, we split the data (50% train, 50% test), trained and tested the model, and calculated the MAE. We ran this 100 times for each regressor and took the overall average to show which regressor preformed the best with all the features. The results from these tests are in Table 1.

From the results we can see that *decision trees* give the lowest MAE compared to the other regressors. We also noticed that *decision trees* ran the train/test cycle much faster than any other regressor. Based on these results, we decided to use the *decision tree* regressor for the rest of our tests.

**Table 1.** The average error for the each regressors over 100 runs with 50–50 split

|            | SVR   | KRR   | DT    | AB    |
|------------|-------|-------|-------|-------|
| Avg. error | 26.4° | 55.2° | 16.2° | 22.1° |

## 4.2 Moving Average Max Value

One of the first attempts for formulating a reliable predictor was to use the *moving average max value* (MAMV). We considered using this feature by itself as a naive approach. We predict as follows: whichever index the *moving average max value* falls on is the predicted angle (in degrees). For our tests, we used a moving average with size 41 (MAMV-41), which was ranked the best using `SelectKBest`, for smoothing the angle data. Since no training was required, we used all the rotations to calculate the MAE. As seen in Table 2 the MAE was 57.1°. Figure 6 shows the errors for each rotation, marking the inside and outside rotations, as well.

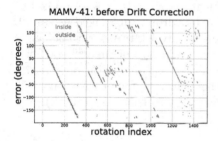

**Fig. 5.** Errors before drift correction. Lines are runs

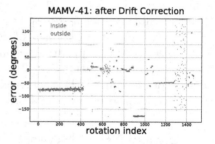

**Fig. 6.** Errors after drift correction. Lines are runs

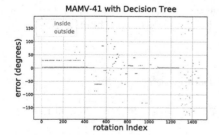

**Fig. 7.** Errors for MAMV-41 with Decision Tree learning. Even/odd for test/train split

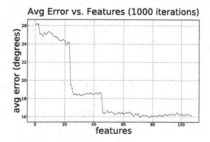

**Fig. 8.** MAE vs. ranked features from `SelectKBest` using *Decision tree* over 1000 runs

Our next step was to use the *decision trees* with the MAMV-41 feature. We applied a 50/50 train/test split in the data and calculated the MAE for each run. We averaged and reported the MAE for all runs. The average MAE over 1000 runs was 25.9° (Table 2). Figure 7 shows a graph of errors for the train/test split where the odd index rotations were for training and the even index rotations were for testing.

## 4.3 SelectKBest

As mentioned before, we used `SelectKBest` to rank all the features. In order to get stable rankings, we ran this 100 times and averaged those ranks. Once we had the ranked list of features, we created a "feature profile" by iteratively adding the next best feature, running train/test with the *decision tree* regressor for that set of features, and recording the MAE. We repeated this process 1000 times and the results are shown in (Fig. 8). It is to be noted that the error does not change considerably for a large number of consecutive features but there are steep drops in the error around certain features. This is because many features are similar and using them for the prediction task does not change the error significantly. The first plateau consists mostly of MAMV features since they were ranked the best among all the other features. The first major drop is at ranked feature 24, which marks the start of the set of continuous wavelet transform coefficients (CWT). The second major drop is cause by the addition of the $2^{nd}$ coefficient from Welch's transformation [24] (2-WT) at ranked feature 46. Beyond that, no significant decrease in MAE is achieved by the inclusion of another feature. The best average MAE over the whole profile is 15.7° at 78 features (Table 2).

## 4.4 RFECV and Neural Network

We ran `RFECV` with a decision tree regressor using a different random state every time. `RFECV` consistently returned three features: MAMV-23, MAMV-41, and 2-WT. Using these three features, we trained and tested on the data with a 50/50 split 10000 times with the decision tree regressor. The average MAE was 11.0° (Table 2). Between `RFECV` and `SelectKBest`, there are four unique features which stand out among the rest. To be thorough, we found the average MAE for all groups of three and four from these four features. None of them were better than the original three features from `RFECV`.

Table 2. Comparison of average error among predictor methods

|  | MAMV-41 | MAMV-41 (DT) | SelectKBest | RFECV | Neural net |
|---|---|---|---|---|---|
| Avg. MAE | ±57.1° | ±25.9° | ±15.7° | ±11.0° | ±15.7° |

For the neural network approach, we used all the features which were produced from the feature selection phase. We settled on a $108 \Rightarrow 32 \Rightarrow 4 \Rightarrow 1$

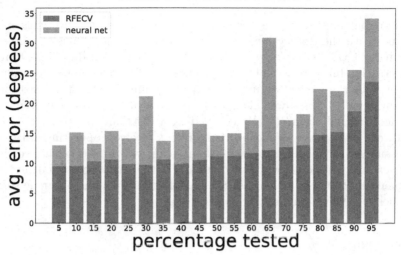

**Fig. 9.** Neural net vs. RFECV performance. The x-axis represents the percentage of the dataset tested with the other partition being used for training (for example 5% tested means 95% of the dataset was used for training).

layering. The average MAE over 100 runs with a 50/50 train/test split was 15.7° (Table 2). In order to show how the neural network stacked against feature selection, we performed an experiment showing each method's performance versus a range of train/test splits. Figure 9 shows that RFECV with it's three features performed better than our neural network at all splits.

### 4.5  Discussion

From our results in Table 2, we determined that RFECV was the best feature selector. The amount of time it takes to filter out significant features is comparable to SelectKBest, but RFECV produces fewer features which leads to lower training and testing times. Figure 9 shows that RFECV beats neural networks consistently for a range of train/test splits. There are a couple of possible reasons why RFECV performed better. The SelectKBest strategy ranks each feature independent of the other features, which means similar features will have similar ranks. As features are iteratively added, many consecutive features will be redundant. This is evident in Fig. 8 where the addition of similar features cause very little change in MAE creating plateaus in the plot. Our SelectKBest method was, in a way, good at finding some prominent features (where massive drops in MAE occurred), but not in the way we intended whereas RFECV was better at ranking diverse features.

## 5    Conclusion and Future Work

The main contribution of this paper is to show that using pure data mining techniques with the RSSI values, one can achieve good accuracies in direction-finding using COTS directional receivers. There are several directions that can be pursued for future work: (1) How accurately can cell phones locations be analyzed with the current setup? (2) Can we minimize the total size of our receive system? (3) How well does this system work for different frequencies and ranges around them? (4) When used in a distributed setting, how much accuracy can one achieve for localization, given $k$ receivers operating at the same time (assuming the distances between them is known). In the journal version of this paper, we will show how to theoretically solve this problem, but extensive experimental results are lacking. In the near future, we plan to pursue some of these questions using our existing hardware setup.

## References

1. Bahl, V., Padmanabhan, V.: RADAR: an in-building RF-based user location and tracking system. Institute of Electrical and Electronics Engineers, Inc., March 2000. https://www.microsoft.com/en-us/research/publication/radar-an-in-building-rf-based-user-location-and-tracking-system/
2. Blossom, E.: GNU radio: tools for exploring the radio frequency spectrum. Linux J. **2004**(122), 4 (2004). http://dl.acm.org/citation.cfm?id=993247.993251
3. Breiman, L., Friedman, J., Olshen, R., Stone, C.: Classification and Regression Trees. Wadsworth, Belmont (1984)
4. Capkun, S., Hamdi, M., Hubaux, J.P.: GPS-free positioning in mobile ad hoc networks. Clust. Comput. **5**(2), 157–167 (2002)
5. Christ, M., Kempa-Liehr, A.W., Feindt, M.: Distributed and parallel time series feature extraction for industrial big data applications. arXiv preprint arXiv:1610.07717 (2016)
6. Finders, D.: Introduction into theory of direction finding (2017). http://telekomunikacije.etf.bg.ac.rs/predmeti/ot3tm2/nastava/df.pdf. Accessed 28 Feb 2017
7. Freund, Y., Schapire, R.E.: A decision-theoretic linearization of on-line learning and an application to boosting (1995)
8. Gething, P.: Radio Direction-Finding: And the Resolution of Multicomponent Wave-Fields. IEE Electromagnetic Waves Series. Peter Peregrinus, London (1978). https://books.google.com/books?id=BCcIAQAAIAAJ
9. Graefenstein, J., Albert, A., Biber, P., Schilling, A.: Wireless node localization based on RSSI using a rotating antenna on a mobile robot. In: 2009 6th Workshop on Positioning, Navigation and Communication, pp. 253–259, March 2009
10. Graefenstein, J., Albert, A., Biber, P., Schilling, A.: Wireless node localization based on RSSI using a rotating antenna on a mobile robot. In: 6th Workshop on Positioning, Navigation and Communication (WPNC 2009), pp. 253–259. IEEE (2009)
11. Guyon, I., Weston, J., Barnhill, S., Vapnik, V.: Gene selection for cancer classification using support vector machines. Mach. Learn. **46**, 389–422 (2002)

12. Hamilton, J.D.: Time Series Analysis, vol. 2. Princeton University Press, Princeton (1994)
13. Huang, W., Xiong, Y., Li, X.Y., Lin, H., Mao, X., Yang, P., Liu, Y., Wang, X.: Swadloon: direction finding and indoor localization using acoustic signal by shaking smartphones. IEEE Trans. Mob. Comput. **14**(10), 2145–2157 (2015). http://dx.doi.org/10.1109/TMC.2014.2377717
14. Ito, S., Kawaguchi, N.: Orientation estimation method using divergence of signal strength distribution. In: Third International Conference on Networked Sensing Systems, pp. 180–187 (2006)
15. Kolster, F.A., Dunmore, F.W.: The radio direction finder and its application to navigation, Washington (1922). ISBN: 978-1-333-95286-0
16. Moell, J., Curlee, T.: Transmitter Hunting: Radio Direction Finding Simplified. TAB Book, McGraw-Hill Education (1987). https://books.google.com/books?id=RfzF2-fHJ6MC
17. Murphy, K.P.: Machine Learning: A Probabilistic Perspective. MIT Press, Cambridge (2012)
18. Porat, B., Friedlander, B.: Direction finding algorithms based on high-order statistics. IEEE Trans. Signal Process. **39**(9), 2016–2024 (1991)
19. Roy, R., Kailath, T.: ESPRIT-estimation of signal parameters via rotational invariance techniques. IEEE Trans. Acoust. Speech Signal Process. **37**(7), 984–995 (1989)
20. Satoh, H., Ito, S., Kawaguchi, N.: Position estimation of wireless access point using directional antennas. In: Strang, T., Linnhoff-Popien, C. (eds.) LoCA 2005. LNCS, vol. 3479, pp. 144–156. Springer, Heidelberg (2005). https://doi.org/10.1007/11426646_14
21. Savarese, C., Rabaey, J.M., Beutel, J.: Location in distributed ad-hoc wireless sensor networks. In: Proceedings of the IEEE International Conference on Acoustics, Speech, and Signal Processing (ICASSP 2001), vol. 4, pp. 2037–2040. IEEE (2001)
22. Schmidt, R.: Multiple emitter location and signal parameter estimation. IEEE Trans. Antennas Propag. **34**(3), 276–280 (1986)
23. Ward, T., Pasiliao, E.L., Shea, J.M., Wong, T.F.: Autonomous navigation to an RF source in multipath environments. In: 2016 IEEE Military Communications Conference (MILCOM 2016), pp. 186–191, November 2016
24. Welch, P.: The use of fast fourier transform for the estimation of power spectra: a method based on time averaging over short, modified periodograms. IEEE Trans. Audio Electroacoust. **15**(2), 70–73 (1967)
25. Wikipedia: Direction finding – Wikipedia, the free encyclopedia (2016). https://en.wikipedia.org/wiki/Direction_finding. Accessed 20 Dec 2016
26. Zhu, J., Zou, H., Rosset, S., Hastie, T.: Mutli-class AdaBoost (2009)
27. Zhuo, L., Dan, S., Yougang, G., Yaqin, S., Junjian, B., Zhiliang, T.: The distinction among electromagnetic radiation source models based on directivity with support vector machines. In: 2014 International Symposium on Electromagnetic Compatibility, Tokyo (EMC 2014/Tokyo), pp. 617–620. IEEE (2014)

# Sequential Keystroke Behavioral Biometrics for Mobile User Identification via Multi-view Deep Learning

Lichao Sun[1]([✉]), Yuqi Wang[3], Bokai Cao[1], Philip S. Yu[1], Witawas Srisa-an[2], and Alex D. Leow[1]

[1] University of Illinois at Chicago, Chicago, IL 60607, USA
{lsun29,caobokai,psyu}@uic.edu, aleow@psych.uic.edu
[2] University of Nebraska–Lincoln, Lincoln, NE 68588, USA
witty@cse.unl.edu
[3] Hong Kong Polytechnic University, Kowloon, Hong Kong
csyqwang@comp.polyu.edu.hk

**Abstract.** With the rapid growth in smartphone usage, more organizations begin to focus on providing better services for mobile users. User identification can help these organizations to identify their customers and then cater services that have been customized for them. Currently, the use of cookies is the most common form to identify users. However, cookies are not easily transportable (e.g., when a user uses a different login account, cookies do not follow the user). This limitation motivates the need to use behavior biometric for user identification. In this paper, we propose DEEPSERVICE, a new technique that can identify mobile users based on user's keystroke information captured by a special keyboard or web browser. Our evaluation results indicate that DEEPSERVICE is highly accurate in identifying mobile users (over 93% accuracy). The technique is also efficient and only takes less than 1 ms to perform identification.

## 1  Introduction

Smart mobile devices are now an integral part of daily life; they are our main interface to cyber-world. We use them for on-line shopping, education, entertainment, and financial transactions. As such, it is not surprising that companies are working hard to improve their mobile services to gain competitive advantages. Accurately and non-intrusively identifying users across applications and devices is one of the building blocks for better mobile experiences, since not only companies can attract users based on their characteristics from various perspectives, but also users can enjoy the personalized services without much effort [15].

User identification is a fundamental, but yet an open problem in mobile computing. Traditional approaches resort to user account information or browsing history. However, such information can pose security and privacy risks, and it is not robust as can be easily changed, e.g., the user changes to a new device or using a different application. Monitoring biometric information including a

© Springer International Publishing AG 2017
Y. Altun et al. (Eds.): ECML PKDD 2017, Part III, LNAI 10536, pp. 228–240, 2017.
https://doi.org/10.1007/978-3-319-71273-4_19

user's typing behaviors tends to produce consistent results over time while being less disruptive to user's experience. Furthermore, there are different kinds of sensors on mobile devices, meaning rich biometric information of users can be simultaneously collected. Thus, monitoring biometric information appears to be quite promising for mobile user identification.

To date, only a few studies have utilized biometric information for mobile user identification on web browser [1, 19]. Important questions such as what kind of biometric information can be used, how does one capture user characteristics from the biometric, and what accuracy of the mobile user identification can be achieved, are largely unexplored. Although there are some researches on mobile user authentication through biometrics [8, 20], authentication is a simplified version of identification, and directly employing authentication would either be infeasible or lead to low accuracy. This work focuses on mobile user identification, and could also be applied to authentication.

In this paper, we collect information from basic keystroke and the accelerometer on the phone, and then propose DEEPSERVICE, a multi-view deep learning method, to utilize this information. To the best of our knowledge, this is the first time multi-view deep learning is applied to mobile user identification. Through several empirical experiments, we showed that the proposed method is able to capture the user characteristics and identify users with high accuracy.

Our contributions are summarized as follows.

1. We propose DEEPSERVICE, a multi-view deep learning method, to utilize easy to collect user biometrics for accurate user identification.
2. We conduct several experiments to demonstrate the effectiveness and superiority of the proposed method against various baseline methods.
3. We give several analyses and insights through the experiments.

The rest of this paper is organized as follows. Section 2 provides background information on deep learning, and reviews prior research efforts related to this work. Section 3 introduces DEEPSERVICE and describes the design and implementation details. Section 4 reports the results of our empirical evaluation on the performance of DEEPSERVICE with respect to other learning techniques. The last section concludes this work and discusses future work.

## 2   Background and Related Work

In this section, we provide additional background information on deep learning structure, and review prior research efforts related to our proposed work.

### 2.1   Background on Deep Learning Structure

Deep learning is a branch of machine learning based on a set of algorithms that attempt to model high level abstractions in data, which is also called deep structured learning, deep neural network learning or deep machine learning. Deep learning is a concept and a framework instead of a particular method.

There are two main branches in deep learning: Recurrent Neural Network (RNN) and Convolutional Neural Networks (CNN). CNN is frequently used in computer vision areas and RNN is applied to solve sequential problems such as nature language process. The simplest form of an RNN is shown as follows:

$$h_k = \phi(W x_k + U h_{k-1})$$

where $h_k$ is the hidden state, $W$ and $U$ are parameters need be learned, and $\phi(.)$ is the is a nonlinear transformation function such as tanh or sigmoid.

Long Short Term Memory network (LSTM) is a special case of RNN, capable of learning long-term dependencies [13]. Specifically, RNN only captures the relationship between recent keystroke information and uses it for prediction. LSTM, on the other hand, can capture long-term dependencies. Consider trying to predict the tapping information in the following text "I plan to visit China ... I need find a place to get some Chinese currency". The word "Chinese" is relevant with respect to the word "China", but the distance between these two words is long. To capture information of the long-term dependencies, we need to use LSTM instead of the standard RNN model.

While LSTM can be effective, it is a complex deep learning structure that can result in high overhead. Gated Recurrent Unit (GRU) is a special case of LSTM but with simpler structures (e.g., using less parameters) [6]. In many problem domains including ours, GRU can produce similar results to LSTM. In some cases, it can even produce better results than LSTM. In this work, we implemented Gated Recurrent Unit (GRU).

Also note that with GRU, it is quite straightforward for each unit to remember the existence of a specific pattern in the input stream over a long series of time steps comparing to LSTM. Any information and patterns will be overwritten by update gate due to its importance.

We can build single-view single-task deep learning model by using GRU as shown in Fig. 2(b). We choose any one view of the dataset such as the view of alphabet as used in this study. We use the normalized dataset as the input of GRU. GRU will produce a final output vector which can help us to do user Identification. A typical GRU is formulated as:

$$z_t = \sigma_g(W_z x_t + U_z h_{t-1})$$
$$r_t = \sigma_g(W_r x_t + U_r h_{t-1})$$
$$\tilde{h}_t = tanh(W x_t + U(r_t \odot h_{t-1}))$$
$$h_t = z_t \tilde{h}_t + (1 - z_t) h_{t-1}$$

where $\odot$ is an element-wise multiplication. $\sigma_g$ is the sigmoid and equals $1/(1 + e^{-x})$. $z_t$ is the update gate which decides how much the unit updates its activation or content. $r_t$ is reset gate of GRU, allowing it to forget the previously computed state.

In Sect. 3, we extend the single-view technique to develop DEEPSERVICE, a multi-view multi-class framework.

## 2.2 Related Work

Most previous works focus on user authorization, rather than identification, based on biometrics. For example, there are multiple approaches to get physiological information that include facial features and iris features [7,9]. This physiological information can also be used for identification, but it requires extra permission from the users. Our method uses behavioral biometrics to identify the users without any cookies or other personal information.

Recently, more research work on continuous authorization problem has emerged for mobile users. Some prior efforts also focus on studying touchscreen gestures [20] or behavioral biometric behaviors such as reading, walking, driving and tapping [3,4]. There are research efforts focusing on offering better security services for mobile users. However, their security models have to be installed on the mobile devices. They then perform binary classifications to detect unauthorized behaviors. Our work focuses on building a general user identification model, which can also be deployed on the web, local devices or even network routers. Our work also focuses on improving users' experience through customized services including providing recommendations and relevant advertisements.

Recently, some research groups focus on mobile user identification based on web browsing information [1,19]. Abramson and Gore try to identify users' web browsing behaviors by analyzing cookies information. However, our model has been designed to target harder problems without using trail of information such as cookies or browsing history. We, instead, use behavioral biometrics to identify users. Information needed can be easily collected from web browser using Javascript.

# 3    DeepService: A Multi-view Multi-class Framework for User Identification on Mobile Devices

DEEPSERVICE is a multi-view and multi-class identification framework via a deep structure. It contains three main steps to identify each user from several users. This process is shown in Fig. 1 and summarized below:

1. In the first step, we collect sequential tapping information and accelerometer information from 40 volunteers who have used our provided smartphones for 8 weeks. We retrieve such sequential data in a real-time manner.
2. In the second step, we prepare the collected information as multi-view data for the problem of user identification.
3. In the third step, we model the multi-view data via a deep structure to perform multi-class learning.
4. In the last step, we compare the performance of the proposed approach with the traditional machine learning techniques for multi-class identification such as support vector machine and random forest. This step is discussed in Sect. 4.

Next, we describe each of the first three steps in turn.

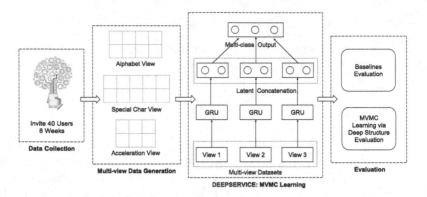

**Fig. 1.** Framework of DEEPSERVICE

## 3.1   Data Collection

First, we describe the data collection process. Our study involves 40 volunteers. The main selection criterion is based on their prior experience with using smartphones. All selected candidates have used smartphones for at least 11 years (some have used smartphones for 18 years). In terms of age, the youngest participant is 30 years old and the oldest one is 63 years old.

Each volunteer was given a same smartphone with the custom software keyboard. Out of the 40 volunteers, we find that 26 of them (17 females and 9 males) have used the provided phones at least 20 times in 8 weeks. The data generated by these 26 volunteers is the one we ended up using in this study. The most active participant has used the phone 4702 times while the least active participant has used the phone only 29 times.

## 3.2   Data Processing

When users type on the smartphone keyboard either locally or on a web browser, our custom keyboard would collect the meta-information associated with the users' typing behaviors, including duration of a keystroke, time since last keystroke, and distance from last keystroke, as well as the accelerometer values along three directions. Due to privacy concerns, the actual characters typed by users are not collected. However, we do collect the categorical information of each keystroke, *e.g.*, alphanumeric characters, special characters, space, backspace. Note that such information can easily be collected from web browser using Javascript as well.

In the data collection process, there are inevitable missing data. For example, when the first time a user uses the phone, the feature *time_since_last_key* is undefined. We replace these missing values with 0. After the complement of missing values, we normalize all the features to the range of $[0, 1]$.

A typical usage of keyboard would likely result in a session consisting more than one keystroke. For example, a simple message such as *"How are you?"*

involves sequential keystrokes as well as multiple types of inputs (alphabets and special characters). In this study, one instance represents one usage session of the phone by the user. A session instance $s_{ij}$ represents the $j$-th session of the $i$-th user in the data set, which consists of three different types of sequential data. Let's denote $s_{ij} = \{c_{ij}^{(1)}, c_{ij}^{(2)}, c_{ij}^{(3)}\}$ where $c_{ij}^{(1)}$ is the time series of alphabet keystrokes, $c_{ij}^{(2)}$ is the time series of special character keystrokes, and $c_{ij}^{(3)}$ is the time series of accelerometer values. It is difficult to align the sequential features in different views because of different timestamps and sampling rates. For example, accelerometer values are much denser than special character keystrokes. Therefore, it is intuitive to treat $c_{ij}^{(1)}$, $c_{ij}^{(2)}$, and $c_{ij}^{(3)}$ as multi-view time series that together compose the complementary information for user identification.

### 3.3  Multi-view Multi-class Deep Learning (MVMC)

Now, we discuss the approach to apply deep learning for constructing the user identification model. The approach is based on Multi-view Multi-class (MVMC) learning with a deep structure.

As mentioned previously, we employ three different views. Each view $V_i$ contains different number of features and different number of samples. In Fig. 2(a), we can use fusion method to combine datasets of different views. However, due to the different number of features, and the number of records in each view of each session, it is hard to build a single-view dataset from many other views. Hence, instead of concatenating different views into one view, we choose to use them separately. This is done to avoid losing information as in the case when multiple views are combined to create a view. One major information that we want to preserve is the sequence of keystrokes. By using multi-view, we are able to maintain each view separately but then use multiple views to make predictions [5,17]. Recently, various methods have been proposed for this purpose [10–12].

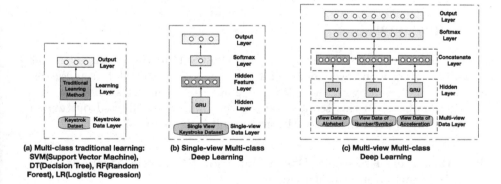

(a) Multi-class traditional learning:
   SVM(Support Vector Machine),
   DT(Decision Tree), RF(Random
   Forest), LR(Logistic Regression)

(b) Single-view Multi-class
   Deep Learning

(c) Multi-view Multi-class
   Deep Learning

**Fig. 2.** A comparison of different frameworks of learning models: left: (a) traditional learning methods; middle (b) single-view multi-class; (c) multi-view multi-class

Before we generate multi-view multi-class learning, we first create single-view multi-class learning as shown in Fig. 2(b) (previously discussed in Sect. 2). Through that model, we can prove the multi-view can help us to improve the performance for identification and we can determine which view most contributes to the identification process. First, we separate the data set into multiple views. In this case, we have a view of alphabets, a view of numbers and symbols, and a view of tapping acceleration. Then, we use GRU and Bidirectional Recurrent Neural Network (we refer to the combination of these two approaches as GRU-BRNN) to build the hidden layers for each view. In the output layer, we use softmax function to perform multiple classifications. Finally, we can evaluate the performance on each view.

We describe the framework that uses multi-view and multi-class learning with deep structure in Fig. 2(c). Comparing to single-view multi-class, multi-view multi-class is a more general model. After we use GRU-BRNN to build the hidden layers for each view. We then concatenate the last layer information from each GRU-BRNN model of each view. We use the last concatenated layer, which contains all information from different views, for identification.

As GRU extracts a latent feature representation out of each time series, the notions of sequence length and sampling time points are removed from the latent space. This avoids the problem of dealing directly with the heterogeneity of the time series from each view. The difference between multi-view multi-class and single-view multi-class learning is that we use the multiple views of the dataset and we use the latent information of each view for prediction, which can improve performance over using only a single-view dataset. We can consider single-view multi-class as a special case of multi-view multi-class learning.

Note that in deep learning, different optimization functions can greatly influence the training speed and the final performance. There are several optimizers such as RMSprop and Adam [16]. In this work, we use an improved version of Adam called Nesterov Adam (Nadam) which is a RMSprop with Nesterov momentum.

## 4    Experiment

To examine the performance of the proposed DEEPSERVICE on identifying mobile users. Our experiments were done on a large-scale real-world data set. We also compared the results with those from several state-of-the-art shallow machine learning methods. In this section, we describe how we conducted our experiments. We then present the experimental results and analysis.

### 4.1    Baselines: Keystroke-Based Behavior Biometric Methods for Continuous Identification

In previous work on keystroke-based continuous identification with machine learning techniques [2,4,14], Support Vector Machine, Decision Tree, and Random Forest are widely used for continuous identification.

*Logistic Regression (LR)*: LR is a linear model with sigmoid function for classification. It is an efficient algorithm which can handle both dense and sparse input.

*Linear Support Vector Machine (LSVM)*: LSVM is widely used in many previous authorization and identification works [2,4,14]. LSVM is a linear model that finds the best hyperplane by maximizing the margin between multiple classes.

*Random Forest/Decision Tree*: Other learning methods such as Random Forest and Decision Tree have not yet been adopted in many behavior biometric work for continuous identification by keystrokes information only. However, Decision Tree is a interpretable classification model for binary classification. It is a tree structure, and features form patterns are nodes in the tree. Random Forest is an ensemble learning method for classification that builds many decision trees during training and combines their outputs for the final prediction. Previous work [18] shows that tree structure methods can work more efficiently than SVM, and can do better binary classification comparing to SVM. We use different traditional learning methods on our data set as the baselines.

## 4.2 Evaluation of DeepService Framework

Since the number of session usage for every user is different, we use four performance measures to evaluate unbalanced results: Recall, Precision F1 Score (F-Measure), and Accuracy. They are defined as:

$$Recall = \frac{TP}{TP+FN} \qquad\qquad F1 = \frac{2*Precision*Recall}{Precision+Recall}$$

$$Precision = \frac{TN}{TN+FP} \qquad Accuracy = \frac{TP+TN}{TP+TN+FP+FN}$$

Next, we report the performance of DEEPSERVICE using our data set. We measured precision, recall and accuracy, f1 score (f-measure) for different models. Based on the results, we can make the following conclusions.

- DEEPSERVICE can identify between two known people at almost 100% accuracy in our experiment. This proves that our data set contains valuable biometric information to distinguish and identify users.
- DEEPSERVICE can do identification with only acceleration records, even when user is not using the keyboard.
- DEEPSERVICE is effective at identifying a large number of users simultaneously either locally or on a web browser.

## 4.3 User Pattern Analysis

In this section, we evaluate the feature patterns for different users. In Fig. 3, we shows the feature patterns analysis of top 5 active users in multi-view.

In the view of Alphabet graphs, each user tends to have unique patterns with respect to the duration, the time since last key, and the number of keystrokes in

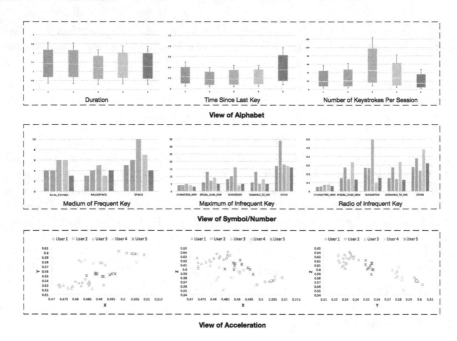

**Fig. 3.** Multi-view pattern analysis of top 5 active user: left is *user1* and right is *user5*

each session. For example, *user3* prefers to use more keystroke in every session with quicker tapping speed than other users.

In the view of Symbol/Number graphs, we have 8 different features. We separate these features into two groups: frequent keys and infrequent keys. A frequent key is defined as a key that is used more than twice per session, otherwise the key is an infrequent keys. (A user tends to use infrequent feature once per session.). We show the medium number of keystroke per session of frequent keys such as auto correct, backspace and space. We also show the range and radio of infrequent keys per session of the top five active users. For example, *user4* frequently uses auto_correct, but she infrequently uses backspace.

In view of Acceleration graphs, we show the correlation of different directions of acceleration. From the last graph, we find that the top 5 active users can be well separated, which proves acceleration can help to identify the users well. It is also proved in our experiments. In both user pattern analysis and experiment, we find view of acceleration can do better identification than other two views.

## 4.4   Identifying Users

DEEPSERVICE can also perform continuous identification. Before we expand to multi-class identification, we first implement a binary-class identification based on multi-view deep learning which is a special case of MVMC learning identification.

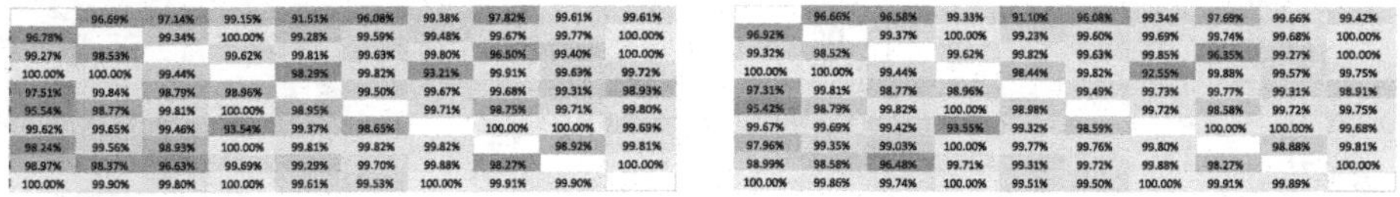

**Fig. 4.** Heatmap of multi-view binary-class identification: left is F1 score; right is accuracy

In Fig. 4, we can see that DEEPSERVICE can do well identification between any two users with 98.97% f1 score and 99.1% accuracy in average. For example, private smartphone usually would not be shared by different people. However, sometimes the private phone could be shared between two people, such as husband and wife. DEEPSERVICE can well separate any two people in this case.

For more general scenarios, we expand binary-identification to N active class (user) identification and use MVMC learning to figure out who is using the phone either locally or on a web browser. Table 1 reports our results.

If we increase the total number of the users in our model, it means we want to identify more people at the same time. For example, if our model is used on a home-router, it may need to identify only members of a family (3 to 10 people) at once. If we, instead, want to identify people working in a small office, we may need to identify more than 10 users. However, it is possible that a larger number of users would degrade the average performance of user identification. This is due to more variation of shared biometric patterns that introduce ambiguity into the system. That's the main reason we want to use multi-view data for user identification, since different users are unlikely to share similar pattens across all views.

**Table 1.** Results of DEEPSERVICE and Baselines

| Method | 5 | | 10 | | 26 | |
|---|---|---|---|---|---|---|
| | Accuracy | F1 | Accuracy | F1 | Accuracy | F1 |
| LR | 66.88% | 66.85% | 44.25% | 45.31% | 27.44% | 30.26% |
| SVM | 68.18% | 68.13% | 44.39% | 45.12% | 30.33% | 31.90% |
| Decision Tree | 68.21% | 67.50% | 53.50% | 52.85% | 43.37% | 42.42% |
| RandomForest | 87.59% | 87.42% | 77.05% | 76.59% | 67.87% | 66.31% |
| Deep Single View | 82.64% | 82.48% | 78.27% | 78.33% | 61.26% | 63.11% |
| DEEPSERVICE | **93.50%** | **93.51%** | **87.35%** | **87.69%** | **82.73%** | **83.25%** |

Table 1 and Fig. 5 report accuracy and F1 values of all learning techniques investigated in this paper. As shown, DEEPSERVICE can identify a user without any cookies and account information. Instead, it simply uses the user's sequential keystroke and accelerometer information. Our approach (DS as shown in Fig. 5) consistently outperforms other approaches listed in Table 1. Moreover, as we increase the number of users, the performance (accuracy and F1) degrades less than those of other approaches.

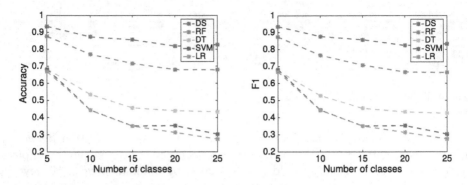

**Fig. 5.** Results with incremental number of classes (users)

In our experiments, we also experimented with using a single-view with deep learning model, and found that the accelerometer view do better identification than other two views. However, when we used information from all three views with MVMC learning, we achieved best performance when compared against the results of other baseline approaches.

### 4.5   Efficiency

To evaluate efficiency of our system, we employ a 15″ Macbook Pro 15 with 2.5 GHz Intel Core i7 and 16 GB of 1600 MHz DDR3 memory, and NVIDIA GeForce GT 750M with 2 GB of video memory. DEEPSERVICE is not the fastest model (decision tree is faster), but it only takes about 0.657 ms per session which shows its feasibility of real-world usage.

## 5   Conclusion and Future Work

We have shown that DEEPSERVICE can be used effectively to identify multiple users. Even though we only use the accelerometer in this work, our results show that more views of dataset can improve the identification performance.

In the future, we want to implement DEEPSERVICE as a tool to help company or government to identify their customers more accurately in the real life. The tool can be implemented on the web or the router. Meanwhile, we will incorporate more sensors, which can be activated from the web browser, to further increase the capability and performance of the DEEPSERVICE.

**Acknowledgements.** This work is supported in part by NSF through grants IIS-1526499, and CNS-1626432, and NSFC 61672313.

# References

1. Abramson, M., Gore, S.: Associative patterns of web browsing behavior. In: 2013 AAAI Fall Symposium Series (2013)
2. Alghamdi, S.J., Elrefaei, L.A.: Dynamic user verification using touch keystroke based on medians vector proximity. In: 2015 7th International Conference on Computational Intelligence, Communication Systems and Networks (CICSyN), pp. 121–126. IEEE (2015)
3. Bo, C., Jian, X., Li, X.-Y., Mao, X., Wang, Y., Li, F.: You're driving and texting: detecting drivers using personal smart phones by leveraging inertial sensors. In: Proceedings of the 19th Annual International Conference on Mobile Computing & Networking, pp. 199–202. ACM (2013)
4. Bo, C., Zhang, L., Jung, T., Han, J., Li, X.-Y., Wang, Y.: Continuous user identification via touch and movement behavioral biometrics. In: 2014 IEEE 33rd International Performance Computing and Communications Conference (IPCCC), pp. 1–8. IEEE (2014)
5. Cao, B., He, L., Wei, X., Xing, M., Yu, P.S., Klumpp, H., Leow, A.D.: t-BNE: tensor-based brain network embedding. SIAM (2017)
6. Chung, J., Gulcehre, C., Cho, K., Bengio, Y.: Empirical evaluation of gated recurrent neural networks on sequence modeling. arXiv preprint arXiv:1412.3555 (2014)
7. de Martin-Roche, D., Sanchez-Avila, C., Sanchez-Reillo, R.: Iris recognition for biometric identification using dyadic wavelet transform zero-crossing. In: 2001 IEEE 35th International Carnahan Conference on Security Technology, pp. 272–277. IEEE (2001)
8. Feng, T., Liu, Z., Kwon, K.-A., Shi, W., Carbunar, B., Jiang, Y., Nguyen, N.: Continuous mobile authentication using touchscreen gestures. In: 2012 IEEE Conference on Technologies for Homeland Security (HST), pp. 451–456. IEEE (2012)
9. Goh, A., Ngo, D.C.L.: Computation of cryptographic keys from face biometrics. In: Lioy, A., Mazzocchi, D. (eds.) CMS 2003. LNCS, vol. 2828, pp. 1–13. Springer, Heidelberg (2003). https://doi.org/10.1007/978-3-540-45184-6_1
10. He, L., Kong, X., Yu, P.S., Yang, X., Ragin, A.B., Hao, Z.: Dusk: a dual structure-preserving kernel for supervised tensor learning with applications to neuroimages. In: Proceedings of the 2014 SIAM International Conference on Data Mining, pp. 127–135. SIAM (2014)
11. He, L., Lu, C.-T., Ding, H., Wang, S., Shen, L., Yu, P.S., Ragin, A.B.: Multi-way multi-level kernel modeling for neuroimaging classification. In: Proceedings of the IEEE Conference on Computer Vision and Pattern Recognition (2017)
12. He, L., Lu, C.-T., Ma, G., Wang, S., Shen, L., Yu, P.S., Ragin, A.B.: Kernelized support tensor machines. In: Proceedings of the 34th International Conference on Machine Learning (2017)
13. Hochreiter, S., Schmidhuber, J.: Long short-term memory. Neural Comput. **9**(8), 1735–1780 (1997)
14. Miluzzo, E., Varshavsky, A., Balakrishnan, S., Choudhury, R.R.: TapPrints: your finger taps have fingerprints. In: Proceedings of the 10th International Conference on Mobile Systems, Applications, and Services, pp. 323–336. ACM (2012)
15. Pedregosa, F., Varoquaux, G., Gramfort, A., Michel, V., Thirion, B., Grisel, O., Blondel, M., Prettenhofer, P., Weiss, R., Dubourg, V., Vanderplas, J., Passos, A., Cournapeau, D., Brucher, M., Perrot, M., Duchesnay, E.: Scikit-learn: machine learning in Python. J. Mach. Learn. Res. **12**, 2825–2830 (2011)

16. Ruder, S.: An overview of gradient descent optimization algorithms. arXiv preprint arXiv:1609.04747 (2016)
17. Shao, W., He, L., Yu, P.S.: Clustering on multi-source incomplete data via tensor modeling and factorization. In: Cao, T., Lim, E.-P., Zhou, Z.-H., Ho, T.-B., Cheung, D., Motoda, H. (eds.) PAKDD 2015. LNCS (LNAI), vol. 9078, pp. 485–497. Springer, Cham (2015). https://doi.org/10.1007/978-3-319-18032-8_38
18. Sun, L., Li, Z., Yan, Q., Srisa-an, W.: SigPID: significant permission identification for android malware detection (2016)
19. Zhang, H., Yan, Z., Yang, J., Tapia, E.M., Crandall, D.J.: mFingerprint: privacy-preserving user modeling with multimodal mobile device footprints. In: Kennedy, W.G., Agarwal, N., Yang, S.J. (eds.) SBP 2014. LNCS, vol. 8393, pp. 195–203. Springer, Cham (2014). https://doi.org/10.1007/978-3-319-05579-4_24
20. Zhao, X., Feng, T., Shi, W.: Continuous mobile authentication using a novel graphic touch gesture feature. In: 2013 IEEE Sixth International Conference on Biometrics: Theory, Applications and Systems (BTAS), pp. 1–6. IEEE (2013)

# Session-Based Fraud Detection in Online E-Commerce Transactions Using Recurrent Neural Networks

Shuhao Wang[1]([✉]), Cancheng Liu[2], Xiang Gao[2], Hongtao Qu[2], and Wei Xu[1]

[1] Tsinghua University, Beijing 100084, China
wsh@physics.so
[2] JD Finance, Beijing 100176, China

**Abstract.** Transaction frauds impose serious threats onto e-commerce. We present CLUE, a novel deep-learning-based transaction fraud detection system we design and deploy at JD.com, one of the largest e-commerce platforms in China with over 220 million active users. CLUE captures detailed information on users' click actions using neural-network based embedding, and models sequences of such clicks using the recurrent neural network. Furthermore, CLUE provides application-specific design optimizations including imbalanced learning, real-time detection, and incremental model update. Using real production data for over eight months, we show that CLUE achieves over 3x improvement over the existing fraud detection approaches.

**Keywords:** Fraud detection · Web mining · Recurrent neural network

## 1 Introduction

Retail e-commerce sales are still quickly expanding. A large online e-commerce website serves millions of users' requests per day. Unfortunately, frauds in e-commerce have been increasing with legitimate user traffic, putting both the financial and public image of e-commerce at risk [5].

Two common forms of frauds in e-commerce websites are *account hijacking* and *card faking* [9]: Fraudsters can steal a user's account on the website to use her account balance, or use a stolen or fake credit card to register a new account. Either case causes losses for both the website and its users. Thus, it is urgent to build effective fraud detection systems to stop such behavior.

Researchers have proposed different approaches to detect fraud [2] using various approaches from rule-based systems to machine learning models like decision tree, support vector machine (SVM), logistic regression, and neural network. All these models use aggregated *features*, such as the total amount of items a user has viewed over the last month, yet many frauds are only detectable by using individual actions instead of aggregates. Also, as fraudulent behaviors change over time to avoid detection, simple features or rules become obsolete quickly.

© Springer International Publishing AG 2017
Y. Altun et al. (Eds.): ECML PKDD 2017, Part III, LNAI 10536, pp. 241–252, 2017.
https://doi.org/10.1007/978-3-319-71273-4_20

Thus, it is essential for a fraud detection system to (1) capture users' behaviors in a way that is as detailed as possible; and (2) choose algorithms to detect the frauds from the vast amount of data. The algorithm must tolerate the dynamics and noise over a long period of time. Previous experience shows that machine learning algorithms outperform rule-based ones [2].

One of the most important piece of information for fraud detection is a user's browsing behavior, or the *sequence* of a user's *clicks* within a session. Statistically, the behaviors of the fraudsters are different from legitimate users. Real users browse items following a certain pattern. They are likely to browse a lot of items similar to the one they have bought for research. In contrast, fraudsters behave more uniformly (e.g. go directly to the items they want to buy, which are usually virtual items, such as #1 in Fig. 1), or randomly (e.g. browse unrelated items before buying, such as #2). Thus, it is important to capture the sequence of each user's clicks, while automatically detect the abnormal behavior patterns.

| Fraudulent User #1 | Fraudulent User #2 |
| --- | --- |
| Visit JD.com | Visit JD.com |
| Search `Game Card` | Search `Apple` |
| Visit `Shanda Game Card (10,000 Game Points)` | Visit `Apple iPad Pro 9.7-inch (128G WLAN, Rose gold)` |
| Checkout `Shanda Game Card (10,000 Game Points)` | Visit `Apple MacBook Air 13.3-inch (128G)` |
| | Visit `Apple iPhone 7 Plus (128G, Gold)` |
| | ... |
| | Visit `Apple iMac 21.5-inch` |
| | Checkout `Apple iPad Pro 9.7-inch (128G WLAN, Rose gold)` |

**Fig. 1.** Examples of fraudulent user browsing behaviors.

We describe our experience with CLUE, a fraud detection system we have built and deployed at JD.com. JD is one of the largest e-commerce platforms in China serving millions of transactions per day, achieving an annual gross merchandise volume (GMV) of nearly 100 billion USD. CLUE is part of a larger fraud detection system in the company. CLUE complements, instead of replacing, other risk management systems. Thus, CLUE only focuses on users' purchase sessions, while leaving the analysis on users' registration, login, payment risk detections, and so on, to other existing systems.

CLUE uses two deep learning methods to capture the users' behavior: Firstly, we use Item2Vec [3], a technique similar to Word2Vec [14], to learn to embed the details of each click (e.g. the item being browsed) into a compact vector representation; Secondly, we use a recurrent neural network (RNN) to capture the sequence of clicks, revealing the browsing behaviors on the time-domain.

In practice, there are three challenges in the fraud detection applications:

(1) The number of fraudulent behaviors is far less than the legitimate ones [2,15], resulting in a highly imbalanced dataset. To capture the degree of imbalance, we define the *risk ratio* as the number of the portion of fraudulent

transactions in all transactions. The typical risk ratio in previous studies is as small as 0.1% [4]. We use a combination of under-sampling legitimate sessions and thresholding [16] to solve the problem.

(2) As the user browsing behaviors, both legitimate and fraudulent, change over time, we observe significant *concept drift* phenomenon [2]. To continuously fine-tune our model, we have built a mechanism that automatically fine-tunes the model with new data points incrementally.

(3) There are tens of millions of user sessions per day. It is challenging to scale the deep learning computation. Our training process is based on TensorFlow [1], using graphics processing units (GPUs) and data parallelism to accelerate computation. The serving module leverages TensorFlow Serving framework, providing real-time analysis of millions of transactions per day.

In summary, our major contributions are:

1. We propose a novel approach to capture detailed user behavior in purchasing sessions for fraud detection in e-commerce websites. Using a RNN-based approach, we can directly model user sessions using intuitive yet comprehensive features.
2. Although the session-modeling approach is general, we optimize it for the fraud detection application scenario. Specifically, we optimize for highly imbalanced datasets, as well as the concept drift problem caused by the ever-changing user behaviors.
3. Last but not least, we have deployed CLUE on JD.com serving over 220 million active users, achieving real-time detection of fraudulent transactions.

## 2    Data and Feature Extraction

In this section, we describe the feature extraction process of turning raw click logs into sequences representing user purchase sessions.

The inputs to CLUE are raw web server logs from standard log collection pipelines. The server log includes standard fields like the requested URL, browser name, client operating system, etc. For this analysis, we remove all personally identifiable information (PII) to protect users' privacy.

We use the web server *session ID* to group logs into user sessions and only focus on sessions with an *order ID*. We label the fraudulent orders using the business department's *case database* that records all fraudulent case reports. In this work, we ignore all sessions that do not lead to an order.

**Feature Extraction Overview.**    The key feature that we capture is the sequence of a user's browsing behavior. Specifically, we capture the behavior using a sequence that consists of a number of clicks within the same session. As we only care about purchasing sessions, the final action in a session is always a *checkout* click. Figure 2 illustrates four sample sessions. Note that there are a different number of clicks per session, so we only use the last $k$ clicks for each session. For short sessions with less than $k$ clicks, we add empty clicks after the checkout (practically, we pad non-existing sessions with zeros to make the session sequences the same length). In CLUE, we use a $k = 50$ that is more than enough to capture the entire sessions in most of the cases [11].

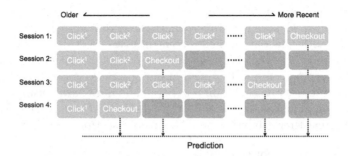

**Fig. 2.** Session padding illustration.

**Encoding Common Fields of a Click.** The standard fields in click logs are straightforward to encode in a feature vector. For example, we include numerical data fields like *dwell time* (i.e. the time a user spends on a particular page) and page loading time. We encode the fields with categorical types using one-hot encoding. These types include the browser language, text encoding settings, client operating systems, device types and so on. Specifically, for the source IP field, we first look up the IP address in an IP geo-location database and encode the location data as categorical data.

**Encoding the URL Information.** We mainly focus on two URL types: "list.jd.com/*" and "item.jd.com/*", respectively. As there are only dozens of merchandise categories, we encode the category using one-hot encoding.

The difficulty is with the items, as there are hundreds of millions of items on JD.com. One-hot encoding will result in a sparse vector with hundreds of millions of dimensions per click. Even worse, the one-hot encoding eliminates the correlations among separate items. Thus, we adopt the Item2Vec [3] to encode items. Item2Vec is a variation of Word2Vec [14], we regard each item as a "word", while regarding each session as a "sentence". So the items that commonly appear at the same positions of a session are embedded into vectors with smaller Euclidean distance.

We observe that the visit frequency follows a steep power-law distribution. If we choose to cover 90% of all the items in the click history, we only need 25 dimensions for Item2Vec, a significant saving on data size. We embed all other 10% items that appear rarely as the same constant vector.

In summary, we embed a URL into three parts, the type, category and item.

## 3    RNN Based Fraudulent Session Detection

Our detection is based on sequences of clicks in a session (as Fig. 2 shows). We feed the clicks of the same session into the model in the time order, and we want to output a *risk score* at the last click (i.e. the checkout action) for each session, indicating how suspicious the session is.

To do so, we need a model that can capture a sequence of actions. We find recurrent neural network (RNN) a good fit. We feed each click to the corresponding time slot of the RNN, and the RNN finally outputs the risk score. Figure 3 illustrates the RNN structure and its input/output. In the following, we use "depth" and "width" to denote the layer number and the number of hidden units per layer, respectively. By default, we use LSTM in CLUE to characterize the long-term dependency of the prediction on the previous clicks. In Sect. 5, we also compare the performance of the GRU alternative.

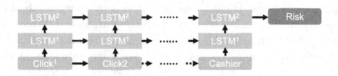

**Fig. 3.** Illustration of the RNN with LSTM cells.

**Dealing with Imbalanced Datasets.** One practical problem in fraud detection is the highly imbalanced dataset. In CLUE, we employ both data and model level approaches [10].

On the data level, we under-sample the legitimate sessions by random skipping, boosting the risk ratio to around 0.5%. After under-sampling, the dataset contains 1.6 million sessions, among which 8,000 are labeled as fraudulent. We use about 6% (about 100,000 sessions) of the dataset as the validation set. We choose test sets, with the risk ratio of 0.1%, from the next continuous time period (e.g. two weeks of data), which is outside of the 1.6 million sessions.

On the model level, we leverage the thresholding approach [16] to implement cost-sensitive learning. By choosing the threshold from the range [0, 1], we can obtain an application specific punishment level imposed on the model for misclassifying minority classes (false negatives) vs. misclassifying the normal class (false positives).

**Model Update.** To save time, we use incremental data to fine-tune the current model. Our experience shows that the incremental update works both efficiently and achieves comparable accuracy as the full update. The quality assurance module is used to guarantee the performance of the switched model is always superior to the current model (see Fig. 4).

## 4   System Architecture and Operation

We have deployed CLUE in real production, analyzing millions of transactions per day. From an engineering point of view, we have the following design goals: (1) *Scalability*: CLUE should scale with the growth of the number of transactions; (2) *Real-time*: we need to detect suspicious sessions before the checkout completes in a synchronous way, giving the business logic a chance to intercept potential frauds; and (3) *Maintainability*: we must be able to keep the model up-to-date over time, while not adding too much training overhead or model switching cost.

## 4.1 Training - Serving Architecture

To meet the goals, we design the CLUE architecture with four tightly coupled components, as Fig. 4 shows.

**Data Input.** We import raw access logs from the centralized log storage into an internal session database within CLUE using standard ETL (i.e. Extract, Transform and Load) tools. During the import process, we sort the logs into different sessions. Then we join the sessions with the purchase database to filter out those sessions without an order ID. Then we obtain the manual labels whether a session is fraudulent or not. We connect to the case database at the business units storing all fraud transaction complaints. We join with the case database (using the order ID) to label those known fraudulent sessions.

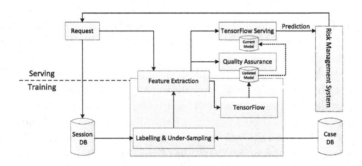

**Fig. 4.** The system architecture of CLUE.

**Model Training.** We perform under-sampling to balance the fraudulent and normal classes. Then the data preprocessing module performs all the feature extraction, including the URL encoding. Note that the item embedding model is trained offline. We then pass the preprocessed data to the TensorFlow-based deep learning module to train the RNN model.

**Online Serving.** After training and model validation, we transfer the trained RNN model to the TensorFlow Serving module for production serving. Requests containing session data from the business department are preprocessed using the same feature extraction module and then fed into the TensorFlow Serving system for prediction. Meanwhile, we persist the session data into the session database for further model updates.

**Model Update.** We perform periodic incremental updates to the model. It uses the latest updated model as the initial parameter and fine-tunes it with the incremental session data. Once the fine-tuned model is ready, and passes the model quality test, it is passed to TensorFlow Serving for production deployment.

## 4.2   Implementation Details

For the RNN Training, we set the initial learning rate to 0.001 and let it exponentially decay by 0.5 at every 5,000 iterations. We adopt Adam [12] for optimization with TensorFlow default configurations. The training process terminates when the loss on the validation set stops decreasing. We use TensorFlow as the deep learning framework [1] and leverage its built-in RNN network. Because of the highly imbalanced dataset, we raise the batch size to 512. The typical training duration is 12 hours (roughly 6–8 epochs).

## 5   Performance Evaluation

We first present our general detection performance on real production data. Then we evaluate the effects of different design choices of CLUE, and their effects on the model performance, such as different RNN structure, embeddings, and RNN cells. We also compare the RNN-based detection method with other features and learning methods. Finally, we evaluate the model update results.

### 5.1   Performance on Real Production Data

The best performance is achieved by an RNN model with 4 layers and 64 units per layer with LSTM cells. It is the configuration we use in production. We evaluate other RNN structures in Sect. 5.2. Compared with traditional machine learning approaches (see Fig. 5(b)), CLUE achieves over 3x improvement over the existing system. Integrating CLUE with existing risk management systems for eight months in production, we have observed that CLUE has brought a significant improvement of the system performance.

### 5.2   Effects of Different RNN Structures

Our model outputs a numerical probability of a session being fraudulent. We use a threshold $T$ to provide a tradeoff between precision and recall [16]. Varying $T$ between $[0, 1]$, we can get the precision of our model corresponding to a particular recall. Figure 5(a) shows the performance of RNNs with different widths using the Precision-Recall (P-R) curve. Throughout the evaluation, we use 4 layers for the RNN model.

The previous study points out that wider neural networks usually provide better memorization abilities, while deeper ones are good at generalization [7]. We want to evaluate the width and depth of the RNN structure to the fraud detection performance. We use $\alpha$-$\beta$ RNN to denote a RNN structure with $\alpha$ layers and $\beta$ hidden units per layer.

Table 1 provides the fraud detection precision with 30% recall, using different $\alpha$ and $\beta$. We see that given a fixed width, the performance improves with the depth increases. However, once the depth becomes too large, overfitting occurs. We find a 4-64 RNN performs the best, outperforming wider models such as

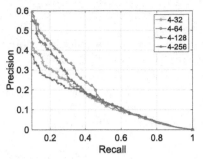

(a) Performance of 4 layer RNNs with different widths.

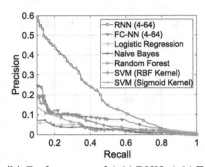

(b) Performance of 4-64 RNN, 4-64 FC-NN and traditional machine learning approaches.

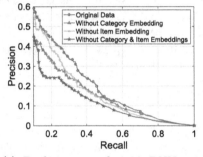

(c) Performance of 4-64 RNN with-/without item & category embeddings.

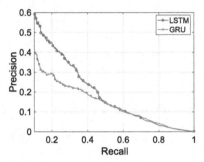

(d) Performance of LSTM and GRU cells, the RNN structure is 4-64.

**Fig. 5.** Experiment results.

**Table 1.** Precision of different RNN structures under the recall of 30%

| #Layer/#Unit | 32 | 64 | 128 | 256 |
|---|---|---|---|---|
| 1 | 19.3% | 23.1% | 24.3% | 25.1% |
| 2 | 23.4% | 23.6% | 26.1% | 27.2% |
| 3 | 24.7% | 24.6% | 29.0% | 27.8% |
| 4 | 24.8% | **33.8%** | 26.4% | 20.8% |

4–128 and 4–256. We believe the reason is that, given the relatively small number of fraudulent sample, the 4–128 RNN model begins to overfit. Also, we can see that the generalization ability of a model seems more important than its memorization ability in our application.

## 5.3   Compare with Traditional Methods

We show that CLUE performs much better than traditional features and learning methods including logistic regression, naive Bayes, SVM, and random forest.

Furthermore, we have investigated the performance of fully connected neural networks (FC-NNs). To leverage these methods, we need to leverage traditional feature engineering approaches by combining the time-dependent browsing history data into a fixed length vector. We follow many related researches and use *bag of words*, i.e., count the number of page views of different types of URLs in a session, and summarize the total dwell time of these URLs into the feature vector. Note that with bag-of-word, we cannot leverage the category and item embedding approaches as introduced in Sect. 2, but we only use one-hot encoding for these data. We plot the results in Fig. 5(b).

FC-NN performs better than other traditional machine learning methods, indicating that with abundant data, deep learning is not only straightforward to apply but also performs better. Meanwhile, with more detailed feature extraction and time-dependent learning, RNNs perform better than FC-NNs. Therefore, we can infer that our performance improvement comes from two aspects: (1) By using more training data, deep learning (FC-NN) outperforms traditional machine learning approaches; (2) RNN further improves the results over FC-NN as it captures both sequence information, and it allows us to use detailed category and item embeddings in the model.

### 5.4   Effects of Key Design Choices

Here we evaluate the effectiveness of various choices in CLUE's feature extraction and learning.

**Category and Item Embeddings.** To show that category and item data are essential features for fraud detection, we remove these features from our click data representation. Figure 5(c) shows that the detection accuracy is significantly lower without such information. Clearly, fraudulent users are different from normal users not because they are performing different clicks, but the real difference is *which* item they click on and in what *order*.

**Using GRU as RNN Cell.** Except for the LSTM, GRU is also an important RNN cell type that deals with the issues in vanilla RNN. Here we investigate the performance of RNNs with GRU cells. In Fig. 5(d), we compare the performance of different RNN cells. We find the LSTM shows better performance.

### 5.5   Model Update

To show the model update effectiveness using historical data, we consider a four consecutive equal-length time periods $P_1, \ldots, P_5$. The proportion of the number of sessions contained in these periods is roughly 2:2:1:1:1. We perform two sets of experiments, and both show the effectiveness of our model update methods.

First, we evaluate the performance of using history up to $P_n$ to predict $P_{n+1}$. We compare the performance of three update strategies: *incremental update*, *full update* (i.e. retrain the model from scratch using all history before the testing period) and *no update*. We use data from $P_1$ to train the initial model. Then we compute the precision (setting the recall to 30%) for the following four time periods using these three strategies. Figure 6(a) presents the results. We can see that incremental update achieves similar performance as the full model update, while the no update strategy performs the worst. It is also interesting that $P_2$ actually works worse than $P_3$ and beyond. We believe it is because the training data from $P_1$ is too few to produce a reasonable model.

Second, we show the performance of using different amounts of training data to predict the last time period, $P_5$. From Fig. 6(b), we can see that using either the full model update or incremental update, adding more data significantly improves the prediction results. It is not only because we are adding more data, but also because we use training data that are closer to time $P_5$, and thus tends to have more similar distributions.

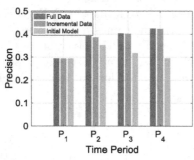
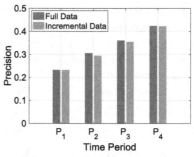

(a) Precision for the next time period.    (b) Precision for the last time period.

**Fig. 6.** The precision of the models trained with incremental and full data for the (a) next time period and (b) last time period, the recall is fixed to be 30%.

## 6    Related Work

**Fraud Detection.** Researchers have investigated fraud detection for a long time. Existing work focuses on credit card [8,13], insurance [17], advertisement [19], and online banking [6] fraud detections. The approaches used in these work include rule-based, graph-based, traditional machine learning, convolutional neural network (CNN) approaches [2,8,20]. They have two drawbacks: (1) These models have difficulties in dealing with time-dependent sequence data; and (2) The model can only take aggregated features (like a count), which directly leads to the loss of the detailed information about individual operations. Our system extracts user browsing histories with detailed feature encodings, and it is able to deal with high-dimensional complex time-dependent data using RNN.

**Recurrent Neural Networks.** Out of the natural language domain, researchers have RNNs to model user behaviors in similar web server logs, especially in session-based recommendation tasks [11,18,21]. To our knowledge, our work is the first application of the RNN-based model in fraud detection. Fraud detection is more challenging in that (1) there are too many items to consider, and thus we cannot use the one-hot encoding in these works; and (2) the frauds are so rare, causing a highly imbalanced dataset.

# 7 Conclusion and Future Work

Frauds are intrinsically difficult to analyze, as they are engineered to avoid detection. Luckily, we are able to observe millions of transactions per day, and thus accumulate enough fraud samples to train an extremely detailed RNN model that captures not only the detailed click information but also the exact sequences. With proper handling of imbalanced learning, concept drift, and real-time serving problems, we show that our features and model, seemingly detailed and expensive to compute, actually scale to support the transaction volumes we have, while providing an accuracy never achieved by traditional methods based on aggregate features. Moreover, our approach is straightforward, without too much ad hoc feature engineering, showing another benefit of using RNN.

As future work, we can further improve the performance of CLUE by building a richer history of a user, including non-purchasing sessions. We are also going to apply the RNN-based representation of sessions into other tasks like recommendation or merchandising.

**Acknowledgement.** We would like to thank our colleagues at JD for their help during this research. This research is supported in part by the National Natural Science Foundation of China (NSFC) grant 61532001, Tsinghua Initiative Research Program Grant 20151080475, MOE Online Education Research Center (Quantong Fund) grant 2017ZD203, and gift funds from Huawei.

# References

1. Abadi, M., Agarwal, A., Barham, P., Brevdo, E., Chen, Z., Citro, C., Corrado, G.S., Davis, A., Dean, J., Devin, M., et al.: TensorFlow: large-scale machine learning on heterogeneous distributed systems. arXiv:1603.04467 (2016)
2. Abdallah, A., Maarof, M.A., Zainal, A.: Fraud detection system: a survey. J. Netw. Comput. Appl. **68**, 90–113 (2016)
3. Barkan, O., Koenigstein, N.: Item2Vec: neural item embedding for collaborative filtering. In: IEEE 26th International Workshop on Machine Learning for Signal Processing, pp. 1–6. IEEE (2016)
4. Bellinger, C., Drummond, C., Japkowicz, N.: Beyond the boundaries of SMOTE. In: Frasconi, P., Landwehr, N., Manco, G., Vreeken, J. (eds.) ECML PKDD 2016. LNCS (LNAI), vol. 9851, pp. 248–263. Springer, Cham (2016). https://doi.org/10.1007/978-3-319-46128-1_16
5. Bianchi, C., Andrews, L.: Risk, trust, and consumer online purchasing behaviour: a chilean perspective. Int. Mark. Rev. **29**(3), 253–275 (2012)
6. Carminati, M., Caron, R., Maggi, F., Zanero, S., Epifani, I.: BankSealer: a decision support system for online banking fraud analysis and investigation. Comput. Secur. **53**, 175–186 (2015)
7. Cheng, H.T., Koc, L., Harmsen, J., Shaked, T., Chandra, T., Aradhye, H., Anderson, G., Corrado, G., Chai, W., Ispir, M., et al.: Wide & deep learning for recommender systems. In: Proceedings of the 1st Workshop on Deep Learning for Recommender Systems, pp. 7–10. ACM (2016)

8. Fu, K., Cheng, D., Tu, Y., Zhang, L.: Credit card fraud detection using convolutional neural networks. In: Hirose, A., Ozawa, S., Doya, K., Ikeda, K., Lee, M., Liu, D. (eds.) ICONIP 2016. LNCS, vol. 9949, pp. 483–490. Springer, Cham (2016). https://doi.org/10.1007/978-3-319-46675-0_53

9. Glover, S., Benbasat, I.: A comprehensive model of perceived risk of e-commerce transactions. Int. J. Electron. Commer. 15(2), 47–78 (2010)

10. He, H., Garcia, E.A.: Learning from imbalanced data. IEEE Trans. Knowl. Data Eng. 21(9), 1263–1284 (2009)

11. Hidasi, B., Quadrana, M., Karatzoglou, A., Tikk, D.: Parallel recurrent neural network architectures for feature-rich session-based recommendations. In: Proceedings of the 10th ACM Conference on Recommender Systems, pp. 241–248 (2016)

12. Kingma, D.P., Ba, J.: Adam: A method for stochastic optimization. arXiv:1412.6980 (2014)

13. Lim, W.Y., Sachan, A., Thing, V.: Conditional weighted transaction aggregation for credit card fraud detection. Adv. Inf. Commun. Technol. 433, 3–16 (2014)

14. Mikolov, T., Sutskever, I., Chen, K., Corrado, G.S., Dean, J.: Distributed representations of words and phrases and their compositionality. In: Advances in Neural Information Processing Systems, pp. 3111–3119 (2013)

15. Phua, C., Alahakoon, D., Lee, V.: Minority report in fraud detection: classification of skewed data. ACM SIGKDD Explor. Newslett. 6(1), 50–59 (2004)

16. Sheng, V.S., Ling, C.X.: Thresholding for making classifiers cost-sensitive. In: National Conference on Artificial Intelligence, pp. 476–481 (2006)

17. Shi, Y., Sun, C., Li, Q., Cui, L., Yu, H., Miao, C.: A fraud resilient medical insurance claim system. In: Thirtieth AAAI Conference, pp. 4393–4394 (2016)

18. Tan, Y.K., Xu, X., Liu, Y.: Improved recurrent neural networks for session-based recommendations. In: Proceedings of the 1st Workshop on Deep Learning for Recommender Systems, pp. 17–22. ACM (2016)

19. Tian, T., Zhu, J., Xia, F., Zhuang, X., Zhang, T.: Crowd fraud detection in internet advertising. In: International Conference on World Wide Web, pp. 1100–1110 (2015)

20. Tseng, V.S., Ying, J.C., Huang, C.W., Kao, Y., Chen, K.T.: FrauDetector: a graph-mining-based framework for fraudulent phone call detection. In: ACM SIGKDD International Conference on Knowledge Discovery and Data Mining, pp. 2157–2166 (2015)

21. Wu, S., Ren, W., Yu, C., Chen, G., Zhang, D., Zhu, J.: Personal recommendation using deep recurrent neural networks in NetEase. In: International Conference on Data Engineering, pp. 1218–1229 (2016)

# SINAS: Suspect Investigation Using Offenders' Activity Space

Mohammad A. Tayebi[1]([✉]), Uwe Glässer[1], Patricia L. Brantingham[2], and Hamed Yaghoubi Shahir[1]

[1] School of Computing Science, Simon Fraser University, Burnaby, Canada
{tayebi,glaesser,syaghoub}@cs.sfu.ca
[2] School of Criminology, Simon Fraser University, Burnaby, Canada
pbrantin@sfu.ca

**Abstract.** Suspect investigation as a critical function of policing determines the truth about how a crime occurred, as far as it can be found. Understanding of the environmental elements in the causes of a crime incidence inevitably improves the suspect investigation process. Crime pattern theory concludes that offenders, rather than venture into unknown territories, frequently commit opportunistic and serial violent crimes by taking advantage of opportunities they encounter in places they are most familiar with as part of their activity space. In this paper, we present a suspect investigation method, called SINAS, which learns the activity space of offenders using an extended version of the random walk method based on crime pattern theory, and then recommends the top-$K$ potential suspects for a committed crime. Our experiments on a large real-world crime dataset show that SINAS outperforms the baseline suspect investigation methods we used for the experimental evaluation.

## 1 Introduction

Crime is a purposive deviant behavior that is an integrated result of different social, economical and environmental factors. Crime imposes substantial costs on society at individual, community, and national levels.

An important policing task is investigating committed or reported crimes—known as criminal or *suspect investigation*. Spatial studies of crime, and more specifically environmental criminology, play an essential role in criminal intelligence [1,5–8].

Modeling spatial aspects of criminal behavior can be seen as an intractable case of the human mobility problem [2,4]. This is mainly because available information about the spatial life of offenders is usually limited to police arrest data on their home and crime locations. Further, spatial displacement of crime is a common phenomenon, meaning offenders shift their crime locations.

In this paper, we propose an approach to Suspect INvestigation using offenders' Activity Space, called SINAS. It first learns activity space of offenders based on crime pattern theory, using existing crime records. Then, given the location

© Springer International Publishing AG 2017
Y. Altun et al. (Eds.): ECML PKDD 2017, Part III, LNAI 10536, pp. 253–265, 2017.
https://doi.org/10.1007/978-3-319-71273-4_21

of a newly occured crime, SINAS ranks known offenders based on their activity space that influences offenders' criminal activity, and finally it recommends the top-$K$ suspects of that crime with the highest probability. Our experiments on a large real-world crime dataset show that SINAS outperforms the baseline suspect investigation methods significantly.

Section 2 explores related work, and Sect. 3 presents the fundamental concepts. Next, Sect. 4 introduces the proposed model, and Sect. 5 discusses the experimental evaluation and the results. Section 6 concludes the paper.

## 2    Background and Related Work

People do not move randomly across urban landscapes. For the most part, they commute between a handful of routinely visited places such as home, work, and their favorite places. With each and every trip, they get more familiar with, and gain new knowledge about, these places and everything along the way. A person will eventually be at ease with a place and it becomes part of their activity space (see Fig. 1). Nodes and Paths are two main components of an activity space. The (activity) nodes are the locations that the person

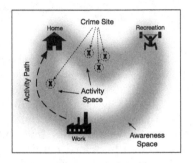

**Fig. 1.** Activity space

frequents (e.g., workplace, residence, recreation). These are the endpoints of a journey. The (activity) paths connect the nodes and represent the person's path of travel between nodes. Activity space of offenders is explored in several studies. The geographic profiling method by Rossmo [8] is widely recognized for inferring the activity space of an offender to determine the likely home location based on their crime locations. This method assumes that offenders select targets and commit crimes near their homes. Frank [3] proposes an approach to infer the activity paths of all offenders in a region based on their crime and home locations. Assuming the home location as the center of an offender's movements, the orientation of activity paths of each individual offender is calculated so as to determine the major directions, relative to their home location, into which they tend to move to commit crimes.

Based on criminological theories, several studies propose mathematical models for spatial and temporal characteristics of crime to predict future crimes. For instance, in [7], the authors use a point-pattern-based transition density model for crime space-event prediction. This model computes the likelihood of a criminal incident occurring at a specified location based on previous incidents. In [9], the authors model the emergence and dynamics of crime hotspots by using a two-dimensional lattice model for residential burglary, where each location is assigned a dynamic attractiveness value, and the behavior of each offender is modeled with a random walk process.

Our proposed method, SINAS, addresses the problem of recommending most likely suspects based on historical spatial information, which is a problem that none of the existing methods is able to address. Thus, there is a challenge in evaluating our experimental results. The model in [7] predicts only the time and location of crimes at an aggregate level. The method proposed in [8] discovers offender home locations based on their crime locations. Finally, the method in [3] finds locations which are centers of interest for committing crime.

# 3  Crime Data Characteristics

We evaluate the efficacy of our approach on a crime dataset representing five years (2001–2006) of police arrest-data for the Province of British Columbia, comprising several million data records, each refers to a reported offence[1]. Our experiments consider all subjects in four main categories: *charged*, *chargeable*, *charge recommended*, and *suspect*. Being in one of these categories means that the police is serious enough about the subject's involvement in a crime as to warrant calling them 'offender'. Here, we concentrate on crimes in Metro Vancouver (population: 2.46 M), with different regions connected through a road network composed of road segments having an average length of 0.2 km (see Table 1).

**Table 1.** Statistical properties of the used dataset

| Property | Value | Property | Value |
|---|---|---|---|
| #crimes | 125,927 | #offenders | 189,675 |
| #offenders with more then 1 crime | 25,162 | #co-offending links | 68,577 |
| #co-offenders in co-offending network | 17,181 | Avg. node degree in co-offending network | 4 |
| #road-segments | 64,108 | Avg. crime per road segment | 2 |

Figures 2a and b illustrate the distribution function of crime incidents per offender and per road segment respectively. Both distributions have heavy-tailed pattern. 83% of the offenders committed only one crime, while less than 1% of the offenders committed 10 or more crimes. Further, 38% of the road segments are linked to at least one crime and 9% of the road segments are linked to 10 or more crimes. Half of all the crimes occurred in only 1% of all road segments, and a total of 25% of all the crimes occurred in only 100 road segments. The average home to crime location distance of 80%, 63% and 40% of all offenders is less than 10 km, 5 km and 2 km, respectively. The average crime location distance of 73%, 52% and 26% of all offenders is less than 10 km, 5 km and 2 km, respectively. One can assume that frequent offenders are generally mobile and have several home locations identified in their records. 41% of the offenders who committed more than one crime have more than one home location.

---

[1] This data was made available for research purpose by Royal Canadian Mounted Police (RCMP) and retrieved from the Police Information Retrieval System (PIRS).

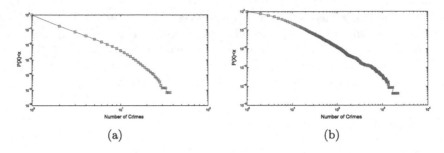

**Fig. 2.** Distribution of: (a) crimes per offender; (b) crimes per road segment

## 3.1   Fundamental Concepts and Definitions

This section introduces the fundamental concepts, definitions, and notations.

**Offender.** Let $V$ be a set of offenders and $C$ be a set of crimes. Each crime event $e \in C$ involves a non-empty subset of criminal offenders $U \subseteq V$. $C_i$ is the set of crimes committed by offender $u_i$. With each crime incident we associate a type of crime, a time when the crime occurred as well as longitude and latitude coordinates of the crime location and home location of involved offenders.

**Co-offending Network.** A *co-offending network* is an undirected graph $G(V, E)$. Each node represents a known offender $u_i \in V$. Offenders $u$ and $v$ are connected, $u_i, u_k \in V$ and $(u_i, u_k) \in E$, if they are known to have committed one or more offences together, and are not connected otherwise. $\Gamma_i$ denotes the set of neighbors of offender $u_i$ in the co-offending network.

**Road Network.** Intuitively, a road network can be decomposed into *road segments*, each of which starts and ends at an intersection. We use the *dual* representation where the role of roads and intersections is reversed. All physical locations along the same road segment are mapped to the same node. Formally, a road network is an undirected graph $R(L, Q)$, where $L$ is a set of nodes, each representing a single road segment. Road segments $l_j$ and $l_k$ are connected, $\{l_j, l_k\} \in Q$, if they have an adjacent intersection in common. Crime locations within a studied geographic boundary are mapped to the closest road segment. Henceforth, the term "road" is used to refer to a road segment.

**Road Features.** A vector $\bar{y}_j$ denotes the features of the road $l_j$ including road length $d_j$, and *road attractiveness* features vector $\bar{a}_j$. Further, $\bar{a}_j$ is a vector of size $m$ where the value of the $k^{th}$ entry of $\bar{a}_j$ corresponds to the total number of crimes of type $k$ committed previously at $l_j$. $\Pi_j$ denotes the set of neighbors of road $l_j$ in the road network. $\Delta \subset L$ denotes a set of roads with the highest crime rate, called crime *hotspots*. $D_{l_j, l_k}$ is the shortest path distance of road $l_j$ from road $l_k$, and $f_j$ denotes the total number of crimes at road $l_j$.

**Anchor Locations.** $L_i$ is the set of roads at which offender $u_i$ has been observed, including all of his known home and crime locations. $f_{i,j}$ and $t_{i,j}$

respectively denote the frequency and the last time $u_i$ was at anchor location $l_j$. *Offender trend* is given by a vector $\bar{x}_i$ of size $m$ which indicates the crime trend of $u_i$ as extracted from his criminal history. That is, the value of the $k^{th}$ entry of $\bar{x}_i$ corresponds to the number of crimes of type $k$ committed by offender $u_i$.

## 3.2 Problem Scope

Crime analysis captures a broad spectrum of facets pertaining to different needs and using different analytical methods. For instance, intelligence analysis aims at recognizing relationships between criminal network actors to identify and arrest offenders. It typically starts with a known crime problem or identified co-offending network, then uses these resources to collect, analyze and compile information about a predetermined target. An important problem is to identify most likely perpetrators of previous crimes. Criminal profiling approaches contribute to criminal intelligence using offenders' characteristics. Also, methods like geographic profiling build on environmental criminology theories and use information related to the environment of offenders and crimes.

**Problem Definition—Suspect Investigation:** In the following, we formally address the problem of suspect investigation. Assume there is a collection of crime records, $C$, from past crime events where each element in $C$ uniquely identifies a single crime incident. When a new crime incident $e$ occurs, police investigates suspects who potentially committed $e$ based on the existing information, that is, anchor locations (home and crime locations) of every offender in $C$ mapped on a road network, the type of crimes they committed, and also the (known) co-offending network $G$ extracted from $C$. The problem definition in abstract formal terms is as follows:

> Given a crime dataset $C$ and new crime incident $e$ at location $l_e$, the goal is to recommend the top-K suspects for $e$ with the highest probability.

Geographic profiling addresses a similar problem of detecting home locations of suspects of a crime incident, given a series of past crimes. However, the novelty of our approach is two-fold: *(1)* it directly targets the identity of offenders rather than their home locations; and *(2)* the input of our approach is a single crime incident, while the input of geographic profiling is a series of crimes.

# 4    SINAS Method

## 4.1    Learning Activity Space

A *random walk* over a graph is a stochastic process in which the initial state is known and the next state is decided using a transition probability matrix that identifies the probability of moving from a node to another node of the graph. Under certain conditions, the random walk process converges to a stationary distribution assigning an importance value to each node of the graph. The random walk method can be modified to satisfy the locality aspect of crimes, which states that offenders do not attempt to move far from their anchor locations. For instance, the random walk method works locally if the likelihood of terminating the walk increases with the distance from the anchor locations.

In our proposed model, starting from an anchor location the offender explores the city through the underlying road network. At each road, he decides whether to proceed to a neighboring road or return to one of his anchor locations. The random walk process continues until it converges to a steady state which reflects the probability of visiting a road by the offender. This probability can be relevant to the offender's exposure to a crime opportunity.

For learning the activity space of an offender, we need to understand his daily life and routines. However, in the crime dataset, generally we miss the *paths* completely and the *nodes* partially. Thus, we improve our incomplete knowledge about offenders with available information in the dataset by defining two different sets of anchor locations: *(1) main anchor locations*, denoted by $\mathcal{L}_i$ for offender $u_i$, is an extension of the offender anchor locations by adding his co-offenders' anchor locations with the assumption that friends in the co-offending network are likely to share the same locations; and *(2) intermediate anchor locations*, denoted by $\mathcal{I}_i$ for offender $u_i$, is the roads closest to the set of his main anchor locations, using a Gaussian model (see Sect. 4.1–Starting Probabilities for details).

An offender starts his random walk either from a main or intermediate anchor location. Given that the actual trajectories in an offender's journey to crime are unknown, SINAS guides offender movements in directions with a higher chance of committing a crime. This is done by taking into account different aspects that influence the offender movement directionality in computing the transition probabilities in a random walk.

**Random Walk Process:** For each single offender $u_i$, we perform a series of random walks on the road network $R(L, Q)$. The random walk process starts from one of the anchor locations of $u_i$ with predefined probabilities (see Sect. 4.1–Starting Probabilities) and traverses the road network to locate a criminal opportunity. At each step $k$ of the random walk, the offender is at a certain road $l_j$ and makes one of two possible decisions: *(1)* With probability $\alpha$, he decides to return to an anchor location and not look for a criminal opportunity this time, choosing an anchor location as follows: *(a)* with probability $\beta$, he decides to return to a main anchor location $l \in \mathcal{L}_i$, and *(b)* with probability $1 - \beta$, he returns to an intermediate anchor location $l \in \mathcal{I}_i$; and *(2)* With probability $1 - \alpha$ he continues looking for a crime opportunity. If he continues his random walk then he has two options in each step of the walk: *(a)* with probability $\theta(u_i, l_j, k)$, he stops the random walk, which means the offender commits a crime at road $l_j$, and *(b)* with probability $1 - \theta(u_i, l_j, k)$ he continues the random walk, moving to another road which is a direct neighbor of $l_j$.

To continue the random walk at road $l_j$, we select a direct neighbor road from $\Pi_{l_j}$. Function $\phi$ computes the transition probability from a roadsegment to one of its neighbor road segments (see Sect. 4.1–Movement Directionality for details). The probability of selecting road segment $l_r$ in the next step is:

$$P(l_j \rightarrow l_r) = \frac{\phi(l_r)}{\sum\limits_{l_p \in \pi_{l_j}} \phi(l_p)} \qquad (1)$$

The probability of being at road $l_r$ at step $k+1$ given that the offender was at road $l_j$ at step $k$ is shown in Eq. 2, where $X_{u_i,k}$ is the random variable for $u_i$ being at road $l_r$ in step $k$. We terminate the random walks when $||F^{m+1}|| - ||F^m|| \leq \epsilon$, where $F^m = (F(u_i, l_1) \dots F(u_i, l_{|L|}))^T$ is the results for $u_i$ after $m$ random walks. For some offenders the random walks do not converge, in which case we terminate the overall process at $m > 10000$.

$$P(X_{u_i,k+1} = l_r | X_{u_i,k} = l_j) = (1 - \alpha)(1 - \theta_{l_j,k}) \times P(l_j \rightarrow l_r) \qquad (2)$$

**Starting Probabilities:** The model distinguishes two types of starting nodes. *(1) Main anchor locations* are all anchor locations of a single offender and his co-offenders: $\mathcal{L}_i = L_i \cup \{l_j : l_j \in L_v, v \in \Gamma_u\}$. The rationale is that offenders who have collaborated in the past likely may have shared information on anchor locations in their activity space, an aspect that possibly affects their choice of future crime locations. In computing the starting probability of each anchor location, the two primary factors are the *frequency* and the *last time* an offender visited an anchor location. The probability that offender $u_i$ starts his random walk from $l_j$ is shown in Eq. 3, where $t$ is the current time, and $\rho$ is the parameter controlling the effect of the timing.

$$S(i,j) = \frac{f_{i,j} \times e^{\frac{-(t-t_{i,j})}{\rho}}}{\sum\limits_{l_k \in \mathcal{L}_i} f_{i,k} \times e^{\frac{-(t-t_{i,k})}{\rho}}} \qquad (3)$$

*(2) Intermediate anchor locations* are the closest locations to main anchor locations. Human mobility models use Gaussian distributions to analyze human movement around a particular point such as home or work location [4]. We assume that offender movement around their main anchor locations follows a Gaussian distribution. Each main anchor location of offender $u_i$ is used as the center, and the probability of $u_i$ being located in a road is modeled with a Gaussian distribution. Given road $l$, the probability of $u_i$ residing at $l$ is:

$$S(i,l) = \sum\limits_{l_j \in \mathcal{L}_i} \frac{f_{i,j}}{\sum\limits_{l_k \in \mathcal{L}_i} f_{i,k}} \frac{\mathcal{N}(l|\mu_{l_j}, \Sigma_{l_j})}{\sum\limits_{l_k \in \mathcal{L}_i} \mathcal{N}(l|\mu_{l_k}, \Sigma_{l_k})} \qquad (4)$$

Here $l$ is a road which does not belong to the set of main anchor locations. $\mathcal{N}(l|\mu_{l_j}, \Sigma_{l_j})$ is a Gaussian distribution for visiting a road when $u_i$ is at anchor location $l_j$, with $\mu_{l_j}$ and $\Sigma_{l_j}$ as mean and covariance. We consider the normalized activity frequency of $u_i$ at $l_j$, meaning that a main anchor location with higher activity frequency has higher importance. For offender $u_i$, the roads with the highest probability of being an intermediate anchor location are added to the set $\mathcal{I}_i$ as additional starting nodes besides the main anchor locations.

**Movement Directionality:** The creation of the main attractor nodes and paths are developed through normal mobility shaped by the urban backcloth or urban environment. Each individual has normal, routine pathways or commuting/mobility routes that are unique. However, the environment where we live influences our actions and movements. Highways, streets and road networks in general guide us to our destinations such as home, workplace, recreation center, and business establishments. In the aggregate, individuals routes overlap or intersect in time and space. These areas of overlap often have rush hours and congestion at intersections or mass transit stops associated with handling large numbers of people. These high activity locations can become crime attractors and crime generators when there are enough suitable targets in those locations. Crime attractors and generators affect directionality of offenders' movement.

One can conclude that starting from an anchor location the probability of offender movement toward crime attractors and generators is higher. To address this fact, in the random walk process, the transition probability is computed based on the proximity of a road to the crime hotspots and the importance of each crime hotspot, which is proportional to the number of crimes committed there. Function $\phi(l_j)$ is used in computing the transition probability (see Sect. 4.1–Random Walk Process) of moving offender $u_i$ from $l_k$ to $l_j$, where $f_n$ is the number of crimes committed at $l_n$. $D_{j,n}$ is the distance of road $l_j$ from the hotspot $l_n \in \Delta$, which is equal to the length of shortest path between two roads.

$$\phi(l_j) = \sum_{n=1}^{|\Delta|} f_n \times \frac{1}{D_{j,n}} \tag{5}$$

**Stopping Criteria:** The probability of stopping the random walk for an offender at a given road corresponds to the probability of this offender committing a crime in that road segment. Two factors influence the stopping probability of offender $u_i$ in the road $l_j$. The first one relates to the similarity of the crime trend of offender $u_i$ and the criminal attractiveness of road $l_j$, where higher similarity means a higher chance of stopping $u_i$ at $l_j$. The second factor is the distance of $l_j$ from the starting point measured in the number of steps $(k)$ from the starting point. To satisfy the locality aspect of crimes, the probability of continuing the random walk decrease while getting farther from the starting point. Thus, the stopping probability (Eq. 6) is inversely proportional to $k$. Also, $sim(i, j)$ denotes the cosine similarity of crime trend of $u_i$ and the attractiveness of $l_j$.

$$\theta(u_i, l_j, k) = sim(i, j) \times \frac{1}{1 + e^{\frac{-k}{2}}} \tag{6}$$

### 4.2   Suspect Recommendation

The crime location is neither even nor random, however, there is an underlying spatial pattern in it. Environmental criminology theories [1] suggest that crimes occur in predictable ways, at offenders' *awareness space* which includes their activity space. To recommend the most likely suspects of a new crime incident based on the learnt offenders' activity space, we rank offenders based on the proximity of the crime location to their activity spaces. An offender is considered as a 'potential suspect' if the crime location is close enough to the activity space of this offender. This approach is based on a crime pattern theory stating that future crime locations are within offenders' activity space which is dependent to their activity nodes and paths. To influence offenders' characteristics, we consider the history of the offenders including the *types* and *number* of their committed crimes. The probability of offender $u_i$ commits a *new* crime $e$ is computed in Eq. 7, where $\mathcal{T}(C_i, e)$ is a boolean function that returns *one* if in the crime records of offender $u_i$, $C_i$, there is a crime event with the same type as crime $e$, and *zero* otherwise. $\omega$ is a parameter that controls the influence of function $\mathcal{T}$. $|C_i|$ is the number of crimes that offender $u_i$ committed previously. $F(u_i, l_k)$ is the probability of $l_k$ being in activity space of offender $u_i$. $D_{l_k, l_e}$ is the distance between roads $l_k$ and $l_e$.

$$Z(u_i, e) = \omega \mathcal{T}(C_i, e) \times |C_i| \times \sum_{k=1}^{n} F(u_i, l_k) \times D_{l_k, l_e} \tag{7}$$

## 5   Experimental Evaluation

We divide the crime dataset chronologically into train and test data. The train data, used to learn the activity space of offenders, includes all crimes that happened in the first 54 months, and the test data includes the remaining six months. The crimes in the test data committed by known offenders are used for suspect investigation. SINAS recommends the top-$K$ suspects most likely to commit a new occurred crime. $K$ is set to 50 in our experiments, but relative results for other values of $K$ are also consistent. We use the recall measure (i.e., the percentage of crimes in which the offender who committed that crime appears in the list of top-$K$ recommended suspects) to evaluate the quality of methods. Before discussing the results, it is crucial to specify the experimental setting.

*(1)* On the one hand, if a new crime occurs in a location where an offender has been observed previously (as his anchor location) then the probability that the same offender is involved in the new crime is higher, and this fact makes the investigation process easier. Formally speaking: assume a crime $e$ committed by offender $u_i$ at $l_j$. If $l_j \in L_i$, then we consider $e$ as an *easy case*; and, if $l_j \notin L_i$, then we consider $e$ as a *hard case*. We therefore define two scenarios: *easy scenario* which includes all (union of easy and hard) cases and *hard scenario* which only includes hard cases. We compare the performance of SINAS for both easy and hard scenarios.

*(2)* On the other hand, repeat offenders are responsible for large percentage of committed crimes and there has long been an interest in the behavior of repeat offenders since controlling these groups of offenders can significantly reduce the overall crime level. Therefore, we distinguish two groups of offenders: repeat

offenders with 10 or more crimes and non-repeat offenders with less than 10 crimes. We compare the performance of SINAS for repeat offenders with all (union of repeat and non-repeat) offenders.

## 5.1 Baseline Methods

As discussed in Sect. 2, there is no suspect investigation method using offenders' spatial information to the best of our knowledge; however, we evaluate the SINAS performance in comparison with some baseline methods. We also perform experiments on different settings of SINAS to learn the meaningfulness of its three principal elements: *(1)* the probabilistic aspect of offenders' activity space, *(2)* offenders' crime types, and *(3)* the frequency of crimes committed by offenders. Following is the list of comparison partners in our experiments.

**SINAS-PPN** takes the probabilistic aspect of offenders' activity space and their crime types into account while ignoring the frequency of committed crimes.

**SINAS-PNP** takes the probabilistic aspect of offenders' activity space and frequency of committed crimes into account while ignoring the type of crimes.

**SINAS-NPP** ignores the probabilistic aspect of offenders' activity space but considers type and frequency of crimes committed by them.

**SINAS** takes all available information including the probabilistic aspect of offenders' activity space, frequency and type of committed crimes into account.

**CrimeFrequency** ranks offenders based on the number of crimes they have committed and includes top-$K$ offenders with the highest crime number in the recommendation list. The intuition behind this method is that *repeat offenders* are more probable to be involved in a new occurred crime.

**Proximity** uses a distance-decay function to compute the proximity of offenders' anchor locations from the location of a new crime. It considers the frequency of being an offender in each of his anchor locations as a factor of their importance. Proximity is comparable to the geographic profiling approach [8].

**Random** recommends suspects randomly from the pool of known offenders.

## 5.2 Experimental Results

Table 2 shows the performance of different variations of SINAS and the other baseline methods for both easy and hard scenarios. For both scenarios, SINAS outperforms the other baseline methods and significantly outperforms Proximity and CrimeFrequency. Interestingly, CrimeFrequency has a good performance in the easy scenario. In the easy scenario that crimes with known locations for

**Table 2.** Recall (%) of different suspect investigation methods (K = 50) for hard and easy scenarios considering all offenders (repeat and non-repeat)

|  | SINAS-PPN | SINAS-PNP | SINAS-NPP | SINAS | CrimeFrequency | Proximity | Random |
|---|---|---|---|---|---|---|---|
| Hard scenario | 3.8 | 5.1 | 4.2 | 5.4 | 1.5 | 3.7 | 0.002 |
| Easy scenario | 10.4 | 11.9 | 5.3 | 12.1 | 1.3 | 10.3 | 0.002 |

offenders are included, Proximity gets the advantage of having those locations exactly in the offender's anchor locations, and therefore it is able to successfully recommend the suspects. CrimeFrequency has the weakest performance but still works much better than Random recommendation method.

In our experimental setting, the number of potential suspects is about 25,000 and $K = 50$. SINAS recall is more than 5% and 12% respectively in the hard and easy scenarios. Contrary to geographic profiling which receives a series of crime locations as an input and criminal profiling which may have rich information about suspects to reduce the search space, SINAS only uses the location of a single crime, and we thus believe this result is a significant contribution to the difficult task of suspect investigation.

Looking at the experimental results of the SINAS variations, we notice all three elements of SINAS contribute to the method performance. Offenders' crime frequency has the most contribution and offenders' crime type has the least influence on the SINAS performance. As already discussed, a large percentage of crimes are committed by repeat offenders and taking this fact into account significantly improves the SINAS performance. As described in Sect. 3, in only half of the repeat offenders we observe strong patterns in their criminal trend. Recognizing complex and latent patterns in criminal activities to serve the suspect investigation task needs a more thorough study of offenders' trend which is beyond the scope of this paper.

Figure 3a shows the performance of SINAS for different values of $K$. As expected, the recall value is increased by increasing $K$, reaching to 9% and 16% in hard and easy scenarios for $K = 100$. Considering the major cost of investigation process for the law enforcement more specifically in serious crimes, using greater values of $K$ to reduce the search space and optimize the spent cost

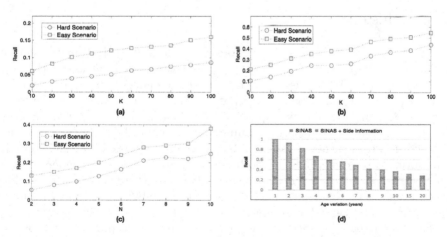

Fig. 3. SINAS performance in the easy and hard scenarios for: (a) different values of $K$, (b) repeat offenders respect to different values of $K$, and (c) group of offenders with greater than or equal $N$ crimes ($K = 50$); (d) SINAS performance for repeat offenders in the hard scenario considering offender's age range ($K = 50$)

and time is reasonable. Figure 3b shows the performance of SINAS for the repeat offenders with respect to different values of $K$. For $K = 50$, SINAS has the recall of 25% and 38% in the hard and easy scenarios. For the repeat offenders that we know more about their spatial activities, the SINAS performance is about two times greater than the method performance for all offenders.

For studying the SINAS performance for repeat and not-repeat offenders, we categorize offenders based on the number of crimes they have committed. Figure 3c shows the SINAS recall for each of these groups of offenders. As depicted, the SINAS performance increases linearly by increasing the number of crimes of the corresponding group, meaning that suspect investigation for a group of offenders who committed more crimes is generally more successful.

SINAS and criminal profiling approaches can be used as complementary tools for suspect investigation. Assume that for a new occurred crime, the police is able to guess the age of the offender based on evidence and witness interviews. Using this piece of information reduces the search space and increases the chance of success. In the following, we discuss the experimental results of applying SINAS on this subset of offenders instead of all offenders. Figure 3d shows the result for this suspect intelligence scenario ($K = 50$), where the x-axis shows the exactness of our knowledge about the age (#years) of the offender. In other words, if the offender exact age is $a$, then the value $b$ on x-axis means SINAS considers offenders with ages in the interval of $[a - b, a + b]$. As shown, having more precise information on the offender's age contributes more to the intelligence process.

With $b = 1$, SINAS is able to investigate all crimes successfully, and even $b = 20$ improves the SINAS performance compared to the situation of having no side information. This result shows the importance of side information in the suspect investigation process.

## 6   Conclusions

This paper proposes the SINAS method for suspect investigation by analyzing the activity space of offenders. It utilizes an extended version of the random walk method and learns the activity space of offenders based on a widely accepted criminological theory, crime pattern theory. Our experimental results show: (1) learning the activity space of offenders from their spatial life contributes to high-quality suspect recommendation; (2) utilizing offenders' criminal trend improves suspect recommendation. Not only does SINAS significantly outperform baseline methods for both repeat and non-repeat offenders, but it also has more satisfying results for repeat offenders where there is more information available on their spatial activities; and (3) SINAS and criminal profiling approaches can be viewed as complementary tools for suspect investigation.

Data mining-based suspect investigation is a multi-step process that has significant operational challenges in practice. Three main steps of this process—question formulation, data preparation, and data mining—have been addressed in our proposed method. However, the ultimate steps, deployment and efficacy evaluation, are beyond the scope of this paper. Making a difference in real-world situations, calls for an iterative process where law enforcement and policymakers act on analytics inferred from data mining-based suspect investigation methods at the strategic, tactical and operational levels.

# References

1. Brantingham, P.J., Brantingham, P.L.: Environmental Criminology. Sage Publications, Beverly Hills (1981)
2. Brockmann, D., Hufnagel, L., Geisel, T.: The scaling laws of human travel. Nature **439**(7075), 462–465 (2006)
3. Frank, R., Kinney, B.: How many ways do offenders travel - evaluating the activity paths of offenders. In: Proceedings of the 2012 European Intelligence and Security Informatics Conference (EISIC 2012), pp. 99–106 (2012)
4. Gonzalez, M.C., Hidalgo, C.A., Barabasi, A.: Understanding individual human mobility patterns. Nature **453**(7196), 779–782 (2008)
5. Gorr, W., Harries, R.: Introduction to crime forecasting. Int. J. Forecast. **19**(4), 551–555 (2003)
6. Harries, K.: Mapping crime principle and practice. U.S. Department of Justice, Office of Justice Programs, National Institute of Justice (1999)
7. Liu, H., Brown, D.E.: Criminal incident prediction using a point-pattern-based density model. Int. J. Forecast. **19**(4), 603–622 (2003)
8. Rossmo, D.K.: Geographic Profiling. CRC Press, Boca Raton (2000)
9. Short, M.B., D'orsogna, M.R., Pasour, V.B., Tita, G.E., Brantingham, P.J., Bertozzi, A.L., Chayes, L.B.: A statistical model of criminal behavior. Math. Models Methods Appl. Sci. **18**(Suppl. 01), 1249–1267 (2008)

# Stance Classification of Tweets Using Skip Char Ngrams

Yaakov HaCohen-kerner[✉], Ziv Ido, and Ronen Ya'akobov

Department of Computer Science, Jerusalem College of Technology, 9116001 Jerusalem, Israel
`kerner@jct.ac.il`, `ziv0798@gmail.com`, `ronenya4321@gmail.com`

**Abstract.** In this research, we focus on automatic supervised stance classification of tweets. Given test datasets of tweets from five various topics, we try to classify the stance of the tweet authors as either in FAVOR of the target, AGAINST it, or NONE. We apply eight variants of seven supervised machine learning methods and three filtering methods using the WEKA platform. The macro-average results obtained by our algorithm are significantly better than the state-of-art results reported by the best macro-average results achieved in the SemEval 2016 Task 6-A for all the five released datasets. In contrast to the competitors of the SemEval 2016 Task 6-A, who did not use any char skip ngrams but rather used thousands of ngrams and hundreds of word embedding features, our algorithm uses a few tens of features mainly character-based features where most of them are skip char ngram features.

**Keywords:** Skip character ngrams · Skip word ngrams · Social data
Short texts · Stance classification · Supervised machine learning · Tweets

## 1 Introduction

Sentiment analysis is the computational study of people's opinions, appraisals, attitudes, and emotions toward entities, individuals, issues, events, topics and their attributes [1]. Stance classification is a sub-domain of sentiment analysis. Stance classification is defined as the task of automatically determining from text whether the text author is in favor of, against, or neutral towards the given target. This task is challenging due to the fact that the available social data contains on the one hand, informal language, e.g., emojis, hashtags, misspellings, onomatopoeia, replicated characters, and slang words and on the other hand, personalized language.

Stance detection is becoming more and more important in many fields. For instance, stance studies can be helpful in detecting electoral issues and understanding how public stance is shaped [2]. Furthermore, stance detection is critical in situations in which a quick detection is needed, such as disaster detection and violence detection [3].

During the last fourteen years, there has been active research concerning stance detection. Most studies focus on debates in online social and political public forums [4–7], congressional debates [8–10], and company-internal discussions [11, 12].

In this study, we explore another field, the field of stance detection in tweets. Twitter as one of the leading social networks presents challenges to the research community since tweets are short, informal, and contain many misspellings, shortenings, and slang

© Springer International Publishing AG 2017
Y. Altun et al. (Eds.): ECML PKDD 2017, Part III, LNAI 10536, pp. 266–278, 2017.
https://doi.org/10.1007/978-3-319-71273-4_22

words. To perform the stance classification tasks we use the popular char/word unigrams/bigrams/trigrams features. Furthermore, we use hashtags, orthographic, and sentiment features that are assumed to contain important social information. We also use char/word skip ngram features.

Skip ngrams are more general than ngrams because their components (usually characters or words) need not be consecutive in the text under consideration, but may leave gaps that are skipped over [13]. The idea behind skip ngram features is to generate features that occur more frequently, which allow overcoming, at least partially, problems such as noise (e.g., misspellings) and sparse data (i.e., most of the data is fairly rare), by considering various skip steps. For the char sequence ABCDE, as an example, in addition to the traditional bigrams AB, BC, CD, and DE, we can define the following skip-bigrams with the skip step of "one": AC, BD, and CE. The main disadvantage of the skip ngram features (for various string and skip lengths) is that their number is relatively high.

The main contribution of this study is the implementation of successful stance classification tasks for short text corpora based mainly on a limited number of char ngrams features in general and char skip ngrams in particular. To the best of our knowledge, we are the first to perform such successful stance classification. The macro-average results obtained by our algorithm are significantly better than the best macro-average results achieved in the SemEval 2016 Task 6-A [14] for all the five released datasets of tweets in the supervised framework.

The rest of this paper is as follows: Sect. 2 presents relevant background on stance classification and skip ngrams. Section 3 describes the applied feature sets. Section 4 presents the examined corpus, the experimental results, and their analysis. Finally, Sect. 5 summarizes the research and suggests future directions.

## 2 Relevant Background

### 2.1 Stance Classification

A shared task held in NLPCC-ICCPOL 2016 [15] focuses on stance detection in Chinese microblogs. The submitted systems were expected to automatically determine whether the author of a Chinese microblog is in favor of the given target, against the given target, or whether neither inference is likely. The authors point that different from regular tasks on sentiment analysis, the microblog text may or may not contain the target of interest, and the opinion expressed may or may not be towards the target of interest. The supervised task, which detects stance towards five targets of interest, has had sixteen team participants. The highest F-score obtained was 0.7106.

The organizers of the SemEval 2016 Task 6-A [14] released five datasets of tweets in the supervised framework. The goal of this task was to classify stance towards five targets: "Atheism", "Climate Change is a Real Concern", "Feminist Movement", "Hillary Clinton", and "Legalization of Abortion" while taking into account that the targets may not explicitly occur in the text. This corpus is the corpus we used in this study. The best results achieved in this task will be compared to our results.

## 2.2  Skip Ngrams

Guthrie et al. [13] examine the use of skip-grams to overcome the data sparsity problem, which refers to the fact that language is a system of rare events, so varied and complex, that even using an extremely large corpus, we can never accurately model all possible strings of words. The authors examine skip-gram modelling using one to four skips with various amount of training data and test against similar documents as well as documents generated from a machine translation system. Their results demonstrate that skip-gram modelling can be more effective in covering trigrams than increasing the size of the training corpus.

Jans et al. [16] were the first to apply skip-grams to predict script events. Their models (1) identify representative event chains from a source text, (2) gather statistics from the event chains, and (3) choose ranking functions for predicting new script events. Predicting script events using 1-skip bigrams and 2-skip bigrams outperform using regular ngrams on various datasets. They estimate that the reason for these findings is that the skipgrams provide many more event pairs and by that better capture statistics about narrative event chains than regular ngrams do.

Sidorov et al. [17] introduce the concept of syntactic ngrams (sn-grams), which enables the use of syntactic information. In sn-grams, neighbors are defined by syntactic relations in syntactic trees. The authors perform experiments for an authorship attribution task (a corpus of 39 documents by three authors) using SVM, NB, and J48 for several profile sizes. The results show that the sn-gram technique outperforms the traditional word ngrams, POS tags, and character features. The best results (accuracy of 100%) were achieved by Sn-grams with the SVM classifier.

Fernández et al. [18] perform supervised sentiment analysis in Twitter. They show that employing skip-grams instead of single words or ngrams improves the results for five datasets including Twitter and SMS datasets. This fact suggests that the skip-grams approach is promising.

Dhondt et al. [19] improve the classification of abstracts from English patent texts using a combination of unigrams and PoS filtered skip-grams. Skip-grams with zero (bigrams) up to two skips were found to be efficient informative phrases and especially noun-noun and adjective-noun combinations make up the most important features for patent classification.

## 3  The Features

In this research, we implement 36,339 features divided into 18 feature sets. Some of these feature sets (e.g., quantitative and orthographic) have been already implemented in previous classification studies [20, 21]. Table 1 presents general details about these feature sets. In a case, where less features are found for a certain feature set than the number assigned to this set then this set contains the number of found features.

The hashtag set contains the following 105 features: frequencies of the top 100 occurring hastags normalized by the # of words in the tweet, # of hashtags in the tweet normalized by the # of the words in the tweet, # of occurrences of 27 positive NRC [22] sentiment words used in hashtags normalized by the # of the words in the tweet, # of occurrences of 51 negative NRC sentiment words used in hashtags [22] normalized by the # of the words in the tweet, # of occurrences of 14,459 positive NRC words used in hashtags normalized by the # of the words in the tweet, and the # of occurrences of 27,812 negative NRC words used in hashtags normalized by the # of the words in the tweet.

**Table 1.** General details about the feature sets.

| # of feature set | Name of feature set | # of features | # of feature set | Name of feature set | # of features |
|---|---|---|---|---|---|
| 1 | hashtag | 105 | 10 | PoS Tags | 36 |
| 2 | sentiment | 6 | 11 | character unigrams | 1000 |
| 3 | quantitative | 5 | 12 | character bigrams | 1000 |
| 4 | emojis | 21 | 13 | character trigrams | 1000 |
| 5 | orthographic | 122 | 14 | word unigrams | 1000 |
| 6 | long words | 11 | 15 | word bigrams | 1000 |
| 7 | stop words | 11 | 16 | word trigrams | 1000 |
| 8 | onomatopoeia | 11 | 17 | skip character ngrams | 15000 |
| 9 | slang | 11 | 18 | skip word ngrams | 15000 |

The sentiment set contains the following 6 features: normalized count of positive/negative sentiment emotion words according to the NRC lexicon [22], normalized counts of positive/negative sentiment words according to the Bing–Liu lexicon [23], and normalized count of positive/negative sentiment words according to the MPQA lexicon [24].

The quantitative set contains the following 5 features: # of characters in the tweet, # of words in the tweet, the average length in characters of a word in the tweet, # of sentences in the tweet, and the average length in words of a sentence in the tweet.

The emoji set contains the following 21 features: the # of emojis in the tweet normalized by the # of the characters in the tweet and frequencies of the top 20 occurring emojis normalized by the # of words in the tweet.

The orthographic set contains 122 features. Due to space limitation, we shall present some of them as follows: # of question marks/# of exclamation marks in the tweet/# of pairs of apostrophes in the tweet/# of legitimate pairs of brackets normalized by the # of characters in the tweet.

The "long words" set contains the following 11 features: # of elongated words (i.e., words that at least one of their letters repeats more than 3 times) normalized by the # of the words in the tweet, and frequencies of the top 10 occurring long words (words that their length is more than 10 characters) normalized by the # of the words in the tweet.

The stop words set contains the following 11 features: # of stop words in the tweet normalized by the # of words in the tweet and frequencies of the top 10 occurring stop words normalized by the # of words in the tweet.

The onomatopoeia set contains the following 11 features: # of onomatopoeia words in the tweet normalized by the # of words in the tweet and frequencies of the top 10 occurring onomatopoeia words normalized by the # of words in the tweet.

The slang set contains the following 11 features: # of slang words in the tweet normalized by the # of words in the tweet and frequencies of the top 10 occurring slang words normalized by the # of words in the tweet.

The PoS Tags set contains frequencies of the 36 PoS tags (see the 'Penn Treebank Project' [25]) normalized by the # of PoS tags in the tweet implemented by the Stanford Log-linear Part-Of-Speech Tagger [26] described in Klein and Manning [27].

Each one of the character unigrams/bigrams/trigrams sets includes the frequencies of the top 1000 occurring character unigrams/bigrams/trigrams normalized by the suitable # of character series in the tweet. Each one of the word unigrams/bigrams/trigrams sets includes the frequencies of the top 1000 occurring word unigrams/bigrams/trigrams normalized by the suitable # of word series in the tweet.

The skip character n-grams set is divided into 15 feature subsets (and the same for the skip word n-grams set). The features included in these 30 sub-sets (1000 features for each sub-set) are defined for all possible combinations of continuous character/word series (of 3–7 characters/words) that enable skip steps (of 2–6 characters/words, respectively). We defined 30,000 features for these 30 sub-sets because we wanted to have 1000 features for each sub-set (similar to what was defined for the character/word unigrams/bigrams/trigrams sets). We did not know which combinations of character/word series and skips will be successful; therefore we decided to define 30 possible combinations. The main reason why we enabled such a big number of features is because we assume that some of these skip ngram features might be very useful to overcome problems that characterized tweets such as noise (e.g., misspellings) and sparse data (i.e., most of the data is fairly rare).

## 4    The Experimental Setup

The examined corpus is the corpus of the SemEval 2016 Task 6-A [14] mentioned above. It includes tweets divided to 5 datasets: Legalization of Abortion, Hillary Clinton, Feminist Movement, Climate Change, and Atheism. Each one of the topics contains tweets with stance class that be labeled into one of three possibilities: FAVOR, AGAINST and NONE. Table 2 presents the distribution of stances in the five supervised datasets. To enable reproducibility, in the next paragraphs, we detail the algorithm, the experiments and their results (in addition to the details in the previous section).

Table 2. Distribution of stances in the five supervised datasets.

| Target | # total | % of instances in train set | | | | % of instances in test set | | | |
|--------|---------|---------|-------|---------|---------|--------|-------|---------|---------|
| | | # train | Favor | Against | Neither | # test | Favor | Against | Neither |
| Abortion | 933 | 653 | 18.5 | 54.4 | 27.1 | 280 | 16.4 | 67.5 | 16.1 |
| Climate | 564 | 395 | 53.7 | 3.8 | 42.5 | 169 | 72.8 | 6.5 | 20.7 |
| Feminist | 949 | 664 | 31.6 | 49.4 | 19.0 | 285 | 20.4 | 64.2 | 15.4 |
| Clinton | 984 | 689 | 17.1 | 57.0 | 25.8 | 295 | 15.3 | 58.3 | 26.4 |
| Atheism | 733 | 513 | 17.9 | 59.3 | 22.8 | 220 | 14.5 | 72.7 | 12.7 |
| All | 4163 | 2914 | 25.8 | 47.9 | 26.3 | 1249 | 24.3 | 57.3 | 18.4 |

Basic baseline accuracy results for each one of the five datasets are computed using all the features (more advanced baseline accuracy results are the state-of-art results reported by the best macro-average results achieved in the SemEval 2016 Task 6-A). We performed extensive experiments using the WEKA platform [28, 29]. Using the same training and test sets as used by the SemEval 2016 Task 6-A, we applied eight variants of seven supervised machine learning (ML) methods with their default parameters, parameter tuning, and 10-fold cross-validation tests, three filter feature selection methods, and seven performance metrics, as follows:

For each dataset (Atheism, Climate Change, Feminist Movement, Hillary Clinton, and Legalization of Abortion), we perform the following steps:

1. Compute all the features from the training dataset
2. Apply the eight variants of the seven supervised ML methods (SMO with two different kernels, LibSVM, J48, Random Forest (RF), Bayes Networks, Naïve Bayes (NB), and Simple Logistics) using all the features to measure the baseline accuracy results.
3. Filter out non-relevant features using three filtering methods (Info Gain [30], Chi-square, and the Correlation Feature Selection method (CFS) [31].
4. Re-apply the eight variants of the seven supervised ML methods using the filtered features for all the 3 filtering methods.
5. Compute the accuracy, precision, recall, and F-Measure values obtained by the top three ML methods while performing various types of parameter tuning, e.g. increasing the # of iterations in RF, performing two experiments for SMO with two different kernels, the default Poly-Kernel, and the normalized Poly-Kernel. We saw that changing the kernel type to these two specific kernels resulted in better results. For LibSVM, we changed the kernel to be the linear kernel, the C value to 0.5 instead value of 1, and we tuned the 'normalize' and 'probabilityEstimates' options.
6. Given the test data, we apply the best ML method (according to the accuracy results) on the features filtered-in by the CFS selection method (found as the best feature selection method), and compute the following seven performance metrics: accuracy, precision, recall, F-Measure, ROC area, PRC area, and the macro-average result.

Due to space limitations, we present in the following sub-section detailed results for only one of the datasets of Task 6-A of SemEval 2016: legalization of abortion. A summary of the results for all the five datasets will be presented after that.

## 4.1 Results for the Legalization of Abortion Dataset

We applied the ML methods described above on all the features and on the filtered features using the three filtering methods. The accuracy results of the baseline version and three versions using the filtered features (Info Gain, Chi–square, and CFS) for each tested ML method for the training dataset of the Legalization of Abortion are presented in Fig. 1.

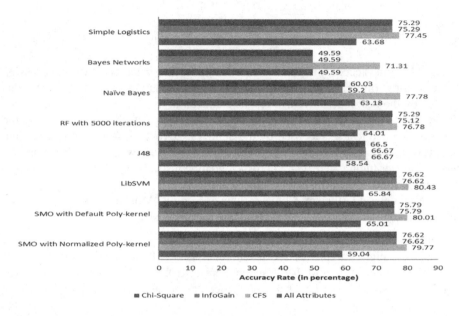

**Fig. 1.** Accuracy rates of the baseline and the filtered features for the abortion dataset.

From Fig. 1, we can see that for all ML methods, the best accuracy results are achieved using the CFS feature selection method. Moreover, in most cases the CFS results are significantly higher than the results obtained by the baseline version that uses all the features. We decided to perform additional experiments with the top three ML methods and to check other measures in addition to accuracy. In Fig. 2, we see the accuracy, precision, recall and F-Measure results of the top three ML methods.

The three accuracy results in Fig. 2 in descending order are: LibSVM with optimized setting (80.43%), SMO with the poly-kernel (80.01%), and SMO with the normalized poly-kernel (79.77%). We can see that the values of the other measures for these three ML methods are also rather similar. There are no significant differences between the results of the three ML methods. Nevertheless, the LibSVM ML method obtained the highest results for all the four measures. Using the filtered features, LibSVM achieved a 14.59% increase over the basic baseline.

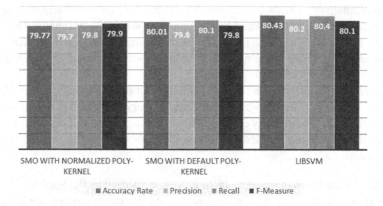

**Fig. 2.** Accuracy, precision, recall and F-measure results of the top three ML methods for the abortion dataset.

The application of the CFS feature selection method on all the features lead to a reduced set of 167 features. 125 features (81%) belong to the skip char ngram feature sets, 30 features (18%) are char ngrams (unigram, bigram and trigram) feature sets, and only 2 features are word unigrams.

**Test Data Results.** The test data for the Legalization of abortion dataset contains 280 tweets. 189 tweets (68%) are likely AGAINST the target, 46 tweets (16%) are likely in FAVOR of the target, and 45 (16%) are NONE of the above.

Based on the results obtained for the training test, we applied the CFS method (the best feature selection method) on the test data, and then we applied the LibSVM method with optimized parameters (the best ML method). The application of the CFS method on all the features lead to a reduced set of 100 features. Again, the dominant feature sets are the feature sets that belong to the char skip ngram features, with 77% of the features. Moreover, almost all the selected features (93%) are character-based features (char skip ngrams, char unigrams, char bigrams, and char trigrams).

The application of the LibSVM method with optimized parameters on the 100 filtered features lead to the following results: accuracy (86.43), Precision (86.2), Recall (86.4), F-Measure (86.3), ROC area (0.93), and PRC area (0.91). The values of the ROC area and the PRC area indicate excellent classification performance.

To estimate the relative importance of each feature, we further applied the InfoGain feature selection method on the 100 filtered features. Analysis of the top 25 ranked features showed that 24 features are character-based. Of these 24 features, 18 features are skip char ngram features (9 skip char bigram features, 8 skip char trigram features, and 1 skip char quadgram feature), and 6 features are from the char ngram feature sets (3 char bigrams and 3 char trigrams). Only one feature is a word unigram.

Examples for a few of those top 25 ranked features are as follows. A skip char trigram feature "wmn", which represents words such as "woman", "women", and "women's" and hastags such as "#women" and "#womenforwomen". A skip char bigram feature "lf", which represents words such as "life", "prolife", and "pro-life" and hastags such as "#everylifematters" and "#ProLifeYouth". A char bigram "wo", which is common to some frequent relevant words and hastags such as "woman", "women", "women's", "#women", and "work", and also of non-relevant frequent words such as "would". The only word unigram, which is among the top ranked features is "men", a group of people which also has what to tweet about abortion.

The main conclusion from these results is that most of the top features are character ngrams and skip character ngrams. These features serve as generalized features that include within them semantically close words and hastags, and their declensions. These features allow to overcome problems such as noise and sparse data and enable successful classification.

**Comparison to the Contest Results.** In the contest, organized by the SemEval 2016 Task 6-A [14] for all the test datasets, the organizers used the macro-average measure as the evaluation metric for the task. The macro-average (also called *Favg*) is defined as:

$$Favg = (Ffavor + Fagainst) / 2 \tag{1}$$

where *Ffavor* and *Fagainst* are defined as follows:

$$Ffavor = 2PfavorRfavor / (Pfavor + Rfavor) \tag{2}$$

$$Fagainst = 2PagainstRagainst / (Pagainst + Ragainst) \tag{3}$$

Our results were: *Fagainst* = 90.7, *Ffavor* = 73.8, and *Favg* = 82.25. The score of 82.25 is significantly higher than the *Favg* results of all the 19 competitors, including the best *Favg* result (66.42) obtained by the baseline SVM-ngrams team using all the possible word ngrams (this team was not a part of the official competition) and the best *Favg* result (63.32) achieved by the DeepStance team (a part of the official competition) using ngrams, word embedding vectors, sentiment analysis features such as those drawn from sentiment lexicons [32], and stance bearing hashtags.

In contrast to the *Favg* scores of many of the competitors of the SemEval 2016 Task 6-A, that were obtained using thousands of ngrams and hundreds of word embedding features, our *Favg* score is significantly better mainly probably due to the use of the CFS feature selection method and the use of only 100 derived features where 93 of them are character-based features and 77 of them are skip char ngram features.

## 4.2   Summary of the Results for All Five Datasets

Table 3 presents a summary of the results of our algorithm for all the five test datasets and Table 4 presents a comparison of the *Favg* values and an analysis of our features.

General findings that can be derived from Table 3 are: (1) The best ML methods are the two SVM's versions and Naïve Bayes; (2) The best filtering method is CFS; (3) The number of the filtered features is relatively very small (between 53 to 111); and (4) The values of all measures are relatively high (around 85% and up) for all test datasets.

**Table 3.** Summary of the results of our algorithm for all the five test datasets.

| Data Set | Best ML method | Best filtering method | # of filtered features | Acc | Prc | Rec | F-M | ROC area | PRC area |
|---|---|---|---|---|---|---|---|---|---|
| Abortion | LibSVM | CFS | 100 | 86.43 | 86.2 | 86.4 | 86.3 | 0.93 | 0.91 |
| Climate | SMO norm. pol-kernel | CFS | 53 | 86.39 | 85.1 | 86.4 | 85.75 | 0.82 | 0.79 |
| Feminist | SMO default pol-kernel | CFS | 102 | 83.51 | 83.9 | 83.5 | 83.7 | 0.82 | 0.75 |
| Clinton | NB | CFS | 111 | 85.42 | 86.5 | 85.4 | 85.95 | 0.93 | 0.88 |
| Atheism | NB | CFS | 74 | 79.55 | 85.6 | 79.5 | 82.44 | 0.93 | 0.91 |

**Table 4.** Comparison of the *Favg* values and an analysis of our features.

| Data Set | % of skip char ngrams | % of char ngrams | Best team in Task 6-A | | *Favg* of our system |
|---|---|---|---|---|---|
| | | | Team | *Favg* | |
| Abortion | 77.0% | 93.0% | SVM-ngrams | 66.42 | 82.25 |
| Climate | 77.4% | 94.3% | IDI@NTNU | 54.86 | 65.1 |
| Feminist | 75.5% | 94.1% | MITRE | 62.09 | 79.45 |
| Clinton | 82.9% | 96.4% | TakcLab | 67.12 | 77.8 |
| Atheism | 78.4% | 93.2% | TakeLab | 67.25 | 80.95 |
| Average | 78.24% | 94.2% | – | 63.55 | 77.11 |

General findings that can be drawn from Table 4 are: (1) The average rate of the skip char ngram features is around 78%; (2) The average rate of all the character-based features is around 94%; and (3) The average value of our *Favg* (77.11) is significantly higher than the average value of *Favg* of the best teams in the five experiments (63.55).

On the one hand, it is not surprising that the best classification results are successful with char ngrams features (around 94% of the features) because tweets are much more characterized by characters than by words, tweets are known as relatively short (up to 140 characters), and they contain also various hashtags, typos, shortcuts, slang words, onomatopoeia, and emojis.

On the other hand, it is relatively surprising that the skip character ngrams (around 78% of the features) contribute the most to the success of the classification tasks. The skip character ngrams that can be regarded as a type of generalized ngrams (because they enable gaps that are skipped over) have been discovered as "anti-noise" features that perform very well in a noisy environment such as twitter corpora.

As mentioned before by Guthrie et al. [13], skip-grams enable to overcome the data sparsity problem (i.e., the text corpus is composed of rare text units) for machine translation tasks even for an extremely large corpus. Based on our experiments, skip character ngrams do not only enable to overcome the data sparsity problem (which characterizes short text corpora) but also help to overcome noisy problems (e.g., misspellings, onoma-topoeia, replicated characters, and slang words), which also characterize short text corpora.

## 5   Summary, Conclusions and Future Work

In this study, we present an implementation of stance classification tasks based mainly on a limited number of features, which contain mainly char ngrams features in general and char skip ngrams in particular. To the best of our knowledge, we are the first to perform successful stance classification using mainly skip character ngrams.

The macro-average results obtained by our algorithm are significantly higher than the state-of-art results reported by the best macro-average results achieved in the SemEval 2016 Task 6-A [14] for all the five released datasets of tweets in the framework of task-A (the supervised framework).

In contrast to the competitors of the SemEval 2016 Task 6-A, that did not use any char skip ngrams but rather used thousands of ngrams and hundreds of word embedding features, our algorithm uses a limited number of features (53–111) derived by the CFS selection method, mainly character-based features where most of them are skip char ngram features.

Our experiments show that two feature sets are very helpful for stance classification of tweets: (1) char ngrams features in general probably because tweets are much more characterized by characters than by words, tweets are relatively short (up to 140 char-acters), and contain also various typos, shortcuts, hashtags, slang words, onomatopoeia, and emojis and (2) skip character ngrams in particular probably because they serve as generalized ngrams that allow to overcome problems such as noise and sparse data.

In order to examine the usefulness of character ngrams in general and skip character ngrams in particular we suggest the following future research proposals: conducting additional experiments for larger social corpora of various types of short text files written in various languages based on more feature sets and applying additional supervised ML methods such as deep learning methods.

**Acknowledgments.**   The authors thank three anonymous reviewers for their help and fruitful comments.

## References

1. Liu, B., Zhang, L.: A survey of opinion mining and sentiment analysis. In: Aggarwal, C., Zhai, C. (eds.) Mining Text Data, pp. 415–463. Springer, Boston (2012). https://doi.org/10.1007/978-1-4614-3223-4_13
2. Mohammad, S.M., Zhu, X., Kiritchenko, S., Martin, J.: Sentiment, emotion, purpose, and style in electoral tweets. Inf. Process. Manage. **51**(4), 480–499 (2015)
3. Basave, C., He, A.E., He, Y., Liu, K., Zhao, J.: A weakly supervised bayesian model for violence detection in social media (2013)

4. Somasundaran, S., Wiebe, J.: Recognizing stances in ideological on-line debates. In: Proceedings of the NAACL HLT 2010 Workshop on Computational Approaches to Analysis and Generation of Emotion in Text, pp. 116–124. Association for Computational Linguistics (2010)
5. Murakami, A., Raymond, R.: Support or oppose? Classifying positions in online debates from reply activities and opinion expressions. In: Proceedings of the 23rd International Conference on Computational Linguistics: Posters, pp. 869–875. Association for Computational Linguistics (2010)
6. Anand, P., Walker, M., Abbott, R., Tree, J.E.F., Bowmani, R., Minor, M.: Cats rule and dogs drool!: classifying stance in online debate. In Proceedings of the 2nd Workshop on Computational Approaches to Subjectivity and Sentiment Analysis, pp. 1–9. Association for Computational Linguistics (2011)
7. Sridhar, D., Foulds, J., Huang, B., Getoor, L., Walker, M.: Joint models of disagreement and stance in online debate. In: Annual Meeting of the Association for Computational Linguistics (2015)
8. Thomas, M., Pang, B., Lee, L.: Get out the vote: determining support or opposition from Congressional floor-debate transcripts. In: Proceedings of the 2006 Conference on Empirical Methods in Natural Language Processing, pp. 327–335. Association for Computational Linguistics (2006)
9. Yessenalina, A., Yue, Y., Cardie, C.: Multi-level structured models for document-level sentiment classification. In: Proceedings of EMNLP, pp. 1046–1056 (2010)
10. Burfoot, C., Bird, S., Baldwin, T.: Collective classification of congressional floor-debate transcripts. In: Proceedings of the 49th Annual Meeting of the Association for Computational Linguistics: Human Language Technologies-Volume 1, pp. 1506–1515. Association for Computational Linguistics (2011)
11. Agrawal, R., Rajagopalan, S., Srikant, R., Xu, Y.: Mining newsgroups using networks arising from social behavior. In: Proceedings of WWW, pp. 529–535 (2003)
12. Rajendran, P., Bollegala, D., Parsons, S.: Contextual stance classification of opinions: a step towards enthymeme reconstruction in online reviews. In: Proceedings of the 3rd Workshop on Argument Mining, pp. 31–39. Association for Computational Linguistics, Berlin (2016)
13. Guthrie, D., Allison, B., Liu, W., Guthrie, L., Wilks, Y.: A closer look at skip-gram modelling. In: Proceedings of the 5th International Conference on Language Resources and Evaluation (LREC-2006), pp. 1222–1225 (2006)
14. Mohammad, S.M., Kiritchenko, S., Sobhani, P., Zhu, X., Cherry, C.: SemEval-2016 task 6: detecting stance in tweets. In: Proceedings of SemEval, pp. 31–41 (2016)
15. Xu, R., Zhou, Yu., Wu, D., Gui, L., Du, J., Xue, Y.: Overview of NLPCC Shared Task 4: stance detection in Chinese microblogs. In: Lin, C.-Y., Xue, N., Zhao, D., Huang, X., Feng, Y. (eds.) ICCPOL/NLPCC -2016. LNCS (LNAI), vol. 10102, pp. 907–916. Springer, Cham (2016). https://doi.org/10.1007/978-3-319-50496-4_85
16. Jans, B., Bethard, S., Vulić, I., Moens, M.F.: Skip n-grams and ranking functions for predicting script events. In: Proceedings of the 13th Conference of the European Chapter of the Association for Computational Linguistics, pp. 336–344. Association for Computational Linguistics (2012)
17. Sidorov, G., Velasquez, F., Stamatatos, E., Gelbukh, A., Chanona-Hernández, L.: Syntactic n-grams as machine learning features for natural language processing. Expert Syst. Appl. **41**(3), 853–860 (2014)
18. Fernández, J., Gutiérrez, Y., Gómez, J.M., Martınez-Barco, P.: GPLSI: supervised sentiment analysis in twitter using skipgrams. In: Proceedings of the 8th International Workshop on Semantic Evaluation (SemEval 2014), number SemEval, pp. 294–299 (2014)

19. Dhondt, E., Verberne, S., Weber, N., Koster, C., Boves, L.: Using skipgrams and pos-based feature selection for patent classification. Comput. Linguist. Neth. J. **2**, 52–70 (2012)
20. HaCohen-Kerner, Y., Beck, H., Yehudai, E., Rosenstein, M., Mughaz, D.: Cuisine: classification using stylistic feature sets and/or name-based feature sets. J. Am. Soc. Inform. Sci. Technol. **61**(8), 1644–1657 (2010)
21. HaCohen-Kerner, Y., Beck, H., Yehudai, E., Mughaz, D.: Stylistic feature sets as classifiers of documents according to their historical period and ethnic origin. Appl. Artif. Intell. **24**(9), 847–862 (2010)
22. Mohammad, S.M., Kiritchenko, S., Zhu, X.: National Research Council Canada (NRC) Hashtag unigram Lexicon (2013). http://saifmohammad.com/WebPages/SCL.html. Accessed 18 Apr 2017
23. https://www.cs.uic.edu/~liub/FBS/sentiment-analysis.html. Accessed 18 Apr 2017
24. http://mpqa.cs.pitt.edu/lexicons/subj_lexicon/. Accessed 18 Apr 2017
25. https://www.ling.upenn.edu/courses/Fall_2003/ling001/penn_treebank_pos.html. Accessed 18 Apr 2017
26. http://nlp.stanford.edu/software/tagger.shtml. Accessed 18 Apr 2017
27. Klein, D., Manning, C.D.: Accurate unlexicalized parsing. In: Proceedings of the 41st Annual Meeting on Association for Computational Linguistics, vol. 1, pp. 423–430. Association for Computational Linguistics (2003)
28. Witten, I.H., Frank, E.: Data Mining: Practical Machine Learning Tools and Techniques, 2nd edn. Morgan Kaufmann Series in Data Management Systems. Morgan Kaufmann, San Mateo (2005)
29. Hall, M.: Correlation-based feature selection for machine learning. Doctoral dissertation, The University of Waikato (1999)
30. Yang, Y., Pedersen, J.O.: A comparative study on feature selection in text categorization. In: Icml, pp. 412–420 (1997)
31. Hall, M., Frank, E., Holmes, G., Pfahringer, B., Reutemann, P., Witten, I.H.: The WEKA data mining software. ACM SIGKDD Explor. Newsl. **11**(1), 10 (2009)
32. Kiritchenko, S., Zhu, X., Mohammad, S.M.: Sentiment analysis of short informal texts. J. Artif. Intell. Res. **50**, 723–762 (2014)

# Structural Semantic Models for Automatic Analysis of Urban Areas

Gianni Barlacchi[1,2(✉)], Alberto Rossi[3], Bruno Lepri[3],
and Alessandro Moschitti[1,4]

[1] University of Trento, Trento, Italy
`gianni.barlacchi@gmail.com, amoschitti@gmail.com`
[2] SKIL - Telecom Italia, Trento, Italy
[3] Bruno Kessler Foundation (FBK), Trento, Italy
`{alrossi,lepri}@fbk.eu`
[4] Qatar Computing Research Institute, HBKU, Doha, Qatar

**Abstract.** The growing availability of data from cities (e.g., traffic flow, human mobility and geographical data) open new opportunities for predicting and thus optimizing human activities. For example, the automatic analysis of land use enables the possibility of better administrating a city in terms of resources and provided services. However, such analysis requires specific information, which is often not available for privacy concerns. In this paper, we propose a novel machine learning representation based on the available public information to classify the most predominant land use of an urban area, which is a very common task in urban computing. In particular, in addition to standard feature vectors, we encode geo-social data from Location-Based Social Networks (LBSNs) into a conceptual tree structure that we call Geo-Tree. Then, we use such representation in kernel machines, which can thus perform accurate classification exploiting hierarchical substructure of concepts as features. Our extensive comparative study on the areas of New York and its boroughs shows that Tree Kernels applied to Geo-Trees are very effective improving the state of the art up to 18% in Macro-F1.

## 1   Introduction

The demographic trend clearly shows an increasing concentration of people in huge cities. By 2030, 9% of the world population is expected to live in just 41 *mega-cities*, each one with more than 10M inhabitants. Thus, the growing availability of data [2] makes it possible to discover new interesting aspects about cities and its life at a fine unprecedented granularity.

A fundamental challenge that policy makers and urban planners are dealing with is *land use classification*, which plays an important role for infrastructure planning and development, real-estate evaluations, and authorizations of business permits. More in detail, policy makers and urban planners need to associate different urban areas with specific human activities (e.g., residential, industrial, business, nightlife and others). However, traditional survey-based approaches

© Springer International Publishing AG 2017
Y. Altun et al. (Eds.): ECML PKDD 2017, Part III, LNAI 10536, pp. 279–291, 2017.
https://doi.org/10.1007/978-3-319-71273-4_23

to classify areas are time consuming and very costly to be applied to modern huge cities. Therefore, automatic approaches using novel sources of data (e.g., data from mobile phones, LBSNs, etc.) have been proposed. For example, [19] designed supervised and unsupervised approaches to infer New York City (NYC) land use from *check-in*. A check-in usually consists of latitude and longitude coordinates associated with additional metadata such as the venue where the user checked-in, comments and photos. Such data can be extracted from LBSNs like Foursquare[1], a social network application that provides the number and type of activities present in the target area (e.g., *Arts & Entertainment, Nightlife Spot*, etc.). The approach basically used feature vectors, mainly consisting of the number of check-in with the associated activity inferred from the Foursquare category of the place (e.g., *eating* if the check-in is done in a *restaurant*). As Gold Standard, the authors used data provided by the NYC Department of City Planning in 2013 mapped on a grid of $200 \times 200$ m.

In this paper, we represent geographical areas in two different ways: (i) as a bag-of-concepts (BOC), e.g., *Arts and Entertainment, College and University, Event, Food* extracted from the Foursquare description of the area; and (ii) as the same concepts above organized in a tree, reflecting the hierarchical category structure of Foursquare activities. We designed kernels combining BOC vectors with Tree Kernels (TKs) [6,9,10,17] applied to concept trees and used them in Support Vector Machines (SVMs). This way, our model (i) can learn complex structural and semantic patterns encoded in our hierarchical conceptualization of an area and (ii) highly improves the accuracy of standard classification methods based on BOC. Our GeoTK represents an interesting novelty as we show that TKs not only can capture semantic information from natural language text, e.g., as shown for semantic role labeling [12] and question answering [3,15], but they can also convey conceptual features from the hierarchy above to perform semantic inference, such as deciding which is the major activity of a land. Our approach is largely applicable as (i) it can use any hierarchical category structure for POIs categories (e.g., OpenStreet Map POIs data); and (ii) many cities offer open access to their land use data.

Finally, we carry out a study with different granularities of the areas to be analyzed. This also enables to analyze the trade-off between the precision in targeting the area of interest and the accuracy with which we carry out the estimation. More in detail, we divide the NYC area in squares with edges of 50, 100, 200 and 250 m and, for each cell, we classify its most predominant land use class (e.g., Residential, Commercial, Manufacturing, etc.). Our extensive experimentation, including a comparative study as well as the use of several machine learning models, shows that GeoTKs are very effective and improve the state of the art up to 18% in Macro-F1.

The reminder of this paper is organized as follows, Sect. 2 introduces the related work, Sect. 3 describes the task and the related data, Sect. 4 presents

---

[1] https://foursquare.com.

our hierarchical tree representation and our GeoTK. Then, Sect. 5 illustrates the evaluation of our approach, and finally Sect. 6 derives some conclusions.

## 2   Related Work

Several works have targeted land use inference by means of different sources of information. For example, [18] built a framework that, using human mobility patterns derived from taxicab trajectories and Point Of Interests (POIs), classifies the functionality of an area for the city of Beijing. The model is similar to the one used for topic discovery in a textual document, where the functionality of an area is the topic, the region is the document, and POIs and mobility patterns are metadata and words, respectively. Specifically, [18] have used an advanced model combining Latent Dirichlet Allocation (LDA) with Dirichlet Multinomial Regression (DMR), in order to insert also information coming from the POIs (metadata). Hence, for each region, after the parameter estimation with DMR, they have a vector representing the intensity of each topic. This vector is then used to aggregate formal regions having similar functions by k-means clustering.

Similarly, [1] proposed a spatio-temporal approach for the detection of functional regions. They exploited three different clustering algorithms by using different set of features extracted from Foursquare's POIs and check-in activities in Manhattan (New York). This task permits to better understand how the functionality of a city's region changes over time. Other works have used geo-tagged data from social networks: for example, [8] used tweets as input data to predict the land use of a certain area of Manhattan. Moreover, they try to infer POIs from tweets' patterns clustering the surface with Self-Organizing-Map, then characterizing each region with a specific tweet pattern and finally using k-means to infer land use. Again, [19] have used check-in data to compare unsupervised and supervised approaches to land use inference.

Finally, some works have also used Call Detail Records (CDRs) [7,8,13,16], which are typically used by mobile phone operators for billing purposes. This data registers the time and type of the communication (e.g., incoming calls, Internet, outgoing SMS), and the radio base station handling the communication. For example, [16] have used CDRs jointly with a Random Forest classifier to build a time-varying land use classification for the city of Boston. The intuition behind this work is to mine a time-variant relation between movement patterns and land use. In particular, they perform a Random Forest prediction and then they compare it with the predictions obtained for the neighboring regions, applying a sort of consensus validation (e.g., they modify the prediction if a certain number of neighbors belong to a different uniform function). This way, they model different land uses for different temporal slots of the day.

Compared to the state of the art, the main novelties introduced by our work are the following: (i) we model the hierarchical semantic information of Foursquare using GeoTK, thus adding powerful structural features in our classification models; and (ii) we study how the size of the grid impacts on the accuracy of different models, thus investigating the trade-off between granularity of the analysis and accuracy. It should be also noted that, in contrast to previous work, GeoTK does not rely on external resources (e.g., mobile phone data) or heavy features engineering in addition to the structural kernel model.

# 3    Datasets

We use the shape file of New York provided by the NYC government[2]. This file is publicly available and contains the entire shape of New York divided in the 5 boroughs: Manhattan, Brooklyn, Staten Island, Bronx, and Queens. Then, we build a grid over the entire city in order to enable our classification task. The goal is to infer the land use of a region given a target label and a feature representation of the region. In the next subsections, we describe (i) the land use data and labels utilized by our approach, and (ii) the Foursquare's POIs used to obtain a feature representation of the land of a region.

## 3.1    Land Use

In our study, we use MapPluto, a freely available dataset provided by the NYC government, which contains precise geo-referenced information for each city's borough. For example, it provides the precise category and shape for each building in the city (Fig. 1). More in detail, it contains the following land use categories: (i) *One and Two Family Buildings*, (ii) *Multi-Family Walk-Up Buildings*, (iii) *Multi-Family Elevator Buildings*, (iv) *Mixed Residential and Commercial Buildings*, (v) *Commercial and Office Buildings*, (vi) *Industrial and Manufacturing Buildings*, (vii) *Transportation and Utility*, (viii) *Public Facilities and Institutions*, (ix) *Open Space and Outdoor Recreation*, (x) *Parking Facilities*, and (xi) *Vacant Land*. Land use information is very fine-grained, and in most cases there is only one land use assigned to one building, thus making it very difficult to determine the land use with just POI information. A reasonable trade-off between classification accuracy and the desired area granularity consists in segmenting the regions in squared cells: each cell will refer to more than one land use but we consider the predominant class as its primary use.

## 3.2    Foursquare's Point of Interests

We extracted 206,602 POIs from the entire NYC. As for the land use data, we have several sources of information, but we focused on the ten macro-categories of the POIs, each one specialized in maximum four levels of detail. These levels follow a hierarchical structure[3], where each level of a category has a finite number of subcategories as node children. For instance, the first level of POIs main categories is constituted by: (i) *Arts and Entertainment*, (ii) *College and University*, (iii) *Event*, (iv) *Food*, (v) *Nightlife Spot*, (vi) *Outdoors and Recreation*,

---

[2] http://www1.nyc.gov/site/planning/data-maps/open-data/districts-download-metadata.page.

[3] https://developer.foursquare.com/categorytree.

**Fig. 1.** Example of land use distribution in New York City.

(vii) *Professional and Other Places*, (viii) *Residence*, (ix) *Shop and Service*, and (x) *Travel and Transport*. The second level includes 437 categories whereas the third level contains a smaller number of categories, 345.

## 4    Semantic Structural Models for Land Use Analysis

Previous works [4,13,14] have mainly used features extracted from LBSNs (e.g., Foursquare's POIs) in the XGboost algorithm [5]. However, these feature vectors have several limitations such as (i) the small amount of information available for the target area and (ii) their inherent scalar nature, which does not capture the existence and the type of relations between different POIs. Here, we propose a much powerful approach based on TKs applied to a semantic structure based on the hierarchical organization of the Foursquare categories.

### 4.1    Bag-of-Concepts

The most straightforward way to represent an area by means of Foursquare data is to use its POIs. Every venue is hierarchically categorized (e.g., *Professional and Other Places → Medical Center → Doctor's office*) and the categories are used to produce an aggregated representation of the area. We use this feature representation by aggregating all the venues together, namely we count the macro-level category (e.g., *Food*) in all the POIs that we found in any cell grid. This way, we generate the Bag-Of-Concepts (BOC) feature vectors, counting the number of each activity under each macro-category.

## 4.2   Hierarchical Tree Representation of Foursquare POIs

Every LBSN (e.g., Foursquare) has its own hierarchy of categories, which is used to characterize each location and activity (e.g., restaurants or shops) in the database. Thus, each POI in Foursquare is associated with a hierarchical path, which semantically describes the type of location/activity (e.g., for *Chinese Restaurant*, we have the path *Food* → *Asian Restaurant* → *Chinese Restaurant*). The path is much more informative than just the target POI name, as it provides feature combinations following the structure and the node proximity information, e.g., *Food & Asian Restaurant* or *Asian Restaurant & Chinese Restaurant* are valid features whereas *Food & Chinese Restaurant* is not.

In this work, we propose, a tree structure, Geo-Tree (GT), where its nodes are Foursquare categories and the edges among them are the same provided in the hierarchical category tree of Foursquare. Our structure is basically composed of all paths associated with the POIs that we find in the target grid cell. Precisely, we connect all these paths in a new root node. This way, the first level of root children corresponds to the most general category in the list (e.g., *Arts & Entertainment, Event, Food*, etc.), the second level of our tree corresponds to the second level of the hierarchical tree of Foursquare, and so on. The terminal nodes are the finest-grained descriptions in terms of category about the area (e.g., *College Baseball Diamond* or *Southwestern French Restaurant*). For example, Fig. 2 illustrates the semantic structure of a grid cell obtained by combining all the categories' chains of each venue. Given such representation, we can encode all its substructures in kernel machines using TKs as described in the next section.

**Fig. 2.** Example of Geo-Tree built according to the hierarchical categorization of Foursquare venues.

## 4.3   Geographical Tree Kernels (GeoTK)

Structural kernels are very effective means for automatic feature engineering [11]. In kernel machines both learning and classification algorithms only depend on the evaluation of inner products between instances, which correspond to compute similarity scores. In several cases, the similarity scores can be efficiently and implicitly handled by kernel functions by exploiting the following dual formulation of the classification function: $\sum_{i=1..l} y_i \alpha_i K(o_i, o) + b$, where $o_i$ are the training objects, $o$ is the classification example, $K(o_i, o)$ is a kernel function that implicitly defines the mapping from the objects to feature vectors $x_i$. In case of tree kernels, $K$ determines the shape of the substructures describing trees.

## 4.4   Tree Kernels

In the majority of machine learning approaches, data examples are transformed in feature vectors, which in turn are used in dot products for carrying out both learning and classification steps. Kernel Machines (KMs) allow for replacing the dot product with kernel functions, which compute the dot product directly from examples (i.e., they avoid the transformation of examples in vectors).

Given two input trees, TKs evaluate the number of substructures, also called fragments, that they have in common. More formally, let $\mathcal{F} = \{f_1, f_2, \ldots, f_{\mathcal{F}}\}$ be the space of all possible tree fragments and $\chi_i(n)$ an indicator function such that it is equal to 1 if the target $f_1$ is rooted in $n$, equal to 0 otherwise. TKs over $T_1$ and $T_2$ are defined by $TK(T_1, T_2) = \sum_{n_1 \in N_{T_1}} \sum_{n_2 \in N_{T_2}} \Delta(n_1, n_2)$, where $N_{T_1}$ e $N_{T_2}$ are the set of nodes of $T_1$ and $T_2$ and

$$\Delta(n_1, n_2) = \sum_{i=1}^{\mathcal{F}} \chi_i(n_1)\chi_i(n_2) \tag{1}$$

represents the number of common fragments rooted at nodes $n_1$ and $n_2$. The number and the type of fragments generated depends on the type of the used tree kernel functions, which, in turn, depends on $\Delta(n_1, n_2)$.

**Syntactic Tree Kernels (STK).** Its computation is carried out by using $\Delta_{STK}(n_1, n_2)$ in Eq. 1 defined as follows (in a syntactic tree, each node can be associated with a production rule):

  (i) if the productions at $n_1$ and $n_2$ are different $\Delta_{STK}(n_1, n_2) = 0$;
 (ii) if the productions at $n_1$ and $n_2$ are the same, and $n_1$ and $n_2$ have
      only leaf children then $\Delta_{STK}(n_1, n_2) = \lambda$; and
(iii) if the productions at $n_1$ and $n_2$ are the same, and $n_1$ and $n_2$ are
      not pre-terminals then $\Delta_{STK}(n_1, n_2) = \lambda \prod_{j=1}^{l(n_1)}(1 + \Delta_{STK}(c_{n_1}^j, c_{n_2}^j))$,

where $l(n_1)$ is the number of children of $n_1$ and $c_n^j$ is the $j$-th child of the node $n$. Note that, since the productions are the same, $l(n_1) = l(n_2)$ and the computational complexity of STK is $O(|N_{T_1}||N_{T_2}|)$ but the average running time tends to be linear, i.e., $O(|N_{T_1}| + |N_{T_2}|)$, for natural language syntactic trees [10].

Finally, by adding the following step:

  (0) if the nodes $n_1$ and $n_2$ are the same then $\Delta_{STK}(n_1, n_2) = \lambda$,

also the individual nodes will be counted by $\Delta_{STK}$. We call this kernel STK$_b$.

**The Partial Tree Kernel (PTK).** [10] generalizes a large class of tree kernels as it computes one of the most general tree substructure spaces. Given two trees, $PTK$ considers any connected subset of nodes as possible features of the substructure space. Its computation is carried out by Eq. 1 using the following $\Delta_{PTK}$ function:

if the labels of $n_1$ and $n_2$ are different $\Delta_{PTK}(n_1, n_2) = 0$;

else $\Delta_{PTK}(n_1, n_2) = \mu\left(\lambda^2 + \sum_{I_1, I_2, l(I_1) = l(I_2)} \lambda^{d(I_1) + d(I_2)} \prod_{j=1}^{l(I_1)} \Delta_{PTK}(c_{n_1}(I_{1j}), c_{n_2}(I_{2j}))\right),$

where $\mu, \lambda \in [0, 1]$ are two decay factors, $I_1$ and $I_2$ are two sequences of indices, which index subsequences of children $u$, $I = (i_1, ..., i_{|u|})$, in sequences of children $s$, $1 \leq i_1 < ... < i_{|u|} \leq |s|$, i.e., such that $u = s_{i_1}..s_{i_{|u|}}$, and $d(I) = i_{|u|} - i_1 + 1$ is the distance between the first and last child.

When the PTK is applied to the semantic Geo-Tree of Fig. 2, it can generate effective fragments, e.g., those in Fig. 3.

**Fig. 3.** Some of the exponential fragment features from the tree of Fig. 2

**Combination of TKs and Feature Vectors.** Our TKs do not consider the frequency[4] of the POIs present in a given grid cell. Thus, it may be useful to enrich the feature space with further information that can be encoded in the model using a feature vector. To this end, we need to use a kernel that combines tree structures and feature vectors. More specifically, given two geographical areas, $x^a$ and $x^b$, we define a combination as: $K(x^a, x^b) = TK(\mathbf{t}^a, \mathbf{t}^b) + KV(\mathbf{v}^a, \mathbf{v}^b)$, where $TK$ is any structural kernel function applied to tree representations, $\mathbf{t}^a$ and $\mathbf{t}^b$ of the geographical areas and $KV$ is a kernel applied to the feature vectors, $\mathbf{v}^a$ and $\mathbf{v}^b$, extracted from $x^a$ and $x^b$ using any data source available (e.g., text, social media, mobile phone and census data).

## 5   Experiments

We test the effectiveness of our approach on the *land use classification* task, where the goal is to assign to each area the predominant land use class as performed in previous work by [16,19]. We first test several models on Manhattan using several grid sizes, then we focus on evaluating the best models on all NYC boroughs and finally, we use the best models on the entire NYC, also enabling comparisons with previous work.

---

[4] It is possible to add the frequency in the kernel computation but for our study we preferred to have a completely different representation from previous typical frequency-based approaches.

## 5.1   Experimental Setup

We performed our experiments on the data from NYC boroughs, evaluating grids of various dimensions: $50 \times 50$, $100 \times 100$, $200 \times 200$ and $250 \times 250$ m. We applied a pre-processing step in order to filter out cells for which it is not possible to perform land use classification. In particular, from each grid, we removed the cells (i) that cover areas without a specified land use (e.g., cell in the sea) and (ii) for which we do not have POIs (e.g., cells from Central Park). For each grid, we created training, validation and test sets, randomly sampling 60%, 20%, 20% of the cells, respectively. We labelled the dataset following the same category aggregation strategy proposed by [19], who assigned the predominant land use class to each grid cell. Note that given the categories described in Sect. 3.1, we merged (i) *One & Two Family Buildings*, (ii) *Multi-Family Walk-Up Buildings* and (iii) *Multi-Family Elevator Buildings* into a single general *Residential* category. Then, we also aggregated (i) *Industrial & Manufacturing*, (ii) *Public Facilities & Institutions*, (iii) *Parking Facilities* and (iv) *Vacant Land* into a new category called *Other*. Thus, the aggregated dataset contains six different classes: (i) *Residential*, (ii) *Commercial and Office Buildings*, (iii) *Mixed Residential and Commercial Buildings*, (iv) *Open Space and Outdoor Recreation*, (v) *Transportation and Utility*, (vi) *Other*. The names and distribution of examples in training and test sets (for the grid of $200 \times 200$) are shown in Table 1. Compared to the original categorization, this new taxonomy has a lower granularity, thus facilitating the identification of the predominant class in each cell.

**Table 1.** Distribution of land use classes in the training and test set for NYC.

| Size | Commercial | Mixed | Open space | Other | Residential | Transportation | Total |
|------|-----------|-------|-----------|-------|-------------|----------------|-------|
| Train | 394 | 225 | 1220 | 1622 | 6248 | 538 | 10247 |
| Test | 175 | 85 | 534 | 615 | 2330 | 214 | 3953 |

To train our models, we adopted SVM-Light-TK[5], which allow us to use structural kernels [10] in SVM-light[6]. We experimented with linear, polynomial and radial basis function kernels applied to standard feature vectors. We measured the performance of our classifier with Accuracy, Macro-Precision, Macro-Recall and Macro-F1 (Macro indicates the average over all categories).

## 5.2   Results for Land Use Classification

We trained multi-class classifiers using common learning algorithm such as Logistic Regression (LogReg), XGboost [5], and SVM using linear, polynomial and radial basis function kernel, named SVM-{Lin, Poly, Rbf}, respectively, and our structural semantic models, indicated with STK, $STK_b$ and PTK. We also combined kernels with a simple summation, e.g., PTK+Poly indicates an SVM using such kernel combination.

---

[5] http://disi.unitn.it/moschitti/Tree-Kernel.htm.
[6] http://svmlight.joachims.org/.

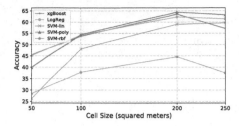
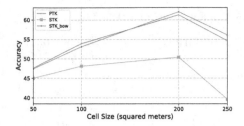

**Fig. 4.** Accuracy of common machine learning models on different cell sizes in Manhattan.

**Fig. 5.** Accuracy of GeoTKs according to different cell sizes of Manhattan.

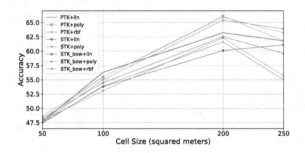

**Fig. 6.** Accuracy of kernel combinations using BOC vectors and GeoTKs according to different cell sizes of Manhattan.

We first tested our models individually just on Manhattan using different grid sizes. Figures 4 and 5 show the accuracy of the multi-classifier for different models according to different granularity of the sampling grid. We note that SVM-Poly, XGboost and LogReg show comparable accuracy. PTK and $STK_b$ perform a little bit less than the feature vector models. Interestingly, the kernel combinations in Fig. 6 provide the best results. This is an important finding as XGboost is acknowledged to be the state of the art for land use classification. Additionally, when the size of the grid cell becomes larger, the accuracy of TKs degrades faster than the one of kernels based on feature vectors, mainly because the conceptual tree becomes too large. After the preliminary experiments above, we selected the most accurate models on Manhattan and tested them on the other boroughs of NYC. Table 2 shows that TKs are more accurate than vectors-based models and the combinations further improve both models.

In the final experiments, we tested our best models on the entire NYC with a grid of $200 \times 200$. We first tuned the following parameters on a validation set: (i) the decay factors $\mu$ and $\lambda$ for TK, (ii) $C$ value for all the SVM approaches, and the specific parameters, i.e., degree in *poly* and $\gamma$ in RBF kernels, (iii) the important and the parameters of XGBoost such as the maximum depth of the tree and the minimum sum of weights of all observations in a child node.

**Table 2.** Accuracy of the best models for each New York borough and cell size.

| Area | Cell | XGBoost | SVM-poly | PTK | PTK+poly | STK | STK+poly | STK$_b$ | STK$_b$+poly |
|---|---|---|---|---|---|---|---|---|---|
| Manhattan | 50 | 45.0 | 39.9 | 47.6 | 48.0 | 45.0 | 47.5 | 47.4 | 48.6 |
| | 100 | 54.0 | 54.4 | 53.9 | 55.5 | 48.1 | 55.0 | 53.1 | 55.5 |
| | 200 | 63.0 | 64.4 | 61.3 | 66.1 | 50.4 | 65.4 | 62.1 | 65.9 |
| | 250 | 57.0 | 63.2 | 54.6 | 61.8 | 39.6 | 63.9 | 56.1 | 63.2 |
| Bronx | 50 | 43.0 | 30.9 | 44.9 | 44.9 | 42.2 | 43.4 | 42.4 | 43.2 |
| | 100 | 50.0 | 43.7 | 53.2 | 54.1 | 51.2 | 53.2 | 54.7 | 54.0 |
| | 200 | 59.0 | 56.4 | 62.6 | 60.6 | 56.4 | 60.4 | 61.8 | 61.8 |
| | 250 | 59.0 | 58.6 | 63.5 | 64.9 | 59.3 | 59.6 | 63.0 | 65.2 |
| Brooklyn | 50 | 49.0 | 44.2 | 51.3 | 51.6 | 48.7 | 51.3 | 51.4 | 52.2 |
| | 100 | 61.0 | 61.0 | 63.1 | 63.5 | 62.4 | 62.9 | 63.1 | 63.2 |
| | 200 | 71.0 | 71.5 | 72.9 | 73.6 | 70.1 | 73.2 | 73.3 | 73.8 |
| | 250 | 70.0 | 68.9 | 71.3 | 72.6 | 67.9 | 70.3 | 70.6 | 71.4 |
| Queens | 50 | 48.0 | 32.4 | 51.5 | 51.5 | 50.2 | 51.0 | 50.5 | 50.3 |
| | 100 | 58.0 | 57.2 | 61.4 | 61.3 | 59.8 | 60.6 | 61.6 | 61.7 |
| | 200 | 67.0 | 66.5 | 70.5 | 71.3 | 69.3 | 69.9 | 70.4 | 71.0 |
| | 250 | 68.0 | 68.3 | 72.9 | 73.1 | 70.1 | 72.2 | 72.4 | 73.6 |
| StatenIsland | 50 | 51.0 | 38.63 | 54.4 | 55.2 | 52.8 | 54.6 | 53.8 | 54.9 |
| | 100 | 57.0 | 56.73 | 58.1 | 58.7 | 53.6 | 57.4 | 56.0 | 58.1 |
| | 200 | 60.0 | 60.0 | 61.8 | 61.1 | 60.2 | 60.0 | 61.3 | 60.9 |
| | 250 | 66.0 | 64.87 | 67.4 | 66.3 | 66.0 | 67.2 | 67.9 | 67.4 |

Table 3 shows the results in terms of Accuracy, Macro F1, Macro-Precision and Macro-Recall. The model *baseline* is obtained by always classifying an example with the label *Residential*, which is the most frequent. We note that: (i) all the feature vector and TK combinations show high accuracy, demonstrating the superiority of GeoTK over all the other models. (ii) $STK_b$+poly (polynomial kernel of degree 2) achieved the highest accuracy, improving over XGBoost up to 4.2 and 6.5 absolute percent points in accuracy and F1, respectively: these correspond to an improvement up to 18% over the state of the art.

Finally, Zhan et al. [19] is the result obtained on the same dataset using check-in data from Foursquare. Although an exact comparison cannot be carried out for possible differences in the experiment setting (e.g., Foursquare data changing over time), we note that our model is 1.8 absolute percentage points better.

**Table 3.** Classification results on New York City.

| Model | Acc. | F1 | Prec. | Rec. |
|---|---|---|---|---|
| Baseline | 58.9 | 12.4 | 0.98 | 16.6 |
| XGBoost | 63.2 | 36.1 | 57.9 | 31.9 |
| SVM-poly | 62.1 | 27.4 | 51.3 | 25.9 |
| **$STK_b$+poly** | **67.4** | **42.6** | **63.9** | **37.4** |
| PTK+poly | 66.9 | 41.4 | 63.8 | 36.2 |
| $STK_b$ | 66.6 | 38.1 | 52.8 | 33.9 |
| PTK | 65.9 | 37.2 | 58.7 | 33.0 |
| STK+poly | 65.5 | 37.3 | 54.5 | 33.3 |
| STK | 62.7 | 25.9 | 41.5 | 24.7 |
| Zhan et al. | 65.6 | – | – | – |

# 6   Conclusions

In this paper, we have introduced a novel semantic representation of POIs to better exploit geo-social data in order to deal with the *primary land use classification* of an urban area. This gives the urban planners and policy makers the possibility to better administrate and renew a city in terms of infrastructures, resources and services. Specifically, we encode data from LBSNs into a tree structure, the Geo-Tree and we used such representations in kernel machines. The latter can thus perform accurate classification exploiting hierarchical substructure of concepts as features. Our extensive comparative study on the areas of New York and its boroughs shows that TKs applied to Geo-Trees are very effective, improving the state of the art up to 18% in Macro-F1.

**Acknowledgments.** This work has been partially supported by the EC project CogNet, 671625 (H2020-ICT-2014-2, Research and Innovation action).

# References

1. Assem, H., Xu, L., Buda, T.S., O'Sullivan, D.: Spatio-temporal clustering approach for detecting functional regions in cities. In: ICTAI, pp. 370–377. IEEE (2016)
2. Barlacchi, G., De Nadai, M., Larcher, R., Casella, A., Chitic, C., Torrisi, G., Antonelli, F., Vespignani, A., Pentland, A., Lepri, B.: A multi-source dataset of urban life in the city of Milan and the Province of Trentino. Sci. Data **2**, 150055 (2015)
3. Barlacchi, G., Nicosia, M., Moschitti, A.: Sacry: syntax-based automatic crossword puzzle resolution system. In: ACL-IJCNLP 2015, p. 79 (2015)
4. Calabrese, F., Di Lorenzo, G., Ratti, C.: Human mobility prediction based on individual and collective geographical preferences. In: ITSC, pp. 312–317. IEEE (2010)
5. Chen, T., Guestrin, C.: Xgboost: a scalable tree boosting system. In: KDD, pp. 785–794. ACM, New York (2016)
6. Collins, M., Duffy, N.: New ranking algorithms for parsing and tagging: kernels over discrete structures, and the voted perceptron. In: ACL (2002)
7. De Nadai, M., Staiano, J., Larcher, R., Sebe, N., Quercia, D., Lepri, B.: The death and life of great Italian cities: a mobile phone data perspective. In: Proceedings of the 25th International Conference on World Wide Web, pp. 413–423. International World Wide Web Conferences Steering Committee (2016)
8. Frias-Martinez, V., Soto, V., Hohwald, H., Frias-Martinez, E.: Characterizing urban landscapes using geolocated tweets. In: SocialCom, pp. 239–248. IEEE (2012)
9. Gärtner, T.: A survey of kernels for structured data. ACM SIGKDD Explor. Newsl. **5**(1), 49–58 (2003)
10. Moschitti, A.: Efficient convolution kernels for dependency and constituent syntactic trees. In: Fürnkranz, J., Scheffer, T., Spiliopoulou, M. (eds.) ECML 2006. LNCS (LNAI), vol. 4212, pp. 318–329. Springer, Heidelberg (2006). https://doi.org/10.1007/11871842_32
11. Moschitti, A.: Making tree kernels practical for natural language learning. In: EACL, vol. 113, p. 24 (2006)
12. Moschitti, A., Pighin, D., Basili, R.: Tree kernels for semantic role labeling. Comput. Linguist. **34**(2), 193–224 (2008)
13. Noulas, A., Mascolo, C., Frias-Martinez, E.: Exploiting foursquare and cellular data to infer user activity in urban environments. In: MDM, vol. 1, pp. 167–176. IEEE (2013)
14. Noulas, A., Scellato, S., Mascolo, C., Pontil, M.: Exploiting semantic annotations for clustering geographic areas and users in location-based social networks. Soc. Mob. Web **11**(2) (2011)
15. Severyn, A., Moschitti, A.: Automatic feature engineering for answer selection and extraction. EMNLP **13**, 458–467 (2013)
16. Toole, J.L., Ulm, M., González, M.C., Bauer, D.: Inferring land use from mobile phone activity. In: SIGKDD International Workshop on Urban Computing, pp. 1–8. ACM (2012)
17. Vishwanathan, S.V.N., Smola, A.J.: Fast kernels for string and tree matching. In: Becker, S., Thrun, S., Obermayer, K. (eds.) NIPS, pp. 569–576. MIT Press (2002)
18. Yuan, J., Zheng, Y., Xie, X.: Discovering regions of different functions in a city using human mobility and POIs. In: KDD, pp. 186–194. ACM (2012)
19. Zhan, X., Ukkusuri, S.V., Zhu, F.: Inferring urban land use using large-scale social media check-in data. Netw. Spat. Econ. **14**(3–4), 647–667 (2014)

# Taking It for a Test Drive: A Hybrid Spatio-Temporal Model for Wildlife Poaching Prediction Evaluated Through a Controlled Field Test

Shahrzad Gholami[1(✉)], Benjamin Ford[1], Fei Fang[2], Andrew Plumptre[3], Milind Tambe[1], Margaret Driciru[4], Fred Wanyama[4], Aggrey Rwetsiba[4], Mustapha Nsubaga[5], and Joshua Mabonga[5]

[1] University of Southern California, Los Angeles, USA
{sgholami,benjamif,tambe}@usc.edu
[2] Harvard University, Boston, MA 02138, USA
fangf07@seas.harvard.edu
[3] Wildlife Conservation Society, New York City, NY 10460, USA
aplumptre@wcs.org
[4] Uganda Wildlife Authority, Kampala, Uganda
{margaret.driciru,fred.wanyama,aggrey.rwetsiba}@ugandawildlife.org
[5] Wildlife Conservation Society, Kampala, Uganda
{mnsubuga,jmabonga}@wcs.org

**Abstract.** Worldwide, conservation agencies employ rangers to protect conservation areas from poachers. However, agencies lack the manpower to have rangers effectively patrol these vast areas frequently. While past work has modeled poachers' behavior so as to aid rangers in planning future patrols, those models' predictions were not validated by extensive field tests. In this paper, we present a hybrid spatio-temporal model that predicts poaching threat levels and results from a five-month field test of our model in Uganda's Queen Elizabeth Protected Area (QEPA). To our knowledge, this is the first time that a predictive model has been evaluated through such an extensive field test in this domain. We present two major contributions. First, our hybrid model consists of two components: (i) an ensemble model which can work with the limited data common to this domain and (ii) a spatio-temporal model to boost the ensemble's predictions when sufficient data are available. When evaluated on real-world historical data from QEPA, our hybrid model achieves significantly better performance than previous approaches with either temporally-aware dynamic Bayesian networks or an ensemble of spatially-aware models. Second, in collaboration with the Wildlife Conservation Society and Uganda Wildlife Authority, we present results from a five-month controlled experiment *where rangers patrolled over 450 sq km across QEPA*. We demonstrate that our model successfully predicted (1) where snaring activity would occur and (2) where it would not occur; in areas where we predicted a high rate of snaring activity, rangers found more snares

---

Shahrzad Gholami and Benjamin Ford are both first authors of this paper.

© Springer International Publishing AG 2017
Y. Altun et al. (Eds.): ECML PKDD 2017, Part III, LNAI 10536, pp. 292–304, 2017.
https://doi.org/10.1007/978-3-319-71273-4_24

and snared animals than in areas of lower predicted activity. These findings demonstrate that (1) our model's predictions are selective, (2) our model's superior laboratory performance extends to the real world, and (3) these predictive models can aid rangers in focusing their efforts to prevent wildlife poaching and save animals.

**Keywords:** Predictive models · Ensemble techniques
Graphical models · Field test evaluation · Wildlife protection
Wildlife poaching

# 1   Introduction

Wildlife poaching continues to be a global problem as key species are hunted toward extinction. For example, the latest African census showed a 30% decline in elephant populations between 2007 and 2014 [1]. Wildlife conservation areas have been established to protect these species from poachers, and these areas are protected by park rangers. These areas are vast, and rangers do not have sufficient resources to patrol everywhere with high intensity and frequency.

At many sites now, rangers patrol and collect data related to snares they confiscate, poachers they arrest, and other observations. Given rangers' resource constraints, patrol managers could benefit from tools that analyze these data and provide future poaching predictions. However, this domain presents unique challenges. First, this domain's real-world data are few, extremely noisy, and incomplete. To illustrate, one of rangers' primary patrol goals is to find wire snares, which are deployed by poachers to catch animals. However, these snares are usually well-hidden (e.g., in dense grass), and thus rangers may not find these snares and (incorrectly) label an area as not having any snares. Second, poaching activity changes over time, and predictive models must account for this temporal component. Third, because poaching happens in the real world, there are mutual spatial and neighborhood effects that influence poaching activity. Finally, while field tests are crucial in determining a model's efficacy in the world, the difficulties involved in organizing and executing field tests often precludes them.

Previous works in this domain have modeled poaching behavior with real-world data. Based on data from a Queen Elizabeth Protected Area (QEPA) dataset, [6] introduced a two-layered temporal graphical model, CAPTURE, while [4] constructed an ensemble of decision trees, INTERCEPT, that accounted for spatial relationships. However, these works did not (1) account for both spatial and temporal components nor (2) validate their models via extensive field testing.

In this paper, we provide the following contributions. (1) We introduce a new hybrid model that enhances an ensemble's broad predictive power with a spatio-temporal model's adaptive capabilities. Because spatio-temporal models require a lot of data, this model works in two stages. First, predictions are made with an ensemble of decision trees. Second, in areas where there are sufficient data, the ensemble's prediction is boosted via a spatio-temporal model.

(2) In collaboration with the Wildlife Conservation Society and the Uganda Wildlife Authority, we designed and deployed a large, controlled experiment to QEPA. Across 27 areas we designated across QEPA, rangers patrolled approximately 452 km over the course of five months; to our knowledge, this is the largest controlled experiment and field test of Machine Learning-based predictive models in this domain. In this experiment, we tested our model's selectiveness: is our model able to differentiate between areas of high and low poaching activity?

In experimental results, (1) we demonstrate our model's superior performance over the state-of-the-art [4] and thus the importance of spatio-temporal modeling. (2) During our field test, rangers found over three times more snaring activity in areas where we predicted higher poaching activity. When accounting for differences in ranger coverage, rangers found twelve times the number of findings per kilometer walked in those areas. These results demonstrate that (i) our model is selective in its predictions and (ii) our model's superior predictive performance in the laboratory extends to the real world.

## 2   Background and Related Work

Spatio-temporal models have been used for prediction tasks in image and video processing. Markov Random Fields (MRF) were used by [11,12] to capture spatio-temporal dependencies in remotely sensed data and moving object detection, respectively.

Critchlow et al. [2] analyzed spatio-temporal patterns in illegal activity in Uganda's Queen Elizabeth Protected Area (QEPA) using Bayesian hierarchical models. With real-world data, they demonstrated the importance of considering the spatial and temporal changes that occur in illegal activities. However, in this work and other similar works with spatio-temporal models [8,9], no standard metrics were provided to evaluate the models' predictive performance (e.g., precision, recall). As such, it is impossible to compare our predictive models' performance to theirs. While [3] was a field test of [2]'s work, [8,9] do not conduct field tests to validate their predictions in the real-world.

In the Machine Learning literature, [6] introduced a two-layered temporal Bayesian Network predictive model (CAPTURE) that was also evaluated on real-world data from QEPA. CAPTURE, however, assumes one global set of parameters for all of QEPA which ignores local differences in poachers' behavior. Additionally, the first layer, which predicts poaching attacks, relies on the current year's patrolling effort which makes it impossible to predict future attacks (since patrols haven't happened yet). While CAPTURE includes temporal elements in its model, it does not include spatial components and thus cannot capture neighborhood specific phenomena. In contrast to CAPTURE, [4] presented a behavior model, INTERCEPT, based on an ensemble of decision trees and was demonstrated to outperform CAPTURE. While their model accounted for spatial correlations, it did not include a temporal component. In contrast to these predictive models, our model addresses both spatial and temporal components.

It is vital to validate predictive models in the real world, and both [3,4] have conducted field tests in QEPA. [4] conducted a one month field test in QEPA and demonstrated promising results for predictive analytics in this domain. Unlike the field test we conducted, however, that was a preliminary field test and was not a controlled experiment. On the other hand, [3] conducted a controlled experiment where their goal, by selecting three areas for rangers to patrol, was to maximize the number of observations sighted per kilometer walked by the rangers. Their test successfully demonstrated a significant increase in illegal activity detection at two of the areas, but they did not provide comparable evaluation metrics for their predictive model. Also, our field test was much larger in scale, involving 27 patrol posts compared to their 9 posts.

## 3    Wildlife Crime Dataset: Features and Challenges

This study's wildlife crime dataset is from Uganda's Queen Elizabeth Protected Area (QEPA), an area containing a wildlife conservation park and two wildlife reserves, which spans about 2,520 km$^2$. There are 37 patrol posts situated across QEPA from which Uganda Wildlife Authority (UWA) rangers conduct patrols to apprehend poachers, remove any snares or traps, monitor wildlife, and record signs of illegal activity. Along with the amount of patrolling effort in each area, the dataset contains 14 years (2003–2016) of the type, location, and date of wildlife crime activities.

Rangers lack the manpower to patrol everywhere all the time, and thus illegal activity may be undetected in unpatrolled areas. Patrolling is an imperfect process, and there is considerable uncertainty in the dataset's negative data points (i.e., areas being labeled as having no illegal activity); rangers may patrol an area and label it as having no snares when, in fact, a snare was well-hidden and undetected. These factors contribute to the dataset's already large class imbalance; there are many more negative data points than there are positive points (crime detected). It is thus necessary to consider models that estimate hidden variables (e.g., whether an area has been attacked) and also to evaluate predictive models with metrics that account for this uncertainty, such as those in the Positive and Unlabeled Learning (PU Learning) literature [5]. We divide QEPA into 1 km$^2$ grid cells (a total of 2,522 cells), and we refer to these cells as targets. Each target is associated with several static geospatial features such as terrain (e.g., slope), distance values (e.g., distance to border), and animal density. Each target is also associated with dynamic features such as how often an area has been patrolled (i.e., coverage) and observed illegal activities (e.g., snares) (Fig. 1).

## 4    Models and Algorithms

### 4.1    Prediction by Graphical Models

**Markov Random Field (MRF).** To predict poaching activity, each target, at time step $t \in \{t_1, ..., t_m\}$, is represented by coordinates $i$ and $j$ within the boundary of QEPA. In Fig. 2(a), we demonstrate a three-dimensional network

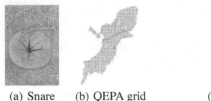

(a) Snare     (b) QEPA grid

**Fig. 1.** Photo credit: UWA ranger

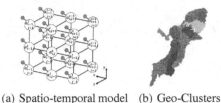

(a) Spatio-temporal model   (b) Geo-Clusters

**Fig. 2.** Geo-clusters and graphical model

for spatio-temporal modeling of poaching events over all targets. Connections between nodes represent the mutual spatial influence of neighboring targets and also the temporal dependence between recurring poaching incidents at a target. $a_{i,j}^t$ represents poaching incidents at time step $t$ and target $i, j$. Mutual spatial influences are modeled through first-order neighbors (i.e., $a_{i,j}^t$ connects to $a_{i\pm1,j}^t$, $a_{i,j\pm1}^t$ and $a_{i,j}^{t-1}$) and second-order neighbors (i.e., $a_{i,j}^t$ connects to $a_{i\pm1,j\pm1}^t$); for simplicity, the latter is not shown on the model's lattice. Each random variable takes a value in its state space, in this paper, $\mathcal{L} = \{0, 1\}$.

To avoid index overload, henceforth, nodes are indexed by serial numbers, $\mathcal{S} = \{1, 2, ..., N\}$ when we refer to the three-dimensional network. We introduce two random fields, indexed by $\mathcal{S}$, with their configurations: $\mathcal{A} = \{a = (a_1, ..., a_N) | a_i \in \mathcal{L}, i \in \mathcal{S}\}$, which indicates an *actual* poaching attack occurred at targets over the period of study, and $\mathcal{O} = \{o = (o_1, ..., o_N) | o_i \in \mathcal{L}, i \in \mathcal{S}\}$ indicates a *detected* poaching attack at targets over the period of study. Due to the imperfect detection of poaching activities, the former represents the hidden variables, and the latter is the known observed data collected by rangers, shown by the gray-filled nodes in Fig. 2(a). Targets are related to one another via a neighborhood system, $\mathcal{N}_n$, which is the set of nodes neighboring $n$ and $n \notin \mathcal{N}_n$. This neighborhood system considers all spatial and temporal neighbors. We define neighborhood attackability as the fraction of neighbors that the model predicts to be attacked: $u_{\mathcal{N}_n} = \sum_{n \in \mathcal{N}_n} a_n / |\mathcal{N}_n|$.

The probability, $P(a_i | u_{\mathcal{N}_n}, \boldsymbol{\alpha})$, of a poaching incident at each target $n$ at time step $t$ is represented in Eq. 1, where $\boldsymbol{\alpha}$ is a vector of parameters weighting the most important variables that influence poaching; $\boldsymbol{Z}$ represents the vector of time-invariant ecological covariates associated with each target (e.g., animal density, slope, forest cover, net primary productivity, distance from patrol post, town and rivers [2, 7]). The model's temporal dimension is reflected through not only the backward dependence of each $a_n$, which influences the computation of $u_{\mathcal{N}_n}$, but also in the past patrol coverage at target $n$, denoted by $c_n^{t-1}$, which models the delayed deterrence effect of patrolling efforts.

$$p(a_n = 1 | u_{\mathcal{N}_n}, \boldsymbol{\alpha}) = \frac{e^{-\boldsymbol{\alpha}[\boldsymbol{Z}, u_{\mathcal{N}_n}, c_n^{t-1}, 1]^{\mathsf{T}}}}{1 + e^{-\boldsymbol{\alpha}[\boldsymbol{Z}, u_{\mathcal{N}_n}, c_n^{t-1}, 1]^{\mathsf{T}}}} \tag{1}$$

Given $a_n$, $o_n$ follows a conditional probability distribution proposed in Eq. 2, which represents the probability of rangers detecting a poaching attack at target $n$.

The first column of the matrix denotes the probability of not detecting or detecting attacks if an attack has not happened, which is constrained to 1 or 0 respectively. In other words, it is impossible to detect an attack when an attack has not happened. The second column of the matrix represents the probability of not detecting or detecting attacks in the form of a logistic function if an attack has happened. Since it is less rational for poachers to place snares close to patrol posts and more convenient for rangers to detect poaching signs near the patrol posts, we assumed $dp_n$ (distance from patrol post) and $c_n^t$ (patrol coverage devoted to target $n$ at time $t$) are the major variables influencing rangers' detection capabilities. Detectability at each target is represented in Eq. 2, where $\beta$ is a vector of parameters that weight these variables.

$$p(o_n|a_n) = \begin{bmatrix} p(o_n = 0|a_n = 0) & p(o_n = 0|a_n = 1, \beta) \\ p(o_n = 1|a_n = 0) & p(o_n = 1|a_n = 1, \beta) \end{bmatrix} = \begin{bmatrix} 1, & \dfrac{1}{1 + e^{-\beta[dp_n, c_n^t, 1]^\top}} \\ 0, & \dfrac{e^{-\beta[dp_n, c_n^t, 1]^\top}}{1 + e^{-\beta[dp_n, c_n^t, 1]^\top}} \end{bmatrix} \quad (2)$$

We assume that $(o, a)$ is pairwise independent, meaning $p(o, a) = \prod_{n \in \mathcal{S}} p(o_n, a_n)$.

**EM Algorithm to Infer on MRF.** We use the Expectation-Maximization (EM) algorithm to estimate the MRF model's parameters $\theta = \{\alpha, \beta\}$. For completeness, we provide details about how we apply the EM algorithm to our model. Given a joint distribution $p(o, a|\theta)$ over observed variables $o$ and hidden variables $a$, governed by parameters $\theta$, EM aims to maximize the likelihood function $p(o|\theta)$ with respect to $\theta$. To start the algorithm, an initial setting for the parameters $\theta^{old}$ is chosen. At E-step, $p(a|o, \theta^{old})$ is evaluated, particularly, for each node in MRF model:

$$p(a_n|o_n, \theta^{old}) = \frac{p(o_n|a_n, \beta^{old}) . p(a_n|u_{\mathcal{N}_n}^{old}, \alpha^{old})}{p(o_n)} \quad (3)$$

M-step calculates $\theta^{new}$, according to the expectation of the complete log likelihood, $\log p(o, a|\theta)$, given in Eq. 4.

$$\theta^{new} = \arg \max_{\theta} \sum_{a_n \in \mathcal{L}} p(a|o, \theta^{old}) . \log p(o, a|\theta) \quad (4)$$

To facilitate calculation of the log of the joint probability distribution, $\log p(o, a|\theta)$, we introduce an approximation that makes use of $u_{\mathcal{N}_n}^{old}$, represented in Eq. 5.

$$\log p(o, a|\theta) = \sum_{n \in \mathcal{S}} \sum_{a_n \in \mathcal{L}} \log p(o_n|a_n, \beta) + \log p(a_n|u_{\mathcal{N}_n}^{old}, \alpha) \quad (5)$$

Then, if convergence of the log likelihood is not satisfied, $\theta^{old} \leftarrow \theta^{new}$, and repeat.

**Dataset Preparation for MRF.** To split the data into training and test sets, we divided the real-world dataset into year-long time steps. We trained the model's parameters $\theta = \{\alpha, \beta\}$ on historical data sampled through time steps $(t_1, ..., t_m)$ for all targets within the boundary. These parameters were used to predict poaching activity at time step $t_{m+1}$, which represents the test set for evaluation purposes. The trade-off between adding years' data (performance) vs. computational costs led us to use three years ($m = 3$). The model was thus trained over targets that were patrolled throughout the training time period $(t_1, t_2, t_3)$. We examined three training sets: 2011–2013, 2012–2014, and 2013–2015 for which the test sets are from 2014, 2015, and 2016, respectively.

Capturing temporal trends requires a sufficient amount of data to be collected regularly across time steps for each target. Due to the large amount of missing inspections and uncertainty in the collected data, this model focuses on learning poaching activity only over regions that have been continually monitored in the past, according to Definition 1. We denote this subset of targets as $\mathcal{S}_c$.

**Definition 1.** *Continually vs. occasionally monitoring: A target $i, j$ is continually monitored if all elements of the coverage sequence are positive; $c_{i,j}^{t_k} > 0, \forall k = 1, ..., m$ where $m$ is the number of time steps. Otherwise, it is occasionally monitored.*

Experiments with MRF were conducted in various ways on each data set. We refer to (a) a *global* model with spatial effects as **GLB-S**, which consists of a single set of parameters $\theta$ for the whole QEPA, and (b) a *global* model without spatial effects (i.e., the parameter that corresponds to $u_{\mathcal{N}_n}$ is set to 0) as **GLB**. The spatio-temporal model is designed to account for temporal and spatial trends in poaching activities. However, since learning those trends and capturing spatial effects are impacted by the variance in local poachers' behaviors, we also examined (c) a *geo-clustered* model which consists of multiple sets of local parameters throughout QEPA with spatial effects, referred to as **GCL-S**, and also (d) a *geo-clustered* model without spatial effects (i.e., the parameter that corresponds to $u_{\mathcal{N}_n}$ is set to 0) referred to as **GCL**.

Figure 2(b) shows the geo-clusters generated by Gaussian Mixture Models (GMM), which classifies the targets based on the geo-spatial features, $Z$, along with the targets' coordinates, $(x_{i,j}, y_{i,j})$, into 22 clusters. The number of geo-clusters, 22, are intended to be close to the number of patrol posts in QEPA such that each cluster contains one or two nearby patrol posts. With that being considered, not only are local poachers' behaviors described by a distinct set of parameters, but also the data collection conditions, over the targets within each cluster, are maintained to be nearly uniform.

## 4.2   Prediction by Ensemble Models

A **Bagging ensemble model** or Bootstrap **agg**regation technique, called Bagging, is a type of ensemble learning which bags some weak learners, such as decision trees, on a dataset by generating many bootstrap duplicates of the dataset and learning decision trees on them. Each of the bootstrap duplicates are obtained by randomly choosing M observations out of M with replacement, where M denotes the training dataset size. Finally, the predicted response of the ensemble is computed by taking an average over predictions from its individual decision trees. To learn a Bagging ensemble, we used the *fitensemble* function of MATLAB 2017a. **Dataset preparation** for the Bagging ensemble model is designed to find the targets that are liable to be attacked [4]. A target is assumed to be attackable if it has ever been attacked; if any observations occurred in the entire training period for a given target, that target is labeled as attackable. For this model, the best training period contained 5 years of data.

### 4.3 Hybrid of MRF and Bagging Ensemble

Since the amount and regularity of data collected by rangers varies across regions of QEPA, predictive models perform differently in different regions. As such, we propose using different models to predict over them; first, we used a Bagging ensemble model, and then improved the predictions in some regions using the spatio-temporal model. For global models, we used MRF for all continually monitored targets. However, for geo-clustered models, for targets in the continually monitored subset, $\mathcal{S}_c^q$, (where temporally-aware models can be used practically), the MRF model's performance varied widely across geo-clusters according to our experiments. $q$ indicates clusters and $1 \leq q \leq 22$. Thus, for each $q$, if the average Catch Per Unit Effort (CPUE), outlined by Definition 2, is relatively large, we use the MRF model for $\mathcal{S}_c^q$. In Conservation Biology, CPUE is an indirect measure of poaching activity abundance. A larger average CPUE for each cluster corresponds to more frequent poaching activity and thus more data for that cluster. Consequently, using more complex spatio-temporal models in those clusters becomes more reasonable.

**Definition 2. *Average CPUE* is** $\sum_{n \in \mathcal{S}_c^q} o_n / \sum_{n \in \mathcal{S}_c^q} c_n^t$ *in cluster q.*

To compute CPUE, effort corresponds to the amount of coverage (i.e., 1 unit = 1 km walked) in a given target, and catch corresponds to the number of observations. Hence, for $1 \leq q \leq 22$, we will boost selectively according to the average CPUE value; some clusters may not be boosted by MRF, and we would only use Bagging ensemble model for making predictions on them. Experiments on historical data show that selecting 15% of the geo-clusters with the highest average CPUE results in the best performance for the entire hybrid model (discussed in the Evaluation Section).

## 5 Evaluations and Discussions

### 5.1 Evaluation Metrics

The imperfect detection of poaching activities in wildlife conservation areas leads to uncertainty in the negative class labels of data samples [4]. It is thus vital to evaluate prediction results based on metrics which account for this inherent uncertainty. In addition to standard metrics in Machine Learning (e.g., precision, recall, F1) which are used to evaluate models on datasets with no uncertainty in the underlying ground truth, we also use the L&L metric introduced in [5], which is a metric specifically designed for models learned on Positive and Unlabeled datasets. L&L is defined as $L\&L = \frac{r^2}{Pr[f(Te)=1]}$, where $r$ denotes the recall and $Pr[f(Te) = 1]$ denotes the probability of a classifier $f$ making a positive class label prediction.

## 5.2   Experiments with Real-World Data

Evaluation of models' attack predictions are demonstrated in Tables 1 and 2. Precision and recall are denoted by Prec. and Rec. in the tables. To compare models' performances, we used several baseline methods, (i) Positive Baseline, **PB**; a model that predicts poaching attacks to occur in all targets, (ii) Random Baseline, **RB**; a model which flips a coin to decide its prediction, (iii) Training Label Baseline, **TL**; a model which predicts a target as attacked if it has been ever attacked in the training data. We also present the results for Support Vector Machines, **SVM**, and AdaBoost methods, **AD**, which are well-known Machine Learning techniques, along with results for the best performing predictive model on the QEPA dataset, INTERCEPT, **INT**, [4]. Results for the Bagging ensemble technique, **BG**, and RUSBoost, **RUS**, a hybrid sampling/boosting algorithm for learning from datasets with class imbalance [10], are also presented. In all tables, **BGG\*** stands for the best performing model among all variations of the hybrid model, which will be discussed in detail later. Table 1 demonstrates that **BGG\*** outperformed all other existing models in terms of L&L and also F1.

**Table 1.** Comparing all models' performances with the best performing BGG model.

| Year | 2014 | | | | | 2015 | | | | | 2016 | | | | |
|---|---|---|---|---|---|---|---|---|---|---|---|---|---|---|---|
| Mdl | PB | RB | TL | SVM | BGG* | PB | RB | TL | SVM | BGG* | PB | RB | TL | SVM | BGG* |
| Prec. | 0.06 | 0.05 | 0.26 | 0.24 | 0.65 | 0.10 | 0.08 | 0.39 | 0.4 | 0.69 | 0.10 | 0.09 | 0.45 | 0.45 | 0.74 |
| Rec. | 1.00 | 0.46 | 0.86 | 0.3 | 0.54 | 1.00 | 0.43 | 0.78 | 0.15 | 0.62 | 1.00 | 0.44 | 0.75 | 0.23 | 0.66 |
| F1 | 0.10 | 0.09 | 0.4 | 0.27 | 0.59 | 0.18 | 0.14 | 0.52 | 0.22 | 0.65 | 0.18 | 0.14 | 0.56 | 0.30 | 0.69 |
| L&L | 1.00 | 0.43 | 4.09 | 1.33 | 6.44 | 1.00 | 0.37 | 3.05 | 0.62 | 4.32 | 1.00 | 0.38 | 3.4 | 1.03 | 4.88 |
| Mdl | RUS | AD | BG | INT | BGG* | RUS | AD | BG | INT | BGG* | RUS | AD | BG | INT | BGG* |
| Prec. | 0.12 | 0.33 | 0.62 | 0.37 | 0.65 | 0.2 | 0.52 | 0.71 | 0.63 | 0.69 | 0.19 | 0.53 | 0.76 | 0.40 | 0.74 |
| Rec. | 0.51 | 0.47 | 0.54 | 0.45 | 0.54 | 0.51 | 0.5 | 0.53 | 0.41 | 0.62 | 0.65 | 0.54 | 0.62 | 0.66 | 0.66 |
| F1 | 0.19 | 0.39 | 0.58 | 0.41 | 0.59 | 0.29 | 0.51 | 0.61 | 0.49 | 0.65 | 0.29 | 0.53 | 0.68 | 0.51 | 0.69 |
| L&L | 1.12 | 2.86 | 6.18 | 5.83 | 6.44 | 1.03 | 2.61 | 3.83 | 3.46 | 4.32 | 1.25 | 2.84 | 4.75 | 2.23 | 4.88 |

Table 2 provides a detailed comparison of all variations of our hybrid models, **BGG** (i.e., when different MRF models are used). When **GCL-S** is used, we get the best performing model in terms of L&L score, which is denoted as **BGG\***.

**Table 2.** Performances of hybrid models with variations of MRF (BGG models)

| Year | 2014 | | | | 2015 | | | | 2016 | | | |
|---|---|---|---|---|---|---|---|---|---|---|---|---|
| Model | GLB | GLB-S | GCL | GCL-S | GLB | GLB-S | GCL | GCL-S | GLB | GLB-S | GCL | GCL-S |
| Prec. | 0.12 | 0.12 | 0.63 | 0.65 | 0.19 | 0.19 | 0.69 | 0.69 | 0.18 | 0.19 | 0.72 | 0.74 |
| Recall | 0.58 | 0.65 | 0.54 | 0.54 | 0.52 | 0.58 | 0.65 | 0.62 | 0.50 | 0.46 | 0.66 | 0.66 |
| F1 | 0.20 | 0.20 | 0.58 | 0.59 | 0.28 | 0.29 | 0.65 | 0.65 | 0.27 | 0.27 | 0.69 | 0.69 |
| L&L | 1.28 | 1.44 | 6.31 | 6.44 | 0.99 | 1.14 | 4.32 | 4.32 | 0.91 | 0.91 | 4.79 | 4.88 |

The poor results of learning a global set of parameters emphasize the fact that poachers' behavior and patterns are not identical throughout QEPA and should be modeled accordingly.

Our experiments demonstrated that the performance of the MRF model within $S_c^q$ varies across different geo-clusters and is related to the CPUE value for each cluster, $q$. Figure 3(a) displays an improvement in L&L score for the **BGG\*** model compared to **BG** vs. varying the percentile of geo-clusters used for boosting. Experiments with the 2014 test set show that choosing the 85th percentile of geo-clusters for boosting with MRF, according to CPUE, (i.e., selecting 15% of the geo-clusters, with highest CPUE), results in the best prediction performance. The 85th percentile is shown by vertical lines in Figures where the **BGG\*** model outperformed the **BG** model. We used a similar percentile value for experiments with the MRF model on test sets of 2015 and 2016. Figure 3(b) and (c) confirm the efficiency of choosing an 85th percentile value.

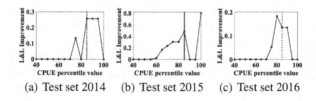

(a) Test set 2014     (b) Test set 2015     (c) Test set 2016

**Fig. 3.** L&L improvement vs. CPUE percentile value; BGG\* compared to BG

## 6   QEPA Field Test

While our model demonstrated superior predictive performance on historical data, it is important to test these models in the field.

The initial field test we conducted in [4], in collaboration with the Wildlife Conservation Society (WCS) and the Uganda Wildlife Authority (UWA), was the first of its kind in the Machine Learning (ML) community and showed promising improvements over previous patrolling regimes. Due to the difficulty of organizing such a field test, its implications

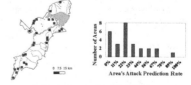

(a) Patrolled areas (b) Prediction rates

**Fig. 4.** Patrol area statistics

were limited: only two 9-km$^2$ areas (18 km$^2$) of QEPA were patrolled by rangers over a month. Because of its success, however, WCS and UWA graciously agreed to a larger scale, controlled experiment: also in 9 km$^2$ areas, but rangers patrolled 27 of these areas (243 km$^2$, spread across QEPA) over five months; this is the largest to-date field test of ML-based predictive models in this domain. We show the areas in Fig. 4(a). Note that rangers patrolled these areas in addition to other areas of QEPA as part of their normal duties.

This experiment's goal was to determine the selectiveness of our model's snare attack predictions: does our model correctly predict both where there are and are not snare attacks? We define attack prediction rate as the proportion of targets (a 1 km by 1 km cell) in a patrol area (3 by 3 cells) that are predicted to be attacked. We considered two experiment groups that corresponded to our model's attack prediction rates from November 2016 to March 2017: High (group 1) and Low (group 2). Areas that had an attack prediction rate of 50% or greater were considered to be in a high area (group 1); areas with less than a 50% rate were in group 2. For example, if the model predicted five out of nine targets to be attacked in an area, that area was in group 1. Due to the importance of QEPA for elephant conservation, we do not show which areas belong to which experiment group in Fig. 4(a) so that we do not provide data to ivory poachers.

To start, we exhaustively generated all patrol areas such that (1) each patrol area was 3 × 3 km$^2$, (2) no point in the patrol area was more than 5 km away from the nearest ranger patrol post, and (3) no patrol area was patrolled too frequently or infrequently in past years (to ensure that the training data associated with all areas was of similar quality); in all, 544 areas were generated across QEPA. Then, using the model's attack predictions, each area was assigned to an experiment group. Because we were not able to test all 544 areas, we selected a subset such that no two areas overlapped with each other and no more than two areas were selected for each patrol post (due to manpower constraints). In total, 5 areas in group 1 and 22 areas in group 2 were chosen. Note that this composition arose due to the preponderance of group 2 areas (see Table 3). We provide a breakdown of the areas' exact attack prediction rates in Fig. 4(b); areas with rates below 56% (5/9) were in group 2, and for example, there were 8 areas in group 2 with a rate of 22% (2/9). Finally, when we provided patrols to the rangers, *experiment group memberships were hidden to prevent effects where knowledge of predicted poaching activity would influence their patrolling patterns and detection rates.*

**Table 3.** Patrol area group memberships

| Experiment group | Exhaustive patrol area groups | Final patrol area groups |
|---|---|---|
| High (1) | 50 (9%) | 5 (19%) |
| Low (2) | 494 (91%) | 22 (81%) |

## 6.1    Field Test Results and Discussion

The field test data we received was in the same format as the historical data. However, because rangers needed to physically walk to these patrol areas, we received additional data that we have omitted from this analysis; observations made outside of a designated patrol area were not counted. Because we only predicted where snaring activity would occur, we have also omitted other observation types made during the experiment (e.g., illegal cattle grazing). We present results from this five-month field test in Table 4. To provide additional context for these results, we also computed QEPA's park-wide historical CPUE (from November 2015 to March 2016): 0.04.

**Table 4.** Field test results: observations

| Experiment group | Observation count (%) | Mean count (std) | Effort (%) | CPUE |
|---|---|---|---|---|
| High (1) | 15 (79%) | 3 (5.20) | 129.54 (29%) | 0.12 |
| Low (2) | 4 (21%) | 0.18 (0.50) | 322.33 (71%) | 0.01 |

Areas with a high attack prediction rate (group 1) had significantly more snare sightings than areas with low attack prediction rates (15 vs. 4). This is despite there being far fewer group 1 areas than group 2 areas (5 vs. 22); on average, group 1 areas had 3 snare observations whereas group 2 areas had 0.18 observations. It is worth noting the large standard deviation for the mean observation counts; the standard deviation of 5.2, for the mean of 3, signifies that not all areas had snare observations. Indeed, two out of five areas in group 1 had snare observations. However, this also applies to group 2's areas: only 3 out of 22 areas had snare observations.

We present Catch per Unit Effort (CPUE) results in Table 4. When accounting for differences in areas' effort, group 1 areas had a CPUE that was over ten times that of group 2 areas. Moreover, when compared to QEPA's park-wide historical CPUE of 0.04, it is clear that our model successfully differentiated between areas of high and low snaring activity. The results of this large-scale field test, the first of its kind for ML models in this domain, demonstrated that our model's superior predictive performance in the laboratory extends to the real world.

## 7    Conclusion

In this paper, we presented a hybrid spatio-temporal model to predict wildlife poaching threat levels. Additionally, we validated our model via an extensive five-month field test in Queen Elizabeth Protected Area (QEPA) where rangers patrolled over $450\,km^2$ across QEPA—the largest field-test to-date of Machine Learning-based models in this domain. On real-world historical data from QEPA, our hybrid model achieves significantly better performance than prior work. On the data collected from our field test, we demonstrated that our model successfully differentiated between areas of high and low snaring activity. These findings demonstrated that our model's predictions are selective and also that its superior laboratory performance extends to the real world. Based on these promising results, future work will focus on deploying these models as part of a software package to UWA to aid in planning future anti-poaching patrols.

**Acknowledgments.** This research was supported by MURI grant W911NF-11-1-0332, NSF grant with Cornell University 72954-10598 and partially supported by Harvard Center for Research on Computation and Society fellowship. We are grateful to the Wildlife Conservation Society and the Uganda Wildlife Authority for supporting data collection in QEPA. We also thank Donnabell Dmello for her help in data processing.

# References

1. Great Elephant Census: The great elephant census—a Paul G. Allen project. Press Release, August 2016
2. Critchlow, R., Plumptre, A., Driciru, M., Rwetsiba, A., Stokes, E., Tumwesigye, C., Wanyama, F., Beale, C.: Spatiotemporal trends of illegal activities from ranger-collected data in a Ugandan National Park. Conserv. Biol. **29**(5), 1458–1470 (2015)
3. Critchlow, R., Plumptre, A.J., Alidria, B., Nsubuga, M., Driciru, M., Rwetsiba, A., Wanyama, F., Beale, C.M.: Improving law-enforcement effectiveness and efficiency in protected areas using ranger-collected monitoring data. Conserv. Lett. **10**(5), 572–580 (2017). Wiley Online Library
4. Kar, D., Ford, B., Gholami, S., Fang, F., Plumptre, A., Tambe, M., Driciru, M., Wanyama, F., Rwetsiba, A., Nsubaga, M., et al.: Cloudy with a chance of poaching: adversary behavior modeling and forecasting with real-world poaching data. In: Proceedings of the 16th Conference on Autonomous Agents and MultiAgent Systems, pp. 159–167 (2017)
5. Lee, W.S., Liu, B.: Learning with positive and unlabeled examples using weighted logistic regression. In: ICML, vol. 3 (2003)
6. Nguyen, T.H., Sinha, A., Gholami, S., Plumptre, A., Joppa, L., Tambe, M., Driciru, M., Wanyama, F., Rwetsiba, A., Critchlow, R., et al.: CAPTURE: a new predictive anti-poaching tool for wildlife protection. In: AAMAS, pp. 767–775 (2016)
7. O'Kelly, H.J.: Monitoring Conservation Threats, Interventions, and Impacts on Wildlife in a Cambodian Tropical Forest, p. 149. Imperial College, London (2013)
8. Rashidi, P., Wang, T., Skidmore, A., Mehdipoor, H., Darvishzadeh, R., Ngene, S., Vrieling, A., Toxopeus, A.G.: Elephant poaching risk assessed using spatial and non-spatial Bayesian models. Ecol. Model. **338**, 60–68 (2016)
9. Rashidi, P., Wang, T., Skidmore, A., Vrieling, A., Darvishzadeh, R., Toxopeus, B., Ngene, S., Omondi, P.: Spatial and spatiotemporal clustering methods for detecting elephant poaching hotspots. Ecol. Model. **297**, 180–186 (2015)
10. Seiffert, C., Khoshgoftaar, T.M., Van Hulse, J., Napolitano, A.: Rusboost: a hybrid approach to alleviating class imbalance. IEEE SMC-A Syst. Hum. **40**(1), 185–197 (2010)
11. Solberg, A.H.S., Taxt, T., Jain, A.K.: A Markov random field model for classification of multisource satellite imagery. IEEE TGRS **34**(1), 100–113 (1996)
12. Yin, Z., Collins, R.: Belief propagation in a 3d spatio-temporal MRF for moving object detection. In: IEEE CVPR, pp. 1–8. IEEE (2007)

# Unsupervised Signature Extraction
# from Forensic Logs

Stefan Thaler[1(✉)], Vlado Menkovski[1], and Milan Petkovic[1,2]

[1] Technical University of Eindhoven,
Den Dolech 12, 5600 MB Eindhoven, Netherlands
{s.m.thaler,v.menkovski}@tue.nl
[2] Philips Research Laboratories,
High Tech Campus 34, Eindhoven, Netherlands
milan.petkovic@philips.com

**Abstract.** Signature extraction is a key part of forensic log analysis. It involves recognizing patterns in log lines such that log lines that originated from the same line of code are grouped together. A log signature consists of immutable parts and mutable parts. The immutable parts define the signature, and the mutable parts are typically variable parameter values. In practice, the number of log lines and signatures can be quite large, and the task of detecting and aligning immutable parts of the logs to extract the signatures becomes a significant challenge. We propose a novel method based on a neural language model that outperforms the current state-of-the-art on signature extraction. We use an RNN auto-encoder to create an embedding of the log lines. Log lines embedded in such a way can be clustered to extract the signatures in an unsupervised manner.

**Keywords:** Information forensic · RNN auto-encoder
Neural language model · Log clustering · Signature extraction

## 1 Introduction

An important step of an information forensic investigation is log analysis. Logs contain valuable information for reconstructing incidents that have happened on a computer system. In this context, a log line is a sequence of tokens that give information about the state of a process that created this log line. The tokens of each log lines are partially natural language and partially structured data. Tokens may be words, numbers, variables or punctuation characters such as brackets, colons or dots.

Log signatures are the print statements that produce the log lines. Log signatures have fixed parts and may have variable parts. Fixed parts consist of a sequence of tokens of arbitrary length that uniquely identify signatures. The variable parts may also be of arbitrary length and variable parts in log lines that originate from the same signature differ.

© Springer International Publishing AG 2017
Y. Altun et al. (Eds.): ECML PKDD 2017, Part III, LNAI 10536, pp. 305–316, 2017.
https://doi.org/10.1007/978-3-319-71273-4_25

The goal of a forensic investigator is to uncover a sequence of events from a forensic log that reveal a security incident. The sequence of events describes the actions that the users of this computer system took. Traces of these actions are typically stored in logs. Similar events may have different log lines associated with them because the variable part reports the state of the system at that time, which makes finding such events difficult. Knowing the log signatures of a log enable a forensic investigator to group together log lines that belong to the same event, even though the log lines differ. Finding such signatures is challenging because of the unknown number of signatures and the unknown number and position of fixed parts. Signature extraction is the process of finding a set of log signatures given a set of log lines.

State-of-the-art approaches identify log signatures based on the position and frequency of tokens [1,12,23]. These approaches typically assume that frequent words define the fixed parts of the signature. This assumption holds if the ratio of log lines per signature in the analyzed log is high. This can be the case for many application logs where the tokens of fixed signature parts are repeated with a high frequency. However, in information forensics, logs commonly have many signatures but few log lines per signature. In this case, the number of occurrence of tokens of variable parts may be higher than fixed tokens, which can cause a confusion of which tokens are fixed and which ones are variable. Confusing fixed tokens with variable ones leads to signatures that match too few log lines, and mixing variable tokens with fixed ones will result into signatures that will match too few log lines.

To address the challenge of signature extraction from forensic logs, we propose to use a method that takes contextual information about the tokens into account. Our approach is inspired by recent advances in the NLP domain, where sequence-to-sequence models have been successfully used to capture natural language [3,9].

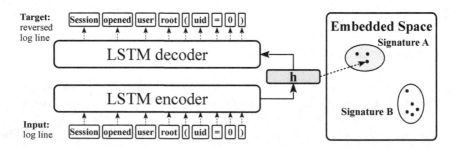

**Fig. 1.** We first embed forensic log lines using an RNN auto-encoder. We then cluster the embedded log lines and assign them to a signature.

Typically, sequence-to-sequence models consist of two recurrent neural networks (RNNs), an encoder network and a decoder network. The encoder network learns to represent an input sequence, and the decoder learns to construct the output sequence from this representation. We use such a model to learn an encoding function that encodes the log line, and a decoding function that learns to reconstruct the reversed input log line from this representation. Figure 1 depicts this idea. Based on the findings in the NLP domain, we assume that this embedding function takes into account contextual information, and embeds similar log lines close to each other in the embedded space. We then cluster the embedded log lines and use the clusters as signature assignment.

In detail, the main contributions of our paper are:

- We propose a method, LSTM-AE+C, that uses an RNN auto-encoder to create a log line embedding space. We then cluster the log lines in the embedding space to determine the signature assignment. We detail this method in Sect. 2.
- We demonstrate on our own and two public datasets that LSTM-AE+C outperforms two state-of-the-art approaches for signature extraction. We detail the experiment setup in Sect. 3 and discuss the results after that.

## 2    Method

Our method LSTM-AE+C for signature extraction of forensic logs can be divided into two steps. First, we train a sequence-to-sequence auto-encoder network to learn an embedded space for log lines. Sequence-to-sequence neural networks for natural language translation have been introduced by Sutskever et al. [19] and widely applied since then. We use a similar model, however, instead of using it in a sequence-to-sequence manner, we use it as auto-encoder that reconstructs the input sequence. Secondly, we cluster the embedded log lines to extract the signatures.

We depict a schematic overview of our model in Fig. 2. To learn an embedding we train the LSTM auto-encoder to reconstruct each input log line. To do that, the encoder part of the auto-encoder needs to encode the log line into a fixed size vector that is fed into the decoder. The fixed size of the vector limits the capacity of the auto-encoder and provides for a regularization that restricts the

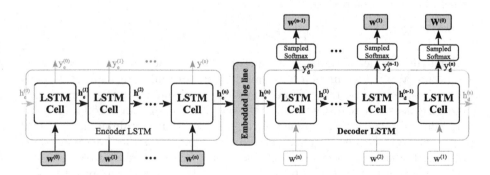

**Fig. 2.** We use a sequence-to-sequence LSTM auto-encoder to learn embeddings for our log lines.

auto-encoder from learning an identity function. We use that representation as embedding for the log lines. In the remaining section we will first detail the components of our model and their relationships to each other, then detail the learning objective and finally describe how we extract signatures.

## 2.1   Model

The input to our model is log lines. We treat log lines as a sequence of tokens of length $n$, where a token can be a word, variable part or delimiter. The set of unique tokens is our input vocabulary, where each token in the vocabulary gets a unique id.

Since the number of such tokens in a log can be potentially very large, we learn a dense representation for the tokens of our log lines. To get these dense representations, we use a token embedding matrix $E^{(v \times u)}$, where $v$ is the unique number of tokens that we have in our token vocabulary and $u$ is the number of hidden units of the encoder network. The index of each row of $E$ is also the position of $v$ in the vocabulary. We denote an in $E$ embedded token as $w$.

Next, want to learn the log line embedding. To do so, we learn an encoder function $ENC$ using an LSTM [7], which is a variant of a recurrent neural network. We chose an LSTM for both the encoder and decoder, because it addresses the vanishing gradient problem. $h_e(t)$ is the hidden encoder state at time step $t$, and $y_e^{(t)}$ is the encoder output at time step t. $w_e^{(t)}$ is the embedded input word for time step $t$. We use the encoding state and and input word at each time step to calculate the next state and the next output. We discard all output $s$ of the encoder. We use the final hidden state $h_e^{(n)}$ to embed our log lines. $h_e^{(n)}$ also serves as initial hidden state for our decoder network.

$$(y_e^{(t)}, h_e^{(t+1)}) = ENC(w_e^{(t)}, h_e^{(t)})$$

Our decoder function $DEC$ is trained to learn to reconstruct the reverse sequence $S'$ given the last hidden state $h_e^{(n)}$ of our encoding network. The structure of the network is identical to the encoder, except we feed the network the reverse sequence of embedded tokens as input.

$$(y_d^{(t)}, h_d^{(t+1)}) = DEC(w_d^{(t)}, h_d^{(t)})$$

From the decoder outputs $y_d^{(t)}$ we predict the reverse sequence of tokens $S'$. Calculating a softmax function for a large token vocabulary is computationally very expensive because the softmax function is calculated as the sum over all potential classes. Therefore, we predict the output tokens of our decoder sequence using sampled softmax [8]. Sampled softmax is a candidate training method that approximates the desired softmax function by solving a task that does not require predicting the correct token from all token. Instead, the task sampled softmax solves is to identify the correct token from a randomly drawn sample of tokens.

## 2.2   Objective

To embed our log lines in an embedded space, the model needs to maximize the probability of predicting a reversed sequence of tokens $S'$ given a sequence of tokens $S$. In other words, we want to train the encoding network to learn an embedding that contains enough information for the decoder to reconstruct it. However, as an effect of the regularization, we expect the model to use the structure to use the structure of the log lines to create a more efficient representation, that in turn allows us to extract the signatures.

$$\theta^* = arg \max_{\theta} \sum_{(S,S')} \log p(S'|S; \theta)$$

$\theta$ are the parameters of our model, $S$ represents a log line, and $S'$ represents a reversed log line.

$S$ is a sequence of tokens of arbitrary length. To model the joint probability over $S'_0, \ldots, S'_{t-1}$ given $S$ and $\theta$, it is common to use the chain rule for probabilities.

$$\log p(S'|S, \theta) = \sum_{t=0}^{n} \log p(S'_t|S, \theta, S'_0, \ldots, S'_{t-1})$$

When training the network, $S$ and $S'$ are the inputs and the targets of one training example. We calculate the sum of equation 2.2 per batch using RMSProp [22]. We detail the hyper parameters of the training process in Sect. 3.3.

## 2.3   Extracting Signatures

After the training of our auto-encoder model is complete, we use encoding network to generate the embedded vector. We expect that due to the lregularization structurally similar log lines will be embedded close to each other, which enables us to use a clustering algorithm to group log lines which belong to the same signature.

Since forensic logs may be very large, we cluster the embedded log lines using the BIRCH algorithm [27]. BIRCH is an iterative algorithm that dynamically builds a height-balanced cluster feature tree. The algorithm has an almost linear runtime complexity on the number of training examples, and it does not require the whole dataset to be stored in memory. These two properties make the algorithm well suited for applications on large datasets.

## 3   Experiments

We compare our method to LogCluster [23] and IPLoM [11]. Many algorithms have been designed to be applied to a special type of application log, where the number of signatures is known up front. However, in an information forensic context the forensic logs that are being analyzed stem from an unknown system, which means that the number of signatures is not known up front. Therefore it is important that IPLoM and LogCluster do not require a fixed amount of clusters as a hyper parameter. Furthermore, in a study by He et al. [6], IPLoM and SLCT were amongst the best-performing signature extraction algorithms. LogCluster is the improved version of SLTC that addresses multiple shortcomings of SLCT. We thus assume the LogCluster would have outperformed SLCT in He's evaluation. For IPLoM, we use the implementation provided by [6]. For LogCluster we use the implementation provided by the author online[1]. We implemented our own method LSTM-AE+C in Tensorflow version 1.0.1. Our experiments are available on GitHub[2].

### 3.1    Evaluation Metrics

To assess our approach, we treat the log signature extraction problem as log clustering problem because log clustering and log signature extraction are related problems [20]. The key difference between clustering and signature extraction is, that the goal of a clustering approach is the find the best clusters according to some metric, whereas the goal of signature extraction is to find the right set of signatures. This set of signatures does not have to be the best set of clusters.

We evaluate the quality of the retrieved clusters of all our evaluated approaches with two metrics: the V-Measure [15] and the adjusted mutual information [24]. The V-Measure is the harmonic mean between the homogeneity and the completeness of clusters. It is based on the conditional entropy of the clusters. The adjusted mutual information describes the mutual information of different cluster labels, adjusted for chance. It is normalized to the size of the clusters. Both approaches are independent of permutations on the true and predicted labels. The values of the V-Measure and the adjusted mutual information can range from 0.0 to 1.0. In both cases, 1.0 means perfect clusters and 0.0 means random label assignment.

Additionally, we assess the cluster quality for clusters retrieved with LSTM-AE+C using the Silhouette score. The Silhouette score measures the tightness and separation of clusters and only depends on the partition of the data [16]. It ranges between $-1.0$ and $1.0$, where a negative score means many wrong cluster assignments and 1.0 means perfect clustering.

We validate the stability of our approaches using 10-fold, randomly sub-sampled 10000 log lines [10]. In Sect. 3.4, we report the average scores for our metrics and their standard deviation.

### 3.2    Datasets

We use three logs to evaluate and compare our method: a forensic log that we extracted from a virtual machine hard drive and the system logs of two

---

[1] https://ristov.github.io/logcluster/.
[2] https://github.com/stefanthaler/2017-ecml-forensic-unsupervised.

**Table 1.** The log file statistics are as follows

| Log name | Lines | Signatures | Unique tokens |
|----------|-------|------------|---------------|
| Forensic | 11.023 | 852 | 4.114 |
| BlueGene/L | 474.796 | 355 | 114.495 |
| Spirit2 | 716.577 | 691 | 59.972 |

high-performance cluster computers, BlueGene/L(BGL) and Spirit [13]. An overview over the log statistics is presented in Table 1.

We created our forensic log by extracting it from a Ubuntu 16.04 system image disk using the open source log2timeline tool[3]. We manually created the signatures for this dataset by looking at the Ubuntu source code. The difference between a forensic log and a system log is that a forensic log contains information from multiple log files on the examined system, whereas a system log only contains the logs that were reported by the system daemon. The system log is part of the forensic log, but it also contains other logs, which typically leads to more complexity in such log files.

BlueGene/L(BGL) was a high-performance cluster that was installed in the Lawrence Livermore National Labs. The publicly available system log was recorded during March 6th and April 1st in 2006. It consists of 4.747.963 log lines in total. In our experiments, we use a stratified sample which has 474.796 log lines. We manually extracted the signatures for this log file.

Spirit was a high-performance cluster that was installed in the Sandia National Labs. The publicly available system log was recorded during January 1st and the 11th of July in 2006. It consists of 272.298.969 log lines in total. In our experiments, we use a stratified sample which has 716.577 lines. We also extracted the signatures for this log by hand.

The BlueGene/L and the Spirit logs are publicly available and can be downloaded from the Usenix webpage[4]. We publish our dataset on GitHub[5].

For all three log files, we removed fixed position data such as timestamps or dates at the beginning of each log message. In the case of our forensic log we completely removed these columns. In the case of the other two logs, we replaced the fixed elements with special token, such as TIME_STAMP. We added this preprocessing because it reduces the sequence complexity, but it does not reduce the quality of the extracted signatures.

### 3.3  Hyper Parameters and Training Details

IPLoM supports the following parameters: File support threshold (FST), which controls the number of clusters found; partition support threshold (PST), which limits the backtracking of the algorithm; upper bound (UB) and lower bound

---

[3] https://github.com/log2timeline/.

[4] https://www.usenix.org/cfdr-data.

[5] https://github.com/stefanthaler/2017-ecml-forensic-unsupervised.

(LB) which control when to split a cluster and cluster goodness threshold (CGT) [11]. We evaluate IPLoM by performing a grid search on the following parameter ranges: FST between 1 and 20 in 1 steps, PST of 0.05, UB between 0.5 and 0.9 in 0.1 steps, LB, between 0.1–0.5 in 0.1 steps and CGT between 0.3 and 0.6 in 0.1 steps. We chose the parameters according to the guidelines of the original paper.

LogCluster supports two main parameters: support threshold (ST), which controls the minimum amount of patterns and the word frequency (WF), which sets the frequency of words within a log line. We evaluate LogCluster by performing grid search using the following parameter ranges: ST between 1 and 3000 in and WF of 0.3, 0.6 and 0.9.

We generate each input token sequence by splitting a log line at each special character. Furthermore, we add a special token at the beginning and the end of the sequence that marks the beginning and the end of a sequence. Within a batch, sequences are zero-padded to the longest sequence in this batch, and zero inputs are ignored during training.

All embeddings and LSTM cells had 256 units. Both encoder and decoder network had a 1-layer LSTM. We trained all our LSTM auto-encoders for ten epochs using RMSProp [22]. We used a learning rate of 0.02 and decayed the learning rate by 0.95 after every epoch. Each training batch had 200 examples and the maximum length number of steps to unroll the LSTM auto-encoder was 200. We used 500 samples to calculate the sampled softmax loss. We used dropout on the decoder network outputs [17] to prevent overfitting and to regularize our network to learn independent representations. Finally, we clip the gradients of our LSTM encoder and LSTM decoder at 0.5 to avoid exploding gradients [14].

The hyper parameters and the architecture of our model were empirically determined. We tried LSTMs with attention mechanism [3], batch normalization, multiple layers of LSTMs, and more units. However, these measures had little effect on the quality of the clusters; therefore we chose the simplest possible architecture. We used the same architecture and hyper parameters for all our experiments.

The second step in our method is to cluster the embedded log lines to find signatures. We cluster the embedded log lines using the BIRCH cluster algorithm [27]. We performed the clustering using grid search on distance thresholds between 1 and 50 in 0.5 steps, and a branching factor of either 15, 30 or 50.

## 3.4   Results

We report the results of our experiments in Table 2. Each value reports the best performing hyper parameter settings. Each score is the average of 10-fold random sub-sampling followed by the standard deviation of this average. We do not report on the Silhouette score for LogCluster and IPLoM because both algorithms do not provide a means to calculate the distance between different log lines.

**Table 2.** Log clustering evaluation, best averages and standard deviation.

| Log file | Approach | V-Measure | Adj. Mut. Inf. | Silhouette |
|---|---|---|---|---|
| Forensic | LogCluster [23] | 0.904 ±0.000 | 0.581 ±0.000 | N/A |
| | IPLoM [11] | 0.825 ± 0.001 | 0.609 ± 0.001 | N/A |
| | LSTM-AE+C (Ours) | **0.935** ± 0.002 | **0.864** ± 0.004 | 0.705 ± 0.001 |
| BlueGene/L | LogCluster [23] | 0.592 ± 0.004 | 0.225 ± 0.005 | N/A |
| | IPLoM [11] | 0.828 ± 0.003 | 0.760 ± 0.005 | N/A |
| | LSTM-AE+C (ours) | **0.948** ± 0.005 | **0.900** ± 0.001 | 0.827 ± 0.002 |
| Spirit | LogCluster [23] | 0.829 ± 0.002 | 0.677 ± 0.004 | N/A |
| | IPLoM [11] | 0.920 ± 0.004 | 0.895 ± 0.003 | N/A |
| | LSTM-AE+C (ours) | **0.930** ± 0.010 | **0.902** ± 0.008 | 0.815 ± 0.004 |

## 3.5   Discussion of Results

As can be seen from Table 2, our approach significantly outperforms the two
word-frequency based baseline approaches on the three datasets, both regarding
V-Measure and Adjusted Mutual Information. The standard deviation is below
0.005 in all reported experiments, which indicates that clustering is consistently
stable over the datasets.

For all three log files we obtain a Silhouette score of greater than 0.70, which
indicates that the cluster algorithm has found a strong structure in the embedded log lines. The weakest structure has been found in the Forensic log. We
hypothesize that the high signature-to-log-line ratio in this log causes the lower
Silhouette score.

Finding the optimal number of clusters for a clustering or signature extraction
approach is a well-known problem. We do not address the topic of finding the
optimal number of signatures in this paper, but it is a fundamental research
topic in many methods for finding the optimal number of clusters have been
proposed, for example [18,21].

## 4   Related Work

Log signature extraction has been studied to achieve a variety of goals such
as anomaly and fault detection in logs [5], pattern detection [1,12,23], profile
building [23], or compression of logs [12,20].

Most of the approaches use word-position or word-frequency based heuristics
to extract signatures from logs. Tang et al. propose to use frequent word-bigrams
to obtain signatures [20]. Fu et al. propose to use a weighted word-edit distance
function to extract signatures [5]. Makanju et al. use the log line length as well as
word frequencies to extract signatures [11]. Vaarandi et al. use word frequencies
and word correlation scores to determine the fixed parts of log lines and thereby
the signatures [23]. Xu et al. propose a method that is not base on statistical

features of the log lines. Instead, they propose to create to extract the signatures from the source code [26].

Recently, RNN sequence-to-sequence models have been successfully applied for neural language modeling and statistical machine translation tasks [2,3,19]. Apart from that, Johnson et al. demonstrated on a large scale that sequence-to-sequence models can be used to allow translation between languages even if explicit training data from source to target language is not available [9].

Auto-encoders have been successfully applied to clustering tasks, such as clustering text and images [25]. Variational recurrent auto-encoders have been used to cluster music snippets [4].

## 5  Conclusion and Future Work

We have presented the LSTM-AE+C a method for clustering forensic logs according to their log signatures. Knowing that log lines belong to the same signature enables a forensic investigator to run more sophisticated analysis on a forensic log, for example, to reconstruct security incidents. Our method uses two components: an LSTM encoder and a hierarchical clustering algorithm. The LSTM encoder is trained as part of an auto-encoder on a log in an unsupervised fashion, and then the clustering algorithm assigns embedded log lines to their signature.

Experiments on three different datasets show that this method outperforms two state-of-the-art algorithms on clustering log lines based on their signatures both in V-Measure and adjusted mutual information. Moreover, we find that the Silhouette score of all found clusters by our method are greater than 0.70, which indicates strongly structured clusters.

One potential way of improving this method is to add a regularization term that aids the auto-encoder in embedding the clustering. Adding a regularization term could be a possible way to inject domain knowledge in the learning process and therefore increase the quality of the learned representation. For example, one could penalize the reconstruction loss of likely variable parts such as memory addresses, numbers or dates less.

Furthermore, we intend to investigate whether the attention mechanism of attentive LSTMs could be used to identify mutable and fixed parts of log lines. Finally, another future direction to our approach is to extract signatures that are human-interpretable. One potential way of addressing this is by using the decoder network to sample log lines from the embedding space.

**Acknowledgment.** This work has been partially funded by the Dutch national program COMMIT under the Big Data Veracity project.

# References

1. Aharon, M., Barash, G., Cohen, I., Mordechai, E.: One graph is worth a thousand logs: uncovering hidden structures in massive system event logs. In: Buntine, W., Grobelnik, M., Mladenić, D., Shawe-Taylor, J. (eds.) ECML PKDD 2009. LNCS (LNAI), vol. 5781, pp. 227–243. Springer, Heidelberg (2009). https://doi.org/10.1007/978-3-642-04180-8_32
2. Cho, K., van Merrienboer, B., Gulcehre, C., Bahdanau, D., Bougares, F., Schwenk, H., Bengio, Y.: Learning phrase representations using RNN encoder-decoder for statistical machine translation. In: Proceedings of the 2014 Conference on Empirical Methods in Natural Language Processing (EMNLP), pp. 1724–1734 (2014). http://arxiv.org/abs/1406.1078
3. Bahdana, D., Bahdanau, D., Cho, K., Bengio, Y.: Neural machine translation by jointly learning to align and translate. In: ICLR 2015, pp. 1–15 (2014). http://arxiv.org/abs/1409.0473v3
4. Fabius, O., van Amersfoort, J.R.: Variational recurrent auto-encoders. arXiv preprint arXiv:1412.6581 (2014)
5. Fu, Q., Lou, J.g., Wang, Y., Li, J.: Execution anomaly detection in distributed systems through unstructured log analysis. In: ICDM, vol. 9, pp. 149–158 (2009)
6. He, P., Zhu, J., He, S., Li, J., Lyu, M.R.: An evaluation study on log parsing and its use in log mining. In: Proceedings - 46th Annual IEEE/IFIP International Conference on Dependable Systems and Networks (DSN 2016), pp. 654–661 (2016)
7. Hochreiter, S., Schmidhuber, J.: Long short-term memory. Neural Comput. **9**(8), 1735–1780 (1997)
8. Jean, S., Cho, K., Memisevic, R., Bengio, Y.: On using very large target vocabulary for neural machine translation (2014). http://arxiv.org/abs/1412.2007
9. Johnson, M., Schuster, M., Le, Q.V., Krikun, M., Wu, Y., Chen, Z., Thorat, N., Viégas, F., Wattenberg, M., Corrado, G., et al.: Google's multilingual neural machine translation system: enabling zero-shot translation. arXiv preprint arXiv:1611.04558 (2016)
10. Lange, T., Roth, V., Braun, M.L., Buhmann, J.M.: Stability-based validation of clustering solutions. Neural Comput. **16**(6), 1299–1323 (2004)
11. Makanju, A., Zincir-Heywood, A.N., Milios, E.E.: A lightweight algorithm for message type extraction in system application logs. IEEE Trans. Knowl. Data Eng. **24**(11), 1921–1936 (2012)
12. Makanju, A.A.O., Zincir-Heywood, A.N., Milios, E.E.: Clustering event logs using iterative partitioning. In: Proceedings of the 15th ACM SIGKDD International Conference on Knowledge Discovery and Data Mining (KDD 2009), p. 1255. ACM (2009)
13. Oliner, A.J., Stearley, J.: What supercomputers say: a study of five system logs today's menu motivation data seven insights recommendations. In: DSN, pp. 575–584. IEEE (2007)
14. Pascanu, R., Mikolov, T., Bengio, Y.: On the difficulty of training recurrent neural networks. In: ICML, vol. 28, no. 3, pp. 1310–1318 (2013)
15. Rosenberg, A., Hirschberg, J.: V-Measure: a conditional entropy-based external cluster evaluation measure. In: EMNLP-CoNLL, vol. 7, pp. 410–420 (2007)
16. Rousseeuw, P.J.: Silhouettes: a graphical aid to the interpretation and validation of cluster analysis. J. Comput. Appl. Math. **20**, 53–65 (1987)
17. Srivastava, N., Hinton, G.E., Krizhevsky, A., Sutskever, I., Salakhutdinov, R.: Dropout: a simple way to prevent neural networks from overfitting. J. Mach. Learn. Res. **15**(1), 1929–1958 (2014)

18. Sugar, C.A., James, G.M.: Finding the number of clusters in a dataset: an information-theoretic approach. J. Am. Statist. Assoc. **98**(463), 750–763 (2003)
19. Sutskever, I., Vinyals, O., Le, Q.V.: Sequence to sequence learning with neural networks. In: NIPS, pp. 1–9 (2014)
20. Tang, L., Li, T., Perng, C.S.: LogSig: generating system events from raw textual logs. In: CIKM, pp. 785–794. ACM (2011)
21. Tibshirani, R., Walther, G., Hastie, T.: Estimating the number of clusters in a data set via the gap statistic. J. R. Statist. Soc. Ser. B (Statist. Methodol.) **63**(2), 411–423 (2001)
22. Tieleman, T., Hinton, G.: Lecture 6.5-rmsprop: divide the gradient by a running average of its recent magnitude. COURSERA Neural Netw. Mach. Learn. **4**(2), 26–31 (2012)
23. Vaarandi, R., Pihelgas, M.: LogCluster - a data clustering and pattern mining algorithm for event logs. In: 12th International Conference on Network and Service Management (CNSM 2015), pp. 1–8. IEEE Computer Society (2015)
24. Vinh, N.X., Epps, J., Bailey, J.: Information theoretic measures for clusterings comparison: variants, properties, normalization and correction for chance. J. Mach. Learn. Res. **11**, 2837–2854 (2010)
25. Xie, J., Girshick, R., Farhadi, A.: Unsupervised deep embedding for clustering analysis. arXiv preprint arXiv:1511.06335 (2015)
26. Xu, W., Huang, L., Fox, A., Patterson, D., Jordan, M.I., Huang, L., Fox, A., Patterson, D., Jordan, M.I.: Detecting large-scale system problems by mining console logs. In: 22nd ACM Symposium on Operating Systems Principles, pp. 117–131. ACM (2009)
27. Zhang, T., Ramakrishnan, R., Livny, M.: BIRCH: an efficient data clustering method for very large databases. In: ACM SIGMOD Record, vol. 25, pp. 103–114. ACM (1996)

# Urban Water Flow and Water Level Prediction Based on Deep Learning

Haytham Assem[1]([⊠]), Salem Ghariba[2], Gabor Makrai[3], Paul Johnston[4], Laurence Gill[4], and Francesco Pilla[2]

[1] Cognitive Computing Group, Innovation Exchange, IBM, Dublin, Ireland
haythama@ie.ibm.com
[2] Department of Planning and Environmental Policy,
University College Dublin, Dublin, Ireland
{salem.ghariba,francesco.pilla}@ucd.ie
[3] York Centre for Complex Systems Analysis (YCCSA),
University of York, Heslington, York, UK
gabor.makrai@york.ac.uk
[4] Department of Civil, Structural, and Environmental Engineering,
Trinity College Dublin, Dublin, Ireland
{pjhnston,laurence.gill}@tcd.ie

**Abstract.** The future planning, management and prediction of water demand and usage should be preceded by long-term variation analysis for related parameters in order to enhance the process of developing new scenarios whether for surface-water or ground-water resources. This paper aims to provide an appropriate methodology for long-term prediction for the water flow and water level parameters of the Shannon river in Ireland over a 30-year period from 1983–2013 through a framework that is composed of three phases: city wide scale analytics, data fusion, and domain knowledge data analytics phase which is the main focus of the paper that employs a machine learning model based on deep convolutional neural networks (DeepCNNs). We test our proposed deep learning model on three different water stations across the Shannon river and show it out-performs four well-known time-series forecasting models. We finally show how the proposed model simulate the predicted water flow and water level from 2013–2080. Our proposed solution can be very useful for the water authorities for better planning the future allocation of water resources among competing users such as agriculture, demotic and power stations. In addition, it can be used for capturing abnormalities by setting and comparing thresholds to the predicted water flow and water level.

**Keywords:** Deep learning · Water management
Convolutional neural networks · Urban computing

## 1 Introduction

Simulating and forecasting the daily time step for the hydrological parameters especially daily water flow (streamflow) and water level with sort of high

© Springer International Publishing AG 2017
Y. Altun et al. (Eds.): ECML PKDD 2017, Part III, LNAI 10536, pp. 317–329, 2017.
https://doi.org/10.1007/978-3-319-71273-4_26

accuracy on the catchment scale is a key role in the management process of water resource systems. Reliable models and projections can be hugely used as a tool by water authorities in the future allocation of the water resource among competing users such as agriculture, demotic and power stations. Catchment characteristics are important aspects in any hydrological forecasting and modeling process. The performance of modeling and projection methods for single hydrometric station varies according to its catchment climatic zone and characteristics. Karran et al. [11] state that methods that are proven as effective for modeling streamflow in the water abundant regions might be unusable for the dryer catchments, where water scarcity is a reality due to the intermittent nature of streams. Climate characteristics may severely affect the performance of different forecasting methods in different catchments and this area of research still requires much more exploration. The understanding of streamflow and water level dynamics is very important, which is described by various physical mechanisms occurring on a wide range of temporal and spatial scales [20]. Simulating these mechanisms and relations can be executed by physical, conceptual or data-driven models. However physical and conceptual models are the only current ways for providing physical interpretations and illustrations into catchment-scale processes, they have been criticized for being difficult to implement for high-resolution time-scale prediction, in addition to the need too many different types of data sets, which are usually very difficult to obtain. In general, physical and conceptual models are very difficult to run and the more resolution they have, the more data they need, which leads to over parametrize complex models [1].

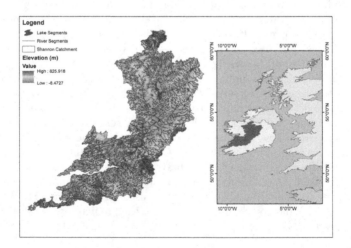

**Fig. 1.** Shannon river catchments and segments.

In this paper, we introduce a water management framework for the aim of providing insights of how to better allocate water resources by providing a highly accurate forecasting model based on deep convolutional neural networks (termed as DeepCNNs in the rest of the paper) for predicting the water flow and water level for the Shannon river in Ireland, the longest river in Ireland at 360.5 km. It drains the Shannon River Basin which has an area of 16,865 km$^2$, one fifth of the area of Ireland. Figure 1 shows Shannon river segments and catchments across Ireland. To the best of our knowledge, this paper is the first to explore and show the effectiveness of the deep learning models in the hydrology domain for long-term projections by employing deep convolutional network model and comparing its performance and showing that it out-performs other well-known time series forecasting models. We organize the paper as follows: Sect. 2 reviews the related work and identify our exact contribution with respect to the state-of-the-art. Section 3 introduces our proposed framework for water management. Section 4 presents the proposed architecture of the deep convolutional neural networks. Section 5 describes the experiments illustrates our results. Finally, we conclude the paper in Sect. 6.

## 2   Related Work

Artificial Neural Networks (ANNs) have been used in hydrology in many applications such as water flow (stream flow) modeling, water quality assessment and suspended sediment load predictions. The first uses for ANN in hydrology is introduced initially in the early 1990s [3], which find the method useful for forecasting process in the hydrological application. The ANN then has been used in many hydrological applications to confirm the usefulness and to model different hydrological parameters, as stream flow. The multi-layer perceptron (MLP) ANN models seem to be the most used ANN algorithms, which are optimized with a back-propagation algorithm, these models are improved the short-term hydrological forecasts. Examples of recent remarkable published applications for the use of ANN in hydrology are as follows [2,12]. Support Vector Machines (SVMs) have been recently adapted to the hydrology applications that is firstly used in 2006 by Khan and Coulibaly in [13], who state that SVR model out performs MLP ANNs in 3–12 month water levels predictions of a lake, then the use of SVM in hydrology has been promoted and recommended in many studies' as described in [4] from the use of flood stages, storm surge prediction, stream flow modeling to even daily evapotranspiration estimations.

The limited ability to process the non-stationary data is the biggest concern of the machine learning techniques applied to the hydrology domain, which leads to the recent application of hybrid models, where the input data are preprocessed for non-stationary characteristics first and then run through the post processing machine learning models to deal with the non-linearity issues. Wavelet transformation combined with machine learning models has been proven to give highly accurate and reliable short-term projections. The most popular hybrid model is the wavelet transform coupled with an artificial neural network (WANN). Kim and Valdés [14] is one of the first hydrological applications of the WANN model, which address the area of forecasting drought in the Conchos River Basin, Mexico, then many following published studies provide the application of WANN in streamflow forecasting and many research areas in hydrological modeling and prediction. In general, all the studies that compare between the ANN and WANN conclude that the WANN models have outperformed the stand alone ANNs [3]. Furthermore, wavelet transform coupled with SVM/SVR (WSVM/WSVR) has been proposed to be used in hydrology applications. To the best of our knowledge, there is a very little research into the application of this hybrid model for streamflow forecasting and there is no application on water level forecasting.

Karran et al. [11] compares the use of four different models, artificial neural networks (ANNs), support vector regression (SVR), wavelet-ANN, and wavelet-SVR for one single station in each watershed of Mediterranean, Oceanic, and Hemiboreal watershed, the results show that SVR based models performed best overall. Kisi et al. [16] have applied the WSVR models with different methods to model monthly streamflow and find that the WSVR models outperformed the stand alone SVR. From the previous state-of-the-art work, we have concluded that the previous mentioned machine learning models (ANNs, SVMs, WANNS, and WSVMs) are the most well-studied and well-known in the field of hydrology. Hence, we build in this paper four baselines employing the previous mentioned models for having a fair comparison for our proposed deep convolutional neural networks across three various water stations. To the best of our knowledge, this paper is the first to adapt Deep Learning technique in the hydrology domain and showing better accuracy across three water stations compared to state-of-the-art models used in the hydrology applications.

## 3    Water Management Framework

In this section, we summarize the three phases for the proposed framework for predicting water flow and water level through multistage analytics process. **(a) City wide scale data analytics:** This phase is composed mainly of two steps, the first step utilize the dynamically spatial distributed water balance model integrating the climate and land use changes. This stage use a wide range of input parameters and grids including seasonally climate variables and changes, land use and its seasonal parameters and future changes, seasonal groundwater depth, soil properties, topography, and slope. The output of this step is several parameters including runoff, recharge, interception, evapotranspiration, soil evaporation, transpiration including total uncertainties or error in the water balance. We utilize *runoff* from this step as an extracted feature to be passed to the data storage (please refer to [6] for the description of the used model). In the second step, we gathered the data for the *temp-max* and *temp-min* from Met Eireann[1] from 1983–2013, the national meteorological service in Ireland. We further simulated the future temperatures from 2013–2080 using statistical down scaling model as described in [7]. **(b) Data Fusion:** In this phase, we follow a stage-based fusion method [22] in which we fused the features extracted from the previous stage with the two observed outputs for water flow and water level from 1983–2013. Furthermore, we normalize and scale the data and store it in a

---

[1] http://www.met.ie.

data-storage for further being processed by the next phase. **(c) Domain knowledge data analytics:** This phase is our main focus for the paper in which we
consume the features stored in the data-storage and train our proposed model
along with the baseline models for the aim of predicting water flow and water
level across three different water stations.

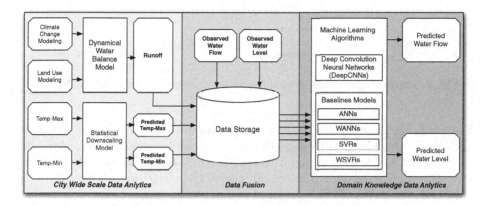

**Fig. 2.** Water management framework.

## 4   Deep Convolutional Neural Networks

In order to design an effective forecasting model for predicting water flow and
water level across several years, we needed to exploit the time series nature of the
data. Intuitively, analyzing the data over a sufficient wide time interval rather
than only including the last reading would potentially lead to more information
for the future water flow and water level. A first approach is that we concatenate
various data samples together and feed them to a machine learning model, this
is what we did in the baseline models which boosts the performance achieved.
To achieve further improvements, we make use of the adequacy of convolutional
neural networks for such type of data [18]. We propose the following architecture, each input sample consists of 10 consecutive readings concatenated together
(10 worked best on our datasets). Each of the three input features (Temp-max,
Temp-min, and Run-off) is fed to the network to a separate channel. The resulting dataset is a tensor of $N \times T \times D$ dimensions, where $N$ is the number of data
points (the total number of records minus the number of concatenated readings).
$T$ is the length of the concatenated strings of events and $D$ is the number of collected features. Each of the resulting tensor records, of dimensionality $1 \times T \times D$
is processed by a stack of convolution layers as shown in Fig. 3.

The first convolution layer utilizes a set of three-channel convolution filters of
size $l$. We do not employ any pooling mechanisms since the dimensionality of the
data is relatively low. In addition, zero padding was used for preserving the input
data dimensionality. Each of these filters provides a vector of length 10, each of its

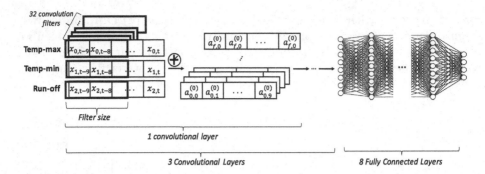

**Fig. 3.** The proposed convolutional neural network architecture (DeepCNNs).

elements further goes to non linear transformation using ReLu [19] as a transfer function. The resulting outputs are further processed by another similar layers of convolutional layers, with as many channels as convolution filters in the previous layer. Given an input record $x$, we can therefore definer the entries output by filter $f$ of convolution layer $l$ at position $i$ as shown in Eq. 1. Finally, the last convolution layer is flattened and further processed through a feedforward fully connected layers.

$$a_{f,i}^{(l)} = \begin{cases} \phi(\sum_{j=0}^{2} \sum_{k=0}^{c-1} w_{fjk}^{(l)} x_{j,i+k-c/2} + b_{fl}), & \text{if } l = 0 \\ \phi(\sum_{j=0}^{n(l-1)-1} \sum_{k=0}^{c-1} w_{fjk}^{(l)} a_{j,i+k-c/2}^{(l-1)} + b_{fl}), & \text{otherwise} \end{cases} \quad (1)$$

where $\phi$ is the non-linear activation function. $x_{j,i}$ is the value of a channel (which corresponds to a feature) $j$ at position $i$ of the input record (if $i$ is negative or greater than 10, then $x_{j,i} = 0$). $w_{fjk}^{(l)}$ is the value of channel $j$ of convolution filter $f$ of layer $l$ at position $k$, and $b_{fl}$ is the bias of filter $f$ at layer $l$. $n(l)$ is the number of convolutions filters at layer $l$.

## 5    Experiments

In this section, we first describe the dataset used in our experiments, then we give an overview on the used baseline models, and finally we show our results discussing the key findings and observations.

### 5.1    Dataset

Following the procedures described in Fig. 2, the resulted datasets stored in the data storage are comprised of five parameters named, max-temp, min-temp, run-off, water flow and water level where the first three represents the featuresof the trained models while the later two represents the outputs of the models. The used parameters can be defined as follows:

- **max-temp, min-temp:** These are the highest and lowest temperatures recorded in °C during each day in the dataset.
- **run-off:** Runoff is described as the part of the water cycle that flows over land as surface water instead of being absorbed into groundwater or evaporating and is measured in $mm$.
- **water flow:** Water flow (streamflow) is the volume of water that moves through a specific point in a stream during a given period of time (one day in our case) and is measured in $m^3/sec$.
- **water level:** This parameter indicates the maximum height reached by the water in the river during the day and is measured in $m$.

The previous parameters in the dataset are for 30-years (1983–2013) resulting in 11,392 samples where each sample represents a day. The datasets formulated related to three different water hydrometric stations named, Inny, lower-shannon, and suck.

## 5.2   Baselines

In this section we describe the baseline models that have been developed for assessing the performance of the proposed deep convolutional neural network. We choose two very popular ordinary machine learning algorithms that has already shown success in hyrdology, Artificial Neural Networks (ANNs) [8] and Support Vector Machines (SVMs) [4]. In addition, we choose two wavelet transformation models that have shown stable outcomes and in particular for the time-series forecasting problems, Wavelet-ANNs (WANNs) and Wavelet-SVMs (WSVMs).

- **ANNs:** We developed three layer feed-forward neural network employing backpropogation algorithm. An automated RapidMinder algorithm proposed in [17] is utilized for optimizing the number of neurons in the hidden layer with setting the number of epochs to 500, learning rate to 0.1 and momentum to 0.1 as well.
- **SVMs:** We developed SVM with non-linear dot kernel which requires two parameters to be configured by the user, namely cost ($C$) and epsilon ($\epsilon$). We set $C$ to 0.0001 and $\epsilon$ to 0.001. The selected combination was adjusted to the most precision that could be acquired through a trial and error process for a more localized optimization for the model parameters.

We used Discrete Wavelet Transforms (DWTs) to decompose the original time series into a time-frequency representation at different scales (wavelet sub-times series). In this type of baselines, we set the level of decomposition to 3, two levels of details and one level of approximations. The signals were decomposed using the redundant trous algorithm [5] in conjunction with the non-symmetric db1 wavelet as the mother function[2]. Three sets of wavelet sub-time series were

---

[2] The using of the átrous algorithm with the db1 wavelet mother function is a result of the optimizing Python Wavelet tool [9].

created, including a low-frequency component (Approximation) that uncovers the signal's trend, and two sets of high-frequency components (Details). The original signal is always recreated by summing the details with the smoothest approximation of the signal. All the input time series are gone through the designed wavelet transform and the resulted sub datasets have been used by the following models:

– **WANNs:** The decomposed time series are fed to the ANN method for the prediction of water flow and water level for one day ahead. The WANNs model employs discrete wavelet transform to overcome the difficulties associated with the conventional ANN model, as the wavelet transform is known to overcome the non-stationary properties of time series.
– **WSVMs:** The WSVR are built in the same way as the WANN model.

## 5.3  Results and Discussion

We followed the previous design for the convolutional neural network in which we performed a random grid search of the hyperparameter space and choose the best performing set. We found that the best performing model is composed of 3 convolutional layers, each of which learns 32 convolution patches of width 5 employing zero padding. After the convolutional layers, we employed 8 fully stacked connected layers. The convolutional layers are regularized using dropout technique [21] with a probability of 0.2 for dropping units. All dense layers employ L2 regularization with $\lambda = 0.000025$. All layers are batch normalized [10] and use ReLu units [19] for activation with an exception for the output layer because it is a regression problem and we are interested in predicting numerical values directly without transform. The efficient ADAM optimization algorithm [15] is used and a mean squared error loss function is optimized with a minibatches of size 10. We set aside 30% from the whole data for testing the performance of the trained model while the 70% rest of the data act as the training dataset. From the training dataset, we select 90% for training each model, and the remaining 10% as the validation set, we used it to export the best model if any improvements on the validation score, we continue the whole process for 200 epochs. In addition, we reduce the learning rate by a factor of 2 once learning stagnates for 20 consecutive epochs. Figures 4, 5 and 6 show the output of the previous training process for the Inny, lower-shannon, and suck water stations respectively where the $x$ axis represents the daily time steps while $y$ axis indicates the output whether it is the water flow (streamflow) or water level. The blue line in the figures indicates the original dataset (ground truth), the green indicates the output of the model on the training dataset, while the red line indicates the output of the model on the test data in which it has not been exposed at all to the model during the training procedures.

We compare our proposed model with the other baseline models described in the previous section. We use the following three evaluation metrics for our comparisons: (a) *Root-mean-square error (RMSE)*: is the most frequently used

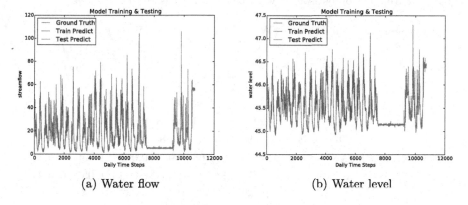

(a) Water flow                          (b) Water level

**Fig. 4.** Inny water station.

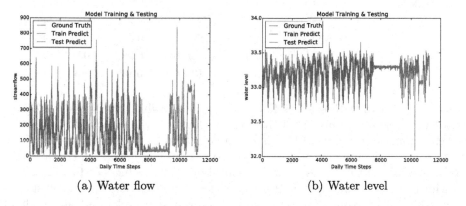

(a) Water flow                          (b) Water level

**Fig. 5.** Lower-shannon water station.

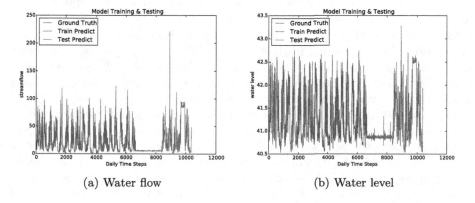

(a) Water flow                          (b) Water level

**Fig. 6.** Suck water station.

**Table 1.** Comparison between baseline models and DeepCNNs.

| Water station | Model | Water flow | | | Water level | | |
|---|---|---|---|---|---|---|---|
| | | RMSE | MAE | $R^2$ | RMSE | MAE | $R^2$ |
| Inny | ANNs | 2.721 | 1.249 | 0.977 | 0.061 | 0.025 | 0.982 |
| | SVMs | 2.712 | 0.956 | 0.977 | 0.06 | 0.022 | 0.983 |
| | WANNs | 2.785 | 1.389 | 0.977 | 0.061 | 0.026 | 0.982 |
| | WSVMs | 2.673 | 0.933 | 0.978 | 0.06 | 0.023 | 0.983 |
| | **DeepCNNs** | **2.14** | **0.92** | **0.98** | **0.05** | **0.02** | **0.99** |
| Lower-Shannon | ANNs | 27.1 | 16.665 | 0.974 | 0.063 | 0.039 | 0.853 |
| | SVMs | 29.782 | 18.191 | 0.969 | 0.066 | 0.037 | 0.842 |
| | WANNs | 27.335 | 16.622 | 0.973 | 0.063 | 0.038 | 0.854 |
| | WSVMs | 30.715 | 19.89 | 0.968 | 0.065 | 0.036 | 0.842 |
| | **DeepCNNs** | **22.30** | **13.43** | **0.98** | **0.05** | **0.03** | **0.87** |
| Suck | ANNs | 4.25 | 2.09 | 0.985 | 0.08 | 0.042 | 0.986 |
| | SVMs | 3.831 | 1.783 | 0.987 | 0.079 | 0.031 | 0.986 |
| | WANNs | 4.252 | 1.954 | 0.985 | 0.079 | 0.039 | 0.986 |
| | WSVMs | 4.075 | 1.469 | 0.985 | 0.08 | 0.031 | 0.986 |
| | **DeepCNNs** | **3.46** | **1.43** | **0.99** | **0.06** | **0.03** | **0.99** |

metric for assessing time-series forecasting models which measures the differences between values predicted by a model and the values actually observed. (b) *Mean absolute error (MAE)*: is a quantity used to measure how close forecasts or predictions are to the eventual outcomes. (c) *Coefficient of determination* $(R^2)$: is a metric that gives an indication about the goodness of fit of a model in which a closer value to 1 indicates a better fitted model. Table 1 illustrates the results of the comparisons between our proposed model and all baselines across the previous described three performance metrics for the three different water stations. Interestingly, we noticed that the proposed deep convolutional neural network model outperforms all baselines across the three different performance metrics. This suggests that predicting water flow and water level in rivers manifests itself in a complex fashion, and motivates further research in the application of deep learning methods to the water management domain. From such comparison, it is observed as well that SVMs is the second best performing model for the Inny and Suck water stations. ANNs is the second best performing model for the lower-shannon water station for both outputs.

Finally, and based on the forecasted/simulated values for the features (Temp-max, Temp-min and run-off) from 2013–2080, we show in Fig. 7a and b the prediction for water flow and water level respectively for the lower-shannon station employing our trained model based on the DeepCNNs. Based on the predictions by our proposed model, it is worth noting from Fig. 7a that there will be a significant increase in the water flow crossing 250 m³ in several days across 2028, 2040 and 2059 while a less but still significant increase across several days in 2047, 2048, 2076 and 2078. It could be observed as well from Fig. 7b that there will be a significant rise of water level crossing 33.4 m in several days in 2021 and bit less in 2032, 2044, 2045 and others as well. These results should be very useful for further being assessed by water authorities for building mitigation plans for the impact of such increase as well as better planning for water allocation across various competing users.

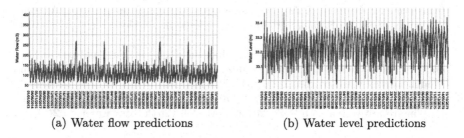

(a) Water flow predictions    (b) Water level predictions

**Fig. 7.** Predictions of water flow and water level for lower-shannon water station from 2013–2080 using the DeepCNNs proposed model.

## 6    Conclusion and Outlook

This paper presents the application of a new data-driven methods for modeling and predicting daily water flow and water level on the catchment scale for the Shannon river in Ireland. We have designed a deep convolutional network architecture to exploit the time-series nature of the data. Using several features captured at real across three various water stations, we have shown that the proposed convolutional network outperforms other four well-known time series forecasting models (ANNs, SVMs, WANNs and WSVMs). The inputs to the models consist of a combination of 30-years daily time series data sets (1983–2013), which can be divided between observed data sets (maximum temperature, minimum temperature, water level and water flow) and simulated data set, runoff. Based on the proposed deep convolutional network model, we further show the predictions of the water flow and water level for the lower-shannon water station from the duration of 2013–2080. Our proposed solution should be very useful for water authorities in the future allocation of water resources among competing users such as agriculture, demotic and power stations. In addition, it could formulate the basis of a decision support system by setting thresholds on water flow and water level predictions for the sake of creating accurate emergency alarms for capturing any expected abnormalities for the Shannon river.

## References

1. Beven, K.J., et al.: Streamflow Generation Processes. IAHS Press, Wallingford (2006)
2. Chattopadhyay, P.B., Rangarajan, R.: Application of ann in sketching spatial non-linearity of unconfined aquifer in agricultural basin. Agric. Water Manag. **133**, 81–91 (2014)

3. Daniell, T.: Neural networks. Applications in hydrology and water resources engineering. In: National Conference Publication - Institute of Engineers. Australia (1991)
4. Deka, P.C., et al.: Support vector machine applications in the field of hydrology: a review. Appl. Soft Comput. **19**, 372–386 (2014)
5. Dutilleux, P.: An implementation of the algorithme àtrous to compute the wavelet transform. In: Combes, J.M., Grossmann, A., Tchamitchian, P. (eds.) Wavelets. IPTI, pp. 298–304. Springer, Heidelberg (1989). https://doi.org/10.1007/978-3-642-75988-8_29
6. Gharbia, S.S., Alfatah, S.A., Gill, L., Johnston, P., Pilla, F.: Land use scenarios and projections simulation using an integrated gis cellular automata algorithms. Model. Earth Syst. Environ. **2**(3), 151 (2016)
7. Gharbia, S.S., Gill, L., Johnston, P., Pilla, F.: Multi-GCM ensembles performance for climate projection on a GIS platform. Model. Earth Syst. Environ. **2**(2), 1–21 (2016)
8. Govindaraju, R.S., Rao, A.R.: Artificial Neural Networks in Hydrology, vol. 36. Springer Science & Business Media, Heidelberg (2013). https://doi.org/10.1007/978-94-015-9341-0
9. Hanke, M., Halchenko, Y.O., Sederberg, P.B., Hanson, S.J., Haxby, J.V., Pollmann, S.: PyMVPA: a python toolbox for multivariate pattern analysis of FMRI data. Neuroinformatics **7**(1), 37–53 (2009)
10. Ioffe, S., Szegedy, C.: Batch normalization: accelerating deep network training by reducing internal covariate shift. arXiv preprint arXiv:1502.03167 (2015)
11. Karran, D.J., Morin, E., Adamowski, J.: Multi-step streamflow forecasting using data-driven non-linear methods in contrasting climate regimes. J. Hydroinformatics **16**(3), 671–689 (2014)
12. Kenabatho, P., Parida, B., Moalafhi, D., Segosebe, T.: Analysis of rainfall and large-scale predictors using a stochastic model and artificial neural network for hydrological applications in Southern Africa. Hydrol. Sci. J. **60**(11), 1943–1955 (2015)
13. Khan, M.S., Coulibaly, P.: Application of support vector machine in lake water level prediction. J. Hydrol. Eng. **11**(3), 199–205 (2006)
14. Kim, T.W., Valdés, J.B.: Nonlinear model for drought forecasting based on a conjunction of wavelet transforms and neural networks. J. Hydrol. Eng. **8**(6), 319–328 (2003)
15. Kingma, D., Ba, J.: ADAM: a method for stochastic optimization. arXiv preprint arXiv:1412.6980 (2014)
16. Kisi, O., Cimen, M.: A wavelet-support vector machine conjunction model for monthly streamflow forecasting. J. Hydrol. **399**(1), 132–140 (2011)
17. Klinkenberg, R.: RapidMiner: Data Mining Use Cases and Business Analytics Applications. Chapman and Hall/CRC, Boca Raton (2013)
18. LeCun, Y., Bengio, Y., et al.: Convolutional networks for images, speech, and time series. In: The Handbook of Brain Theory and Neural Networks, vol. 3361, No. 10 (1995)
19. Nair, V., Hinton, G.E.: Rectified linear units improve restricted Boltzmann machines. In: Proceedings of the 27th International Conference on Machine Learning (ICML 2010), pp. 807–814 (2010)
20. Sivakumar, B.: Forecasting monthly streamflow dynamics in the Western United States: a nonlinear dynamical approach. Environ. Model. Softw. **18**(8), 721–728 (2003)

21. Srivastava, N., Hinton, G.E., Krizhevsky, A., Sutskever, I., Salakhutdinov, R.: Dropout: a simple way to prevent neural networks from overfitting. J. Mach. Learn. Res. **15**(1), 1929–1958 (2014)
22. Zheng, Y.: Methodologies for cross-domain data fusion: an overview. IEEE Trans. Big Data **1**(1), 16–34 (2015)

# Using Machine Learning for Labour Market Intelligence

Roberto Boselli[1,2], Mirko Cesarini[1,2], Fabio Mercorio[1,2(✉)],
and Mario Mezzanzanica[1,2]

[1] Department of Statistics and Quantitative Methods, University of Milano-Bicocca,
Milan, Italy
fabio.mercorio@unimib.it
[2] CRISP Research Centre, University of Milano-Bicocca, Milan, Italy

**Abstract.** The rapid growth of Web usage for advertising job positions provides a great opportunity for real-time labour market monitoring. This is the aim of Labour Market Intelligence (LMI), a field that is becoming increasingly relevant to EU Labour Market policies design and evaluation. The analysis of Web job vacancies, indeed, represents a competitive advantage to labour market stakeholders with respect to classical survey-based analyses, as it allows for reducing the time-to-market of the analysis by moving towards a fact-based decision making model. In this paper, we present our approach for automatically classifying million Web job vacancies on a standard taxonomy of occupations. We show how this problem has been expressed in terms of text classification via machine learning. Then, we provide details about the classification pipelines we evaluated and implemented, along with the outcomes of the validation activities. Finally, we discuss how machine learning contributed to the LMI needs of the European Organisation that supported the project.

**Keywords:** Machine learning · Text classification
Governmental application

## 1 Introduction

In recent years, the European Labour demand conveyed through specialised Web portals and services has grown exponentially. This also contributed to introduce the term "Labour Market Intelligence" (LMI), that refers to the use and design of AI algorithms and frameworks to analyse Labour Market Data for supporting decision making. This is the case of *Web job vacancies*, that are job advertisements containing two main text fields: a *title* and a *full description*. The title shortly summarises the job position, while the full description field usually includes the position details and the relevant skills that the employee should hold.

There is a growing interest in designing and implementing real LMI applications to Web Labour Market data for supporting the policy design and evaluation activities through evidence-based decision-making. In 2010 the European

© Springer International Publishing AG 2017
Y. Altun et al. (Eds.): ECML PKDD 2017, Part III, LNAI 10536, pp. 330–342, 2017.
https://doi.org/10.1007/978-3-319-71273-4_27

Commission has published the communication "A new impetus for European Cooperation in Vocational Education and Training (VET) to support the Europe 2020 strategy",[1] aimed at promoting education systems in general, and VET in particular. In 2016, the European Commission's highlighted the importance of Vocational and Educational activities, as they are "valued for fostering job-specific and transversal skills, facilitating the transition into employment and maintaining and updating the skills of the workforce according to sectorial, regional, and local needs".[2] In 2016, the EU and Eurostat launched the ESSnet Big Data project, involving 22 EU member states with the aim of "integrating big data in the regular production of official statistics, through pilots exploring the potential of selected big data sources and building concrete applications".

The rationale behind all these initiatives is that reasoning over Web job vacancies represents an added value for both *public and private* labour market operators to deeply understand the Labour Market dynamics, occupations, skills, and trends: (*i*) by reducing the time-to-market with respect to classical survey-based analyses (results of official Labour Market surveys actually require up to one year before being available); (*ii*) by overcoming the linguistic boundaries through the use of standard classification systems rather than proprietary ones; (*iii*) by representing the resulting knowledge over several dimensions (e.g., territory, sectors, contracts, etc.) at different level of granularity and (*iv*) by evaluating and comparing international labour markets to support fact-based decision making.

*Contribution.* In this paper we present our approach for classifying Web job vacancies, we designed and realised within a research call-for-tender[3] for the Cedefop EU organisation[4]. Specifically, the goal of this project was twofold: first, the evaluation of the effectiveness of using Web Job vacancies for LMI activities through a feasibility study, second, the realisation of a working prototype that collects and analyses Web job vacancies over 5 Countries (United Kingdom, Ireland, Czech Republic, Italy, and Germany) for obtaining near-real time labour market information. Here we focus on the classification task showing the performances achieved by three classification pipelines we evaluated for realising the system.

We begin by discussing related work in Sect. 2. In Sect. 3 we discuss how the problem of classifying Web job vacancies has been solved through machine learning, providing details on the feature extraction techniques used. Section 4

---

[2] The Commission Communication "A New Skills Agenda for Europe" COM (2016) 381/2, available at https://goo.gl/Shw7bI.

[3] "Real-time Labour Market information on skill requirements: feasibility study and working prototype". Cedefop Reference number AO/RPA/VKVET-NSOFRO/Real-time LMI/010/14. Contract notice 2014/S 141-252026 of 15/07/2014 https://goo.gl/qNjmrn.

[4] Cedefop European agency supports the development of European Vocational Education and Training (VET) policies and contributes to their implementation - http://www.cedefop.europa.eu/.

provides the experimental results about the evaluation of three distinct pipelines employed. Section 5 concludes the paper and describes the ongoing research.

## 2    Related Work

Labour Market Intelligence is an *emerging* cross-disciplinary field of studies that is gaining research interests in both *industrial* and *academic* communities.

*Industries.* Information extraction from unstructured texts in the labour market domain mainly focused on the e-recruitment process (see, e.g., [19]) attempting to support or automate the *resume* management by matching candidate profiles with job descriptions using machine learning approaches [11,30,32]. Concerning companies, their need to automatize Human Resource (HR) department activities is strong; as a consequence, a growing amount of commercial skill-matching products have been developed in the last years, for instance, Burning-Glass, Workday, Pluralsight, EmployInsight, and TextKernel. To date, the only commercial solution that uses international standard taxonomies is Janzz: a Web based platform to match labour demand and supply in both public and private sectors. It also provides APIs access to its knowledge base, but it is not aimed at classifying job vacancies. Worth of mentioning is Google Job Search API, a pay-as-you-go service announced in 2016 for classifying job vacancies through the Google Machine Learning service over O*NET, that is the US standard occupation taxonomy. Though this commercial service is still a closed alpha, it is quite promising and also sheds the light on the needs for reasoning with Web job vacancies using a common taxonomy as baseline.

*Literature.* Since the early 1990s, *text classification* (TC) has been an active research topic. It has been defined as "the activity of labelling natural language texts with thematic categories from a predefined set" [29]. Most popular techniques are based on the *machine learning* paradigm, according to which an automatic text classifier is created by using an inductive process able to learn, from a set of pre-classified documents, the characteristics of the categories of interest.

In the recent literature, text classification has proven to give good results in categorizing many real-life Web-based data such as, for instance, news and social media [15,33], and sentiment analysis [20,25]. To the best of our knowledge, text classifiers have not been applied yet to the classification of Web job vacancies published on several Web sites for analysing the Web job market of a geographical area, and this system is the first example in this direction.

All these approaches are quite relevant and effective, and they also make evidence of the importance of the Web for labour market information. Nonetheless, they differ from our approach in two aspects. First, we aim to classify *job vacancies* according to a target classification system for building a (language independent) knowledge base for analyses purposes, rather than matching resumes on job vacancies. Furthermore, resumes are usually accurately written by candidates whilst Web advertisements are written in a less accurate way, and this quality issue might have unpredictable effects on the information derived from them (see, e.g. [6,9,21,22] for practical applications). Second, the system aims at producing analyses based on standard taxonomies to support the fact-based decision making activities of several stakeholders.

# 3    Text Classification in LMI

*The Need for a Standard Occupations Taxonomy.* The use of proprietary and language-dependent taxonomies can prevent the effective monitoring and evaluation of Labour Market dynamics across national borders. For these reasons, a great effort has been made by International organisations for designing *standard* classifications systems, that would act as a lingua-franca for the Labour Market to overcome the linguistic boundaries as well. One of the most important classification system designed for this purposes is ISCO: The *International Standard Classification of Occupations* has been developed by the International Labour Organization as a four-level classification that represents a standardised way for organising the labour market occupations. In 2014, ISCO has been extended through ESCO: the multilingual classification system of European Skills, Competences, Qualifications and Occupations, that is emerging as the European standard for supporting the whole labour market intelligence over 24 EU languages. Basically, the ESCO data model includes the ISCO hierarchical structure as a whole and extends it through a taxonomy of skills, competences and qualifications.

## 3.1    The Classification Task

Text categorisation aims at assigning a Boolean value to each pair $(d_j, c_i) \in D \times C$ where $D$ is a set of documents and $C$ a set of predefined categories. A *true* value assigned to $(d_j, c_i)$ indicates document $d_j$ to be set under the category $c_i$, while a false value indicates $d_j$ cannot be assigned under $c_i$. In our scenario, we consider a set of job vacancies $\mathcal{J}$ as a collection of documents each of which has to be assigned to one (and only one) ISCO occupation code. We can model this problem as a text classification problem, relying on the definition of [29].

Formally speaking, let $\mathcal{J} = \{J_1, \ldots, J_n\}$ be a set of job vacancies, the classification of $\mathcal{J}$ under the ESCO classification system consists of $|O|$ independent problems of classifying each job vacancy $J \in \mathcal{J}$ under a given ESCO occupation code $o_i$ for $i = 1, \ldots, |O|$. Then, a *classifier* is a function $\psi : \mathcal{J} \times O \to \{0,1\}$ that approximates an unknown target function $\check{\psi} : \mathcal{J} \times O \to \{0,1\}$. Clearly, as we deal with a single-label classifier, $\forall j \in \mathcal{J}$ the following constraint must hold: $\sum_{o \in O} \psi(j, o) = 1$.

In this paper, job vacancies are classified according to the $4^{th}$ level of the ISCO taxonomy (and the corresponding multilingual concepts of the ESCO ontology) as further detailed in the next sections. The choice of the ISCO $4^{th}$ level (also referred as ISCO 4 digits classification) is a trade-off between the granularity of occupations (the more digits the better) and the effort to develop an automatic classifier (the fewer digits the better). The job vacancy classification is translated into a supervised machine learning text classification problem, namely a multiclass single label classification problem i.e., a job offer is classified to one and only one 4 digits ISCO code over a set of 436 available ones.

Within this project we decided to use *titles* for occupation classification. Indeed, in our very preliminary studies [2] we experimentally observed that titles are often concise and highly focused on describing the proposed occupations while other topics are hardly dealt, making titles suitable for the classification task.

### 3.2 Feature Extraction

Two feature extraction methods have been evaluated for classifying job occupation, namely: Bag of Word Approach, and Word2Vec, that we describe in the following.

**Bag of Word Feature Extraction.** Titles were pre-processed according to the following steps: (*i*) html tag removal, (*ii*) html entities and symbol replacement, (*iii*) tokenization, (*iv*) lower case reduction, (*v*) stop words removal (using the stop-words list provided by the NLTK framework [5]), (*vi*) stemming (using the Snowball stemmer), (*vii*) n-grams frequency computation (actually, unigram and bigram frequencies were computed, n-grams which appear less than 4 times or that appear in more than 30% of the documents are discarded, since they are not significant for classification). Each title is pre-processed according to the previous steps and is transformed into a set of n-gram frequencies.

**Word2Vec Feature Extraction.** Each word in a title was replaced by a corresponding vector of an n-dimensional space. We used a vector representation of words belonging to the family of neural language models [3] and specifically we used the Word2Vec [23,24] representation.

In neural language models, every word is mapped to a unique vector, given a word $w$ and its context $k$ (n words nearby $w$), the concatenation or sum of the vectors of the context words is then used as features for prediction of the word $w$ [24]. This problem can be viewed as a machine learning problem where $n$ context words are fed into a neural network that should be trained to predict the corresponding word, according to the Continuous Bag of Words (CBOW) model proposed in [23].

The word vector representations are the coefficient of the internal layers of the neural network, for more details the interested reader can refer to [24]. The word vectors are also called word embeddings.

After the training ends, words with similar meaning are mapped to a similar position in the vector space [23]. For example, "powerful" and "strong" are close to each other, whereas "powerful" and "Paris" are more distant. The word vector differences also carry meaning.

We used the GENSIM [27] implementation of Word2Vec to identify the vector representations of the words. Since Word2Vec requires huge text corpora for producing meaningful vectors, we used all the downloaded job vacancies to train the Word2Vec (the *unlabelled dataset*, about 6 Million of job vacancies, as outlined in Sect. 4.1). The 6 Million job vacancy texts underwent the steps from $(i)$ to $(vi)$ of the processing pipeline described in Subsect. 3.2 before being used for training the Word2Vec model. Actually, the Wod2Vec model was trained using vectors of size equals to 300 using the CBOW training algorithm.

The Word2Vec embeddings were used to process the titles of the labelled dataset introduced in Sect. 4.1 as follows: steps from $(i)$ to $(vi)$ of the processing pipeline described in Subsect. 3.2 were executed on titles. The first 15 tokens of titles where considered (i.e., tokens exceeding the $15^{th}$ were dropped, as the affected titles account for less than 0.2% of total vacancies). Each word in the title was replaced by the corresponding (word) vector, e.g., given a set of n titles each one composed by 15 words, the output of the substitution can be viewed as a 3-dimensional array (e.g., a 3-dimensional matrix or a 3-dimensional tensor) of the shape: [n_documents, 15, word_vector_dimension].

# 4    Experimental Results

This section introduces the evaluation performed on the several classification pipelines and the dataset used.

## 4.1    Datasets

Two datasets have been considered in the experiments outlined in this section:

**Labelled.** A set of 35,936 job vacancies manually labelled using 4 digits ISCO code. Not all the 4 digits ISCO occupations are present in the dataset, only 271 out of 436 ISCO codes were actually found. It is worth to mention that ISCO tries to categorize all possible occupations, but some are hardly found on the Web (e.g., 9624 *Water and firewood collectors*[5]). The interested reader can refer to [13] for further information.

**Unlabelled.** A set of 6,005,916 unlabelled vacancies. The vacancies have been collected for one year scraping 7 web sites focusing on the UK and Irish Job Market. For each vacancy both a title and a full description is available.

The *labelled* dataset was used to train a classifier to be used later to identify the ISCO occupations on the *unlabelled* vacancy dataset. The latter was used to compute the Word2Vec word embeddings.

In the following sections, the classification pipelines we have evaluated are introduced. For evaluation purposes, the labelled dataset was randomly split into train and test (sub)sets containing respectively 75% and 25% of the vacancies. The vacancies of each ISCO code were distributed in the two subsets using the same proportions.

---

[5] Tasks include cutting and collecting wood from forests for sale in market or for own consumption ... drawing water from wells, rivers or ponds, etc. for domestic use.

## 4.2   Classification Pipelines

This subsection will introduce the several classification pipelines which have been evaluated for classifying job vacancies. Each pipeline has parameters whose optimal values have been found performing a Grid Search as detailed in Sect. 4.3.

**BoW - SVM.** The BoW feature extraction pipeline (described in Sect. 3.2) was used on the (labelled) training dataset and the results were used to feed two classifiers, namely Linear SVM and Gaussian SVM, the latter also known as radial basis function (RBF) SVM kernel [8]. They will be called LinearSVM and RBF SVM hereafter.

According to [14], SVM is well suited to the particular properties of texts, namely high dimensional feature spaces, few irrelevant features (dense concept vector), and sparse instance vectors. The parameters evaluated during the grid search are $C \in \{0.01, 0.1, 1, 10, 100\}$ for LinearSVM classifier, while $C \in \{0.01, 0.1, 1\} \times$ Gamma $\in \{0.1, 1, 10\}$ for RBF SVM.

**BoW - Neural Network.** The BoW feature extraction pipeline (described in Sect. 3.2) was also used to feed the fully connected neural networks described below. Each neural network has an input of size 5,820 (the number of features produced by the feature extraction pipeline) and an output of size 271 (the number of ISCO codes in the training set). Each layer (if not otherwise specified) use the Linear Rectifier as non linearity, excluding the last layer which uses Softmax. In the networks described below each fully connected layer is preceded by a batch normalization layer [12], whose purpose is to accelerate the network training (and it doesn't have an effect on the classification performances).

- (FCNN1) is a 4 layer neural network having 2 hidden layers: a batch normalization and a fully connected layer of size 3,000.
- (FCNN2) is a 5 layer neural network, having 4 hidden layers: two fully connected layers respectively of 3,900 and 2,000 neuron size, each one preceded by a batch normalization layer.

**Word2Vec - Convolutional Neural Networks.** Convolutional Neural Networks (CNNs) are a type of Neural Network where the first layers act as filters to identify patterns in the input data set and consequently to work on a more abstract representation of the input. CNNs were originally employed in Computer Vision, the interested reader can refer to [17, 18, 28] for more details. CNNs have been employed to solve text classification tasks [7, 16].

In this paper, we evaluated the convolutional neural network described in Table 1 over the results of the Word2Vec pipeline described in Sect. 3.2.

The first two convolutional layers perform a convolution over the Word2Vec features producing as output respectively the results of 200 and 100 filters. At the end of the latter convolutional layer, each title can be viewed as a matrix of 15 × 100 (respectively, the number of words and the quantity of filter values computed on each word embedding). The FeaturePoolLayer averages the 15 values for each filter, at the end of this layer each title can be viewed as a vector

**Table 1.** The Convolutional Neural Networks Structure. Each layer non linearity is specified in the note (if any). A BatchNormLayer performs a batch normalization

| Network layer | Layer type | Note |
|---|---|---|
| 1 | Input Layer | input shape: [n_documents, 15, word_vector_dimension ] |
| 2 | Conv1DLayer | num_filters = 200, filter_size = 1, stride = 1, pad = 0, nonlinearity = linear rectifier |
| 3 | Conv1DLayer | num_filters = 100, filter_size = 1, stride = 1, pad = 0, nonlinearity = linear rectifier |
| 4 | FeaturePoolLayer | for each filter it computes the mean across the 15 word values |
| 5 | Fully Connected Layer | num_units = 2000, nonlinearity = linear rectifier |
| 6 | BatchNormLayer | |
| 7 | Fully Connected Layer | num_units = 500, nonlinearity = linear rectifier |
| 8 | BatchNormLayer | |
| 9 | Fully Connected Layer | The final layer, nonlinearity = softmax |

of 100 values. Then, two fully connected layers follow of respectively 2, 000 and 500 neurons (each fully connected layer is preceded by a Batch Normalisation Layer). The last layer has as many neurons as the number of ISCO codes available in the training set and employs softmax as non linearity.

The network was trained using Gradient Descent, the network accounts for about $10^6$ weights to be updated during the training. It wasn't necessary splitting the documents into batches during the training since the GPU memory was enough to handle all of them. The initial learning rate was 0.1 and Nesterov Momentum was employed. Early stopping was used to guess when to stop the training.

## 4.3    Experimental Settings

The classification pipelines previously introduced have been evaluated using the train and test datasets on which the labelled dataset was split into. The unlabelled dataset was used to train the Word2Vec model, which was then employed in the feature extraction process over the labelled dataset. The extracted features were used to perform a supervised machine learning process. Each classification pipeline has parameters requiring tuning, therefore a grid search was performed on the train set using a k-fold cross validation (k = 5) to identify the combination of parameters maximizing the F1-score (actually it was used the weighted F1-score). For each classification pipeline, the best combination of parameters was evaluated against the test set. The results are outlined in the remaining of this section.

The classifiers were built using the Scikit-learn [26], Theano [31], and Lasagne [10] frameworks running on an Intel Xeon machine with 32 GB Ram and an NVidia CUDA 4 GB GPU. Considering the BoW Feature Extraction, the Linear SVM classifier parameters and performances are shown in Table 2a, the SVM RBF performances are shown in Table 2c, and the Fully Connected Neural Network classifier performances are shown in Table 2b.

**Table 2.** Classification pipelines parameters and performances. NGram Range focuses on BoW feature extraction: (1,1) is for *Only Unigrams*, (1,2) is for *Both Unigrams and Bigrams*. The F1 score, precision, and recall are the weighted average of the corresponding scores computed on each ISCO code class. In Table (C) only a subsets of the results is shown. Grid search was computed using a 5-fold cross validation on the training set

(a) BoW - LinearSVM

| C | NGram Range | F1-S | Prec | Rec |
|---|---|---|---|---|
| 0.01 | (1, 1) | 0.786 | 0.798 | 0.797 |
| 100 | (1, 1) | 0.797 | 0.806 | 0.793 |
| 100 | (1, 2) | 0.816 | 0.826 | 0.813 |
| 10 | (1, 1) | 0.825 | 0.831 | 0.825 |
| 0.01 | (1, 2) | 0.834 | 0.842 | 0.838 |
| 10 | (1, 2) | 0.835 | 0.842 | 0.833 |
| 1 | (1, 1) | 0.845 | 0.849 | 0.846 |
| 0.1 | (1, 1) | 0.846 | 0.851 | 0.849 |
| 1 | (1, 2) | 0.854 | 0.858 | 0.854 |
| **0.1** | **(1, 2)** | **0.858** | **0.862** | **0.859** |

(b) BoW - Neural Net.

| NGram Range | Net | F1-S | Prec | Rec |
|---|---|---|---|---|
| FCNN1 | (1, 1) | 0.778 | 0.786 | 0.783 |
| FCNN2 | (1, 1) | 0.784 | 0.790 | 0.783 |
| FCNN2 | (1, 2) | 0.801 | 0.809 | 0.799 |
| **FCNN1** | **(1, 2)** | **0.816** | **0.822** | **0.818** |

(c) BoW - RBF SVM

| F1-S | Prec | Rec | C | γ | Ngram Range |
|---|---|---|---|---|---|
| 0.016 | 0.049 | 0.025 | 0.01 | 10 | (1, 1) |
| 0.018 | 0.061 | 0.023 | 0.01 | 10 | (1, 2) |
| 0.020 | 0.033 | 0.029 | 0.01 | 0.1 | (1, 1) |
| 0.027 | 0.049 | 0.036 | 0.01 | 1 | (1, 2) |
| 0.028 | 0.038 | 0.039 | 0.01 | 0.1 | (1, 2) |
| ... | ... | ... | ... | ... | ... |
| 0.718 | 0.849 | 0.644 | 100 | 1 | (1, 1) |
| 0.827 | 0.840 | 0.822 | 100 | 0.1 | (1, 1) |
| 0.831 | 0.845 | 0.825 | 100 | 0.1 | (1, 2) |
| 0.836 | 0.852 | 0.832 | 1 | 0.1 | (1, 1) |
| 0.840 | 0.851 | 0.836 | 10 | 0.1 | (1, 1) |
| 0.841 | 0.854 | 0.836 | 10 | 0.1 | (1, 2) |
| **0.842** | **0.861** | **0.835** | **1** | **0.1** | **(1, 2)** |

**Results Summary.** Table 3 summarises the best parameters for each classification pipeline and outlines the performance computed on the test set.

**Table 3.** Classification pipelines performances computed on the test dataset

| Classification pipeline | Notes | F1 Score | Precision | Recall |
|---|---|---|---|---|
| BoW SVM Linear | C = 0.1, NGram_Range = (1,2) | **0.857** | 0.870 | **0.865** |
| BoW SVM RBF | C = 1, Gamma = 0.1, NGram_Range = (1,2) | 0.849 | **0.878** | 0.856 |
| BoW neural network | Net = FCNN1, NGram_Range = (1,2) | 0.820 | 0.835 | 0.830 |
| W2V CNN | Net = CNN | 0.787 | 0.802 | 0.797 |

The BoW SVM Linear has the best performances, therefore it has been chosen for implementing the occupation classification pipeline for the English language in the prototype. As stated in the literature, text classification can be efficiently solved using linear classifiers [14] like Linear SVM, and the additional complexity of non-linear classification does not tend to pay for itself [1], except for some special data sets. Considering the Word2Vec Convolutional Neural Network, the performances are shown in Table 3, the authors would have expected better results, and the matter calls for further experiments.

## 4.4 Results Validation by EU Organisation

The project provided two main outcomes to the Cedefop EU Agency that supported it. On one side, a *feasibility study*, that has not been addressed in this paper, reporting some best practices identified by labour market experts involved within the project and belonging to the ENRLMM.[6] As a major result, the project provided a *working prototype* that has been deployed at Cedefop in June 2016 and it is currently running on the Cedefop Datacenter. In Fig. 1 we report a snapshot from a demo dashboard that provides an overview of the occupation trends over the period June-September 2015 collecting up to 700 K unique Web job vacancies. To date, the system collected 7+ million job vacancies over the 5 EU countries, and it is among a selection of research projects of Italian universities framed within a Big Data context [4].

Below we report an example of how the classified job vacancies could be used to support LMI monitoring, focusing only on 4-months scraping data. One might closely looks at the differences between countries in terms of labour market demands. Comparing UK against Italy, we can see that "Sales, Marketing and development managers" are the most requested occupations at ESCO (level 2) in the UK over this period whilst, rolling up on ESCO level 1 we can observe that "Professionals" are mainly asked in the "Information and Communication" sector, followed by "Administrative and support service activities" according to the NACE taxonomy. Furthermore, the type of contract is usually specified offering permanent contracts. Differently, the Italian labour market, that has a job vacancy posting rate ten times lower than UK over the same period, is looking at Business Service Agents requested in the "Manufacturing" field offering often temporary contracts.

*Interactive Demos.* Due to the space restrictions, the dashboard in Fig. 1 and some other demo dashboards have been made available online: *Industry and Occupation Dashboard* at https://goo.gl/bdqMkz, *Time-Series Dashboard* at https://goo.gl/wwqjhz, and *Occupations Dashboard* at https://goo.gl/M1E6x9

*Project Results Validation.* Finally, the project results have been discussed and endorsed in a workshop which took place in Thessaloniki, in December 2015.[7] The methodology and the results obtained have been validated as effective by

---

[6] The Network on Regional Labour Market Monitoring, http://www.regional labourmarketmonitoring.net/.

[7] The Workshop agenda and participants list is available at https://goo.gl/71Oc7A.

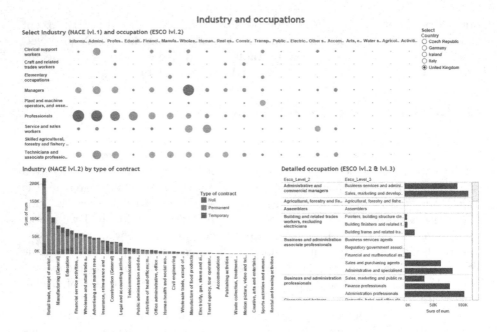

**Fig. 1.** A snapshot from the System Dashboard deployed. Interactive Demo available at https://goo.gl/bdqMkz

leading experts on LMI and key stakeholders. In 2017, we have been granted by Cedefop the extension of the prototype to all the 28 EU Countries.[8]

## 5   Conclusions and Expected Outcomes

In this paper we have described an innovative real-world data system we developed within a European research call-for-tender, granted by a EU organisation, aimed at classifying Web job vacancies through machine learning algorithms. We designed and evaluated several classification pipelines for assigning ISCO occupation codes to job vacancies, focusing on the English language. The classification performances guided the implementation of similar pipelines for different languages. The main outcome of this project is a working prototype actually running on the Cedefop European Agency datacenter, collecting and classifying Web job vacancies from 5 EU Countries. The developed system provides an important contribution to the whole LMI community and it is among the first research projects that employed machine learning algorithms for obtaining near real-time information on Web job vacancies. The results have been validated by EU labour market experts and put the basis of a further call to extend the system to all the EU Countries, which we are currently working on.

---

[8] "Real-time Labour Market information on Skill Requirements: Setting up the EU system for online vacancy analysis AO/DSL/VKVET-GRUSSO/Real-time LMI 2/009/16. Contract notice - 2016/S 134-240996 of 14/07/2016 https://goo.gl/5FZS3E.

# References

1. Aggarwal, C.C., Zhai, C.: A survey of text classification algorithms. In: Aggarwal, C., Zhai, C. (eds.) Mining Text Data, pp. 163–222. Springer, Boston (2012). https://doi.org/10.1007/978-1-4614-3223-4_6

2. Amato, F., Boselli, R., Cesarini, M., Mercorio, F., Mezzanzanica, M., Moscato, V., Persia, F., Picariello, A.: Challenge: processing web texts for classifying job offers. In: 2015 IEEE International Conference on Semantic Computing (ICSC), pp. 460–463 (2015)

3. Bengio, Y., Ducharme, R., Vincent, P., Jauvin, C.: A neural probabilistic language model. J. Mach. Learn. Res. **3**, 1137–1155 (2003)

4. Bergamaschi, S., Carlini, E., Ceci, M., Furletti, B., Giannotti, F., Malerba, D., Mezzanzanica, M., Monreale, A., Pasi, G., Pedreschi, D., et al.: Big data research in Italy: a perspective. Engineering **2**(2), 163–170 (2016)

5. Bird, S., Klein, E., Loper, E.: Natural Language Processing with Python: Analyzing Text with the Natural Language Toolkit. O'Reilly Media Inc., Sebastopol (2009)

6. Boselli, R., Mezzanzanica, M., Cesarini, M., Mercorio, F.: Planning meets data cleansing. In: The 24th International Conference on Automated Planning and Scheduling (ICAPS 2014), pp. 439–443. AAAI (2014)

7. Collobert, R., Weston, J.: A unified architecture for natural language processing: deep neural networks with multitask learning. In: The 25th International Conference on Machine Learning, pp. 160–167. ICML, ACM (2008)

8. Cortes, C., Vapnik, V.: Support-vector networks. Mach. Learn. **20**(3), 273–297 (1995)

9. Dasu, T.: Data glitches: monsters in your data. In: Sadiq, S. (ed.) Handbook of Data Quality, pp. 163–178. Springer, Heidelberg (2013). https://doi.org/10.1007/978-3-642-36257-6_8

10. Dieleman, S., Schlüter, J., Raffel, C., Olson, E., Sønderby, S.K., Nouri, D., et al.: Lasagne: first release, August 2015. https://doi.org/10.5281/zenodo.27878

11. Hong, W., Zheng, S., Wang, H.: Dynamic user profile-based job recommender system. In: Computer Science and Education (ICCSE). IEEE (2013)

12. Ioffe, S., Szegedy, C.: Batch normalization: Accelerating deep network training by reducing internal covariate shift. arXiv preprint arXiv:1502.03167 (2015)

13. International standard classification of occupations (2012). Accessed 11 Nov 2016

14. Joachims, T.: Text categorization with support vector machines: learning with many relevant features. In: Nédellec, C., Rouveirol, C. (eds.) ECML 1998. LNCS, vol. 1398, pp. 137–142. Springer, Heidelberg (1998). https://doi.org/10.1007/BFb0026683

15. Khan, F.H., Bashir, S., Qamar, U.: TOM: Twitter opinion mining framework using hybrid classification scheme. Decis. Support Syst. **57**, 245–257 (2014)

16. Kim, Y.: Convolutional neural networks for sentence classification. arXiv preprint arXiv:1408.5882 (2014)

17. Krizhevsky, A., Sutskever, I., Hinton, G.E.: ImageNet classification with deep convolutional neural networks. In: Advances in Neural Information Processing Systems 25, pp. 1097–1105. Curran Associates, Inc. (2012)

18. Lee, H., Grosse, R., Ranganath, R., Ng, A.Y.: Convolutional deep belief networks for scalable unsupervised learning of hierarchical representations. In: The 26th Annual International Conference on Machine Learning, ICML 2009, pp. 609–616. ACM (2009)

19. Lee, I.: Modeling the benefit of e-recruiting process integration. Decis. Support Syst. **51**(1), 230–239 (2011)
20. Melville, P., Gryc, W., Lawrence, R.D.: Sentiment analysis of blogs by combining lexical knowledge with text classification. In: ACM SIGKDD International Conference on Knowledge Discovery and Data Mining. ACM (2009)
21. Mezzanzanica, M., Boselli, R., Cesarini, M., Mercorio, F.: Data quality sensitivity analysis on aggregate indicators. In: Helfert, M., Francalanci, C., Filipe, J. (eds.) Proceedings of the International Conference on Data Technologies and Applications, Data 2012, pp. 97–108. INSTICC (2012)
22. Mezzanzanica, M., Boselli, R., Cesarini, M., Mercorio, F.: A model-based evaluation of data quality activities in KDD. Inf. Process. Manag. **51**(2), 144–166 (2015)
23. Mikolov, T., Chen, K., Corrado, G., Dean, J.: Efficient estimation of word representations in vector space. arXiv preprint arXiv:1301.3781 (2013)
24. Mikolov, T., Sutskever, I., Chen, K., Corrado, G.S., Dean, J.: Distributed representations of words and phrases and their compositionality. In: Advances in Neural Information Processing Systems, pp. 3111–3119 (2013)
25. Pang, B., Lee, L., Vaithyanathan, S.: Thumbs up?: sentiment classification using machine learning techniques. In: Proceedings of the ACL-02 Conference on Empirical Methods in Natural Language Processing, vol. 10, pp. 79–86. Association for Computational Linguistics (2002)
26. Pedregosa, F., Varoquaux, G., Gramfort, A., Michel, V., Thirion, B., Grisel, O., Blondel, M., Prettenhofer, P., Weiss, R., Dubourg, V., Vanderplas, J., Passos, A., Cournapeau, D., Brucher, M., Perrot, M., Duchesnay, E.: Scikit-learn: machine learning in Python. J. Mach. Learn. Res. **12**, 2825–2830 (2011)
27. Řehůřek, R., Sojka, P.: Software framework for topic modelling with large corpora. In: Proceedings of the LREC 2010 Workshop on New Challenges for NLP Frameworks, pp. 45–50. ELRA, Valletta, May 2010. http://is.muni.cz/publication/884893/en
28. Schmidhuber, J.: Deep learning in neural networks: an overview. Neural Netw. **61**, 85–117 (2015)
29. Sebastiani, F.: Machine learning in automated text categorization. ACM Comput. Surv. (CSUR) **34**(1), 1–47 (2002)
30. Singh, A., Rose, C., Visweswariah, K., Chenthamarakshan, V., Kambhatla, N.: Prospect: a system for screening candidates for recruitment. In: Proceedings of the 19th ACM International Conference on Information and Knowledge Management, pp. 659–668. ACM (2010)
31. Theano Development Team: Theano: A Python framework for fast computation of mathematical expressions. arXiv e-prints abs/1605.02688, May 2016. http://arxiv.org/abs/1605.02688
32. Yi, X., Allan, J., Croft, W.B.: Matching resumes and jobs based on relevance models. In: Proceedings of the 30th Annual International ACM SIGIR Conference on Research and Development in Information Retrieval, pp. 809–810. ACM (2007)
33. Zubiaga, A., Spina, D., Martínez-Unanue, R., Fresno, V.: Real-time classification of twitter trends. JASIST **66**(3), 462–473 (2015)

# Nectar Track

# Activity-Driven Influence Maximization in Social Networks

Rohit Kumar[1,4(✉)], Muhammad Aamir Saleem[1,2], Toon Calders[1,3],
Xike Xie[5], and Torben Bach Pedersen[2]

[1] Université Libre de Bruxelles, Brussels, Belgium
rohit.kumar@ulb.ac.be
[2] Aalborg University, Aalborg, Denmark
[3] Universiteit Antwerpen, Antwerp, Belgium
[4] Universitat Politécnica de Catalunya (BarcelonaTech), Barcelona, Spain
[5] University of Science and Technology of China, Hefei, China

**Abstract.** Interaction networks consist of a static graph with a time-stamped list of edges over which interaction took place. Examples of interaction networks are social networks whose users interact with each other through messages or location-based social networks where people interact by checking in to locations. Previous work on finding influential nodes in such networks mainly concentrate on the static structure imposed by the interactions or are based on fixed models for which parameters are learned using the interactions. In two recent works, however, we proposed an alternative activity data driven approach based on the identification of influence propagation patterns. In the first work, we identify so-called information-channels to model potential pathways for information spread, while the second work exploits how users in a location-based social network check in to locations in order to identify influential locations. To make our algorithms scalable, approximate versions based on sketching techniques from the data streams domain have been developed. Experiments show that in this way it is possible to efficiently find good seed sets for influence propagation in social networks.

## 1 Introduction

Understanding how information propagates in a network has a broad range of applications like viral marketing [6], epidemiology and outdoor marketing [7]. For example, imagine a computer games company that has budget to hand out samples of their new product to 50 gamers, and want to do so in a way that achieves maximal exposure. In that situation the company would like to target those customers that have maximal influence on social media. For this purpose they monitor interactions between gamers, and learn from these interactions which ones are the most influential. Notice that for the company it is also important that the selected people are not only influential, but that their combined influence should be maximal; selecting 50 highly influential gamers in the same sub-community is likely less effective than targeting less influential users but from different communities. This example is a typical instance of

Y. Altun et al. (Eds.): ECML PKDD 2017, Part III, LNAI 10536, pp. 345–348, 2017.
https://doi.org/10.1007/978-3-319-71273-4_28

the *Influence maximization problem* [6]. The common ingredients of an influence maximization problem are: a graph in which the nodes represent users of a social network, an information propagation model, and a target number of seed nodes that need to be identified such that they jointly maximize the influence spread in the network under the given propagation model.

Earlier works in this area studied different propagation models, such as linear threshold (LT) or independent cascade (IC) models [3], the complexity of the influence maximization problem under these models, and efficient heuristic algorithms. For instance, Kempe et al. [3] proved that the influence maximization problem under the LT and IC models is NP-hard and they provided a greedy algorithm to select seed sets using maximum marginal gain. As the model was based on Monte Carlo simulations it was not very scalable for large graphs.

A critical issue in the application of influence maximization algorithms is that of selecting the right propagation model. Most of these propagation models rely on parameters such as the influence a user exerts on his/her neighbors. Therefore, a second important line of work deals with learning these parameters based on observations. For instance, in a social network we could observe that user $a$ liking a post is often followed by user $b$, who is friend of $a$, liking the same post. In such a case it is plausible that user $a$ has a high influence on user $b$, and hence that the parameter expressing the influence of $a$ on $b$ should get a high value. The parameter learning problem is hence, based on a record of activities in the network, estimate the most likely parameter setting for explaining the observed propagation. The resulting optimized model can then be used to address the problem of selecting the best seed nodes. Goyal et al. [2] proposed the first such data based approach to find influential users in a social network. They estimate the influence probability of one user on his/her friend by observing all the different activities the friend follows in a given time window divided by total number of activities done by the user.

All these works share one property: they are based on models and if activity data is used, it is only indirectly to estimate model parameters. Recently, however, new, model-independent and purely data-driven methods have emerged. Our two papers, [7] published at WSDM and [4] published at EDBT should be placed in this category of data-based approaches.

## 2 Data-Driven Information Maximization

In [4] we proposed a new time constrained model to consider real interaction data to identify influence of every node in an interaction network [5]. The central idea in our approach is to mine frequent *information channels* between different nodes and use the presence of an information channel as an indication of possible influence among the nodes. An *information channel*$(ic(u, v))$ is a sequence of interactions between nodes $u$ and $v$ forming a path in the network which respects the time order. As such, an information channel represents a potential way information could have flown in the interaction network. An interaction could be bidirectional, for instance a chat or call between two users where information flows in both directions, or uni-directional where information flows from one user to another, for example in an email interaction or a re-tweet.

Figure 1 illustrates the notion of an information channel. There are interactions from user $a \rightarrow b$ and $c \rightarrow e$ at 9 AM, from $b \rightarrow d$ and $b \rightarrow c$ at 9:05 AM and $d \rightarrow f$ at 9:10 AM. These interactions form an interaction network. There is an information channel $a \rightarrow c$ via the temporal path $a \rightarrow b \rightarrow c$ but there is no information channel from $a \nrightarrow e$ as there is no time respecting path from $a$ to $e$. We define the *duration*$(dur(ic(u, v)))$ of an information channel as the time differ-ence of the first and last interaction on the information channel. For example, the dura-tion of the information channel $a \rightarrow b \rightarrow c$

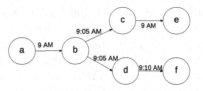

**Fig. 1.** Information channels between different nodes in the network. Every node is a user in a social network and the edges represents an interaction between users.

is 10 min. There could be multiple information channels of different durations between two nodes in a network. The intuition of the information channel notion is that node $u$ could only have sent information to node $v$ if there exists a time respecting series of interactions connecting these two nodes. Therefore, nodes that can reach many other nodes through information channels are more likely to influence other nodes than nodes that have information channels to only few nodes. This notion is captured by the *influence reachability set*. The *Influence reachability set (IRS)* $\sigma(u)$ of a node $u$ in a network $G(V, \mathcal{E})$ is defined as the set of all the nodes to which $u$ has an information channel.

In [4] we presented a one-pass algorithm to find the IRS of all nodes in an interaction network. We developed a time-window based HyperLogLog sketch [1] to compactly store the IRS of all the nodes and provided a greedy algorithm to do influence maximization.

## 3   Finding Influential Locations

Outdoor marketing can also benefit from the same data based approach to maximize influ-ence spread [7]. Recently, with the pervasive-ness of location-aware devices, social network data is often complemented with geographi-cal information, known as location-based social networks (LBSNs). In [7] we study naviga-tion patterns of users based on LBSN data to determine influence of a location on another location. Using the LBSN data we construct an interaction graph with nodes as locations and the edges representing the users traveling between locations. For example, in Fig. 2 there

| loc | Check-ins | | |
|-----|-----------|--|--|
| | Users | | |
| | t=1 | t=2 | t=3 |
| $T_1$ | $b, c, e, f$ | $a, h$ | $f$ |
| $T_2$ | $a, h$ | $f, g$ | $a$ |
| $M_1$ | $g$ | $i$ | $d$ |
| $H_1$ | – | $b, c, d, e$ | $i$ |
| $H_2$ | $d, i$ | – | – |

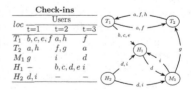

**Fig. 2.** Running example of a LBSN [7]. Nodes in the graph are the locations visited by users a–h. Edges are the movement of user between locations in a time window.

is an edge from location $T_1$ to $T_2$ due to users $a$ and $f$ visiting both locations within one trip.

We define the influence of a location by its capacity to spread its visitors to other locations. The intuition behind this definition is that good locations to seed with messages such as outdoor marketing promotions, are locations from which its visitors go to many other locations thus spreading the message. Thus, location influence indirectly captures the capability of a location to spread a message to other geographical regions. For example, if a company wants to distribute free t-shirts to promote some media campaign in a city, it would get maximum exposure by selecting neighborhoods such that the visitors of these neighborhood spread to maximum other neighborhoods in the city. In [7] we provide an exact on-line algorithm and a more memory-efficient but approximate variant based on the HyperLogLog sketch to maintain a data structure called *Influence Oracle* that allows to greedily find a set of influential locations.

## 4   Conclusion

In both of our works, through simulation experiments, we have shown that the data driven approach is quite accurate in modeling influence spread in the network. We also used time window based variations of the HyperLogLog sketch as an alternative to capture the influence set of every node in the network enabling us to scale our algorithms to very high data volumes.

**Acknowledgement.** This work was supported by the Fonds de la Recherche Scientifique-FNRS under Grant(s) no. T.0183.14 PDR. Xike Xie is supported by the CAS Pioneer Hundred Talents Program and the Fundamental Research Funds for the Central Universities.

## References

1. Flajolet, P., Fusy, É., Gandouet, O., Meunier, F.: Hyperloglog: the analysis of a near-optimal cardinality estimation algorithm. In: DMTCS Proceedings (2008)
2. Goyal, A., Bonchi, F., Lakshmanan, L.V.: A data-based approach to social influence maximization. Proc. VLDB Endowment **5**(1), 73–84 (2012)
3. Kempe, D., Kleinberg, J., Tardos, É.: Maximizing the spread of influence through a social network. In: KDD (2003)
4. Kumar, R., Calders, T.: Information propagation in interaction networks. In: EDBT (2017)
5. Kumar, R., Calders, T., Gionis, A., Tatti, N.: Maintaining sliding-window neighborhood profiles in interaction networks. In: Appice, A., Rodrigues, P.P., Santos Costa, V., Gama, J., Jorge, A., Soares, C. (eds.) ECML PKDD 2015. LNCS (LNAI), vol. 9285, pp. 719–735. Springer, Cham (2015). https://doi.org/10.1007/978-3-319-23525-7_44
6. Richardson, M., Domingos, P.: Mining knowledge-sharing sites for viral marketing. In: KDD (2002)
7. Saleem, M.A., Kumar, R., Calders, T., Xie, X., Pedersen, T.B.: Location influence in location-based social networks. In: WSDM (2017)

# An AI Planning System for Data Cleaning

Roberto Boselli[1,2], Mirko Cesarini[1,2], Fabio Mercorio[1,2(✉)],
and Mario Mezzanzanica[1,2]

[1] Department of Statistics and Quantitative Methods,
University of Milano-Bicocca, Milan, Italy
`fabio.mercorio@unimib.it`
[2] CRISP Research Centre, University of Milano-Bicocca, Milan, Italy

**Abstract.** Data Cleaning represents a crucial and error prone activity
in KDD that might have unpredictable effects on data analytics, affecting
the believability of the whole KDD process. In this paper we describe how
a bridge between AI Planning and Data Quality communities has been
made, by expressing both the data quality and cleaning tasks in terms
of AI planning. We also report a real-life application of our approach.

**Keywords:** AI planning · Data quality · Data cleaning · ETL

## 1 Introduction and Motivation

A challenging issue in data quality is to automatically check the quality of a
source dataset and then to identify cleaning activities, namely a sequence of
actions able to cleanse a dirty dataset. Data quality is a domain-dependent
concept, usually defined as "fitness for use", thus reaching a satisfying level of
data quality strongly depends on the analysis purposes. Focusing on *consistency*,
which can be seen as "the violation of semantic rules defined over a set of data
items" [1], the state-of-the-art solutions mainly rely on functional dependencies
(FDs) and their variants, that are powerful in specifying integrity constraints.
Consistency requirements are usually defined on either a single tuple, two tuples,
or a set of tuples [4]. Though the first two kind of constraints can be modelled
through FDs, the latter one requires reasoning with a (finite but variable in
length) set of data items (e.g., time-related data), and this makes the use of
FD-based approaches uneffective (see, e.g., [4,10]). This is the case of *logitudi-
nal data* (aka *historical or time-series data*), which provide knowledge about a
given subject, object or phenomena observed at multiple sampled time points.
In addition, it is well known that FDs are expressive enough to model static
constraints, which evaluate the current state of the database, but they do not
take into account how the database state has evolved over time [3]. Furthermore,
though FDs enable the detection of errors, they cannot be used as guidance to
fix them [9].

In such a context graphs or tree formalisms are deemed also appropriate to
model the *expected* data behaviour, that formalises how the data should evolve

© Springer International Publishing AG 2017
Y. Altun et al. (Eds.): ECML PKDD 2017, Part III, LNAI 10536, pp. 349–353, 2017.
https://doi.org/10.1007/978-3-319-71273-4_29

over time for being considered as consistent, and this makes the exploration-based technique (as AI Planning) a good candidate for the data quality task. The idea that underlies our work is to cast the problem of checking the consistency of a set of data items as a planning problem. This, in turn, allows using off-the-shelf AI planning tools to perform two separated tasks: (i) to catch inconsistencies and (ii) to synthesise a sequence of actions able to cleanse any (modelled) inconsistencies found in the data. In this paper we summarise results from our recent works on data consistency checking [15] and cleaning [2,14].

*AI Planning at a Glance.* Planning in Artificial Intelligence is about the decision making performed by computer programs when trying to achieve some goal. It requires to synthesise a sequence of actions that will transform a system configuration, step by step, into the desired one (i.e., the goal state). Roughly, planning requires two main elements: (i) *the domain*, i.e., a set of states of the environment $S$ together with the set of actions $A$ specifying the transitions between these states; (ii) *the problem*, that consists of the set of facts whose composition determine an initial state $s_0 \in S$ of the environment, and a set of facts $G \subseteq S$ that models the goals of the planning taks. A solution (aka *plan*) is a bounded sequence of actions $a_1, \ldots, a_n$ that can be applied to reach a goal configuration. Planning formalisms are expressive enough to model complex temporal constraints, then a cleaning approach based on AI planning might allow domain experts to concentrate on *what* quality constraints have to be modelled rather than on *how* to check them. Recently, AI Planning contributed to the trace alignment problem in the context of Business Process Modelling [5].

## 2    A Data Cleaning Approach Framed Within KDD

Our approach requires to map a sequence of events as actions of a planning domain, so that AI planning algorithms can be exploited to find inconsistencies and to fix them. Intuitively, let us consider an events sequence $\epsilon = e_0, e_2, \ldots, e_{n-1}$. Each event $e_i$ will contain a number of observation variables whose evaluation determines a snapshot of the subject's *state*[1] at time point $i$, namely $s_i$. Then, the evaluation of any further event $e_{i+1}$ might change the value of one or more state variables of $s_i$, generating a new state $s_{i+1}$.

We encode the expected subjects' behaviour (the so-called *consistency model*) as a transition system. A consistent trajectory represents a sequence of events that does not violate any consistency constraints. Given a $\epsilon$ event sequence as input, the planner deterministically determines a trajectory $\pi = s_0 e_0 s_1 \ldots s_{n-1} e_{n-1} s_n$ on the finite state system explored (i.e., a *plan*) where each state $s_{i+1}$ results by applying event $e_i$ on $s_i$. Once a model describing the evolution of an event sequence has been defined, we detect quality issues by solving a planning problem where a consistency violation is the goal condition. If a plan is found by a planning system, the event sequence is marked as inconsistent in the original data quality problem. Our system works in three steps (Fig. 1).

---

[1] A value assignment to a set of finite-domain state variables.

**Step 1 [Universal Checker].** We simulate the execution of *all* the event sequences - within a finite-horizon - summarising all the inconsistencies found during the exploration[2] into an object, we call *Universal Checker* (UCK), that represents a taxonomy of the inconsistencies that may affect a data source. The UCK computed can be seen as a list of tuples $(id, s_i, a_i)$, that specifies the inconsistency with $id$ might arise in a state $s_i$ as consequence of applying $a_i$.

**Step 2 [Universal Cleanser].** For any given tuple $(id, s_i, a_i)$ of the Universal Checker, we construct a new planning problem which differs from the previous one in terms of both initial and goal states: (i) the new initial state is $s_i$, that is a consistent state where the event $e_i$ can be applied leading to an inconsistent state $s_{i+1}$; (ii) the new goal is to be able to "execute action $a_i$". Intuitively, a *cleaning action sequence* applied to state $s_i$ transforms it into a state $s_j$ where action $a_i$ can be applied without violating any consistency rules. To this end, the planner *explores* the state space and *collects* all the optimal corrections according to a given criterion. The output of this phase is a *Universal Cleanser*. Informally, it can be seen as a set of policies, computed off-line, able to bring the system to the goal from any state reachable from the initial ones (see, e.g., [8,12]). In our context, the universal cleanser is a lookup table that returns a sequence of actions able to fix an event $e_i$ occurring in a state $s_j$.

**Step 3 [Cleanse the Data].** Given a set of event sequences $D = \{\epsilon_1, \ldots, \epsilon_n\}$ the system uses the planner to verify the consistency of each $\epsilon_i$. If an inconsistency is found, the system retrieves its identifier from the Universal Checker, and then selects the cleaning actions sequence through a look-up on the Universal Cleanser.

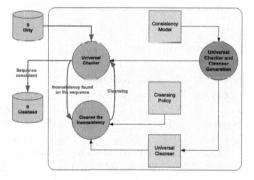

The Universal Cleanser presents two important features that makes it effective in dealing with real data: first, it is synthesised off-line and only summarises cost-optimal action sequences. Clearly, the cost function

**Fig. 1:** A graphical representation of the Consistency Verification and Cleaning Process.

is domain-dependent and usually driven by the purposes of the analysis (we discussed how to select different cleaning alternatives in [13,14]). Second, the UC is *data-independent* as it has been synthesised by considering *all* the (bounded) event sequences, thus *any* data sources conform to the model can be handled. Our approach has been implemented on top of the UPMurphi planner [6,7].

---

[2] Notice that this task can be accomplished by forcing the planner to continue the search even if a goal has been found.

*Real-life Application*[3]. Our approach has been applied to the *mandatory communication*[4] domain, that models labour market data of Italian citizens at regional level. Here, inconsistencies represent career transitions not permitted by the Italian Labour Law. Thanks to our approach, we synthesised both the Universal Checker and Cleanser for the domain (i.e., 342 distinct inconsistencies found and up to 3 cleaning action sequence synthesised for each). The system has been employed within the KDD process that analysed the real career sequences of 214,432 citizens composed of 1,248,751 mandatory notifications. For details about the quality assessment see [15] whilst for cleaning details see [14].

## 3   Concluding Remarks

We presented a general approach that expresses Data quality and cleaning tasks in terms of AI Planning problem, connecting two distinct research areas. Our approach has been formalised and fully-implemented on top of the UPMurphi planner, and applied to a real-life example analysing and cleaning million records concerning labour market movements of Italian citizens.

We are working on (i) including machine-learning algorithms to identify the *most suited* cleaning action, and (ii) applying our approach to build training sets for data cleaning tools based on machine-learning (e.g., [11]).

## References

1. Batini, C., Scannapieco, M.: Data Quality: Concepts, Methodologies and Techniques. Data-Centric Systems and Applications. Springer, New York (2006)
2. Boselli, R., Cesarini, M., Mercorio, F., Mezzanzanica, M.: Planning meets data cleansing. In: The 24th ICAPS. AAAI Press (2014)
3. Chomicki, J.: Efficient checking of temporal integrity constraints using bounded history encoding. ACM Trans. Database Syst. (TODS) **20**(2), 149–186 (1995)
4. Dallachiesa, M., Ebaid, A., Eldawy, A., Elmagarmid, A.K., Ilyas, I.F., Ouzzani, M., Tang, N.: NADEEF: a commodity data cleaning system. In: SIGMOD (2013)
5. De Giacomo, G., Maggi, F.M., Marrella, A., Patrizi, F.: On the disruptive effectiveness of automated planning for LTLf-based trace alignment. In: AAAI (2017)
6. Della Penna, G., Intrigila, B., Magazzeni, D., Mercorio, F.: UPMurphi: a tool for universal planning on PDDL+ problems. In: The 19th ICAPS, pp. 106–113 (2009)
7. Della Penna, G., Intrigila, B., Magazzeni, D., Mercorio, F.: A PDDL+ benchmark problem: the batch chemical plant. In: ICAPS, pp. 222–224. AAAI Press (2010)
8. Della Penna, G., Magazzeni, D., Mercorio, F.: A universal planning system for hybrid domains. Appl. Intell. **36**(4), 932–959 (2012)
9. Fan, W., Li, J., Ma, S., Tang, N., Yu, W.: Towards certain fixes with editing rules and master data. Proc. VLDB Endowment **3**(1–2), 173–184 (2010)

---

[3] This work was partially supported within a Research Project granted by the CRISP Research Centre and Arifl Agency (Regional Agency for Education and Labour.
[4] The Italian Ministry of Labour and Welfare: Annual report about the CO system, available at http://goo.gl/XdALYd last accessed may 2017.

10. Hao, S., Tang, N., Li, G., He, J., Ta, N., Feng, J.: A novel cost-based model for data repairing. IEEE Trans. Knowl. Data Eng. **29**(4) (2017)
11. Krishnan, S., Wang, J., Wu, E., Franklin, M.J., Goldberg, K.: ActiveClean: Interactive data cleaning while learning convex loss models. arXiv preprint arXiv:1601.03797 (2016)
12. Mercorio, F.: Model checking for universal planning in deterministic and non-deterministic domains. AI Commun. **26**(2), 257–259 (2013)
13. Mezzanzanica, M., Boselli, R., Cesarini, M., Mercorio, F.: Data quality sensitivity analysis on aggregate indicators. In: DATA, pp. 97–108 (2012)
14. Mezzanzanica, M., Boselli, R., Cesarini, M., Mercorio, F.: A model-based approach for developing data cleansing solutions. ACM J. Data Inf. Q. **5**(4), 1–28 (2015)
15. Mezzanzanica, M., Boselli, R., Cesarini, M., Mercorio, F.: A model-based evaluation of data quality activities in KDD. Inf. Process. Manag. **51**(2), 144–166 (2015)

# Comparing Hypotheses About Sequential Data: A Bayesian Approach and Its Applications

Florian Lemmerich[1(✉)], Philipp Singer[1,2,3,4], Martin Becker[2],
Lisette Espin-Noboa[1], Dimitar Dimitrov[1], Denis Helic[3], Andreas Hotho[2],
and Markus Strohmaier[1,4]

[1] GESIS - Leibniz Institute for the Social Sciences, Mannheim, Germany
{florian.lemmerich,lisette.espin-noboa,dimitar.dimitrov,
markus.strohmaier}@gesis.org
[2] University of Würzburg, Würzburg, Germany
{becker,hotho}@informatik.uni-wuerzburg.de
[3] Graz University of Technology, Graz, Austria
dhelic@tugraz.at
[4] RWTH Aachen, Aachen, Germany
markus.strohmaier@humtec.rwth-aachen.de

**Abstract.** Sequential data can be found in many settings, e.g., as sequences of visited websites or as location sequences of travellers. To improve the understanding of the underlying mechanisms that generate such sequences, the *HypTrails* approach provides for a novel data analysis method. Based on first-order Markov chain models and Bayesian hypothesis testing, it allows for comparing a set of hypotheses, i.e., beliefs about transitions between states, with respect to their plausibility considering observed data. HypTrails has been successfully employed to study phenomena in the online and the offline world. In this talk, we want to give an introduction to HypTrails and showcase selected real-world applications on urban mobility and reading behavior on Wikipedia.

## 1 Introduction

Today, large collections of data are available in the form of sequences of transitions between discrete states. For example, people move between different locations in a city, users navigate between web pages on the world wide web, or users listen to sequences of songs of a music streaming platform. Analyzing such datasets can leverage the understanding of behavior in these application domains. In typical machine learning and data mining approaches, parameters of a model (e.g., Markov chains) are learned automatically in order to capture the data generation process and make predictions. However, it is then often difficult to interpret the learned parameters or to relate them to basic intuitions and existing theories about the data, specifically if many parameters are involved.

---

This work summarizes a previous publication presenting the HypTrails approach [5] and three selected papers [1–3] that utilize it.

© Springer International Publishing AG 2017
Y. Altun et al. (Eds.): ECML PKDD 2017, Part III, LNAI 10536, pp. 354–357, 2017.
https://doi.org/10.1007/978-3-319-71273-4_30

In a recently introduced line of research, we therefore aim to establish an alternative approach: we develop a method that allows to capture the belief in the generation of sequential data as Bayesian priors over parameters and then compare such *hypotheses* with respect to their plausibility given observed data. In this work, we want to showcase our general approach [5], which we call *HypTrails*, and present some practical applications in various domains [1–3], i.e., sequences of visited locations derived from photos uploaded to Flickr, taxi directions in Manhattan, and navigation of readers in Wikipedia.

## 2  Bayesian Hypotheses Comparison in Sequential Data

For comparing hypotheses about the transition behavior in sequence data, we follow a Bayesian approach. As an underlying model, we utilize first-order Markov chain models. Such models assume a memory-less transition process between discrete states. That means that the probability of the next visited state depends only on the current one. The parameters of this model, i.e., the transition probabilities $p_{ij}$ between the states, can be written as a single matrix.

In HypTrails, we want to compare a set of hypotheses $H_1, \ldots, H_n$ with respect to how well they can explain the generation of the observed data. Each of the hypotheses captures a belief in the transition between the states as derived from theory in the application domain, from other related datasets, or from human intuition. To specify a hypothesis, the user can express a *belief matrix*, in which a high value in a cell $(i, j)$ reflects a belief that transitions between the states $i$ and $j$ are more common. With HypTrails, these belief matrices are then automatically transformed into Bayesian Dirichlet priors over the model parameters (i.e., the transition probabilities in the Markov chain). This transformation can be performed for different concentration parameters $\kappa$. A higher value of $\kappa$ generates a prior that corresponds to a stronger belief in the hypothesis. For each hypothesis $H_i$, and each concentration parameter $\kappa$, we can then compute the marginal likelihood $P(D|H_i)$ of the data given the hypothesis. Given our model, the marginal likelihood can efficiently be computed in closed form. The higher the marginal likelihood of a hypothesis is, the more plausible it appears to be with respect to the observed data. For quantifying the support of one hypothesis over another, we utilize Bayes Factors, a Bayesian alternative to frequentists p-values, which can directly be interpreted with lookup tables [4]. For a set of hypotheses, the marginal likelihoods induce an ordering of the hypotheses with respect to their plausibility given the data. However, the plausibility of hypotheses is always only checked relatively against each other. Therefore, often a simple hypothesis is used as a baseline, e.g., the uniform hypothesis that assumes all transitions to be equally likely.

To compare hypotheses, all priors should be derived using the same belief strength $\kappa$. To make comparisons across different belief strength, HypTrails results are typically visualized as line plots, in which each line corresponds to one hypothesis. The x-axis specifies different values of the concentration parameter $\kappa$, and the y-axis describes the marginal likelihood of a hypothesis, cf. Fig. 1.

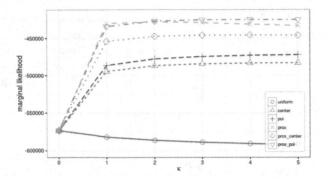

**Fig. 1. Example result of HypTrails (Flickr study, Berlin).** Each line represents one hypothesis. The x-axis defines different concentration parameters (strengths of belief), the y-axis indicates (logs of) marginal likelihoods for each hypothesis. It can be seen that the baseline "uniform" hypotheses is by far the least plausible of these hypothesis, while a mixture of proximity and center hypotheses ("prox-center") and a mixture of proximity and point-of-interest hypotheses ("prox-poi") perform best.

## 3    Applications

Next, we outline three real-world applications of this technique.

### 3.1    Urban Mobility in Flickr

In a first study, we focused on geo-temporal trails derived from Flickr. In particular, we crawled all photos on Flickr with geo-spatial information (i.e., latitude and longitude) from 2010 to 2014 for four major cities (Berlin, London, Los Angeles, and New York). We used a map grid to construct a discrete state space of locations. Then, we created a sequence of locations for each user that uploaded pictures of that city based on the picture locations. On the sequences, we evaluated a variety of hypotheses such as a proximity hypothesis (next location is near the current one), a point-of-interest hypothesis (next location will be at a tourist attraction or transportation hub), a center hypothesis (next location will be close to the city center), and combinations of them. As a result, rankings are mostly consistent across cities. Combinations of proximity and point-of-interest hypotheses are overall most plausible. Figure 1 shows example results for Berlin.

### 3.2    Taxi Usage in Manhattan

In a second study, we investigated again trails of urban mobility. In particular, we studied a dataset of taxi trails in Manhattan[1]. In this study, we used *tracts* (small administrative) units as a state space of locations. Using additional information on these tracts extracted from census data and data from the FourSquare API,

---

[1] http://www.andresmh.com/nyctaxitrips/.

we investigated more than 60 hypotheses such as "taxis drive to tracts with similar ethnic distribution" or "taxis will drive to popular locations w.r.t. check-ins". We also performed spatio-temporal clustering of the sequence data and applied HypTrails on the individual clusters to find behavioral traits that are typical for certain times and places. For instance, we discovered a group of taxi rides to locations with a high density of party venues on weekend nights.

### 3.3  Link Usage in Wikipedia

In another work, we studied transitions between articles in the online encyclopedia Wikipedia. In particular, we were interested in which links on a Wikipedia page get frequently used. For that purpose, we applied HypTrails on a recently published dataset of all transitions between Wikipedia pages for one month[2] using the set of all articles as state space. For constructing hypotheses, we considered hypotheses based on visual features of the links (e.g., "links in the lead paragraph get clicked more often" or "links in the main text get clicked more often"), hypothesis based on text similarity between articles, and hypotheses based on the structure of the link network of Wikipedia articles. As a result, hypotheses that assume people to prefer links at the top and left-hand side, and hypotheses that express a belief in more frequent usage of links towards the periphery of the article network are most plausible.

## 4  Conclusion

In this work, we gave a short introduction into the HypTrails approach that allows to compare the plausibility of hypotheses about the generation of a sequential datasets. Additionally, we described three real-world applications of this technique for studying urban mobility and reading behavior in Wikipedia.

## References

1. Becker, M., Singer, P., Lemmerich, F., Hotho, A., Helic, D., Strohmaier, M.: Photowalking the city: comparing hypotheses about urban photo trails on Flickr. In: International Conference on Social Informatics (SocInfo), pp. 227–244 (2015)
2. Dimitrov, D., Singer, P., Lemmerich, F., Strohmaier, M.: What makes a link successful on Wikipedia? In: International World Wide Web Conference, pp. 917–926 (2017)
3. Espín Noboa, L., Lemmerich, F., Singer, P., Strohmaier, M.: Discovering and characterizing mobility patterns in urban spaces: a study of Manhattan taxi data. In: International Workshop on Location and the Web, pp. 537–542 (2016)
4. Kass, R.E., Raftery, A.E.: Bayes factors. J. Am. Stat. Assoc. **90**(430), 773–795 (1995)
5. Singer, P., Helic, D., Hotho, A., Strohmaier, M.: Hyptrails: a Bayesian approach for comparing hypotheses about human trails on the web. In: International World Wide Web Conference, pp. 1003–1013 (2015)

---

[2] https://datahub.io/dataset/wikipedia-clickstream.

# Data-Driven Approaches for Smart Parking

Fabian Bock[1](✉), Sergio Di Martino[2], and Monika Sester[1]

[1] Institute of Cartography and Geoinformatics, Leibniz University Hannover,
Hannover, Germany
{bock,sester}@ikg.uni-hannover.de
[2] Department of Electrical Engineering and Information Technologies,
University of Naples "Federico II", Naples, Italy
sergio.dimartino@unina.it

**Abstract.** Finding a parking space is a key problem in urban scenarios, often due to the lack of actual parking availability information for drivers. Modern vehicles, able to identify free parking spaces using standard on-board sensors, have been proven to be effective probes to measure parking availability. Nevertheless, spatio-temporal datasets resulting from probe vehicles pose significant challenges to the machine learning and data mining communities, due to volume, noise, and heterogeneous spatio-temporal coverage. In this paper we summarize some of the approaches we proposed to extract new knowledge from this data, with the final goal to reduce the parking search time. First, we present a spatio-temporal analysis of the suitability of taxi movements for parking crowd-sensing. Second, we describe machine learning approaches to automatically generate maps of parking spots and to predict parking availability. Finally, we discuss some open issues for the ML/KDD community.

## 1   Introduction

Very often, in urban scenarios, drivers have to roam at the end of their trips on the search for a parking space, worsening the overall traffic and wasting time and fuel [5]. *Smart Parking* refers to Information and Communication Technology solutions meant to improve parking search by providing information about parking locations and their actual or estimated availability. While it is rather trivial to gather parking availability information for parking facilities, it becomes tricky for on-street parking, where there are mainly two sensing strategies: stationary or mobile collection. The former relies on sensors embedded in the road infrastructure, continuously measuring whether stalls are free or occupied. However, it is too expensive to cover a wider city area with those sensors. The latter mainly exploits participatory or opportunistic crowd-sensing solutions from mobile apps or probe vehicles [6], that can occasionally detect free parking spaces. Mobile sensors are pretty cheap to deploy in comparison to the stationary ones but the quality and the spatio-temporal resolution of the obtainable data streams is lower, posing many challenges for the automatic extraction of useful knowledge.

In this paper, we give an overview of some approaches we used to exploit mobile sensor data for Smart Parking scenarios. In particular, in Sect. 2 we

© Springer International Publishing AG 2017
Y. Altun et al. (Eds.): ECML PKDD 2017, Part III, LNAI 10536, pp. 358–362, 2017.
https://doi.org/10.1007/978-3-319-71273-4_31

summarize a study showing that a small fleet of taxis equipped with standard sensors can provide parking availability information comparable to a large number of stationary sensors [1,2]. Then, in Sect. 3, we show two actual Smart Parking use cases attainable with machine learning techniques on probe vehicle data: (I) identification of parking legality of small road segments [4], and (II) prediction of parking availability [3]. Finally, issues still to be faced by the ML/KDD Community are discussed in Sect. 4.

## 2  Mining Taxi GPS Trajectories to Assess Quality of Crowd-Sensed Parking Data

Probe vehicles are a promising solution to scan parking availability, since series sensors, like side-scanning ultrasonic sensors or windshield-mounted cameras, can be effectively used to determine free parking spaces. Mathur et al. [6] were the first to conduct a preliminary evaluation on the potentiality of taxis as probe vehicles, using a real-world dataset of GPS trajectories in San Francisco, USA. Some simplifications in their assumptions motivated us to investigate more deeply the topic, answering the questions whether the spatio-temporal distribution of a fleet of taxis is suitable for parking crowd-sensing and how many taxis are needed [1,2]. For that, we processed and combined parking availability sensor data from more than 400 road segments with over 3000 parking spaces from the SFpark project[1] in San Francisco, with trajectories of about 500 taxis[2] in the same area, with more than 11 million GPS points over three weeks.

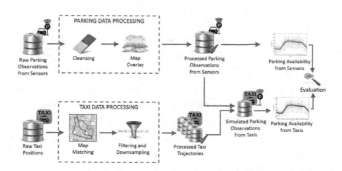

**Fig. 1.** The evaluation pipeline for combining parking data with taxi trajectories [2].

An overview of the processing steps to compare the spatio-temporal characteristics of parking and taxi movements is illustrated in Fig. 1. Both datasets needed to be matched to the same street network, taken from *OpenStreetMap*. Also some non-trivial cleansing and filtering were required to have comparable

---

[1] http://sfpark.org/.
[2] http://crawdad.org/epfl/mobility/20090224/.

datasets. The taxi trajectories were then aggregated to compute a typical weekly behavior per road segment. Assuming that taxis would have observed parking availability each time they traversed a road segment, we calculated a dataset of parking observations achievable by taxis by downsampling the stationary sensor data according to the timestamps of taxi visits per road segment. The actual parking availability was then estimated from the last observation of the taxis.

In a direct comparison of the parking availability information from mobile and stationary sensors, we found that the regular trips of 300 taxis (about 20% of all licensed taxis in San Francisco at that time) were sufficient to cover the SFpark project area in San Francisco with a maximal deviation of ±1 parking spaces with respect to stationary sensors in more than 85% of the cases. This result is remarkable since the taxi coverage revealed strong variability with the time of day, but can be explained by the fact that parking turnover showed a similar time dependence. The time until the next taxi visit was less than 30 min for about 60% of all road segments and time instants. Therefore, we concluded that the spatio-temporal movements of taxis are well suited to crowd-sense parking availability.

## 3   Machine Learning Approaches for On-Street Parking Information

In this section we describe two use cases for Smart Parking we can derive on top of the data stream coming from probe vehicles.

*Learning Parking Legality from Locations of Parked Vehicles.* The location of parking lanes and the legality of parking in a specific spot is the first relevant information for drivers looking for a free space. Parking might be not allowed in front of e.g. garage exits, or even for a full road if the road is narrow. Thus, drivers should focus their search on areas with many parking spaces. As the location of parking spaces is often unknown to non-local drivers and on-street parking maps do not exist in many cities, we developed a crowd-sensing approach to learn the parking legality from the location of parked vehicles at different time instants [4]. For every small road segment unit, several spatial and temporal features were extracted and a binary decision was performed to distinguish legal from illegal parking spots. Multiple classifiers were evaluated on parking availability data, collected on 9 trips with a probe vehicle on more than five kilometers of potential parking spaces. Results show that the random forest classifier achieved the best results. However, also k-means clustering plus a simple classification heuristic performed nearly as good as the first one, without the need for costly training data.

*Predicting Parking Availability.* Based on the parking availability information, a prediction is useful to provide a suggestion to drivers approaching their destination. There exist some data-driven prediction approaches in the literature [5], mostly formulated as a regression problem and only considering input data at a constant frequency. Since most of the drivers are rather interested whether or not there is at least one free parking space in a road segment and what is the corresponding probability, we reformulated the problem as a binary classification task that also is robust for irregular sampling [3]. As features, we used the last observations of the sensors in a road segment as well as the aggregated observations up to a certain distance in the surroundings, the parking capacity, and the time of day and day of week. We evaluated the approach using a random forest classifier with data from the SFpark project. Results show that the binary parking availability of a road segment can be predicted with a F1-score of about 75% for 30 min ahead.

## 4   Conclusions and Open Issues

Crowd-sensing vehicles, measuring the on-street parking availability during their regular trips, represent a new source for large amounts of parking data that promise to mitigate parking search problems. For example, maps of parking spaces can be automatically generated, parking availability predicted, and search recommendations given to frustrated drivers. However, due to the highly irregular spatio-temporal coverage of the generated data, also new research challenges arise for these applications. As the highly irregular sampling needs to be considered, standard time series approaches cannot be applied to predict parking availability. Also, as parking is a very dynamic phenomenon, extracting the trends in parking occupancy from the fluctuating data remains a challenge. Another open question is how learned models can be transferred to other cities. Finally, it is also very relevant to investigate whether additional approaches are necessary for irregular events like concerts or sport matches.

## References

1. Bock, F., Attanasio, Y., Di Martino, S.: Spatio-temporal road coverage of probe vehicles: a case study on crowd-sensing of parking availability with taxis. In: Bregt, A., Sarjakoski, T., van Lammeren, R., Rip, F. (eds.) GIScience 2017. LNGC, pp. 165–184. Springer, Cham (2017). https://doi.org/10.1007/978-3-319-56759-4_10
2. Bock, F., Di Martino, S.: How many probe vehicles do we need to collect on-street parking information? In: 2017 International Conference on Models and Technologies for Intelligent Transportation Systems (MT-ITS). IEEE (2017)
3. Bock, F., Di Martino, S., Sester, M.: What are the potentialities of crowdsourcing for dynamic maps of on-street parking spaces? In: Proceedings of the 9th ACM SIGSPATIAL International Workshop on Computational Transportation Science (IWCTS 2016), pp. 19–24. ACM, New York (2016)
4. Bock, F., Liu, J., Sester, M.: Learning on-street parking maps from position information of parked vehicles. In: Sarjakoski, T., Santos, M.Y., Sarjakoski, L.T. (eds.) Geospatial Data in a Changing World. LNGC, pp. 297–314. Springer, Cham (2016). https://doi.org/10.1007/978-3-319-33783-8_17

5. Lin, T., Rivano, H., Moul, F.L.: A survey of smart parking solutions. IEEE Trans. Intell. Transp. Syst. **PP**(99), 1–25 (2017)

6. Mathur, S., Jin, T., Kasturirangan, N., Chandrasekaran, J., Xue, W., Gruteser, M., Trappe, W.: ParkNet: drive-by sensing of road-side parking statistics. In: Proceedings of the 8th International Conference on Mobile Systems, Applications, and Services, pp. 123–136. ACM, New York (2010)

# Image Representation, Annotation and Retrieval with Predictive Clustering Trees

Ivica Dimitrovski[1]($\boxtimes$), Dragi Kocev[2], Suzana Loskovska[1], and Sašo Džeroski[2]

[1] University of Ss Cyril and Methodius, Skopje, Macedonia
{ivica.dimitrovski,suzana.loshkovska}@finki.ukim.mk
[2] Jožef Stefan Institute, Ljubljana, Slovenia
{Dragi.Kocev,Saso.Dzeroski}@ijs.si

**Abstract.** In this paper, we summarize our work on using the predictive clustering framework for image analysis. More specifically, we have used predictive clustering trees to generate image representations, that can then be used to perform image retrieval and/or image annotation. We have evaluated the proposed method for performing image retrieval on general purpose images [6], and annotation of general purpose images [5], medical images [3] and diatom images [4].

**Keywords:** Image representation · Image retrieval
Image annotation · Multi-target prediction · Predictive clustering

## 1 Introduction

The overwhelming increase in the amount of available visual information, especially digital images, has brought up a pressing need to develop efficient and accurate systems for image representation, retrieval and annotation. Most such systems for image analysis use the bag-of-visual-words representation of images. However, the computational bottleneck in all such systems is the construction of the visual codebook, i.e., obtaining the visual words. This is typically performed by clustering hundreds of thousands or millions of local descriptors, where the resulting clusters correspond to visual words. Each image is then represented by a histogram of the distribution of its local descriptors across the codebook.

The major issue in retrieval systems is that by increasing the sizes of the image databases, the number of local descriptors to be clustered increases rapidly: Thus, using conventional clustering techniques is infeasible. While existing approaches are able to solve the efficiency issue, a part of the discriminative power of the codebook is sacrificed for this. Considering this, we propose to construct the visual codebook by using predictive clustering trees (PCTs) [1], which can be constructed and executed efficiently and have good predictive performance.

PCTs are a generalization of decision trees towards the task of structured output prediction, including multi-target regression, (hierarchical) multi-label

© Springer International Publishing AG 2017
Y. Altun et al. (Eds.): ECML PKDD 2017, Part III, LNAI 10536, pp. 363–367, 2017.
https://doi.org/10.1007/978-3-319-71273-4_32

classification and time series prediction. Moreover, the definition of descriptive, clustering and target attributes is flexible thus facilitating the learning of both unsupervised and supervised trees. Furthermore, to increase the stability of the model, we propose to use random forests of PCTs [7]. We create a random forest of PCTs that represents the codebook, i.e., is used to generate the image representation.

The images represented with the bag-of-visual-words can then be used to perform image retrieval and/or annotation. In the former, the indexing structure for performing the retrieval is the same structure representing the codebook – the random forest of PCTs. We evaluate the proposed bag-of-visual-words approach for image retrieval on five benchmark reference datasets. The results reveal that the proposed method produces a visual codebook with superior discriminative power and thus better retrieval performance while maintaining excellent computational efficiency [6].

Additional complexity of image annotation arises from the complexity of the labels used for annotation: Typically, an image depicts more than one object, hence more than one label should be assigned for that image. Moreover, there might be some structure among the labels, such as a hierarchy of labels. To address this additional complexity, we learn ensembles of PCTs to exploit the potential relations that may exist among the labels. We have evaluated this approach on three tasks: multi-label classification of general purpose images [5], and hierarchical multi-label classification of medical images [3], as well as diatom images [4]. The results of the evaluation show that we achieve state-of-the-art predictive performance.

The remainder of this paper is organized as follows. We next briefly present the predictive clustering framework. We then outline the method for constructing image representations. Finally, we describe the evaluation of this approach, first for image retrieval and then for image annotation.

## 2   Predictive Clustering Framework

Predictive Clustering Trees (PCTs) [1] generalize decision trees and can be used for a variety of learning tasks including different types of prediction and clustering. The PCT framework views a decision tree as a hierarchy of clusters: the top-node of a PCT corresponds to one cluster containing all data, which is recursively partitioned into smaller clusters while moving down the tree. The leaves represent the clusters at the lowest level of the hierarchy and each leaf is labeled with its cluster's prototype (prediction). One of the most important steps in the PCT algorithm is the test selection procedure. For each node, a test is selected by using a heuristic function computed on the training examples. The heuristic used in this algorithm for selecting the attribute tests in the internal nodes is the reduction in variance caused by partitioning the instances. Maximizing the variance reduction maximizes cluster homogeneity and improves predictive performance.

In this work, we used three instantiations of PCTs for the tasks of multi-target regression (MTR), multi-label classification (MLC) and hierarchical multi-label classification (HMC). For the MTR task, the variance is calculated as the sum of the normalized variances of the target variables. For the MLC task, we used the sum of the Gini indices of the labels, while for the HMC task, the variance is calculated by using a weighted Euclidean distance that considers the hierarchy of the labels. The prototype function returns as a prediction the tuple with the mean values of the target variables, calculated by using the training instances that belong to the given leaf.

## 3    PCTs for Image Representation

The proposed method for constructing the visual codebook is as follows. First, we randomly select a subset of the local (SIFT) descriptors from all of the training images [8]. Next, the selected local descriptors constitute the training set used to construct a PCT. For the construction of a PCT, we set the descriptive attributes (i.e., the 128 dimensional vector of the local descriptor) to be also target and clustering attributes. Note that this feature is a unique characteristic of the predictive clustering framework. The PCTs are computationally efficient: it is very fast to both construct them and use them to make predictions. However, tree learning is unstable, i.e., the structure of the learned tree can change substantially for small changes in the training data [2]. To overcome this limitation and to further improve the discriminative power of the indexing structure, we use an ensemble (i.e., random forest) of PCTs. The overall codebook is obtained by concatenating the codebooks from each tree.

## 4    PCTs for Image Retrieval

In the proposed system, a PCT (or a random forest of PCTs) represents the search/indexing structure used to retrieve images similar to query images. Namely, for each image descriptor (i.e., each training example used to construct the PCTs), we keep a unique index/identifier. The identifier consists of the image ID from which the local descriptor was extracted coupled with a descriptor ID. This indexing allows for faster computation of the image similarities.

We have evaluated the proposed improvement of the bag-of-visual words approach on three reference datasets and two additional datasets of 100 K images and 1 M images, comparing it to two state-of-the-art methods based on approximate k-means and extremely randomized tree ensembles. The results from the experimental evaluation reveal the following. First and foremost, our system exhibits better retrieval performance by 6–8% (mean average precision) than both competing methods at the same efficiency. Additionally, the increase of the number of local descriptors and number of PCTs used to create the indexing structure improve the retrieval performance of the system.

# 5   PCTs for Hierarchical Annotation of Images

We first use our system for multi-label annotation of general purpose images [5]. We compare the efficiency and the discriminative power of the proposed approach to the literature standard of using k-means clustering. The results reveal that our approach is much more efficient in terms of computational time (24.4 times faster) and produces a visual codebook with better discriminative power as compared to k-means clustering. Moreover, the difference in predictive performance increases with the average number of labels per image.

Next, we evaluate the performance of ensembles of PCTs for HMC (bagging and random forests) on the task of annotation of medical images using the hierarchy from the DICOM header [3]. The experiments on the IRMA database show that random forests of PCTs for HMC outperform SVMs for flat classification. The average difference is 17 points for the ImageCLEF2007 and 20 points for the ImageCLEF2008 dataset (a point in the hierarchical evaluation measure roughly corresponds to one completely misclassified image). Additionally, the random forests are the fastest method; they are 10 times faster than bagging and 5.5 times faster than the SVMs.

Finally, for the task of hierarchical annotation of diatom images, by using random forests of PCTs for HMC, we obtained the best results on the different variants of the ADIAC database of diatom images [4]: The obtained predictive power of our method was in the range 96–98%. More specifically, we outperformed a variety of methods for annotation that use SVMs, bagged decision trees and neural networks. Finally, we used these annotations in an on-line annotation system to assist taxonomists in identifying a wide range of different diatoms.

**Acknowledgments.** We would like to acknowledge the support of the European Commission through the project MAESTRA - Learning from Massive, Incompletely annotated, and Structured Data (Grant number ICT-2013-612944).

# References

1. Blockeel, H., Raedt, L.D., Ramon, J.: Top-down induction of clustering trees. In: Proceedings of the 15th International Conference on Machine Learning, pp. 55–63. Morgan Kaufmann (1998)
2. Breiman, L.: Random forests. Mach. Learn. **45**(1), 5–32 (2001)
3. Dimitrovski, I., Kocev, D., Loskovska, S., Dzeroski, S.: Hierarchical annotation of medical images. Pattern Recognit. **44**(10–11), 2436–2449 (2011)
4. Dimitrovski, I., Kocev, D., Loskovska, S., Dzeroski, S.: Hierarchical classification of diatom images using ensembles of predictive clustering trees. Ecol. Inform. **7**(1), 19–29 (2012)
5. Dimitrovski, I., Kocev, D., Loskovska, S., Dzeroski, S.: Fast and efficient visual codebook construction for multi-label annotation using predictive clustering trees. Pattern Recognit. Lett. **38**, 38–45 (2014)
6. Dimitrovski, I., Kocev, D., Loskovska, S., Dzeroski, S.: Improving bag-of-visual-words image retrieval with predictive clustering trees. Inf. Sci. **329**, 851–865 (2016)

7. Kocev, D., Vens, C., Struyf, J., Džeroski, S.: Tree ensembles for predicting structured outputs. Pattern Recognit. **46**(3), 817–833 (2013)
8. Lowe, D.G.: Distinctive image features from scale-invariant keypoints. Int. J. Comput. Vis. **60**(2), 91–110 (2004)

# Music Generation Using Bayesian Networks

Tetsuro Kitahara[(⊠)]

College of Humanities and Sciences, Nihon University,
3-25-40, Sakurajosui, Stagaya-ku, Tokyo 156-8550, Japan
kitahara@chs.nihon-u.ac.jp
http://www.kthrlab.jp/

**Abstract.** Music generation has recently become popular as an application of machine learning. To generate polyphonic music, one must consider both *simultaneity* (the vertical consistency) and *sequentiality* (the horizontal consistency). Bayesian networks are suitable to model both simultaneity and sequentiality simultaneously. Here, we present music generation models based on Bayesian networks applied to chord voicing, four-part harmonization, and real-time chord prediction.

## 1 Introduction

Music is widely known as an application domain of machine learning. However, in the beginning of the 21st century, recognition/analysis tasks were actively studied, such as music transcription and genre classification. But recently, the number of studies devoted to music generation has been increasing (e.g., [1]).

When generating polyphonic music, one must consider two-directional consistencies: *simultaneity* (i.e., the vertical or pitch-axis consistency) and *sequentiality* (i.e., the horizontal or time-axis consistency). Our team has investigated music generation models considering both simultaneity and sequentiality using Bayesian networks [2–4]. Here, we present our models applied to chord voicing [2], four-part harmonization [3], and real-time chord prediction [4].

## 2 Assumed Music Structure and Fundamental Model

Suppose that a chord progression $C = [c_1, c_2, \cdots, c_N]$ ($c_i$: chord symbol) exists in a piece of music. Each chord $c_i$ (e.g., Am) is played with a particular voicing $(a_i^{(1)}, a_i^{(2)}, \cdots, a_i^{(K)})$ ($a_i^{(k)}$: note name (a.k.a. pitch class)) (e.g., (C, E, A)). As noted in Introduction, a set of simultaneous notes $(a_i^{(1)}, a_i^{(2)}, \cdots, a_i^{(K)})$ should be harmonically consistent with one other, and each sequence $A^{(k)} = [a_1^{(k)}, a_2^{(k)}, \cdots, a_N^k]$ should be temporally smooth. At the same time, a melody $M = [m_{1,1}, m_{1,2}, \cdots, m_{2,1}, \cdots]$ exists, where $m_{i,j}$ represents the note name of the $j$-th note in the $i$-th chord region. The sequences of chords, voicings, and

This work was supported by JSPS KAKENHI Grant Numbers 16K16180, 16H01744, 16KT0136, and 17H00749.

melody notes are considered to have temporal dependencies within each sequence but also depends on one another, as shown in Fig. 1(a). In fact, this fundamental model is difficult to construct because of variations in the number of melody notes within each chord region. We therefore simplify the model based on restrictions to music structures designed for each music generation task.

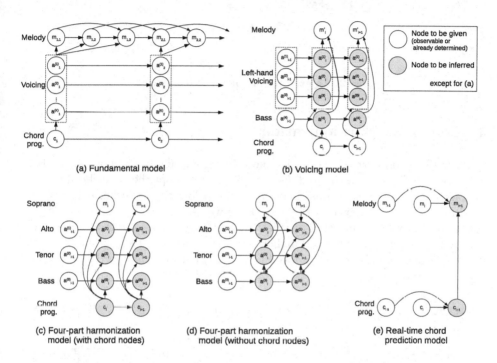

(a) Fundamental model

(b) Voicing model

(c) Four-part harmonization model (with chord nodes)

(d) Four-part harmonization model (without chord nodes)

(e) Real-time chord prediction model

**Fig. 1.** Fundamental model and models specialized to each task

## 3    Chord Voicing

Chord voicing refers to estimating voicings $(A^{(1)}, A^{(2)}, \cdots, A^{(K)})$ according to a given chord progression $C$ and melody $M$. Here we assume $K = 4$ for simplicity. To resolve the difficulty due to variations in the number of melody notes within each chord region, we use a different melody node $m_i' = (r_{i,0}, \cdots, r_{i,11})$ $(0 \leq r_{i,p} \leq 1)$ that represents the relative length of the appearance of each note name. For example, $m_i' = (0.5, 0, 0.25, 0, 0.25, 0, \cdots, 0)$ is given for a melody [E, D, C, C] (with equal duration). The simplified model is shown in Fig. 1(b).

This model is applied sequentially from the beginning to the end of a given piece. Given $c_i$, $m_i'$, and $(a_{i-1}^{(1)}, \cdots, a_{i-1}^{(K)})$, the $i$-th chord voicing $(a_i^{(1)}, \cdots, a_i^{(K)})$ as well as its next voicing $(a_{i+1}^{(1)}, \cdots, a_{i+1}^{(K)})$ is estimated because each voicing should be smoothly connected to the next voicing. $(a_{i+1}^{(1)}, \cdots, a_{i+1}^{(K)})$ will be overridden at the next step because this step is repeated for each increment of $i$.

An example of chord voicing is shown in Fig. 2. The model has been trained with 30 jazz pieces arranged for the electronic organ. Listening tests conducted by music experts revealed that 94.7% of the chord voicings were acceptable.

**Fig. 2.** An example of voicing (excerpted)

## 4   Four-Part Harmonization

Here, we focus on harmonization. Unlike voicing, a sequence of chord symbols is not given—it has to be estimated. For simplicity, we adopt the "one chord for one melody note" assumption. Based on this assumption, the Bayesian network can be simplified to that shown in Fig. 1(c). Here we assume $K = 3$. This problem is called four-part harmonization because the harmony consists of four voices (i.e., soprano, alto, tenor, and bass). Furthermore, we constructed a Bayesian network in which the chord nodes are removed (Fig. 1(d)) because the chord symbols are sometimes too ambiguous.

**Fig. 3.** Example of harmonization (left: model with chord nodes, right: model without chord nodes)

Figure 3 shows an example of harmonization using these two models. Our objective quantitative evaluation reveals that the model shown in Fig. 1(d) generates more temporally smooth harmonies than the model shown in Fig. 1(c) even though harmonizations with the former model tend to contain slightly more dissonant sounds.

## 5   Real-Time Chord Prediction

Finally, we apply our Bayesian network to real-time chord prediction. Music experts can often precisely predict the next chord by listening to the current chord, even if they are not familiar with the piece being played. This ability derives from the fact that chord progressions have strong temporal dependencies; experts have learned these dependencies based on their musical experience. They are therefore able to play an accompaniment to a melody that they are listening to for the first time. The goal here is to achieve a computer system that plays such an accompaniment.

Real-time chord prediction can also be achieved through a simplified version of the fundamental model shown in Fig. 1(a). For simplicity, we estimate only chord symbols, we determine the voicings through a separately designed rule. The model used here is shown in Fig. 1(e). Given a new melody note, its next note is predicted. At the same time, the most likely next chord is inferred based on the current chord and the predicted next note.

An example of chord prediction is shown in Fig. 4. This figure shows that the model appropriately predicts chord progression.

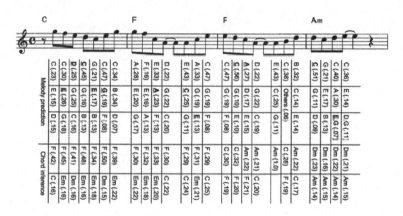

**Fig. 4.** Example of real-time chord prediction results

# 6 Conclusion

We have presented Bayesian network models that achieve different music generation tasks: chord voicing, four-part harmonization, and real-time chord prediction. Bayesian networks are flexible models that are suitable to construct a unified music generation model. In the future, we will apply our model to other types of music generation tasks.

# References

1. Harjeres, G., Pachet, F.: DeepBach: A Steerable Model for Bach Chorales Generation, arXiv:1612.01010 [cs.AI] (2016)
2. Kitahara, T., Katsura, M., Katayose, H., Nagata, N.: Computational model for automatic chord voicing based on Bayesian network. In: ICMPC, pp. 395–398 (2008)

3. Suzuki, S., Kitahara, T.: Four-part harmonization using Bayesian networks: pros and cons of introducing chord nodes. J. New Music Res. **43**(3), 331–353 (2014)
4. Kitahara, T., Totani, N., Tokuami, R., Katayose, H.: BayesianBand: jam session system based on mutual prediction by user and system. In: Natkin, S., Dupire, J. (eds.) ICEC 2009. LNCS, vol. 5709, pp. 179–184. Springer, Heidelberg (2009). https://doi. org/10.1007/978-3-642-04052-8_17

# Phenotype Inference from Text and Genomic Data

Maria Brbić[1], Matija Piškorec[1], Vedrana Vidulin[1], Anita Kriško[2], Tomislav Šmuc[1], and Fran Supek[1,3(✉)]

[1] Ruđer Bošković Institute, Zagreb, Croatia
[2] Mediterranean Institute of Life Sciences, Split, Croatia
[3] Centre for Genomic Regulation, Barcelona, Spain
fran.supek@irb.hr

**Abstract.** We describe ProTraits, a machine learning pipeline that systematically annotates microbes with phenotypes using a large amount of textual data from scientific literature and other online resources, as well as genome sequencing data. Moreover, by relying on a multi-view non-negative matrix factorization approach, ProTraits pipeline is also able to discover novel phenotypic concepts from unstructured text. We present the main components of the developed pipeline and outline challenges for the application to other fields.

**Keywords:** Phenotypic trait · Microbes · Comparative genomics
Late fusion · Text mining · Non-negative matrix factorization

## 1 Introduction

With the development of next-generation DNA sequencing techniques, the number of available microbial genomes has rapidly increased. However, this explosive growth of genomics data is not followed by the phenotypic annotations of organisms, such as growth at extreme temperatures, resistance to radiation, or the ability to cause disease in plants, animals or humans. The systematic annotation of organisms with phenotypic traits is of importance for discovering the associations between genes to phenotypes that would suggest a biological basis for various traits. Existing databases [7,11] rely on manual annotation of organisms, which results in limited coverage. On the other hand, there is a vast amount of unstructured data with phenotype descriptions available in scientific articles and other textual resources. Motivated by this abundance of genomic and of textual data, we developed ProTraits [2] - a machine learning-based pipeline that systematically assigns predictions across large number of organisms and phenotypes. Along with predicting existing phenotypic labels, ProTraits pipeline is also able to define novel phenotypic concepts from unstructured text using a multi-view approach based on non-negative matrix factorization followed by clustering and manual curation. Here, we briefly describe main components of

© Springer International Publishing AG 2017
Y. Altun et al. (Eds.): ECML PKDD 2017, Part III, LNAI 10536, pp. 373–377, 2017.
https://doi.org/10.1007/978-3-319-71273-4_34

our pipeline and present an overview of results. The proposed approach can easily be extended to other fields with the abundant unstructured textual data. The ProTraits database of microbial phenomes is available at http://protraits. irb.hr/.

## 2   Methodology

In this section, we describe the main components of the ProTraits pipeline (Fig. 1): (i) unsupervised phenotype discovery based on multi-view non-negative matrix factorization; (ii) a supervised machine learning framework for phenotype inference from textual and genomic data; (iii) a late-fusion based component for the combination of predictions coming from 11 independent models, and (iv) a user-friendly web interface providing searchable predictions.

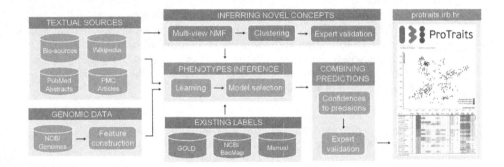

**Fig. 1.** System architecture of the ProTraits pipeline

### 2.1   Initial Data

Text documents describing bacterial and archaeal species were downloaded from six textual resources including Wikipedia, the MicrobeWiki student-edited resource, PubMed abstracts of scientific publications, PubMedCentral full-texts, and an additional set of assorted microbiology resources. The initial set of phenotype assignments was collected from NCBI, BacMap [11] and GOLD databases [7]. The set of biochemical phenotypes was collected manually from individual publications where various microbial species were initially characterized.

### 2.2   Inferring Phenotypic Concepts

We applied non-negative matrix factorization (NMF), commonly used for topic discovery tasks, to each text resource separately to discover novel phenotypic concepts. We then clustered the NMF factors, while requiring that a concept has to be consistently discoverable in at least three text resources. Since the NMF algorithm has a stochastic component, we ran the algorithm multiple times with different random seeds while also varying the number of factors parameter, in order to maximize the diversity of discovered concepts. These groups were then examined by an expert and those describing new phenotypes were retained and used in the same way as labels collected from the existing databases. In total, we discovered 113 non-redundant novel phenotypic concepts.

## 2.3 Phenotype Prediction

In the phenotype prediction task, the learning examples were species and the class label was the presence/absence of a phenotype in that species. A separate model was trained for each of the 424 phenotypes and 10-fold cross-validation used to estimate the accuracy. Once a model was learned, it was applied to the species with unknown phenotypic annotations. To make the functioning of our models more interpretable to biologists, we also provide sets of most important features of all models.

**Predictions from textual data.** We used bag-of-words representation with tf-idf weighting of word frequencies across documents assigned to species in a given text corpus. A Support vector machine (SVM) classifier with a linear kernel was trained on all combinations of text resources and phenotypes.

**Predictions from genome data.** We constructed five different genomic representations for each microbial species: (i) the proteome composition [1,9]; (ii) the gene repertoire encoded as presence/absence of Clusters of Orthologous Groups (COG) gene families [4,6]; (iii) co-occurrence of species across environmental sequencing data sets [3]; (iv) gene neighborhoods [8] encoded as pairwise chromosomal distances between gene family members; and (v) genomic signatures of translation efficiency in gene families [5,10]. Again, we trained models on all combinations of representations and phenotypes. We used the Random Forest (RF) classifier which we found to outperform other tested algorithms.

**Combining predictions.** To combine predictions from different models and provide an interpretable estimate of confidence in each prediction, the confidence scores of each prediction were converted to precisions, based on cross-validation precision-recall curves. Precision scores for organisms in the initially unlabeled set of organisms were calculated via linear interpolation between the neighboring confidence points and then assigned to both positive and negative class for each prediction and further adjusted to account for difference in class sizes, ensuring that the minimum precision of each class is 0, regardless of the number of positive/negative examples. The systematic validation performed by two experts on a random sample of 2,500 predictions showed that the precisions combined using late fusion schemes agree well with human judgment, particularly when requiring agreement of two independent models (either text or genomics-derived).

**Web interface and results.** In summary, ProTraits covers 3,046 microbial organisms and 424 microbial phenotypes. It provides predictions across six textual resources and five independent genomic representations. At the precision threshold higher than 0.9, ProTraits assigns ≈545,000 novel annotations, out of which ≈308,000 are supported in two or more independent predictions. A web interface at http://protraits.irb.hr/ provides precision scores across 11 individual predictors and an integrated score calculated using the two-votes late fusion scheme.

## 3   Challenges and Conclusions

Training separate classifiers for each of the phenotypes does not scale well in terms of computation time required, especially for high-dimensional genomic datasets. However, using existing multi-label classifiers was not straightforward for our datasets since most of the target values were missing. Another challenge was collecting initial labels, as this requires tedious manual curation. While the two existing microbial phenotype databases alleviated this problem in our work, for other important problems in the life sciences, similar databases may not be available. Crucially, the input of field experts has allowed us to validate predictions and inferred concepts, demonstrating that our models are trustworthy.

**Acknowledgments.** This work has been funded by the by the European Union FP7 grants ICT-2013-612944 (MAESTRA) and Croatian Science Foundation grants HRZZ-9623.

## References

1. Brbić, M., Warnecke, T., Kriško, A., Supek, F.: Global shifts in genome and proteome composition are very tightly coupled. Genome Biol. Evol. **7**, 1519–1532 (2015)
2. Brbić, M., Piškorec, M., Vidulin, V., Kriško, A., Šmuc, T., Supek, F.: The landscape of microbial phenotypic traits and associated genes. Nucleic Acids Res. **44**, 10074–10090 (2016)
3. Chaffron, S., Rehrauer, H., Pernthaler, J., von Mering, C.: A global network of coexisting microbes from environmental and whole-genome sequence data. Genome Res. **20**, 947–959 (2010)
4. Feldbauer, R., Schulz, F., Horn, M., Rattei, T.: Prediction of microbial phenotypes based on comparative genomics. BMC Bioinform. **16**, 1–8 (2015)
5. Kriško, A., Copić, T., Gabaldón, T., Lehner, B., Supek, F.: Inferring gene function from evolutionary change in signatures of translation efficiency. Genome Biol. **15**, R44 (2014)
6. MacDonald, N.J., Beiko, R.G.: Efficient learning of microbial genotype-phenotype association rules. Bioinformatics **26**, 1834–1840 (2010)
7. Reddy, T.B.K., Thomas, A.D., Stamatis, D., Bertsch, J., Isbandi, M., Jansson, J., Mallajosyula, J., Pagani, I., Lobos, E.A., Kyrpides, N.C.: The Genomes OnLine Database (GOLD) v. 5: a metadata management system based on a four level (meta)genome project classification. Nucleic Acids Res. **43**, D1099–1106 (2015)
8. Rogozin, I.B., Makarova, K.S., Murvai, J., Czabarka, E., Wolf, Y.I., Tatusov, R.L., Szekely, L.A., Koonin, E.V.: Connected gene neighborhoods in prokaryotic genomes. Nucleic Acids Res. **30**, 2212–2223 (2002)

 9. Smole, Z., Nikolic, N., Supek, F., Šmuc, T., Sbalzarini, I.F., Kriško, A.: Proteome sequence features carry signatures of the environmental niche of prokaryotes. BMC Evol. Biol. 11–26 (2011)
10. Supek, F., Škunca, N., Repar, J., Vlahoviček, K., Šmuc, T.: Translational selection is ubiquitous in prokaryotes. PLoS Genet. **6**, e1001004 (2010)
11. Stothard, P., Van Domselaar, G., Shrivastava, S., Guo, A., O'Neill, B., Cruz, J., Ellison, M., Wishart, D.S.: BacMap: an interactive picture atlas of annotated bacterial genomes. Nucleic Acids Res. **33**, D317–D320 (2005)

# Process-Based Modeling and Design
# of Dynamical Systems

Jovan Tanevski[1](✉), Nikola Simidjievski[1], Ljupčo Todorovski[1,2],
and Sašo Džeroski[1]

[1] Jožef Stefan Institute, Ljubljana, Slovenia
{jovan.tanevski,nikola.simidjievski,saso.dzeroski}@ijs.si
[2] University of Ljubljana, Ljubljana, Slovenia
ljupco.todorovski@fu.uni-lj.si

**Abstract.** Process-based modeling is an approach to constructing explanatory models of dynamical systems from knowledge and data. The knowledge encodes information about potential processes that explain the relationships between the observed system entities. The resulting process-based models provide both an explanatory overview of the system components and closed-form equations that allow for simulating the system behavior. In this paper, we present three recent improvements of the process-based approach: (i) improving predictive performance of process-based models using ensembles, (ii) extending the scope of process-based models towards handling uncertainty and (iii) addressing the task of automated process-based design.

## 1    Introduction

Process-based modeling (PBM) supports knowledge discovery by learning understandable and communicable models of dynamical systems. PBM uses domain-specific knowledge as declarative bias in combination with observed time-series data to address the task of modeling real-world systems. It performs both structure identification and parameter estimation, resulting in a process-based model which specifies a set of differential equations. In turn, such models accurately capture the complex and nonlinear behavior of a dynamical system through time.

Learning models of dynamical systems is a supervised machine learning task: the predictive variables correspond to observed system variables, while the targets correspond to their time derivatives. However, the task bears two specific properties that limit the use of traditional machine learning approaches. First, the resulting models take the form of a set of entities, processes and differential equations, i.e., artifacts used by scientists and engineers to construct explanatory models. On the other hand, machine learning methods operate on classes of predictive models that generalize well over arbitrary data, while keeping the complexity of training and evaluation procedures low. Second, the observed variables are measured at consecutive time points, so the data instances breach the common assumption of their mutual independence.

Y. Altun et al. (Eds.): ECML PKDD 2017, Part III, LNAI 10536, pp. 378–382, 2017.
https://doi.org/10.1007/978-3-319-71273-4_35

The PBM approach relies on the paradigm of computational scientific discovery [3] and more specifically, on approaches to inductive process modeling. On one hand, research in this area has a long tradition and has been applied to a variety of domains [1,2,10,11]. However, while successful, it has been at the margins of mainstream machine learning. On the other hand, the PBM approach has so far focused primarily on applications within a narrow class of problems that emphasize descriptive and deterministic models at output, given a single data type at input. In terms of output, such models are typically simulated and analyzed using the learning data. Therefore, they have a tendency to overfit – rendering them incapable at accurately predicting future system's behavior. Also, these models do not capture the intrinsic uncertainty of the interactions in the system. They always predict exactly the same behavior of the system at output in a deterministic manner: determined only by initial conditions and ignoring the uncertainty in real-world systems. In terms of input, an assumption of the PBM is that time-series of observations are always available and sufficient. This, however, does not hold for problems with limited observability or tasks, such as design, where different types of input are required.

In response, our recent developments of the PBM approach have aimed at bridging the gap between machine learning and domains of application within physical and life sciences. We address the limitations of the PBM approach by broadening the classes of tasks it can address. We build on the tradition of constant performance improvement, but also extend the scope of potential applications. In particular, to improve the performance on the task of predictive modeling, we support the learning of different types of ensembles of process-based models [4–6]. Next, we extended the output to include process-based models that describe stochastic interactions [7]. Finally, in order to address tasks of modeling dynamical systems under limited observability and tasks of design of dynamical systems, we consider different types of input data. Namely, in addition to time-series of observations of system variables we allow for the definition of expected properties of the behavior of the dynamical system [8,9].

## 2   Methods

The PBM learning task takes domain-specific knowledge and time-series data at input (Fig. 1). The resulting model comprises system variables represented as *entities* and their interactions that define the underlying model structure represented as *processes*. This representation allows for straightforward mapping of process-based models into a set of differential equations. The model parameters are fitted to the data using evolutionary optimization methods with the sum-of-squares loss function as the objective. The PBM approach, however, adds an extra layer to the model equations. In particular, the models are constructed using components from a library of domain-knowledge, represented by *template entities and processes*. These templates encode taxonomies of variable and constant properties of the constituents in the dynamical systems as well as the taxonomies of processes (interactions) among them. The (partial) instantiations of

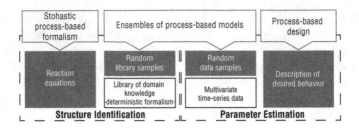

**Fig. 1.** General overview of the three extensions of PBM presented in this paper.

such templates, taken from arbitrary levels of the respective taxonomies, define and constrain the model structure search space for a specific modeling task.

PBM has four distinguishing features. First, it produces *understandable* models, which give clear insight into the structure of a dynamical system building on the traditional mathematical description. The processes relate specific parts of the set of differential equations to understandable real world causal relations between the system's components. Second, process-based models retain the *utility* of traditional mathematical models. They can be readily simulated and analyzed using well established numerical approaches. Third, PBM is *generally applicable* to domains that require models described in terms of equations. Finally, the PBM approach is *modular*. The domain-knowledge library can be instantiated into a number of different modeling components specific to a particular modeling task. It captures the basic modeling principles in a given domain and can be reused for different modeling applications within the same domain.

We report on three extensions of PBM (Fig. 1). To improve the capability to predict future system's behavior, we consider learning of ensembles of process-based models. The constituent base process-based models are learned either from different samples of the measured data [4], random samples of the library of domain knowledge [6] or both [5]. Such sampling approaches have a direct effect on the generalization ability of the ensembles, leading to improved predictive performance. Second, the ensembles of process-based models can provide long-term predictions, relying only on the initial values of the state variables as opposed to traditional ML ensembles (in the context of time-series) that are typically used for short-term prediction.

To capture the intrinsic uncertainty of interactions within real world dynamical systems, we propose an improved finer grained formalism for representing domain knowledge [7]. It encodes the interactions between entities, i.e., processes in the form of reaction equations allowing for both deterministic and stochastic interpretation of process-based models and knowledge.

We extended the input to the PBM approach to different types of data, which allows handling a broader set of tasks ranging from completely data-driven to completely knowledge-driven modeling. In this context, we first strengthen the evaluation bias of modeling tasks with limited observability [9]. We use domain-specific criteria for model selection as part of a general regularized objective function for parameter optimization and model selection. Second, we formulate the novel task of process-based design of dynamical systems [8]. This approach does not take measured data at input, but is completely based on the description of desired properties of the behavior of a dynamical system. We further generalize the task by taking advantage of methods for simultaneous optimization of multiple conflicting objectives (desired properties of the behavior). We use the complete information from the Pareto front of optimal solutions (obtained for every candidate design) to rank the designs and make a well informed selection.

## 3   Significance and Challenges

The methodology for learning ensembles of PBMs extends the scope of the traditional ensemble paradigm in machine learning towards modeling dynamical systems. It improves the generalization power of PBMs, providing more accurate simulation of the future behavior of the modeled systems. The proposed methodology employs four different methods for constructing ensembles of process-based models. Each of these significantly improves the predictive performance (on average up to 60% of relative improvement) over individual models on tasks of modeling population dynamics in three lake ecosystems [4–6].

The extension of the PBM approach towards stochastic process-based models has allowed us to model dynamical systems that are out of the scope of deterministic models. We have demonstrated that the stochastic PBM is capable of reconstructing known, manually constructed models from synthetic and real-world data in the domains of systems biology and epidemiology [7].

The capability of PBM to handle different inputs and multiple modeling objectives has led to important contributions in the domains of systems and synthetic biology. In particular, PBM can address the problem of high structural uncertainty (many candidate model structures) and incomplete data (i.e., limited observability of the system variables). In system biology, our approach can alleviate the model selection problem by strengthening the evaluation bias with introducing domain-specific model selection criteria [9]. In synthetic biology, we can now use PBM to solve the task of automated design. Our results show that PBM is capable of reconstructing known/good designs, as well as proposing novel alternative designs of a synthetic stochastic switch and a synthetic oscillator [8].

Note, finally, that all three extensions of the PBM approach are designed and implemented as independent modular components. Therefore, they are interoperable. They can be, in principle, arbitrarily combined and applied to novel tasks, such as learning ensembles of stochastic process-based models.

Several challenges, that we are aware of and currently working on, remain in PBM. The exhaustive combinatorial search currently in use is computationally inefficient and does not scale well with the number of candidate model structures. It is therefore necessary to integrate methods for heuristic search in our current implementation. An alternative approach to reducing search complexity is to use higher-level constraints on model structures that are more expressive than the current constraints. They can be based on the topological properties of the candidate model structures, or can define a probability distribution over the model structures. Finally, both process-based modeling and design require further evaluation on other related domains, such as neurobiology, systems pharmacology and systems medicine, or on completely new domains. The new applications will most certainly open up new directions for improvement of the PBM approach.

**Acknowledgements.** The authors acknowledge the financial support of the Slovenian Research Agency (research core funding No. P2-0103, No. P5-0093 and project No. N2-0056 Machine Learning for Systems Sciences) and the Ministry of Education, Science and Sport of Slovenia (agreement No. C3330-17-529021).

# References

1. Bridewell, W., Langley, P., Todorovski, L., Džeroski, S.: Inductive process modelling. Mach. Learn. **71**, 109–130 (2008)
2. Džeroski, S., Langley, P., Todorovski, L.: Computational discovery of scientific knowledge. In: Džeroski, S., Todorovski, L. (eds.) Computational Discovery of Scientific Knowledge. LNCS (LNAI), vol. 4660, pp. 1–14. Springer, Heidelberg (2007). https://doi.org/10.1007/978-3-540-73920-3_1
3. Langley, P., Simon, H.A., Bradshaw, G.L., Zytkow, J.M.: Scientific Discovery: Computational Explorations of the Creative Processes. MIT Press, Cambridge (1992)
4. Simidjievski, N., Todorovski, L., Džeroski, S.: Predicting long-term population dynamics with bagging and boosting of process-based models. Expert Syst. Appl. **42**(22), 8484–8496 (2015)
5. Simidjievski, N., Todorovski, L., Džeroski, S.: Learning ensembles of process-based models by bagging of random library samples. In: Calders, T., Ceci, M., Malerba, D. (eds.) DS 2016. LNCS (LNAI), vol. 9956, pp. 245–260. Springer, Cham (2016). https://doi.org/10.1007/978-3-319-46307-0_16
6. Simidjievski, N., Todorovski, L., Džeroski, S.: Modeling dynamic systems with efficient ensembles of process-based models. PLoS One **11**(4), 1–27 (2016)
7. Tanevski, J., Todorovski, L., Džeroski, S.: Learning stochastic process-based models of dynamical systems from knowledge and data. BMC Syst. Biol. **10**(1), 1–30 (2016)
8. Tanevski, J., Todorovski, L., Džeroski, S.: Process-based design of dynamical biological systems. Sci. Rep. **6**(1), 1–13 (2016)
9. Tanevski, J., Todorovski, L., Kalaidzidis, Y., Džeroski, S.: Domain-specific model selection for structural identification of the Rab5-Rab7 dynamics in endocytosis. BMC Syst. Biol. **9**(1), 1–31 (2015)
10. Todorovski, L., Bridewell, W., Shiran, O., Langley, P.: Inducing hierarchical process models in dynamic domains. In: Proceedings of the Twentieth National Conference on Artificial Intelligence, pp. 892–897. AAAI Press (2005)
11. Čerepnalkoski, D., Taškova, K., Todorovski, L., Atanasova, N., Džeroski, S.: The influence of parameter fitting methods on model structure selection in automated modeling of aquatic ecosystems. Ecol. Model. **245**, 136–165 (2012)

# QuickScorer: Efficient Traversal of Large Ensembles of Decision Trees

Claudio Lucchese[1]([✉]), Franco Maria Nardini[1], Salvatore Orlando[1,2],
Raffaele Perego[1], Nicola Tonellotto[1], and Rossano Venturini[1,3]

[1] ISTI–CNR, Pisa, Italy
{claudio.lucchese,francomaria.nardini,raffaele.perego,
nicola.tonellotto}@isti.cnr.it
[2] Ca' Foscari University of Venice, Venice, Italy
orlando@unive.it
[3] University of Pisa, Pisa, Italy
rossano.venturini@unipi.it

**Abstract.** Machine-learnt models based on additive ensembles of binary regression trees are currently deemed the best solution to address complex classification, regression, and ranking tasks. Evaluating these models is a computationally demanding task as it needs to traverse thousands of trees with hundreds of nodes each. The cost of traversing such large forests of trees significantly impacts their application to big and stream input data, when the time budget available for each prediction is limited to guarantee a given processing throughput. Document ranking in Web search is a typical example of this challenging scenario, where the exploitation of tree-based models to score query-document pairs, and finally rank lists of documents for each incoming query, is the state-of-art method for ranking (a.k.a. *Learning-to-Rank*). This paper presents QUICKSCORER, a novel algorithm for the traversal of huge decision trees ensembles that, thanks to a cache- and CPU-aware design, provides a $\sim 9\times$ speedup over best competitors.

**Keywords:** Learning to rank · Ensemble of decision trees · Efficiency

## 1 Introduction

In this paper we discuss QUICKSCORER (QS), an algorithm developed to speedup the application of machine-learnt forests of binary regression trees to score and finally rank lists of candidate documents for each query submitted to a Web search engine. QUICKSCORER was thus developed in the field of *Learning-to-Rank* (LtR) within the IR community. Nowadays, LtR is commonly exploited by Web search engines within their query processing pipeline, by exploiting massive training datasets consisting of collections of query-document pairs, in turn modeled as vectors of hundreds features, annotated with a relevance label.

The interest in exploiting forests of binary regression trees to rank lists of candidate documents is due to the success of gradient boosting tree algorithms [4].

© Springer International Publishing AG 2017
Y. Altun et al. (Eds.): ECML PKDD 2017, Part III, LNAI 10536, pp. 383–387, 2017.
https://doi.org/10.1007/978-3-319-71273-4_36

This kind of algorithms is considered the state-of-the-art LtR solution for addressing complex ranking problems [5]. In search engines, these forests are exploited within a two-stage architecture. While the first stage retrieves a set of possibly relevant documents matching the user query, such expensive LtR-based scorers, optimized for high precision, are exploited in the second stage to *re-rank* the set of candidate documents coming from the first stage. The *time budget* available to re-rank the candidate documents is limited, due to the incoming rate of queries and the users' expectations in terms of response time. Therefore, devising techniques and strategies to speed up document ranking without losing in quality is definitely an urgent research topic in Web search [9].

Strongly motivated by these considerations, the IR community has started to investigate computational optimizations to reduce the scoring time of the most effective LtR rankers based on ensembles of regression trees, by exploiting advanced features of modern CPUs and carefully exploiting memory hierarchies. Among those, the best competitor of QUICKSCORER is vPRED [1].

We argue that QUICKSCORER can also be exploited in different *time-sensitive* scenarios and each time it is needed to use a large forest of binary decision trees, e.g., random forest, for classification/regression purposes and apply it to big and stream data with strict processing throughput requirements.

## 2   QuickScorer

Given a query-document pair $(q, d_i)$, represented by a feature vector $\mathbf{x}$, a LtR model based on an additive ensemble of regression trees predicts a relevance score $s(\mathbf{x})$ used for ranking a set of documents. Typically, a tree ensemble encompasses several binary decision trees, denoted by $\mathcal{T} = \{T_0, T_1, \ldots\}$. Each internal (or branching) node in $T_h$ is associated with a Boolean test over a specific feature $f_\phi \in \mathcal{F}$, and a constant threshold $\gamma \in \mathbb{R}$. Tests are of the form $\mathbf{x}[\phi] \leq \gamma$, and, during the visit, the left branch is taken *iff* the test succeeds. Each leaf node stores the tree prediction, representing the potential contribution of the tree to the final document score. The scoring of $\mathbf{x}$ requires the traversal of all the ensemble's trees and it is computed as a *weighted sum* of all the tree predictions.

Algorithm 1 illustrates QS [3,7]. One important result is that QS computes $s(\mathbf{x})$ by only identifying the branching nodes whose test evaluates to false, called *false nodes*. For each false node detected in $T_h \in \mathcal{T}$, QS updates a bitvector associated with $T_h$, which stores information that is eventually exploited to identify the *exit leaf* of $T_h$ that contributes to the final score $s(\mathbf{x})$. To this end, QS maintains for each tree $T_h \in \mathcal{T}$ a bitvector leafidx[$h$], made of $\Lambda$ bits, one per leaf. Initially, every bit in leafidx[$h$] is set to 1. Moreover, each branching node is associated with a bitvector mask, still of $\Lambda$ bits, identifying the set of unreachable leaves of $T_h$ in case the corresponding test evaluates to false. Whenever a false node is visited, the set of unreachable leaves leafidx[$h$] is updated through a *logical AND* ($\wedge$) with mask. Eventually, the leftmost bit set in leafidx[$h$] identifies the leaf corresponding to the score contribution of $T_h$, stored in the lookup table leafvalues.

**Algorithm 1.** QUICKSCORER

```
1  QUICKSCORER(x,𝒯):
2    foreach Tₕ ∈ 𝒯 do
3      leafidx[h] ← 11...11

4    foreach f_φ ∈ ℱ do // Mask Computation
5      foreach (γ, mask, h) ∈ 𝒩_φ do
6        if x[φ] > γ then
7          leafidx[h] ← leafidx[h] ∧ mask
8        else
9          break

10   score ← 0       // Score Computation
11   foreach Tₕ ∈ 𝒯 do
12     j ← leftmost bit set in leafidx[h]
13     l ← h · Λ + j
14     score ← score + leafvalues[l]

15   return score
```

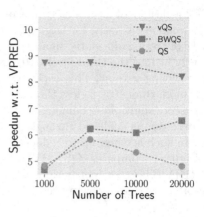

**Fig. 1.** QS performance.

To efficiently identify all the *false nodes* in the ensemble, QS processes the branching nodes of all the trees *feature by feature*. Specifically, for each feature $f_\phi$, QS builds a list $\mathcal{N}_\phi$ of tuples $(\gamma, \text{mask}, h)$, where $\gamma$ is the predicate threshold of a branching node of tree $T_h$ performing a test over the feature $f_\phi$, denoted by $\mathbf{x}[\phi]$, and mask is the pre-computed mask that identifies the leaves of $T_h$ that are un-reachable when the associated test evaluates to false. $\mathcal{N}_\phi$ is statically sorted in ascending order of $\gamma$. Hence, when processing $\mathcal{N}_\phi$ sequentially, as soon as a test evaluates to true, i.e., $\mathbf{x}[\phi] \leq \gamma$, the remaining occurrences surely evaluate to true as well, and their evaluation is thus safely skipped.

We call *mask computation* the first step of the algorithm during which all the bitvectors leafidx[h] are updated, and *score computation* the second step where such bitvectors are used to retrieve tree predictions.

Compared to the classic tree traversal, QUICKSCORER introduces a main novelty. The cost of the traversal does not depend on the average length of the root-to-leaf paths, but rather on the average number of false nodes in the trees of the forest. Experiments on large public datasets with large forests, with 64 leaves per tree and up to 20,000 trees, show that a classic traversal evaluates between 50% and 80% of the branching nodes. This is due to the imbalance of the trees built by state-of-the-art LtR algorithms. On the other hand, on the same datasets, QUICKSCORER always visits less than 30% of the nodes. This results in a largely reduced number of operations and number of memory accesses.

Moreover, QUICKSCORER exploits a cache- and CPU-aware design. For instance, the values of $(\gamma, \text{mask}, h)$ are accessed through a linear scan of the QUICKSCORER data structures, which favours cache prefetching and limits data dependencies. For each feature, QUICKSCORER visits only one true node, thus easing the CPU branch predictor and limiting control dependencies. This makes QUICKSCORER to perform better than competitors also with a special kind of perfectly balanced trees named *oblivious* [6].

The design of QUICKSCORER makes it possible to introduce two further improvements. Firstly, for large LtR models, the forest can be split into multiple *blocks* of trees, sufficiently small to allow the data structure of a single block to entirely fit into the third-level CPU cache. We name BLOCKWISE-QS (BWQS) the resulting variant. This cache-aware algorithm reduces the cache miss ratio from more than 10% to less than 1%. Secondly, the scoring can be vectorized so as to score multiple documents simultaneously. In V-QUICKSCORER (vQS) [8] vectorization is achieved through AVX 2.0 instructions and 256-bits wide registers. In such setting, up to 8 documents can be processed simultaneously.

Figure 1 compares QS, BWQS, and vQS against the best competitor vPRED. The test was performed on a large dataset, with a model with 64 leaves per tree and varying the number of trees of the forest.

## 3   Discussion

In this work, we focused on tree ensembles to tackle the LtR problem. Decision tree ensembles are a popular and effective machine learning tool beyond LtR. Their success is witnessed by the *Kaggle 2015* competitions, where most of the winning solutions exploited MART models, and by the *KDD Cup 2015*, where MART-based algorithms were used by all the top 10 teams [2].

In the LtR scenario, the time budget available for applying a model is limited and must be satisfied. Therefore large models, despite being more accurate, cannot be used because of their high evaluation cost. QS, a novel algorithm for the traversal of decision trees ensembles, is an answer to this problem as it provides $\sim9\times$ speedup over state-of-the-art competitors. Moreover, the need of efficient traversal strategies goes beyond the LtR scenario, for instance when such models are used to classify big data collections. For all these reasons, we believe that QS can help scientists from the data mining community to speed-up the process of evaluating highly effective tree-based models over big and stream datasets.

## References

1. Asadi, N., Lin, J., de Vries, A.P.: Runtime optimizations for tree-based machine learning models. IEEE TKDE **26**(9), 2281–2292 (2014)
2. Chen, T., Guestrin, C.: XGBoost: A scalable tree boosting system. In: Proceedings of SIGKDD, pp. 785–794. ACM (2016)
3. Dato, D., Lucchese, C., Nardini, F.M., Orlando, S., Perego, R., Tonellotto, N., Venturini, R.: Fast ranking with additive ensembles of oblivious and non-oblivious regression trees. ACM TOIS **35**(2), 1–31 (2016)
4. Friedman, J.H.: Greedy function approximation: a gradient boosting machine. Ann. Stat. **29**, 1189–1232 (2000)
5. Gulin, A., Kuralenok, I., Pavlov, D.: Winning The transfer learning track of yahoo!'s learning to rank challenge with YetiRank. In: Yahoo! Learning to Rank Challenge, pp. 63–76 (2011)

6. Langley, P., Sage, S.: Oblivious decision trees and abstract cases. In: Working Notes of the AAAI-94 Workshop on Case-Based Reasoning, pp. 113–117. AAAI Press (1994)
7. Lucchese, C., Nardini, F.M., Orlando, S., Perego, R., Tonellotto, N., Venturini, R.: QuickScorer: a fast algorithm to rank documents with additive ensembles of regression trees. In: Proceedings of SIGIR, pp. 73–82. ACM (2015)
8. Lucchese, C., Nardini, F.M., Orlando, S., Perego, R., Tonellotto, N., Venturini, R.: Exploiting CPU SIMD extensions to speed-up document scoring with tree ensembles. In: Proceedings of SIGIR, pp. 833–836. ACM (2016)
9. Segalovich, I.: Machine learning in search quality at Yandex. Presentation at the Industry Track of SIGIR (2010)

# Recent Advances in Kernel-Based Graph Classification

Nils M. Kriege$^{(\boxtimes)}$ and Christopher Morris

Department of Computer Science, TU Dortmund University, Dortmund, Germany
{nils.kriege,christopher.morris}@tu-dortmund.de

**Abstract.** We review our recent progress in the development of graph kernels. We discuss the hash graph kernel framework, which makes the computation of kernels for graphs with vertices and edges annotated with real-valued information feasible for large data sets. Moreover, we summarize our general investigation of the benefits of explicit graph feature maps in comparison to using the kernel trick. Our experimental studies on real-world data sets suggest that explicit feature maps often provide sufficient classification accuracy while being computed more efficiently. Finally, we describe how to construct valid kernels from optimal assignments to obtain new expressive graph kernels. These make use of the kernel trick to establish one-to-one correspondences. We conclude by a discussion of our results and their implication for the future development of graph kernels.

## 1 Introduction

In various domains such as chemo- and bioinformatics, or social network analysis large amounts of graph structured data is becoming increasingly prevalent. Classification of these graphs remains a challenge as most graph kernels either do not scale to large data sets or are not applicable to all types of graphs. In the following we briefly summarize related work before discussing our recent progress in the development of efficient and expressive graphs kernels.

### 1.1 Related Work

In recent years, various graph kernels have been proposed. Gärtner et al. [5] and Kashima et al. [8] simultaneously developed graph kernels based on random walks, which count the number of walks two graphs have in common. Since then, random walk kernels have been studied intensively, see, e.g., [7,10,13,19,21]. Kernels based on shortest paths were introduced by Borgwardt et al. [1] and are computed by performing 1-step walks on the transformed input graphs, where edges are annotated with shortest-path lengths. A drawback of the approaches mentioned above is their high computational cost. Therefore, a different line of research focuses particularly on scalable graph kernels. These kernels are typically computed by explicit feature maps, see, e.g., [17,18]. This allows to bypass

© Springer International Publishing AG 2017
Y. Altun et al. (Eds.): ECML PKDD 2017, Part III, LNAI 10536, pp. 388–392, 2017.
https://doi.org/10.1007/978-3-319-71273-4_37

the computation of a gram matrix of quadratic size by applying fast linear classifiers [2]. Moreover, graph kernels using assignments have been proposed [4], and were recently applied to geometric embeddings of graphs [6].

# 2  Recent Progress in the Design of Graph Kernels

We give an overview of our recent progress in the development of scalable and expressive graph kernels.

## 2.1  Hash Graph Kernels

In areas such as chemo- or bioinformatics edges and vertices of graphs are often annotated with real-valued information, e.g., physical measurements. It has been shown that these attributes can boost classification accuracies [1,3,9]. Previous graph kernels that can take these attributes into account are relatively slow and employ the kernel trick [1,3,9,15]. Therefore, these approaches do not scale to large graphs and data sets. In order to overcome this, we introduced the *hash graph kernel framework* in [14]. The idea is to iteratively turn the continuous attributes of a graph into discrete labels using randomized hash functions. This allows to apply fast explicit graph feature maps, e.g., [17], which are limited to discrete annotations. In each iteration we sample new hash functions and compute the feature map. Finally, the feature maps of all iterations are combined into one feature map. In order to obtain a meaningful similarity between attributes in $\mathbb{R}^d$, we require that the probability of collision $\Pr[h_1(x) = h_2(y)]$ of two independently chosen random hash functions $h_1, h_2 \colon \mathbb{R}^d \to \mathbb{N}$ equals an adequate kernel on $\mathbb{R}^d$. Equipped with such a hash function, we derived approximation results for several state-of-the-art kernels which can handle continuous information. Moreover, we derived a variant of the Weisfeiler-Lehman subtree kernel which can handle continuous attributes.

Our extensive experimental study showed that instances of the hash graph kernel framework achieve state-of-the-art classification accuracies while being orders of magnitudes faster than kernels that were specifically designed to handle continuous information.

## 2.2  Explicit Graph Feature Maps

Explicit feature maps of kernels for continuous vectorial data are known for many popular kernels like the Gaussian kernel [16] and are heavily applied in practice. These techniques cannot be used to obtain approximation guarantees in the hash graph kernel framework. Therefore, in a different line of work, we developed explicit feature maps with the goal to lift the known approximation results for kernels on continuous data to kernels for graphs annotated with continuous data [11]. More specifically, we investigated how general convolution kernels are composed from base kernels and how to construct corresponding feature maps. We applied our results to widely used graph kernels and analyzed for which kernels and graph properties computation by explicit feature maps is feasible and actually more efficient. We derived approximative, explicit feature maps for state-of-the-art kernels supporting real-valued attributes. Empirically we observed that for graph kernels like GraphHopper [3] and Graph Invariant [15] approximative explicit feature maps achieve a classification accuracy close to the exact methods based on the kernel trick, but required only a fraction of their running time. For the shortest-path kernel [1] on the other hand the approach fails in accordance to our theoretical analysis.

Moreover, we investigated the benefits of employing the kernel trick when the number of features used by a kernel is very large [10,11]. We derived feature maps for random walk and subgraph kernels, and applied them to real-world graphs with discrete labels. Experimentally we observed a phase transition when comparing running time with respect to label diversity, walk lengths and subgraph size, respectively, confirming our theoretical analysis.

### 2.3    Optimal Assignment Kernels

For non-vectorial data, Fröhlich et al. [4] proposed kernels for graphs derived from an optimal assignment between their vertices, where vertex attributes are compared by a base kernel. However, it was shown that the resulting similarity measure is not necessarily a valid kernel [20,21]. Hence, in [12], we studied optimal assignment kernels in more detail and investigated which base kernels lead to valid kernels. We characterized a specific class of kernels and showed that it is equivalent to the kernels obtained from a hierarchical partition of their domain. When such kernels are used as base kernel the optimal assignment (i) yields a valid kernel; and (ii) can be computed in linear time by histogram intersection given the hierarchy. We demonstrated the versatility of our results by deriving novel graph kernels based on optimal assignments, which are shown to improve over their convolution-based counterparts. In particular, we proposed the Weisfeiler-Lehman optimal assignment kernel, which performs favorable compared to state-of-the-art graph kernels on a wide range of data sets.

## 3    Conclusion

We gave an overview about our recent progress in kernel-based graph classification. Our results show that explicit graph feature maps can provide an efficient computational alternative for many known graph kernels and practical applications. This is the case for kernels supporting graphs with continuous attributes and for those limited to discrete labels, even when the number of features is very large. Assignment kernels, on the other hand, are computed by histogram intersection and thereby again employ the kernel trick. This suggests to study the application of non-linear kernels to explicit graph feature maps in more detail as future work.

**Acknowledgement.** We would like to thank the co-authors of our publications [10–12,14]. This research was supported by the German Science Foundation (DFG) within the Collaborative Research Center SFB 876 "Providing Information by Resource-Constrained Data Analysis", project A6.

# References

1. Borgwardt, K.M., Kriegel, H.P.: Shortest-path kernels on graphs. In: IEEE International Conference on Data Mining, pp. 74–81 (2005)
2. Fan, R.E., Chang, K.W., Hsieh, C.J., Wang, X.R., Lin, C.J.: LIBLINEAR: a library for large linear classification. J. Mach. Learn. Res. **9**, 1871–1874 (2008)
3. Feragen, A., Kasenburg, N., Petersen, J., Bruijne, M.D., Borgwardt, K.: Scalable kernels for graphs with continuous attributes. In: Advances in Neural Information Processing Systems, pp. 216–224 (2013)
4. Fröhlich, H., Wegner, J.K., Sieker, F., Zell, A.: Optimal assignment kernels for attributed molecular graphs. In: 22nd International Conference on Machine Learning, pp. 225–232 (2005)
5. Gärtner, T., Flach, P., Wrobel, S.: On graph kernels: hardness results and efficient alternatives. In: Schölkopf, B., Warmuth, M.K. (eds.) COLT-Kernel 2003. LNCS (LNAI), vol. 2777, pp. 129–143. Springer, Heidelberg (2003). https://doi.org/10.1007/978-3-540-45167-9_11
6. Johansson, F.D., Dubhashi, D.: Learning with similarity functions on graphs using matchings of geometric embeddings. In: 21st ACM SIGKDD International Conference on Knowledge Discovery and Data Mining, pp. 467–476 (2015)
7. Kang, U., Tong, H., Sun, J.: Fast random walk graph kernel. In: SIAM International Conference on Data Mining, pp. 828–838 (2012)
8. Kashima, H., Tsuda, K., Inokuchi, A.: Marginalized kernels between labeled graphs. In: 20th International Conference on Machine Learning, pp. 321–328 (2003)
9. Kriege, N., Mutzel, P.: Subgraph matching kernels for attributed graphs. In: 29th International Conference on Machine Learning (2012)
10. Kriege, N., Neumann, M., Kersting, K., Mutzel, M.: Explicit versus implicit graph feature maps: a computational phase transition for walk kernels. In: IEEE International Conference on Data Mining, pp. 881–886 (2014)
11. Kriege, N.M., Neumann, M., Morris, C., Kersting, K., Mutzel, P.: A unifying view of explicit and implicit feature maps for structured data: Systematic studies of graph kernels. CoRR abs/1703.00676 (2017). http://arxiv.org/abs/1703.00676
12. Kriege, N.M., Giscard, P.-L., Wilson, R.C.: On valid optimal assignment kernels and applications to graph classification. In: Advances in Neural Information Processing Systems, pp. 1615–1623 (2016)
13. Mahé, P., Ueda, N., Akutsu, T., Perret, J.L., Vert, J.P.: Extensions of marginalized graph kernels. In: Twenty-First International Conference on Machine Learning, pp. 552–559 (2004)
14. Morris, C., Kriege, N.M., Kersting, K., Mutzel, P.: Faster kernel for graphs with continuous attributes via hashing. In: IEEE International Conference on Data Mining, pp. 1095–1100 (2016)
15. Orsini, F., Frasconi, P., De Raedt, L.: Graph invariant kernels. In: Twenty-Fourth International Joint Conference on Artificial Intelligence, pp. 3756–3762 (2015)
16. Rahimi, A., Recht, B.: Random features for large-scale kernel machines. In: Advances in Neural Information Processing Systems, pp. 1177–1184 (2008)

17. Shervashidze, N., Schweitzer, P., van Leeuwen, E.J., Mehlhorn, K., Borgwardt, K.M.: Weisfeiler-Lehman graph kernels. J. Mach. Learn. Res. **12**, 2539–2561 (2011)
18. Shervashidze, N., Vishwanathan, S.V.N., Petri, T.H., Mehlhorn, K., Borgwardt, K.M.: Efficient graphlet kernels for large graph comparison. In: Twelfth International Conference on Artificial Intelligence and Statistics, pp. 488–495 (2009)
19. Sugiyama, M., Borgwardt, K.M.: Halting in random walk kernels. In: Advances in Neural Information Processing Systems, pp. 1639–1647 (2015)
20. Vert, J.P.: The optimal assignment kernel is not positive definite. CoRR abs/0801.4061 (2008). http://arxiv.org/abs/0801.4061
21. Vishwanathan, S.V.N., Schraudolph, N.N., Kondor, R., Borgwardt, K.M.: Graph kernels. J. Mach. Learn. Res. **11**, 1201–1242 (2010)

# Demo Track

# ASK-the-Expert: Active Learning Based Knowledge Discovery Using the Expert

Kamalika Das[1(✉)], Ilya Avrekh[2], Bryan Matthews[2], Manali Sharma[3],
and Nikunj Oza[4]

[1] USRA, NASA Ames Research Center, Moffett Field, Moffett Field, CA, USA
`kamalika.das@nasa.gov`
[2] SGT Inc., NASA Ames Research Center, Moffett Field, Moffett Field, CA, USA
`{ilya.avrekh-1,bryan.l.matthews}@nasa.gov`
[3] Samsung Semiconductor Inc., San Jose, CA, USA
`manali.s@samsung.com`
[4] NASA Ames Research Center, Moffett Field, Moffett Field, CA, USA
`nikunj.c.oza@nasa.gov`

**Abstract.** Often the manual review of large data sets, either for purposes of labeling unlabeled instances or for classifying meaningful results from uninteresting (but statistically significant) ones is extremely resource intensive, especially in terms of subject matter expert (SME) time. Use of active learning has been shown to diminish this review time significantly. However, since active learning is an iterative process of learning a classifier based on a small number of SME-provided labels at each iteration, the lack of an enabling tool can hinder the process of adoption of these technologies in real-life, in spite of their labor-saving potential. In this demo we present ASK-the-Expert, an interactive tool that allows SMEs to review instances from a data set and provide labels within a single framework. ASK-the-Expert is powered by an active learning algorithm for training a classifier in the backend. We demonstrate this system in the context of an aviation safety application, but the tool can be adopted to work as a simple review and labeling tool as well, without the use of active learning.

**Keywords:** Active learning · Graphical user interface · Review
Labeling

## 1 Introduction

Active learning is an iterative process that requires feedback on instances from a subject matter expert (SME) in an interactive fashion. The idea in active

M. Sharma—This work was done when the author was a student at Illinois Institute of Technology.

The rights of this work are transferred to the extent transferable according to title 17 U.S.C. 105.

Y. Altun et al. (Eds.): ECML PKDD 2017, Part III, LNAI 10536, pp. 395–399, 2017.
https://doi.org/10.1007/978-3-319-71273-4_38

learning is to bootstrap an initial classifier with a few examples from each class that have been labeled by the SME. Traditional active learning approaches select an informative instance from the unlabeled data set and ask SMEs to review the instance and provide a label. This process continues iteratively until a desired level of performance is achieved by the classifier or when the budget (allotted resources) for the SME is exhausted. Much of the research in active learning simulates this interaction between the learner and the SME. In particular, all labels are collected from the SME a priori and during the active learning process, the relevant labeled instances are revealed to the learner, based on its requests at each iteration. The problem of using such retrospective evaluation of an active learning algorithm is twofold. Firstly, the lack of availability of an interactive interface is largely responsible for the generally low adoption of active learning algorithms in practical scenarios. Secondly, the simulated environment fails to achieve the biggest benefit associated with the use of active learning: reduction of SME review time. This is because the SME has to review and label all examples a priori. Therefore, for utilizing active learning frameworks in situations of low availability of labeled data, it is important to have an interactive tool that allows SMEs to review and label instances only when asked by the learner.

## 2    Application and Demo Scenario

A major focus of the commercial aviation community is discovery of unknown safety events in flight operational data through the use of unsupervised anomaly detection algorithms. However, anomalies found using such approaches are abnormal only in the statistical sense, i.e., they may or may not represent an operationally significant event (e.g. represent a real safety concern). After an algorithm produces a list of statistical anomalies, an SME must review the list to identify those that are operationally relevant for further investigation. Usually, less than 1% of the hundreds or thousands of statistical anomalies turn out to be operationally relevant. Therefore, substantial time and effort is spent examining anomalies that are not of interest and it is essential to optimize this review process in order to reduce SME labeling efforts (man hours spent in investigating results). A recently developed active learning method [2] incorporates SME feedback in the form of rationales for classification of flights to build a classifier that can distinguish between uninteresting and operationally significant anomalies with 70% fewer labels compared to manual review and comparable accuracy.

To the best of our knowledge, there exists no published work that describes such software tools for review and annotation of numerical data using active learning. There are some image and video annotation tools that collect labels, such as LabelMe from MIT CSAIL [1]. Additionally there are active learning powered text labeling tools, such as Abstrackr [3] designed specifically for medical experts for citation review and labeling. The major difference between these annotator tools and our tool is the absence of context in our case. Unlike in the case of image or text data where the information is self-contained in the instance being reviewed, in our case, we have to enable the tool to obtain additional contextual information and visualize the feature space on demand. Other domains plagued by label scarcity can also benefit from the adaptation of this tool, with or without the use of an active learning algorithm.

## 3    System Description

In this demo the goal of our annotation interface is to facilitate review of a set of anomalies detected by an unsupervised anomaly detection algorithm and allow labeling of those anomalies as either operationally significant (OS) or not operationally significant (NOS). Our system, as shown in Fig. 1a consists of two components, viz. the coordinator and the annotator.

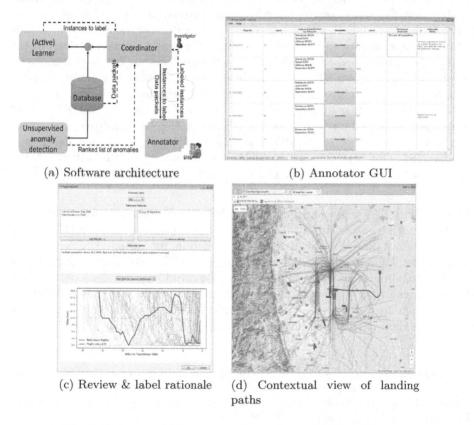

(a) Software architecture

(b) Annotator GUI

(c) Review & label rationale

(d) Contextual view of landing paths

**Fig. 1.** Software architecture and snapshots of ASK-the-Expert

The coordinator has access to the data repository and accepts inputs in the form of a ranked list of anomalies from the unsupervised anomaly detection algorithm. The coordinator is the backbone of the system communicating iteratively with the active learner, gathering information on instances selected for annotation and packing information for transmission to the annotator. Once the annotator collects and sends the labeled instances, the coordinator performs two tasks: (i) resolve labeling conflicts across multiple SMEs through the use of a majority voting scheme or by invoking an investigator review, and (ii) automate the construction of new rationale features as conjunctions and/or disjunctions of raw data features based on the rationale notes entered by the SME in the annotation window. All data exchange between the coordinator and the annotator happens through cloud based storage. The annotator, shown in Fig. 1b is the graphical user interface that the SMEs work with and needs to be installed at the SME end. When the annotator is opened, it checks for new data packets (to be labeled) on the cloud. If new examples need annotation, the annotator window displays the list of examples ranked in the order of importance along with the features identified to be the most anomalous. Clicking on the annotate button next to each example, the SME can delve deeper into that example in order to provide a label for that instance. The functions of the annotator include (i) obtaining examples to be labeled from the cloud and displaying them to the SME, (ii) allowing review of individual features as well as feature interactions (shown in Fig. 1c), and (iii) occasionally providing additional context information by looking at additional data sources (for example, plotting flight paths in the context of other flights landing on the same runway at a certain airport using geographical data from maps, as shown in Fig. 1d). Multiple annotators can be used simultaneously by different SMEs to label the same or different sets of examples. Once the labeled examples are submitted by the annotator, the coordinator collects and consolidates them and sends them back to the learner.

**Demo Plan:** We will demonstrate the ASK-the-Expert tool for an aviation safety case study. Since the data cubes for normal and anomalous flights are proprietary information, the database will be hosted in our laptop. The coordinator tool will be live and running at NASA, gathering the latest set of flights that need to be labeled and uploading them on the cloud. We will demonstrate how the SMEs can review new examples in the context of other flights and provide labels. Their feedback will be sent back to the learner through the coordinator for the next iteration of classifier learning after incorporating new rationale features.

**Acknowledgments.** This work is supported in part by a Center Innovation Fund (CIF) 2017 grant at NASA Ames Research Center and in part by the NASA Aeronautics Mission Directorate. Manali Sharma was supported by the National Science Foundation CAREER award no. IIS-1350337.

# References

1. Russell, B., Torralba, A., Murphy, K., Freeman, W.: LabelMe: a database and web-based tool for image annotation. Int. J. Comput. Vis. **77**(1), 157–173 (2007)
2. Sharma, M., Das, K., Bilgic, M., Matthews, B., Nielsen, D., Oza, N.: Active learning with rationales for identifying operationally significant anomalies in aviation. In: Proceedings of European Conference on Machine Learning and Knowledge Discovery in Databases, ECML-PKDD 2016, pp. 209–225 (2016)
3. Wallace, B., Small, K., Brodley, C., Lau, J., Trikalinos, T.: Deploying an interactive machine learning system in an evidence-based practice center: Abstrackr. In: Proceedings of the 2nd ACM SIGHIT International Health Informatics Symposium, pp. 819–824 (2012)

# Delve: A Data Set Retrieval and Document Analysis System

Uchenna Akujuobi and Xiangliang Zhang[(✉)]

King Abdullah University of Science and Technology, Thuwal, Saudi Arabia
{uchenna.akujuobi,xiangliang.zhang}@kaust.edu.sa

**Abstract.** Academic search engines (e.g., Google scholar or Microsoft academic) provide a medium for retrieving various information on scholarly documents. However, most of these popular scholarly search engines overlook the area of data set retrieval, which should provide information on relevant data sets used for academic research. Due to the increasing volume of publications, it has become a challenging task to locate suitable data sets on a particular research area for benchmarking or evaluations. We propose Delve, a web-based system for data set retrieval and document analysis. This system is different from other scholarly search engines as it provides a medium for both data set retrieval and real time visual exploration and analysis of data sets and documents.

## 1 Introduction

The area of scholarly search engines although sparsely studied, is not a new phenomenon. Search engines provide a new insight into scholarly information searchable on the web, incorporating functionalities to rank and measure academic activities [3]. However, due to the unprecedented rate in the number of scholarly papers published per year [4], researchers often go through an exhaustive step of re-searching and reading through many documents to locate usable data sets (i.e., relevant benchmark/evaluation data sets) that fits their research problem setting. It is therefore, desirable to have a platform where experts and non-experts are able to access not just topic or document information but also relevant data sets, together with the ability to analyze their interconnection. This task can be structured as an information retrieval task [5]. Current systems are designed either for data set search[1] or for scholarly search[2]. One system [1] incorporated the use of data set as a filter agent for their document search results. However, users are often interested in *locating data sets relevant to their search* rather than using the data sets to filter their search.

In Delve[3], we take a different approach by designing a system that allows users to locate both relevant documents and data sets, and also to visualize

---

[1] http://www.re3data.org/.

[2] http://www.scholar.google.com/.

[3] The system can be seen in action at https://youtu.be/bF6PUj8801U.

Y. Altun et al. (Eds.): ECML PKDD 2017, Part III, LNAI 10536, pp. 400–403, 2017.
https://doi.org/10.1007/978-3-319-71273-4_39

and analyze their relationship network. Delve borrows ideas from label propagation [2] algorithm and adopts methods proposed in ParsCit [6] for text mining. Our system also provides a simple and easy-to-use interface built on the d3.js[4] framework which facilitates visualization and analysis of papers and data sets.

## 2   System Design

Our data set was constructed with an initial focus on academic documents published in 17 different conferences and journals between **2001** to **2015**, including ICDE, KDD, TKDE, VLDB, CIKM, NIPS, ICML, ICDM, PKDD, SDM, WSDM, AAAI, IJCAI, DMKD, WWW, KAIS and TKDD. Using the Microsoft graph data set[5], we then extended these documents, adding their references and the references of their references (up to 2 hops away). In total, we currently have **2,116,429** academic publications from more than 1000 different conferences and journals.

**Data Set and Document Analysis.** Our system is built on the citation graph of these more than 2 million papers. Formally, in a directed citation graph $G = \{V, E\}$, two nodes $v_i$ and $v_j$ are linked by edge $E(v_i; v_j)$ if $v_i$ cites $v_j$. Since the system is designed for data set relevant retrieval, an edge $E(v_i; v_j)$ between $v_i$ and $v_j$ can be labeled as: 1 - if $v_i$ cites $v_j$ because $v_i$ uses the data set available/used in $v_j$; and 0 - otherwise. Then based on the labels, we can extract the data set labeled citations. The initial labeling work was conducted by crowd-sourcing on papers and data sets cited by papers published in ICDE, KDD, ICDM, SDM and TKDE from 2001 to 2014. These labels (accounting for *5% of the whole graph edges*) have been manually verified to be correct by three qualified participants. Due to the high cost, it is infeasible to label the remaining 95% of edges manually. Therefore, the main challenging task is to develop a correct and yet efficient algorithm to efficiently assign labels to the large amount of unlabeled edges using the limited amount of verified labels. To solve this problem, we developed a semi-supervised learning method *"link label propagation algorithm"* using ideas borrowed from label propagation algorithm [2].

**Label Assignment.** The original label propagation (LP) algorithm predicts labels for nodes, our task is to predict labels for edges. Therefore we restructure the original graph to $G' = \{V', E'\}$ where $V'$ is the set of edges $E$ in graph $G$, and $E'$ is the set of generated edges. The edges $E'$ are generated by linking each edge $E_i$ in $G$ ($V'_i$ in $G'$) to the top 10 similar edges $E_j$ ($V'_j$ in $G'$) that have the same target node as $E_i$ or where the target node of $E_i$ is the source node of $E_j$. To define the similarity between citations, we extract the number of data set keywords[6] from each citation context (i.e. the sentences which encompass the citations). We then defined a Gaussian similarity score between pairs of edges $(E_i, E_j)$ $Sim_{ij} = \exp(-\frac{\|d_i - d_j\|^2}{2\sigma^2})$, where $d_i = \frac{n_d}{n_c}$. $n_d$ is the number of data set

---

[4] https://d3js.org/.
[5] https://academicgraph.blob.core.windows.net/graph-2015-11-06/index.html.
[6] Manually compiled list of data set related words.

related words in the sentences which encompasses the citation depicted as $E_i$, and $n_c$ is the number of such sentences in the source papers. For edges having the same target nodes, we assign a weight of $1 + Sim_{ij}$, and $0.5 + Sim_{ij}$ otherwise.

With the constructed graph $G' = \{V', E'\}$ where a small portion of $V'$ have verified labels, label propagation algorithm is run to propagate the given labels to unlabeled $V'$. We conducted extensive experiments to evaluate our designed method. Our system achieves an average precision of 82%.

## 3   Use Cases

Delve is based on two components: search and online document analysis.

**Search:** This enables users to search on a keyword, author or phrase for both documents and data sets. Delve analyzes this query and presents the user with results (outputs) ranked by relevance. Figure 1 shows the result of the query "multi-label learning". The search result is split into two: data set results and scholarly document results. The data set result is further split into three parts: 1. Matched data sets (data sets matching the search query). 2. Popular data set (data sets used by the papers matching the search query ordered by popularity). 3. Unavailable data sets (currently temporary or permanently inaccessible relevant data sets, e.g., invalid or closed links). Data sets can be either papers where the data sets are described or web links where the data sets are located.

**On-Line Document Analysis:** This function enables a user to analyze a paper by understanding its relationship with other papers and data sets without having to go through the references; searching each of them manually. It can also be used by authors to discover which papers are advisable to cite in their work. A user can either analyze any document in our database or upload a scholarly document file for analysis, e.g., a PDF file. When a document is uploaded for

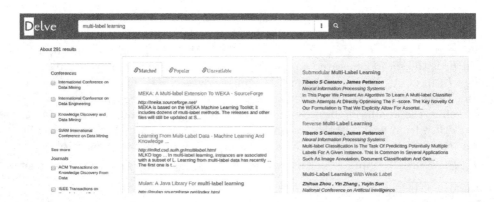

**Fig. 1.** Results from searching for "multi-label learning" in Delve

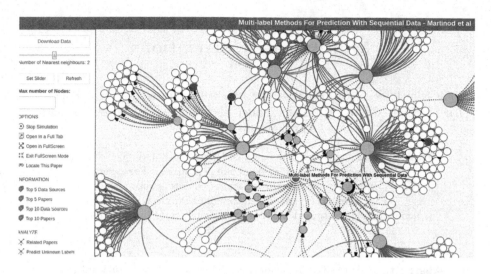

**Fig. 2.** Final output of uploaded file analysis in Delve (Color figure online)

analysis, Delve mines and analyzes the document text, translates the results as a query and displays the result as a visual citation graph, as shown in Fig. 2, which gives the result of analyzing *Multi-label methods for prediction with sequential data* [7]. We would like to point out that this paper is not in our system at the moment of writing this paper. However, based on its references and citations, our system can analyze its relevant papers and visualize the citation relations.

Note that in Fig. 2, the blue edges indicate data set relevant relationships, and the size of the nodes show its importance in the network measured based its citations in the subgraph. Mouse hovering over a node displays the item title and clicking on a node displays more information about the item. In addition, the red edges show a non data set relevant relationship, and broken edges have unknown labels. The unknown labels can be inferred using label propagation.

# References

1. Semantic Scholar: Web. https://www.semanticscholar.org. Accessed 21 Feb 2017
2. Fujiwara, Y., Irie, G.: Efficient label propagation. In: ICML-2014 (2014)
3. Ortega, J.L.: Academic Search Engines: A Quantitative Outlook. Elsevier (2014)
4. National Science Board: Science and engineering indicators. National Science Foundation, Arlington (2012)
5. Sathiaseelan, J.G.R.: A technical study on Information Retrieval using web mining techniques. In: ICIIECS (2015)
6. Councill, I.G., et al.: ParsCit: an open-source CRF reference string parsing package. In: LREC (2008)
7. Read, J., et al.: Multi-label methods for prediction with sequential data. Pattern Recognit. **63**, 45–55 (2017)

# Framework for Exploring and Understanding Multivariate Correlations

Louis Kirsch[✉], Niklas Riekenbrauck, Daniel Thevessen, Marcus Pappik,
Axel Stebner, Julius Kunze, Alexander Meissner, Arvind Kumar Shekar,
and Emmanuel Müller

Hasso Plattner Institute, University of Potsdam, Potsdam, Germany
{louis.kirsch,niklas.riekenbrauck,daniel.thevessen,
marcus.pappik,axel.stebner,julius.kunze,
alexander.meissner}@student.hpi.de,
arvind.shekar@guest.hpi.de, emmanuel.mueller@hpi.de

**Abstract.** Feature selection is an essential step to identify relevant and non-redundant features for target class prediction. In this context, the number of feature combinations grows exponentially with the dimension of the feature space. This hinders the user's understanding of the feature-target relevance and feature-feature redundancy. We propose an interactive Framework for Exploring and Understanding Multivariate Correlations (FEXUM), which embeds these correlations using a force-directed graph. In contrast to existing work, our framework allows the user to explore the correlated feature space and guides in understanding multivariate correlations through interactive visualizations.

## 1 Introduction

The amount of data collected in various applications such as life-sciences, e-commerce and engineering is ever-growing. A common method used to avoid the curse-of-dimensionality and reduce the cost of collecting data is feature selection. In order to provide a smaller yet predictive subset of features, a large variety of existing approaches [4] such as CFS compute the relevance of each feature to the target class, as well as the redundancy between features.

However, the user does not get an overview of all correlations in the dataset. Furthermore, the selection process is non-transparent, as the reason for a feature's relevance or redundancy is not explained by these algorithms. This non-transparency impairs the user's understanding of the data. A high-dimensional dataset may also contain many redundant features, i.e., features exhibiting linear or non-linear dependency. Hence, the first challenge for explaining the feature selection process is to present relevance and redundancy jointly in an informative layout. The second challenge is to guide the user in understanding how features

© Springer International Publishing AG 2017
Y. Altun et al. (Eds.): ECML PKDD 2017, Part III, LNAI 10536, pp. 404–408, 2017.
https://doi.org/10.1007/978-3-319-71273-4_40

are correlated as opposed to merely returning a correlation score. We address these two challenges by contributing FEXUM, a framework that provides:

(1) A visual embedding of feature correlations (relevances and redundancies).
(2) User-reviewable multivariate correlations.

This leads to a more comprehensible selection process in comparison to state-of-the-art tools, reflected in Table 1. While most tools focus on fully-automated statistical selection of features, with FEXUM we aim at explaining traditional black-box algorithms. KNIME is a renowned tool that offers filter-based feature selection using linear correlation and variance measures. However, without customized extensions, it does not address feature redundancy during selection. RapidMiner and Weka take redundancy into account, but do not provide an overview of all feature correlations. Additionally, they do not explain the reason for the relevance of a feature.

**Table 1.** Comparison of feature selection tools

| Tools | Relevance | Redundancy | Correlation overview | Correlation explanation |
|---|---|---|---|---|
| KNIME | ✓ | ✗ | ✗ | ✗ |
| RapidMiner | ✓ | ✓ | ✗ | ✗ |
| Weka | ✓ | ✓ | ✗ | ✗ |
| FEXUM | ✓ | ✓ | ✓ | ✓ |

## 2  FEXUM

FEXUM is an application that allows instant access with a web browser. We achieve this by basing our infrastructure on AngularJS and the Django web framework. To ensure scalability for large datasets, we distribute computations to multiple machines with Celery. The entire framework is open source and available online[1].

### 2.1  Relevance-Redundancy Embedding

As explained in Sect. 1, existing feature ranking methods do not provide a comprehensive overview of correlations that facilitate exploring the dataset. Therefore, our goal is to simultaneously visualize all feature correlations to the target (relevance) and pairwise correlations (redundancy). We allow for arbitrary relevance and redundancy measures. For now, we employ the concept of conditional dependence from [2] to quantify the correlations.

However, it is computationally expensive to calculate the redundancy score for all feature pairs. We propose, hence, to infer the redundancy scores heuristically from random subsets. The pseudo-code for this computation is made available online[2]. Our visualization provides a layout in which a smaller distance of a

---

[1] https://github.com/KDD-OpenSource/fexum.
[2] https://github.com/KDD-OpenSource/fexum-hics/blob/master/FPR.pdf.

feature to the target denotes a greater relevance, while a smaller distance between two features denotes a greater redundancy. We interpret this as a graph in which nodes represent features and weighted edges represent distances. These distances do not obey the triangle inequality and therefore cannot be mapped to metric space. We address this challenge by applying force-directed graph drawing [1]. Our algorithm places features randomly and applies forces proportional to the difference between their current distance and their correlation-defined distance. With these forces, we run a simulation until equilibrium is reached. This method is suitable even for datasets with several hundreds of features. If the correlation measure supports it, the view is updated iteratively, minimizing waiting time for the user. This is the case for our current implementation. As shown in Fig. 1, force-directed graph drawing allows soft clustering of features. Serving as a major advantage, this provides not only a relevance ranking of features, but also an understanding of how features interact with each other in terms of redundancy. This enables the user to freely select one feature from each cluster, potentially in accordance with the user's domain knowledge.

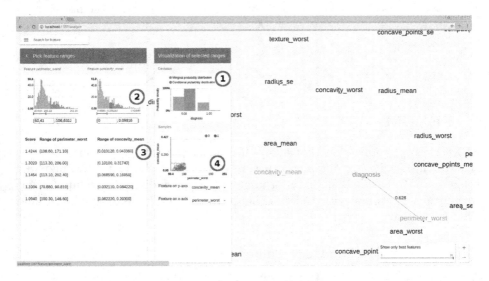

**Fig. 1.** Features drawn using a force-directed graph (right), with the target highlighted in green. An analysis view of two features (left) for inspecting the correlations. (Color figure online)

## 2.2   Understanding Feature Relevance and Redundancy

Having selected a feature set $S \subset X$, where $X = \{x_1, \cdots, x_d\}$ is a $d$-dimensional dataset, the second goal of our framework is to provide insight into its correlations with the target $y$. We propose using the average divergence between the marginal probability of $y$ and the probability of $y$ conditioned on different value ranges of $S$. For every feature $s \in S$, a value range of interest can be chosen.

If a feature $s$ correlates with the target feature $y$, there exists a value range of $s$ which changes the distribution of $y$ in contrast to $y$'s marginal distribution [2].

As shown in Fig. 1, our framework allows specifying the respective value range per feature using value sliders. Therefore, both bivariate and multivariate correlations can be detected. Our framework guides the user with four essential components for understanding correlations with the target.

Both the target's marginal probability distribution and the distribution conditioned on the selected value ranges are rendered in Fig. 1: (1). Changing value ranges updates this plot in real-time, allowing the user to test hypotheses evaluated according to the resulting divergence from the marginal distribution. Identifying the right hypotheses becomes challenging with more features to consider. To address this, value ranges that maximally violate the assumption of statistical independence w.r.t. the target feature are highlighted in a histogram above the sliders in (2). This tells the user which ranges strongly contribute to bivariate correlations. Nevertheless, it is still difficult to find multivariate correlations. Therefore, a table in (3) suggests multiple configurations, where each configuration specifies a value range for each $s \in S$. Each configuration is scored based on the divergence of its probability distribution and only the highest scoring configurations are displayed. Selecting one of these suggestions updates the respective value range sliders and the probability distribution plot. Finally, in case $y$ is categorical, we visualize the data points within the value ranges in our two dimensional scatter plot in (4), each data point colorized according to its respective class.

## 2.3 Demonstration

FEXUM can be used with a wide range of datasets, supplied through upload by the user. While it is currently in use in industry, we will demonstrate our framework on publicly available datasets from medical, social and physical applications. As an example, we now show how our framework enhances feature selection for the Wisconsin Breast Cancer (Diagnostic) dataset [3] in Fig. 1.

In the rendering of our force-directed graph, we observe varied feature relevance scores and clusters of redundant features. In particular, features derived from similar properties such as *radius_mean* and *radius_worst* achieve comparable relevances and are highly redundant to each other. Based on this first impression, we decide to have a closer look at the most relevant feature *perimeter_worst*. We can easily find influential value ranges in the analysis view, because they are highlighted in red in the histogram. The overall relevance score can be corroborated by analyzing several individual value ranges, which can be chosen based on the framework's recommendations or expert knowledge.

Since we support multivariate correlations, the current subset can be iteratively expanded in a similar fashion. As demonstrated, the framework guides in exploration and review of correlations.

# References

1. Fruchterman, T.M., Reingold, E.M.: Graph drawing by force-directed placement. Softw. Pract. Exp. **21**(11), 1129–1164 (1991)
2. Keller, F., Müller, E., Bohm, K.: HiCS: high contrast subspaces for density-based outlier ranking. In: ICDE (2012)
3. Lichman, M.: UCI machine learning repository (2013). http://archive.ics.uci.edu/ml. Accessed 17 Apr 2017
4. Molina, L.C., Belanche, L., Nebot, À.: Feature selection algorithms: a survey and experimental evaluation. In: ICDM (2003)

# *Lit@EVE*: Explainable Recommendation Based on Wikipedia Concept Vectors

M. Atif Qureshi[(✉)] and Derek Greene

Insight Center for Data Analytics, University College Dublin,
Dublin, Republic of Ireland
{muhammad.qureshi,derek.greene}@ucd.ie

**Abstract.** We present an explainable recommendation system for novels and authors, called *Lit@EVE*, which is based on Wikipedia concept vectors. In this system, each novel or author is treated as a concept whose definition is extracted as a concept vector through the application of an explainable word embedding technique called *EVE*. Each dimension of the concept vector is labelled as either a Wikipedia article or a Wikipedia category name, making the vector representation readily interpretable. In order to recommend items, the *Lit@EVE* system uses these vectors to compute similarity scores between a target novel or author and all other candidate items. Finally, the system generates an ordered list of suggested items by showing the most informative features as human-readable labels, thereby making the recommendation explainable.

## 1 Introduction

Recently, considerable attention has been paid to providing meaningful explanations for decisions made by algorithms [5]. On the legislative side, the European Union has approved regulations that requires a "right to explanation" in relation to any user-facing algorithm [2]. This increased emphasis on the need for explainable decision-making algorithms is the first motivation for our work. As further motivation, increasingly recommender systems attempt to offer serendipitous suggestions, where the items being recommended are relevant but also potentially different from those items which they users seen previously [3]. To address both of these motivations, we propose the *Lit@EVE* system, which makes use of Wikipedia articles and categories as a rich source of structured features. Furthermore, to explain the similarity between items, the system makes use of our previously-proposed word embedding algorithm called *EVE* [6]. Word embedding algorithms generate real-valued representation for words or concepts in a vector space, allowing simple comparisons to be made between them by operating over their corresponding vectors. In the case of *EVE*, the dimensions of this space are human-readable, as each dimension represents a single Wikipedia article or category. We demonstrate this approach in the context of recommending books and authors, where *EVE* concept vectors are used to represent both authors and their literary works.

© Springer International Publishing AG 2017
Y. Altun et al. (Eds.): ECML PKDD 2017, Part III, LNAI 10536, pp. 409–413, 2017.
https://doi.org/10.1007/978-3-319-71273-4_41

Recently, Chang et al. [1] described a crowdsourcing-based framework for generating natural language explanations which relies on specific human-generated annotations, whereas our system harnesses the ongoing work of Wikipedia editors, and automatically assigns labels to explain a given recommendation. Moreover, the use of a rich set of Wikipedia articles and categories as features helps to highlight serendipitous aspects of recommended items which are otherwise difficult to discover.

## 2   System Overview

We now present an overview of the *Lit@EVE* system. First, we discuss the dataset used to build our recommender, then we discuss the corresponding *EVE* word embeddings, and finally we show how recommendations are generated using the system.

### 2.1   Dataset

Our dataset is based on the curated "Wikiproject novels"[1] list which contains 49,999 Wikipedia entries (as of 20 April 2017) relating to literature. Many of these entries correspond to novels, although some denote other literary concepts, such as genres, publishers, and tropes. In order to exclusively extract novels, we include only those with a Wikipedia info box that contains an "author" attribute. This filtered set has 18,572 entries corresponding to novels. From the author attribute of each entry, we discovered 2,512 unique authors. Our combined dataset contains both the novel and author entries.

### 2.2   Concept Embeddings

The *EVE* algorithm generates a vector embedding of each word or concept by mapping it to a Wikipedia article[2] [6]. For example, the concept "Harry Potter" is mapped to the Wikipedia article of the novel "Harry Potter". After identifying the concept article, *EVE* generates a vector with dimensions quantifying the association of the mapped article with other Wikipedia articles and categories. In the case of articles, *EVE* exploits the hyperlink structure of Wikipedia. Specifically, associations are calculated as a normalised sum of the number of incoming and outgoing links between the concept article and other Wikipedia articles. Furthermore, a self-citation is also added for the concept article. To quantify associations with Wikipedia categories, *EVE* propagates scores from the concept article to other related Wikipedia categories – e.g., "Harry Potter" has related categories "Fantasy novel series", "Witchcraft in written fiction", etc. Each of the related categories receives a uniform score which is propagated to neighbouring categories (i.e., super and sub categories) by means of a factor

---

[1] https://en.wikipedia.org/wiki/Wikipedia:WikiProject_Novels.
[2] Either an exact match or a best match.

called *jump probability*. The propagation continues until a maximum hop count is reached, which prevents topical drift. The final embedding vector for the concept is constructed from the associations for all articles and categories. For further details on the construction of embedding vectors refer to our paper on *EVE* [6]. We apply this process for all novels and authors in our dataset. The resulting vectors form the input for *Lit@EVE* to generate explainable recommendations.

### 2.3  *Lit@EVE* Recommendations

*Lit@EVE* generates recommendations via a two step process. Firstly, it embeds domain-specific knowledge in the *EVE* vectors, and then it applies a similarity function to these vectors to rank candidate recommendations.

**Domain-specific vector rescaling:** To generate recommendations, we eliminate rare dimensions from the *EVE* vector embeddings for novels and authors and incorporate domain-specific knowledge in the vector embeddings. This is done as follows. First, we calculate the item frequency of each dimension (i.e., the number of novels or authors with a non-zero association for this value). Dimensions with a frequency <3 are eliminated from the model. This limits the dimensionality to 156,553 unique features for novels and authors. Next, we scale the dimensions by the inverse item frequency of each dimension. Furthermore, each association of the Wikipedia hyperlink in the vector representation is scaled by the importance of the Wikipedia hyperlink which is calculated by PageRank score [4]. Finally, the vectors are normalised to unit length.

**Generating recommendations:** The rescaled vectors representing novels and authors are used to generate recommendations. For a target novel or author, we calculate cosine similarities between that item and the rest of the items in the dataset. The candidate list is then sorted by similarity to identify the top recommended items. Each recommended item is explained by the most informative features i.e., the embedding dimensions which maximise the similarity score between the target and recommend item; we select top-n informative features where n equals 10 for this demonstration. The explanation corresponding to the informative feature is the label of that dimension (e.g. "American Horror Novelist").

## 3  User Interface

Figure 1 shows the query-based exploratory interface of *Lit@EVE*. Users may query or select an item (a novel or author) which allows for further exploration through explainable recommendations. Each novel suggested to a user is explained through features such as "Novels Set In Kansas", while each suggested author is also explained with features such as "British Writers". Alternatively, users may opt to browse items strongly associated with features, such as "Fantasy Novels" or "Victorian Novelists". The following use-cases illustrate the various aspects of recommendations generated by *Lit@EVE*:

**Fig. 1.** The *Lit@EVE* interface supports three selection levels – novels, authors, and features.

- Selecting the novel "Harry Potter and the Order of the Phoenix" suggests "The Lord of the Rings" as the recommended novel, with common features such as both being "BILBY Award-winning works", both being "Sequel Novels", and both involving a plot having "Fictional Prisons".
- Selecting the author "Terry Pratchett" offers a list of similar author recommendations e.g. "John Fowles". Both are explained with common features such as "English Humanists", "English Atheists", "20th-century English Novelists".
- Selecting the feature "Nautical Fiction" offers a list of novel recommendations from genres such as "Adventure novel", "Historical Fiction", and "Children's fantasy novel". This may be interesting to a user who is interested in "Nautical Fiction" who would like to browse novels from different genres which incorporate aspects of nautical fiction.

An interesting aspect of the explanations associated with our recommendations is the granularity at which they help users to discover serendipitous aspects around a given novel or author. For instance, in the first use case above, the feature "BILBY Award-winning works" connects diverse works that have won this children's book award, potentially allowing users to make serendipitous discoveries of novels of this type. For further details on the unique aspects of recommendations generated by *Lit@EVE*, we refer the reader to an online video demonstration of the system[3].

**Acknowledgments.** This publication has emanated from research conducted with the support of Science Foundation Ireland (SFI) under Grant Number SFI/12/RC/2289.

# References

1. Chang, S., Harper, F.M., Terveen, L.: Crowd-based personalized natural language explanations for recommendations. In: Proceedings of the 10th ACM Conference on Recommender Systems, pp. 175–182. ACM (2016)
2. Goodman, B., Flaxman, S.: European union regulations on algorithmic decision-making and a "right to explanation". arXiv preprint arXiv:1606.08813 (2016)
3. Kotkov, D., Wang, S., Veijalainen, J.: A survey of serendipity in recommender systems. Knowl.-Based Syst. **111**, 180–192 (2016)
4. Page, L., Brin, S., Motwani, R., Winograd, T.: The PageRank citation ranking: bringing order to the web. Technical report, Stanford InfoLab (1999)
5. Qiu, L., Gao, S., Cheng, W., Guo, J.: Aspect based latent factor model by integrating ratings and reviews for recommender system. Knowl.-Based Syst. **110**, 233–243 (2016)
6. Qureshi, M.A., Greene, D.: EVE: explainable vector based embedding technique using Wikipedia. arXiv preprint arXiv:1702.06891 (2017)

---

[3] http://mlg.ucd.ie/liteve/.

# Monitoring Physical Activity and Mental Stress Using Wrist-Worn Device and a Smartphone

Božidara Cvetković(✉), Martin Gjoreski, Jure Šorn, Pavel Maslov, and Mitja Luštrek

Department of Intelligent Systems, Jožef Stefan Institute,
Jamova 39, 1000 Ljubljana, Slovenia
boza.cvetkovic@ijs.si

**Abstract.** The paper presents a smartphone application for monitoring physical activity and mental stress. The application utilizes sensor data from a wristband and/or a smartphone, which can be worn in various pockets or in a bag in any orientation. The presence and location of the devices are used as contexts for the selection of appropriate machine-learning models for activity recognition and the estimation of human energy expenditure. The stress-monitoring method uses two machine-learning models, the first one relying solely on physiological sensor data and the second one incorporating the output of the activity monitoring and other context information. The evaluation showed that we recognize a wide range of atomic activities with the accuracy of 87%, and that we outperform the state-of-the art consumer devices in the estimation of energy expenditure. In stress monitoring we achieved the accuracy of 92% in a real-life setting.

**Keywords:** Machine-learning · Activity recognition
Estimation of energy expenditure · Mental stress detection
Wrist-worn device · Smartphone

## 1 Introduction

A typical worker in the competitive labor market of developed countries spends long hours in an office (sitting disease) under high mental stress. Since it is acknowledged that a lack of physical activity and mental stress contribute to the development of various diseases, poor mental health and decreased quality of life, it is crucial to increase the self-awareness of the population and provide solutions to improve their lifestyle. Wearable devices and mobile applications with accurate monitoring of physical activity and mental stress modules could offer such solutions.

The popularity of physical activity monitoring is seen in the number of smartphone applications, dedicated devices and smartwatch applications available on the market. The majority of smartphone-only or wristband-only applications are

Y. Altun et al. (Eds.): ECML PKDD 2017, Part III, LNAI 10536, pp. 414–418, 2017.
https://doi.org/10.1007/978-3-319-71273-4_42

either based on step counting, or use a metric called activity counts which correlates motion intensity with the human energy expenditure (EE) using a single regression equation [1]. Such approaches are somewhat effective only for monitoring ambulatory activities. More accurate approaches recognize the user's activity using activity recognition (AR) and utilize it as a machine-learning feature for estimation of EE (activity-based approaches). However, these approaches do not handle the varying location and orientation of the smartphone, which limits their real-life performance.

Monitoring mental stress using commercial and unobtrusive devices is a new and challenging topic, which is why few dedicated devices are available on the market. Until now, the most advanced approach was cStress [2], which utilizes an ECG sensor and is suitable for everyday use. However, the authors proposed replacing the somewhat uncomfortable ECG sensor with a wrist device, and better exploiting the information on the user's context.

We present a mobile application that uses machine learning on smartphone- and wristband sensor data for real-time activity monitoring and mental stress detection. The monitoring automatically adapts to the devices in use and to the orientation and location of the smartphone on the body. The stress detection uses the outputs of the activity monitoring and other information as context to improve the performance.

## 2   System Implementation and Methods with Evaluation

Our system is implemented on standard Android smartphone. It connects to the Microsoft Band 2 wristband over Bluetooth and collects and processes the sensor data from both devices. It perform the activity and mental stress monitoring in real time. The results are shared over MQTT protocol with a web application for visual presentation and demonstration.

### 2.1   Physical Activity Monitoring Method

The physical activity monitoring method is composed of six steps (left side and green-shaded modules of Fig. 1). The inputs are accelerometer and physiological data from a smartphone and/or wristband. The outputs are the recognized

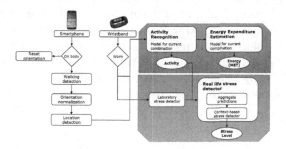

**Fig. 1.** Pipeline for physical activity and stress monitoring. (Color figure online)

activity and the estimated energy expenditure in MET (1 MET is defined as the energy expended at rest, while around 20 MET is expended at extreme exertion). The first step uses heuristics to detect the devices currently present on the user's body. If the smartphone is present, the method anticipates a walking period of 10 s, which is detected using a machine-learning model (second step). The walking segment is used for normalizing the orientation of the smartphone (third step). The normalized data is fed into the location detection machine-learning model, which is trained to recognize whether the smartphone is in the trousers pocket, jacket or a bag (fourth step). The present devices and the recognized location serve as context for the selection of an appropriate machine-learning model for activity recognition. We trained eight models, one for each location and combination of the devices, and one for the smartphone before orientation is normalized. The AR is performed on 2-s data windows and the EE estimation on 10-s data windows. The reader is referred to [3] for details.

The evaluation of the method was performed on dataset of ten volunteers performing a scenario of predefined activities (lying, sitting, standing, walking, Nordic walking, running, cycling, home chores, gardening, etc.). The volunteers were equipped with smartphones, a wristband and an indirect calorimeter for obtaining ground-truth EE. The evaluation was done with the leave-one-subject-out approach. We achieved the AR accuracy of 87%, and the mean absolute error of the EE estimation of 0.64 MET which outperforms the state-of-the-art commercial device Bodymedia (error of 1.03 MET).

## 2.2    Stress Monitoring Method

The mental stress monitoring method is composed of two steps presented in blue-shaded modules of Fig. 1. The first step is a laboratory stress detector, which is a machine-learning model trained to distinguish stressful vs. non-stressful events based on physiological data recorded in a laboratory, where stress was induced by solving mathematical problems under time pressure [4]. The detection is performed on 4-min data. In real life, there are many situations that induce a similar arousal to stress (e.g., exercise), so the laboratory stress detector is inaccurate. The algorithm is enhanced with a context-based stress detector which uses as input the predictions of the laboratory stress detector, as well as the information on the physical activity and other context information (e.g., time of the day, history of predictions), to perform a stress detection every 20 min.

The evaluation of the method was performed on a dataset of 55 days of four volunteers leading their lives as normal. They were equipped with a wristband and a mobile application to label ground-truth stress. The evaluation was done with the leave-one-subject-out approach. We achieved the classification accuracy of 92% and the F-measure of 79% (the results without the context were 17% points worse).

# 3    Demonstration

To demonstrate the performance of the application, the visitor will be offered an Android smartphone and a wristband. He/she will choose the location of the smartphone and weather both devices or only one will be used. The visitor will perform activities of his/her choice and observe the stress level, estimated energy expenditure, recognized activity and location in real time through web application shown in Fig. 2.

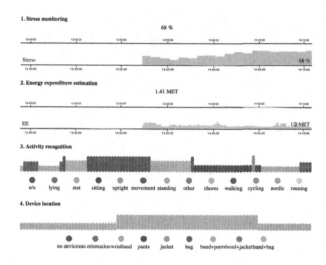

**Fig. 2.** Web application presents the processed data from the smartphone in real time.

# 4    Conclusion

We presented a state-of-the-art application for physical activity and mental stress monitoring, which relies on commercial devices such as many people already use. It is designed to handle real-life situations, and features real-time visual presentation via a web application, which is suitable for demonstration.

# References

1. Crouter, S.E., Kuffel, E., Haas, J.D., Frongillo, E.A., Bassett, D.R.: Refined two-regression model for the ActiGraph accelerometer. Med. Sci. Sport. Exerc. **42**, 1029–1037 (2010)
2. Hovsepian, K., Al'Absi, M., Ertin, E., Kamarck, T., Nakajima, M., Kumar, S.: cStress: towards a gold standard for continuous stress assessment in the mobile environment. In: Proceedings of the ACM International Joint Conference on Pervasive and Ubiquitous Computing (UbiComp 2015), pp. 493–504 (2015)

3. Cvetkovic, B., Szeklicki, R., Janko, V., Lutomski, P., Lustrek, M.: Real-time activity monitoring with a wristband and a smartphone. Inf. Fusion (2017)
4. Gjoreski, M., Gjoreski, H., Luštrek, M., Gams, M.: Continuous stress detection using a wrist device: in laboratory and real life. In: UbiComp Adjunct, pp. 1185–1193 (2016)

# Tetrahedron: Barycentric Measure Visualizer

Dariusz Brzezinski[✉], Jerzy Stefanowski, Robert Susmaga,
and Izabela Szczęch

Institute of Computing Science, Poznan University of Technology,
ul. Piotrowo 2, 60-965 Poznan, Poland
{dbrzezinski,jstefanowski,rsusmaga,iszczech}@cs.put.poznan.pl

**Abstract.** Each machine learning task comes equipped with its own set
of performance measures. For example, there is a plethora of classification measures that assess predictive performance, a myriad of clustering
indices, and equally many rule interestingness measures. Choosing the
right measure requires careful thought, as it can influence model selection and thus the performance of the final machine learning system.
However, analyzing and understanding measure properties is a difficult
task. Here, we present *Tetrahedron*, a web-based visualization tool that
aids the analysis of complete ranges of performance measures based on a
two-by-two contingency matrix. The tool operates in a barycentric coordinate system using a 3D tetrahedron, which can be rotated, zoomed,
cut, parameterized, and animated. The application is capable of visualizing predefined measures (86 currently), as well as helping prototype new
measures by visualizing user-defined formulas.

## 1 Introduction

Classifier selection and evaluation are difficult tasks requiring time and knowledge about the underlying data. One of the most important ingredients when
assessing classifiers is the used *classification performance measure*. An analogous
decision has to be made in association rule mining, where the overwhelming number of generated rules is usually trimmed by a selected *interestingness measure*.
However, many researchers often carry out their experiments with respect to few
selected measures, without discussing their properties and justifying their choice
simply by the measure's popularity.

To aid the analysis of properties of measures based on two-by-two contingency
tables, we put forward *Tetrahedron*, a web-based visualization tool for analyzing *entire ranges* of measure values. The proposed application visualizes 4D
data in 3D using the barycentric coordinate system [1,2]. *Tetrahedron* produces
3D WebGL plots with zooming, rotating, animation, and detailed configuration
capabilities. The presented tool can be used to compare properties of existing
measures, as well as devise new metrics.

## 2 The Visualization Technique

A confusion matrix for binary classification (Table 1) consists of four entries:
*TP*, *FP*, *FN*, *TN*. However, for a dataset of $n$ examples these four entries are

© Springer International Publishing AG 2017
Y. Altun et al. (Eds.): ECML PKDD 2017, Part III, LNAI 10536, pp. 419–422, 2017.
https://doi.org/10.1007/978-3-319-71273-4_43

**Table 1.** Confusion matrix for two-class classification

| Actual \ Predicted | Positive | Negative | total |
|---|---|---|---|
| Positive | $TP$ | $FN$ | $P$ |
| Negative | $FP$ | $TN$ | $N$ |
| total | $\hat{P}$ | $\hat{N}$ | $n$ |

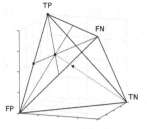

**Fig. 1.** Tetrahedron

sum-constrained, as $n = TP + FP + FN + TN$. Therefore, for a given constant $n$, any three values in the confusion matrix uniquely define the fourth value. This property allows to visualize any classification performance measure based on the two-class confusion matrix using a 4D barycentric coordinate system, tailored to sum-constrained data. The same holds for any $2 \times 2$ matrix, for example, those used to define rule interestingness measures [2].

The *barycentric coordinate system* is a coordinate system in which point locations are specified relatively to hyper-sides of a simplex. A 4D barycentric coordinate system is a tetrahedron, where each dimension is represented as one of the four vertices. Choosing vectors that represent $TP$, $FP$, $FN$, $TN$ as vertices of a regular tetrahedron in a 3D space, one arrives at a barycentric coordinate system depicted in Fig. 1.

In this system, every confusion matrix $\left[\begin{smallmatrix} TP & FN \\ FP & TN \end{smallmatrix}\right]$ is represented as a point of the tetrahedron. Let us illustrate this fact with a few examples. Figure 1 shows a skeleton of a tetrahedron with four exemplary points:

- one located in vertex TP, which represents $\left[\begin{smallmatrix} n & 0 \\ 0 & 0 \end{smallmatrix}\right]$,
- one located in the middle of edge TP–FP, which represents $\left[\begin{smallmatrix} n/2 & 0 \\ n/2 & 0 \end{smallmatrix}\right]$,
- one located in the middle of face $\triangle$TP–FP–FN, which represents $\left[\begin{smallmatrix} n/3 & n/3 \\ n/3 & 0 \end{smallmatrix}\right]$,
- one located in the middle of the tetrahedron, which represents $\left[\begin{smallmatrix} n/4 & n/4 \\ n/4 & n/4 \end{smallmatrix}\right]$.

One way of understanding this representation is to imagine a point in the tetrahedron as the center of mass of the examples in a confusion matrix. If all $n$ examples are true positives, then the entire mass of the predictions is at $TP$ and the point coincides with vertex TP. If all examples are false negatives, the point lies on vertex FN, etc. Generally, whenever $a > b$ ($a, b \in \{TP, FN, FP, TN\}$) then the point is closer to the vertex corresponding to $a$ rather than $b$.

Using the barycentric coordinate system makes it possible to depict the originally 4D data (two-class confusion matrices) as points in 3D. Moreover, an additional variable based on the depicted four values may be rendered as color. In the presented tool, we adapt this procedure to color-code the values of classification performance and rule interestingness measures. A more in-depth description of the visualization and its possible applications can be found in [1,2].

## 3 Tool Overview

The described visualization technique has been implemented as an interactive web-based application. An online version, compatible with all modern web browsers across different client platforms, is publicly available[1]. The application can visualize 86 predefined 4D measures, including 21 classification measures, 16 rule interestingness measures, and 49 general-purpose formulas based on a two-by-two matrix. The user can also visualize custom measures by providing their formula. The main functionalities of the application are:

- **Interactive 3D tetrahedron visualization.** The visualization (Fig. 2a) supports: 86 predefined measures, rotating, zooming, four rendering precisions, saving as an html with WebGL, and exporting images. The user may choose to visualize external views, inner layers, and control point-padding.
- **Cross-sections.** A useful way of visualizing measure values can also be achieved be cutting the tetrahedron with a plane and analyzing the obtained slice. In this application the user can visualize cross-sections (Fig. 2b) which correspond to different class distributions. Interestingly, this particular kind of cross-sections produces a 2D space analogous to that used in ROC charts.
- **Parameter animations.** Several of the application options can be animated. These options can change the visualization parameters automatically in constant intervals creating an animation (Fig. 2c). Such animations can be useful when attempting to analyze: consecutive layers of the tetrahedron, the

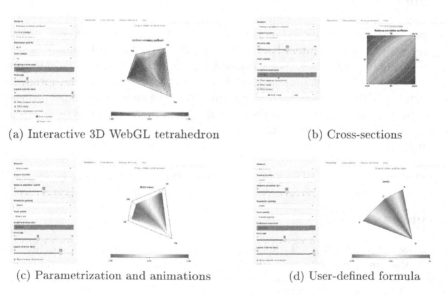

(a) Interactive 3D WebGL tetrahedron

(b) Cross-sections

(c) Parametrization and animations

(d) User-defined formula

**Fig. 2.** Application overview

---

[1] https://dabrze.shinyapps.io/Tetrahedron/. Source codes at: https://github.com/dabrze/tetrahedron (MIT License).

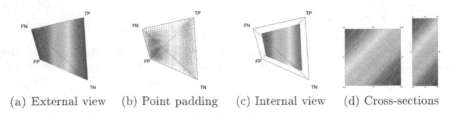

(a) External view     (b) Point padding     (c) Internal view     (d) Cross-sections

**Fig. 3.** Visualizations of classification accuracy (Color figure online)

impact of measure parameters (e.g. the impact of $\beta$ in $F_\beta$-score), or the effect of changing class distributions on cross-sections.

– **Custom measure definition.** It is possible to define a custom measure to be visualized by providing its formula (Fig. 2d).

Since classification accuracy is one of the most intuitive performance measures, let us use it to exemplify visualizations produced by our tool with the default (blue: 0, red: 1) color map. One can notice that confusion matrices with a high number of *FP* and *FN* result in low accuracy (blue), whereas high *TP* and *TN* yield high accuracy (red). Cross-sections for two different class ratios show that on imbalanced data high accuracy can be achieved by trivial majority classifiers. More examples of visual-based analyses can be found in [1,2] (Fig. 3).

## 4  Conclusions

We propose *Tetrahedron*, a web-based visualization tool for analyzing and prototyping measures based on a two-by-two matrix. Its main features include: interactive 3D WebGL barycentric plots, zooming, parameter animation, performing cross-sections, providing custom measure formulas, and saving plots with a single click. Such functionality facilitates visual inspection of various measure properties, such as determining measure monotonicity, symmetries, maximas, or undefined values. Thus, the presented tool can be used to gain further understanding of existing machine learning measures, as well as devise new ones.

**Acknowledgments.** NCN DEC-2013/11/B/ST6/00963, PUT Statutory Funds.

## References

1. Brzezinski, D., Stefanowski, J., Susmaga, R., Szczęch, I.: Visual-based analysis of classification measures with applications to imbalanced data. arXiv:1704.07122
2. Susmaga, R., Szczęch, I.: Can interestingness measures be usefully visualized? Int. J. Appl. Math. Comp. Sci. **25**(2), 323–336 (2015)

# TF Boosted Trees: A Scalable TensorFlow Based Framework for Gradient Boosting

Natalia Ponomareva[1]([✉]), Soroush Radpour[2], Gilbert Hendry[3], Salem Haykal[2], Thomas Colthurst[3], Petr Mitrichev[4], and Alexander Grushetsky[2]

[1] Google, Inc., New York, USA
tfbt-public@google.com
[2] Google, Inc., Mountain View, USA
[3] Google, Inc., Cambridge, USA
[4] Google, Inc., Zurich, Switzerland

**Abstract.** TF Boosted Trees (TFBT) is a new open-sourced framework for the distributed training of gradient boosted trees. It is based on TensorFlow, and its distinguishing features include a novel architecture, automatic loss differentiation, layer-by-layer boosting that results in smaller ensembles and faster prediction, principled multi-class handling, and a number of regularization techniques to prevent overfitting.

**Keywords:** Distributed gradient boosting · TensorFlow

## 1  Introduction

Gradient boosted trees are popular machine learning models. Since their introduction in [3] they have gone on to dominate many competitions on real-world data, including Kaggle and KDDCup [2]. In addition to their excellent accuracy, they are also easy to use, as they deal well with unnormalized, collinear, missing, or outlier-infected data. They can support custom loss functions and are often easier to interpret than neural nets or large linear models. Because of their popularity, there are now many gradient boosted tree implementations, including scikit-learn [7], R gbm [8], Spark MLLib [5], LightGBM [6], XGBoost [2].

In this paper, we introduce another optimized and scalable gradient boosted tree library, **TF Boosted Trees** (**TFBT**), which is built on top of the TensorFlow framework [1]. TFBT incorporates a number of novel algorithmic improvements to the gradient boosting algorithm, including new per-layer boosting procedure which offers improved performance on some problems. TFBT is open source, and available in the main TensorFlow distribution under `contrib/boosted_trees`.

## 2  TFBT Features

In Table 1 we provide a brief comparison between TFBT and some existing libraries. Additionally, TFBT provides the following.

© Springer International Publishing AG 2017
Y. Altun et al. (Eds.): ECML PKDD 2017, Part III, LNAI 10536, pp. 423–427, 2017.
https://doi.org/10.1007/978-3-319-71273-4_44

**Table 1.** Comparison of gradient boosted libraries.

| Lib | D? | Losses | Regularization |
|---|---|---|---|
| scikit-learn | N | $R$: least squares, least absolute dev, huber and quantile. $C$: logistic, Max-Ent and exp | Depth limit, shrinkage, bagging, feature sub-sampling |
| GBM | N | $R$: least squares, least absolute dev, t-distribution, quantile, huber. $C$: logistic, Max-Ent, exp, poisson & right censored observations. Supports *ranking* | Shrinkage, bagging, depth limit, min # of examples per node |
| MLLib | Y | $R$: least squared and least absolute dev. $C$: logistic | Shrinkage, early stopping, depth limit, min # of examples per node, min gain, bagging |
| Light GBM | Y | $R$: least squares, least absolute dev, huber, fair, poisson. $C$: logistic, Max-Ent. Supports *ranking* | Dropout, shrinkage, # leafs limit, feature subsampling, bagging, L1 & L2 |
| XGBoost | Y | $R$: least squares, poisson, gamma, tweedie regression. $C$: logistic, Max-Ent. Supports *ranking* and **custom** | L1 & L2, shrinkage, feature subsampling, dropout, bagging, min child weight and gain, limit on depth and # of nodes, pruning |
| **TFBT** | Y | Any twice differentiable loss from tf.contrib.losses and **custom** losses | L1 & L2, tree complexity, shrinkage, line search for learning rate, dropout, feature subsampling and bagging, limit on depth and min node weight, pre- post- pruning |

*D?* is whether a library supports distributed mode. $R$ stands for regression, $C$ for classification.

**Layer-by-Layer Boosting.** TFBT supports two modes of tree building: *standard* (building sequence of boosted trees in a stochastic gradient fashion) and novel *Layer-by-Layer* boosting, which allows for stronger trees (leading to faster convergence) and deeper models. One weakness of tree-based methods is the fact that only the examples falling under a given partition are used to produce the estimator associated with that leaf, so deeper nodes use statistics calculated from fewer examples. We overcome that limitation by recalculating the gradients and Hessians whenever a new layer is built resulting in stronger trees that better approximate the functional space gradient. This enables deeper nodes to use higher level splits as priors meaning each new layer will have more information and will be able to better adjust for errors from the previous layers. Empirically we found that layer-by-layer boosting generally leads to faster convergence and, with proper regularization, to less overfitting for deeper trees.

**Multiclass Support.** TFBT supports one-vs-rest, as well as 2 variations that reduce the number of required trees by storing per-class scores at each leaf. All other implementations use one-vs-rest (MLLib has no multiclass support).

Since TFBT is implemented in TensorFlow, **TensorFlow specific features** are also available

- Ease of writing **custom loss** functions, as TensorFlow provides automatic differentiation [1] (other packages like XGBoost require the user to provide the first and second order derivatives).
- Ability to easily switch and compare TFBT with other TensorFlow models.
- Ease of debugging with TensorBoard.
- Models can be run on multiple CPUs/GPUs and on multiple platforms, including mobile, and can be easily deployed via TF serving.
- Checkpointing for fault tolerance, incremental training & warm restart.

# 3  TFBT System Design

**Finding Splits.** One of the most computationally intensive parts in boosting is finding the best splits. Both R and scikit-learn work with an exact greedy algorithm for enumerating all possible splits for all features, which does not scale. Other implementations, like XGBoost, work with approximate algorithms to build quantiles of feature values and aggregating gradients and Hessians for each bucket of quantiles. For aggregation, two approaches can be used [4]: either each of the workers works on all the features, and then the statistics are aggregated in Map-Reduce (MLLib) or All-Reduce (XGBoost) fashion, or a parameter server (PS) approach (TencentBoost [4], PSMART [9]) is applied (each worker and PS aggregates statistics only for a subset of features). The All-Reduce versions do not scale to a high-dimensional data and Map-Reduce versions are slow to scale.

**TFBT Architecture.** Our computation model is based on the following needs:

1. Ability to train on datasets that don't fit in workers' memory.
2. Ability to train deeper trees with a larger number of features.
3. Support for different modes of building the trees: standard one-tree-per-batch mode, as well as boosting the tree layer-by-layer.
4. Minimizing parallelization costs. Low cost restarts on stateless workers would allow us to use much cheaper preemptible VMs.

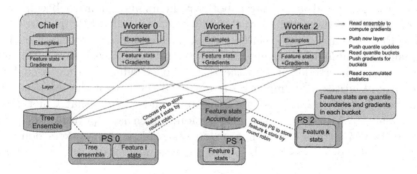

**Fig. 1.** TFBT architecture.

Our design is similar to XGBoost [2], TencentBoost [4] in that we build distributed quantile sketches of feature values and use them to build histograms, to be used later to find the best split. In TencentBoost [4] and PSMART [9] full training data is partitioned and loaded in workers' memory, which can be a problem for larger datasets. To address this we instead work on mini-batches, updating quantiles in an online fashion without loading all the data into the memory. As far as we know, this approach is not implemented anywhere else.

Each worker loads a mini-batch of data, builds a local quantile sketch, pushes it to PS and fetches the bucket boundaries that were built at the previous iteration (Fig. 1). Workers then compute per bucket gradients and Hessians and push them back to the PS. One of the workers, designated as Chief, checks during

**Algorithm 1.** Chief and Workers' work

---

1: **procedure** CALCULATESTATISTICS(PS, MODEL, STAMP, BATCH_DATA, LOSS_FN)
2:     *predictions* ← model.predict(BATCH_DATA)
3:     *quantile_stats* ← calculate_quantile_stats(BATCH_DATA)
4:     push_stats(PS,quantile_stats, stamp)                    ▷ PS updates quantiles
5:     *current_boundaries* ← fetch_latest_boundaries(PS, stamp)
6:     *gradients, hessians* ← calculate_derivatives(predictions,LOSS_FN)
7:     *gradients, hessians* ← aggregate(current_boundaries,gradients, hessians)
8:     push_stats(PS,gradients, hessians, size(BATCH_DATA), stamp)
9: **procedure** DOWORK(PS, LOSS_FN, IS_CHIEF)                 ▷ Runs on workers and 1 chief
10:     **while** true **do**
11:         *BATCH_DATA* ← read_data_batch()
12:         *model* ← fetch_latest_model(PS)
13:         *stamp* ← model.stamp_token
14:         CalculateStatistics(PS,model, stamp, BATCH_DATA,LOSS_FN)
15:         **if** *is_chief* & get_num_examples(PS, stamp) ≥ *N_PER_LAYER* **then**
16:             *next_stamp* ← *stamp* + 1
17:             *stats* ← flush(PS, stamp, next_stamp)        ▷ Update stamp, returns stats
18:             build_layer(PS, model, next_stamp, stats)     ▷ PS updates ensemble

---

each iteration if the PS have accumulated enough statistics for the current layer and if so, starts building the new layer by finding best splits for each of the nodes in the layer. Code that finds the best splits for each feature is executed on the PS that have accumulated the gradient statistics for the feature. The Chief receives the best split for every leaf from the PS and grows a new layer on the tree.

Once the Chief adds a new layer, both gradients and quantiles become stale. To avoid stale updates, we introduce an abstraction called StampedResource - a TensorFlow resource with an int64 stamp. Tree ensemble, as well as gradients and quantile accumulators are all stamped resources with such token. When the worker fetches the model, it gets the stamp token which is then used for all the reads and writes to stamped resources until the end of the iteration. This guarantees that all the updates are consistent and ensures that Chief doesn't need to wait for Workers for synchronization, which is important when using preemptible VMs. Chief checkpoints resources to disk and workers don't hold any state, so if they are restarted, they can load a new mini-batch and continue.

# References

1. Abadi, M., et al.: TensorFlow: large-scale machine learning on heterogeneous distributed systems. In: OSDI (2016)
2. Chen, T., et al.: XGBoost: a scalable tree boosting system. CoRR (2016)
3. Friedman, J.H.: Greedy function approximation: a gradient boosting machine. Ann. Stat. **29**, 1189–1232 (2000)
4. Jiang, J., Jiang, J., Cui, B., Zhang, C.: TencentBoost: a gradient boosting tree system with parameter server. In: 33rd IEEE International Conference on Data Engineering, ICDE 2017, San Diego, CA, USA, 19–22 April 2017, pp. 281–284 (2017). https://doi.org/10.1109/ICDE.2017.87

5. Meng, X., Bradley, J.K., Yavuz, B., Sparks, E.R., Venkataraman, S., Liu, D., Freeman, J., Tsai, D.B., Amde, M., Owen, S., Xin, D., Xin, R., Franklin, M.J., Zadeh, R., Zaharia, M., Talwalkar, A.: MLlib: machine learning in Apache Spark. CoRR (2015). http://arxiv.org/abs/1505.06807
6. Microsoft: Microsoft/dmtk (2013). https://github.com/microsoft/dmtk
7. Pedregosa, F., Varoquaux, G., Gramfort, A., Michel, V., Thirion, B., Grisel, O., Blondel, M., Prettenhofer, P., Weiss, R., Dubourg, V., Vanderplas, J., Passos, A., Cournapeau, D., Brucher, M., Perrot, M., Duchesnay, E.: Scikit-learn: machine learning in Python. J. Mach. Learn. Res. **12**, 2825–2830 (2011)
8. Ridgeway, G.: Generalized boosted models: a guide to the GBM package (2005)
9. Zhou, J., et al.: PSMART: parameter server based multiple additive regression trees system. In: WWW 2017 Companion (2017)

# TrajViz: A Tool for Visualizing Patterns and Anomalies in Trajectory

Yifeng Gao$^{(\boxtimes)}$, Qingzhe Li, Xiaosheng Li, Jessica Lin, and Huzefa Rangwala

George Mason University, Fairfax, USA
{ygao12,qli10,xli22,jessica}@gmu.edu, rangwala@cs.gmu.edu

**Abstract.** Visualizing frequently occurring patterns and potentially unusual behaviors in trajectory can provide valuable insights into activities behind the data. In this paper, we introduce TrajViz, a motif (frequently repeated subsequences) based visualization software that detects patterns and anomalies by inducing "grammars" from discretized spatial trajectories. We consider patterns as a set of sub-trajectories with unknown lengths that are spatially similar to each other. We demonstrate that TrajViz has the capacity to help users visualize anomalies and patterns effectively.

## 1 Introduction

With the rapid growth of tracking technology, a large amount of trajectory data are generated from users' daily activities. Discovering frequently occurring patterns (motifs) and potentially unusual behaviors can be used to summarize the overwhelming amount of trajectories data and obtain meaningful knowledge. In this paper, we present TrajViz, a software that visualizes patterns and anomalies in trajectory datasets. TrajViz extends our previous work in time series motif discovery [1] to sub-trajectory pattern visualization. We consider patterns as a set of sub-trajectories with unknown lengths that are spatially similar to each other. We use a grid-based discretization approach to remove the speed information and adapt a grammar-based motif discovery algorithm, Iterative Sequitur (ItrSequitur), to discover the patterns. We design a user-friendly interface to allow visualization of repeated, as well as unusual sub-trajectories within the datasets.

## 2 Relate Work and Overview of TrajViz

Previously, we introduced a grammar-based motif discovery framework [7], which uses Sequitur [4], a grammar induction algorithm, to find approximate motifs of variable lengths in time series. However, the unique characteristics and challenges associated with spatial trajectory data make it unsuitable and difficult to apply the algorithms directly on trajectory data. In [5], the authors introduced STAVIS, a trajectory analytical system that uses grammar induction to infer variable-length patterns. However, its definition of "pattern" is based on time

© Springer International Publishing AG 2017
Y. Altun et al. (Eds.): ECML PKDD 2017, Part III, LNAI 10536, pp. 428–431, 2017.
https://doi.org/10.1007/978-3-319-71273-4_45

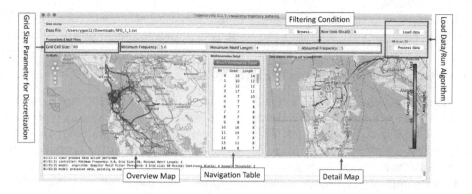

**Fig. 1.** Screenshot of TrajViz and default view for San Franciso Taxi data [6]

series motifs. Therefore, speed variation will significantly affect the quality of patterns discovered. Other work such as [2,9] focuses on either sequential pattern mining based on important locations, or trajectory clustering, both of which are different from the goal of our software.

A screenshot of TrajViz is shown in Fig. 1. TrajViz follows the Visual Information-Seeking Mantra [8]. After processing the data, an overview heat map of pattern density is displayed. User can zoom in to see the detailed map and use domain knowledge to filter out unwanted patterns by setting minimum frequency, minimum continuous blocks length (Minimal Motif Length) and maximum frequency for anomaly detection (Anomaly Frequency). Adjusting these thresholds does not require re-running the discretization and grammar induction steps (introduced in the next subsection). Further details on TrajViz can be found in goo.gl/cKCeDt.

## 3   Our Approach

### 3.1   Discretization

Before we can induce grammars on trajectory data, it is necessary to pre-process the data. We first convert the trajectory data to speed-insensitive symbolic sequences after removing noises from the trajectory dataset. To prepare for discretization, we divide the entire region into an ($\alpha \times \alpha$) equal-frequency grid, where $\alpha$ is the grid size. We assign each grid cell a block ID sequentially from left to right and from top to bottom.

After block IDs are assigned, we use a four-step procedure to convert raw trajectory to a block ID sequence $S_{block}$. First, we up-sample the raw trajectory by using linear interpolation to ensure that the consecutive blocks in $S_{block}$ are spatially adjacent. Then trajectories are converted into block ID sequences based on the order of traversal. Next, we perform further noise removal by removing blocks that are barely covered by the trajectory. Finally, numerosity reduction [3] is adopted to compress the sequence by only recording the first occurrence of consecutively repeating symbol. $S_{block}$ is insensitive to speed variation. This is an important property that allows us to detect spatially-similar sub-trajectories.

## 3.2    Grammar Induction with ItrSequitur

As demonstrated in previous work [7], a context-free grammar summarizes the structure of an input sequence. Intuitively, repeated substrings in $S_{block}$ represent a set of similar sub-trajectories. Therefore, learning a set of grammar rules to identify repeating substrings from $S_{block}$ can discover frequently occurring patterns (sub-trajectories) in trajectory data. Previous work [5] utilizes Sequitur [4], a linear complexity grammar induction approach, to learn the grammar rules. However, Sequitur can only detect patterns if they have identical symbolic representation. In TrajViz, we adapt an iterative version of Sequitur, called ItrSequitur [1], for more robust grammar induction. ItrSequitur iteratively rewrites the input sequence based on the output of Sequitur and re-induces the grammar on the revised sequence until no new grammar can be found. Different from Sequitur, ItrSequitur allows small variation in matching substrings. Therefore, it is robust to noise in the dataset.

## 3.3    Patterns/Anomalies Discovery and Motif Heatmap

TrajViz consolidates the patterns detected by merging patterns that have similar symbolic representations. Top-ranked frequent patterns that satisfy user-defined filtering conditions are listed in the motifs/anomalies table. User can navigate the patterns by clicking through the items in the table; a zoom-in of the selected pattern is then shown on the right panel. Figure 2 shows screenshots of a motif and an anomaly detected. To show the direction of the trajectories, the start points are marked by black circles, and the end points are denoted by black squares.

For each point in a motif, we compute the point density by counting the number of points from other motifs within some distance threshold, and create a motif heatmap. A five-color gradient (blue-cyan-green-yellow-red) is built to

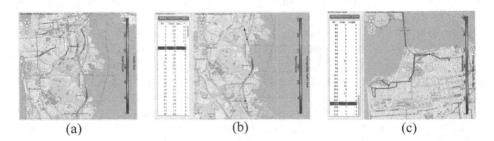

(a)                              (b)                              (c)

**Fig. 2.** Example of patterns detected in San Franciso Taxi Dataset [6] (a) Motif Heatmap (b) A pattern indicates a frequently visited route from the city to airport (c) An unusual (infrequent) round trip route (Color figure online)

linearly map the densities to their specific colors. The most dense points have the red colors while the least dense ones are in blue.

To find anomalies, we create a trajectory rule-density curve by counting the number of grammar rules covering each consecutive pair of block IDs (we consider a pair at a time in order to preserve the direction of the trajectory). The intuition is that, an anomalous subsequence would have zero or very few repetitions, hence low rule-density. TrajViz finds low-density subsequences within a trajectory and marks them as unusual routes (Fig. 2(c)).

# 4    Target Audience

TrajViz provides an efficient, interpretable, and user-interactive mechanism to understand functional activities behind massive trajectory data. TrajViz targets a diverse audience including researchers, practitioners, and scientists who are interested in discovering patterns in trajectory data.

**Acknowledgements.** We would like to thank Ranjeev Mittu at the Naval Research Lab (NRL) for the support and valuable suggestions on our work.

# References

1. Gao, Y., Lin, J., Rangwala, H.: Iterative grammar-based framework for discovering variable-length time series motifs. In: 2016 15th IEEE International Conference on Machine Learning and Applications (ICMLA), pp. 7–12. IEEE (2016)
2. Lee, J.-G., Han, J., Li, X., Gonzalez, H.: Traclass: trajectory classification using hierarchical region-based and trajectory-based clustering. Proc. VLDB Endow. **1**(1), 1081–1094 (2008)
3. Lin, J., Keogh, E., Wei, L., Lonardi, S.: Experiencing sax: a novel symbolic representation of time series. Data Min. Knowl. Disc. **15**(2), 107–144 (2007)
4. Nevill-Manning, C.G., Witten, I.H.: Identifying hierarchical strcture in sequences: a linear-time algorithm. J. Artif. Intell. Res. (JAIR) **7**, 67–82 (1997)
5. Oates, T., Boedihardjo, A.P., Lin, J., Chen, C., Frankenstein, S., Gandhi, S.: Motif discovery in spatial trajectories using grammar inference. In: Proceedings of the 22nd ACM International Conference on Conference on Information & Knowledge Management, pp. 1465–1468. ACM (2013)
6. Piorkowski, M., Sarafijanovic-Djukic, N., Grossglauser, M.: A parsimonious model of mobile partitioned networks with clustering. In: 2009 First International Communication Systems and Networks and Workshops, pp. 1–10. IEEE (2009)
7. Senin, P., Lin, J., Wang, X., Oates, T., Gandhi, S., Boedihardjo, A.P., Chen, C., Frankenstein, S., Lerner, M.: GrammarViz 2.0: a tool for grammar-based pattern discovery in time series. In: Calders, T., Esposito, F., Hüllermeier, E., Meo, R. (eds.) ECML PKDD 2014. LNCS (LNAI), vol. 8726, pp. 468–472. Springer, Heidelberg (2014). https://doi.org/10.1007/978-3-662-44845-8_37
8. Shneiderman, B.: The eyes have it: a task by data type taxonomy for information visualizations. In: IEEE Symposium on Visual Languages, 1996. Proceedings, pp. 336–343. IEEE (1996)
9. Zheng, Y., Zhang, L., Xie, X., Ma, W.-Y.: Mining interesting locations and travel sequences from GPS trajectories. In: Proceedings of the 18th International Conference on World Wide Web, pp. 791–800. ACM (2009)

# TrAnET: Tracking and Analyzing the Evolution of Topics in Information Networks

Livio Bioglio, Ruggero G. Pensa[⊠][iD], and Valentina Rho

Department of Computer Science, University of Turin, Turin, Italy
{livio.bioglio,ruggero.pensa,valentina.rho}@unito.it

**Abstract.** This paper presents a system for tracking and analyzing the evolution and transformation of topics in an information network. The system consists of four main modules for pre-processing, adaptive topic modeling, network creation and temporal network analysis. The core module is built upon an adaptive topic modeling algorithm adopting a sliding time window technique that enables the discovery of ground-breaking ideas as those topics that evolve rapidly in the network.

**Keywords:** Information diffusion · Topic modeling
Citation networks

## 1 Introduction

Information diffusion is an important and widely-studied topic in computational social science and network analytics due to its applications to social media/network analysis, viral marketing campaigns, influence maximization and prediction. An information diffusion process takes place when some nodes (e.g., customers, social profiles, scientific authors) influence some of their neighbors in the network which, in their turn, influence some of their respective neighbors. The definition of "influence" depends on the application. In mouth-to-mouth viral campaign, a user who bought a product at time $t$ influence their neighbors if they buy the same product at time $t + \delta$. In bibliographic networks, author $a$ influences author $b$ when $a$ and $b$ are connected by some relationship (e.g., collaboration, co-authorship, citation) and either $b$ cites one of the papers published by author $a$, or author $b$ publish in the same topic as author $a$ [2].

In this paper we propose a system for topic diffusion analysis based on adaptive and scalable Latent Dirichlet Annotation (LDA [1]) that uses a different notion of influence: for a given topic $x$, author $a$ influences author $b$ when $b$ publish at time $t + \delta$ a paper that cites some papers covering topic $x$ and authored by $a$ at time $t$. Moreover, our focus is on topic evolution rather than on ranking authors, such as in [5]. Our system, in fact, enables the discovery of groundbreaking topics and ideas, which are defined as topics that evolve rapidly in the network. According to our definition, the most interesting topics are those that influence many new research topics, thus stimulating new research ideas. By setting different diffusion

© Springer International Publishing AG 2017
Y. Altun et al. (Eds.): ECML PKDD 2017, Part III, LNAI 10536, pp. 432–436, 2017.
https://doi.org/10.1007/978-3-319-71273-4_46

model parameters, our system enables the flexible analysis of topic evolution and the identification of the most influential authors. The salient features of our system, with respect to other state-of-the-art methods, are: (1) its ability to track the evolution and transformation of topics in time; and (2) its flexibility, enabling multiple types of online and offline analyses.

## 2    System Description

The architecture of the system is presented in Fig. 1. As an input, it takes a corpus consisting of any type of document (including scientific papers, patents, news articles) with explicit references to other previously published documents. First, the documents are pre-processed with NLP techniques that perform tokenization, lemmatization, stopwords removal and term frequency computation in order to prepare the corpus for the topic modeling module. This module adopts a scalable and robust topic modeling library [3] that enables the extraction of an adaptive set of to topics. Thanks to this module, it is possible to assign multiple weighted topics to a document published at time $t + \delta$ according to a topic model computed at the previous instant $t$. Moreover, the topic model can be adapted efficiently to newly inserted documents without recomputing it from scratch. A network creation module is used to extract the bibliographic network from the original corpus. Finally, the evolution of topic is tracked on the bibliographic network by a network analysis module that enables the visualization of several temporal characteristics of topic evolution, and the detection of the most interesting topics according to the evolution speed.

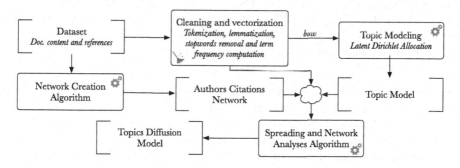

**Fig. 1.** A graphic overview of the overall processing and analysis pipeline.

To perform topic evolution analysis, the spreading model considers several adjustable parameters. For each analysis task we consider: a time scale $[t_0, t_n]$ defining the overall time interval of the analysis; a time window of size $\delta$ and an overlap $\gamma < \delta$ defining a set of time intervals $\{\Delta T_0, \ldots, \Delta T_N\}$ s.t. $\forall i \ T_i = [t_0 + i(\delta - \gamma), t_0 + i(\delta - \gamma) + \delta)$; a set of $K$ topics $\{\tau_1, \ldots, \tau_K\}$ ($K$ being a user-given parameter). Given a topic $\tau_x$, users in the network are activated at time

$\Delta T_0$ if they publish a paper covering topic $\tau_x$ during $\Delta T_0$. Users are activated at time $\Delta T_i$ ($i > 0$) if they cite any paper that contributed to the activation of the users at time $\Delta T_{i-1}$. A paper $p$ is said to cover a topic $\tau_x$ if LDA has assigned $\tau_x$ to paper $p$ with a weight greater than a user-specified threshold.

The whole process is driven within an interactive Jupyter notebook[1]. All modules are implemented in Python. All data are stored in a MongoDB[2] database server. The system runs on Windows, Linux and Mac OS X operating systems using a standard computing platform (e.g., any multi-core Intel Core iX CPU, and 8 GB RAM) and does not require any high-performance GPU architecture.

## 3    Demonstration

**Dataset.** The dataset used for TrAnET demonstration is a subset of the scientific papers citations network. This dataset is created by automatically merging two datasets originally extracted through ArnetMiner [4]: the DBLP and ACM citation networks[3]. The demonstration focuses on papers published from 2000 to 2014 within a set of preselected venues, for a total of about 155,000 papers.

**Text Processing and Topic Extraction.** The input data given to the topic extraction module is obtained as the result of the cleaning and vectorization process performed on the concatenation of paper title and abstract, as described in the previous section. In particular, the cleaning module ignores terms that appears only once in the dataset and in more than 80% of the documents. The topic extraction is performed on the whole dataset using Latent Dirichlet Allocation, searching for 50 topics. The topic model is then used to assign a weighted list of topics to each paper in the dataset. In our demonstration, we consider only topic assignments with weight greater than 0.2.

**Example of Topic Evolution.** To explain how our tool works, the analysis on two representative topics (namely, topics 6 and 34) is shown here: their keywords, sized according to their weight within the topic, are described in Fig. 2a and b, respectively. These topics have been chosen because they are assigned to a comparable number of papers (4498 for topic 6 and 6079 for topic 34) and authors (8430 for topic 6 and 8776 for topic 34). Moreover, they exhibit a very similar publication trend. According to Fig. 2c, which shows the cumulative number of authors that have published for the first time a paper on each topic in each year, the two trends are almost indistinguishable. This result (similar to what can be computed by [2]) shows that these topics have a similar diffusion trend in the bibliographic network. However, there is a strong difference in the evolution speed, as shown in Fig. 2d. Topic 34 (information retrieval) evolves

---

[1] https://jupyter.org/.
[2] https://www.mongodb.com/.
[3] https://aminer.org/citation.

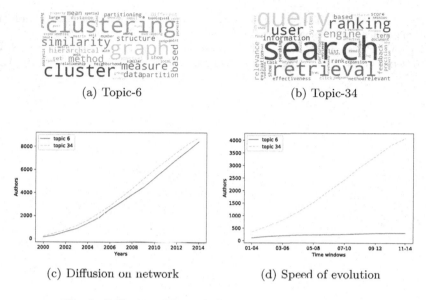

(a) Topic-6                          (b) Topic-34

(c) Diffusion on network            (d) Speed of evolution

**Fig. 2.** Diffusion and word clouds of the selected topics.

more rapidly than topic 6 (clustering). This behavior can be explained by the increasing research efforts in the first field, driven by search engine and social media applications, as well as by Semantic Web technologies. Clustering, in contrast, appears as an evergreen albeit not particularly evolving research field in the time frame considered here. In this experiment, we used $K = 50$, $\delta = 4$ and $\gamma = 3$. By tuning the three parameters suitably, different outcomes will be shown during the demonstration.

The source code and the dataset of the demonstration are available online[4].

**Acknowledgments.** This work is partially funded by project MIMOSA (MultIModal Ontology-driven query system for the heterogeneous data of a SmArtcity, "Progetto di Ateneo Torino_call2014_L2_157", 2015–17).

# References

1. Blei, D.M., Ng, A.Y., Jordan, M.I.: Latent dirichlet allocation. J. Mach. Learn. Res. **3**, 993–1022 (2003)
2. Gui, H., Sun, Y., Han, J., Brova, G.: Modeling topic diffusion in multi-relational bibliographic information networks. In: Proceedings of CIKM 2014, pp. 649–658. ACM (2014)
3. Řehůřek, R., Sojka, P.: Software framework for topic modelling with large corpora. In: Proceedings of the LREC 2010 Workshop on New Challenges for NLP Frameworks, pp. 45–50 (2010)

---

[4] https://github.com/rupensa/tranet.

4. Tang, J., Zhang, J., Yao, L., Li, J., Zhang, L., Su, Z.: ArnetMiner: extraction and mining of academic social networks. In: KDD 2008, pp. 990–998 (2008)
5. Xiong, C., Power, R., Callan, J.: Explicit semantic ranking for academic search via knowledge graph embedding. In: Proceedings of WWW 2017, pp. 1271–1279. ACM (2017)

# WHODID: Web-Based Interface for Human-Assisted Factory Operations in Fault Detection, Identification and Diagnosis

Pierre Blanchart[✉] and Cédric Gouy-Pailler

CEA, LIST, 91191 Gif-sur-Yvette Cedex, France
pierre.blanchart@cea.fr

**Abstract.** We present WHODID: a turnkey intuitive web-based interface for fault detection, identification and diagnosis in production units. Fault detection and identification is an extremely useful feature and is becoming a necessity in modern production units. Moreover, the large deployment of sensors within the stations of a production line has enabled the close monitoring of products being manufactured. In this context, there is a high demand for computer intelligence able to detect and isolate faults inside production lines, and to additionally provide a diagnosis for maintenance on the identified faulty production device, with the purpose of preventing subsequent faults caused by the diagnosed faulty device behavior. We thus introduce a system which has fault detection, isolation, and identification features, for retrospective and on-the-fly monitoring and maintenance of complex dynamical production processes. It provides real-time answers to the questions: "is there a fault?", "where did it happen?", "for what reason?". The method is based on a posteriori analysis of decision sequences in XGBoost tree models, using recurrent neural networks sequential models of tree paths.

The particularity of the presented system is that it is robust to missing or faulty sensor measurements, it does not require any modeling of the underlying, possibly exogenous manufacturing process, and provides fault diagnosis along with confidence level in plain English formulations. The latter can be used as maintenance directions by a human operator in charge of production monitoring and control.

**Keywords:** Production units · Fault detection and identification · Maintenance operator friendly · Tree ensemble · Gradient boosting · LSTM-RNN networks

## 1 Introduction

Modern factories operation and optimization rely on fine-grained monitoring of machines and products. Besides classical purposes such as energy optimization and smart production planning, there is a high demand for systems able to detect and isolate the location of faults occurring in production chains. Thus there has

© Springer International Publishing AG 2017
Y. Altun et al. (Eds.): ECML PKDD 2017, Part III, LNAI 10536, pp. 437–441, 2017.
https://doi.org/10.1007/978-3-319-71273-4_47

been a tremendous effort to design computational intelligences able to represent the underlying dynamics of such complex systems, with the goal of detecting, identifying and possibly explaining the occurrence of faults while the system is in operation. Fault detection and identification is often addressed through an explicit modeling of the system processes using supervised approaches. The first problem with this approach is that it implies learning as many models as there are processing steps, which can be a huge number in modern factories. The second problem comes from the faulty and missing sensor measurements, which, combined with the complex and dynamical nature of some processes make such modeling highly inaccurate and unreliable for fault detection [5]. In our approach, we learn a global fault detection model (FDM) taking all sensor measurements into account for more reliable detection, and we perform a posteriori analysis of this model to perform fault identification and diagnosis. Or course, such an approach is only viable if the global model's decisions are interpretable by any means, and those decisions can be related to the individual physical equipments, e.g. the work stations, for fault isolation/identification. We use XGBoost [1], a gradient boosting tree ensemble classification method, as a FDM since it has proved robustness and even superior performance for such unbalanced two-class classification problems as fault detection. The drawback of such a model is that it does not provide with any direct interpretability of its decision, which is a desirable feature for identification and diagnosis [2,3]. Some approaches cope with this issue by simplifying the learned FDM to make it interpretable [4,6], but degrading the detection performances. In a similar spirit, some models are constrained to be simple enough for interpretability, impacting the detection performance as well [7]. Unlike those, we keep the original FDM and seeks interpretation from directly it using tree path analysis, thus keeping the original FDM performance.

## 2    Fault Detection, Identification and Diagnosis

We train the XGBoost FDM on a large set of engineered features that are related to a physical equipment or a physical entity in the factory such as a station or a production line. Features can be sensor measurements made at stations level, timestamps of products passage in a station, more evolved features such as non-linear projections of sensor measurements, features characterizing the time distribution of faults at a station... XGBoost is particularly suited to the scenario where we are using heterogeneous data, with various dynamics, and possibly many missing/abnormal data. Besides, it is not sensitive to redundant features, making it a very robust approach for fault detection in production industry, where we typically deal with numerical, categorical and timestamps data representing a mix of sensor measurements and feedback from human station operator, and as such very liable to be faulty/redundant or missing.

Identification and diagnosis are then performed in a joint manner by analyzing the trees in the XGBoost model. The idea is to learn sequential models of paths followed by non-faulty data inside the trees. Thus for each node of a tree, we want to have a model able to say what is the most likely path to be followed subsequently by a non-faulty data, i.e. we want to model what is the probability to go to the left branch, to the right branch or to end in a leaf. Those models have a sequential nature since, in a given node, they are conditioned by the path followed from the root to this node. And there is a combinatorial aspect induced by all the possible paths in the tree. We address this aspect by learning recurrent models of tree paths, using long-short term memory recurrent neural networks [8]. Numerical data is used along with the node index, to make the learning problem easier and break the combinatorial aspect, since, numerically speaking, not all tree paths figure in the data: only tree paths potentially existing are learned. We train as many tree path models as there are trees in the XGBoost model, and for each faulty data, we look inside each tree in which node(s) its tree path **diverges** from the "normal" tree path learned from non-faulty data. KL-divergence is used as a measure of divergence in a node between the predicted distribution by our normal path model (probability of "left", "right", "leaf"), and the observed distribution, *i.e.* in which branch the fault data goes. This gives us an indication as to where and why a fault happened, since the faulty data obviously follow paths in the decision trees which at some point diverge from normality. Identification and diagnosis are straightforward to obtain since each node of a decision tree makes direct reference to a feature, and, defines a "normality regime" on this feature thanks to the split value associated to the node. The feature being related to a precise physical equipment, we can easily output as a potential fault identification the concerned equipment, and, as a diagnosis the interval of normality defined by the node split along with the abnormal measure. Such an identification/diagnosis pair can be formulated in plain English and enriched with informations on the sensor(s) measures associated with the node where the divergence was observed. This last part is mostly the responsibility of the industrial actor and has no genericity (Fig. 1).

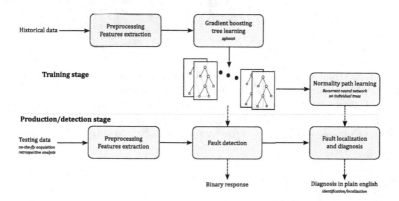

**Fig. 1.** Processing workflow of the fault detection, identification and diagnosis system.

To rank identification diagnosis pairs according to relevance, observed node divergences are aggregated across all the trees of the global defect model by computing in which proportion an individual tree score contributes to the global defect score and reweighing accordingly. It enables a ranking of potential fault diagnosis by decreasing order of relevance. This human readable output then allows an operator in charge of production chain maintenance and control to address the problem in the right place.

## 3   Interface Operation

The interface operation is demonstrated in Fig. 2: the operator selects a production line in the hierarchical view in Fig. 2d and a faulty product in the side menu in Fig. 2a, and obtains a view of the selected line which shows the product parcours through stations along with fault diagnosis shown as tooltips in the stations where a problem was identified (Fig. 2a). A full fault report in plain English is displayed in panel Fig. 2b. The view in Fig. 2c shows algorithmic insights about the model and would not be visible to a production monitoring operator.

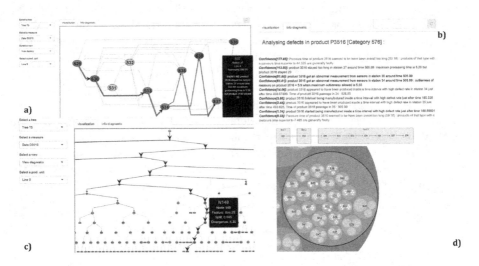

**Fig. 2.** User-Interface overview. (a) Faults (orange stations) are reported on the path of product P3516 through stations in line 3 (red), and detailed in the tip. (b) Full fault diagnosis of P3516 with their respective confidence levels in brackets. (c) Decision path of P3516 (in red) in tree T5, with one node divergence (in orange) referring to a fault in station S29. (d) Hierarchical view of the factory (lines – stations). (Color figure online)

## References

1. Chen, T., Guestrin, C.: XGBoost: a scalable tree boosting system. In: ACM SIGKDD International Conference on Knowledge Discovery and Data Mining (2016)
2. Túlio Ribeiro, M., Singh, S., Guestrin, C.: "Why Should I Trust You?": explaining the predictions of any classifier. In: ACM SIGKDD International Conference on Knowledge Discovery and Data Mining (2016)

3. Lipton, Z. C.: The mythos of model interpretability. In: ICML Workshop on Human Interpretability in Machine Learning (WHI 2016) (2016)
4. Hara, S., Hayashi, K.: Making tree ensembles interpretable. In: ICML Workshop on Human Interpretability in Machine Learning (WHI 2016) (2016)
5. Sobhani-Tehrani, E., Khorasani, K.: Fault Diagnosis of Nonlinear Systems Using a Hybrid Approach. Springer, London (2009). https://doi.org/10.1007/978-0-387-92907-1
6. Gallego-Ortiz, C., Martel, A.L.: Interpreting extracted rules from ensemble of trees: application to computer-aided diagnosis of breast MRI. In: ICML Workshop on Human Interpretability in Machine Learning (WHI 2016) (2016)
7. Letham, B., Rudin, C., McCormick, T.H., Madigan, D.: Interpretable classifiers using rules and Bayesian analysis: building a better stroke prediction model. Ann. Appl. Stat. **9**(3), 1350–1371 (2015)
8. Hochreiter, S., Schmidhuber, J.: Long short-term memory. Neural Comput. **9**(8), 1735–1780 (1997)

# Author Index

Printed in the United States
By Bookmasters